Current Topics in Microbiology and Immunology

Volume 342

Current Topics in Microbiology and Immunology

Previously published volumes
Further volumes can be found at springer.com

Vol. 316: **Pitha, Paula M. (Ed.):**
Interferon: The 50th Anniversary. 2007.
VII, 391 pp. ISBN 978-3-540-71328-9

Vol. 317: **Dessain, Scott K. (Ed.):**
Human Antibody Therapeutics for Viral Disease.
2007. XI, 202 pp. ISBN 978-3-540-72144-4

Vol. 318: **Rodriguez, Moses (Ed.):**
Advances in Multiple Sclerosis and
Experimental Demyelinating Diseases. 2008.
XIV, 376 pp. ISBN 978-3-540-73679-9

Vol. 319: **Manser, Tim (Ed.):**
Specialization and Complementation
of Humoral Immune Responses to Infection.
2008. XII, 174 pp. ISBN 978-3-540-73899-2

Vol. 320: **Paddison, Patrick J.;**
Vogt, Peter K.(Eds.): RNA Interference. 2008.
VIII, 273 pp. ISBN 978-3-540-75156-4

Vol. 321: **Beutler, Bruce (Ed.):**
Immunology, Phenotype First: How Mutations
Have Established New Principles and
Pathways in Immunology. 2008. XIV, 221 pp.
ISBN 978-3-540-75202-8

Vol. 322: **Romeo, Tony (Ed.):**
Bacterial Biofilms. 2008. XII, 299.
ISBN 978-3-540-75417-6

Vol. 323: **Tracy, Steven; Oberste, M. Steven;**
Drescher, Kristen M. (Eds.):
Group B Coxsackieviruses. 2008.
ISBN 978-3-540-75545-6

Vol. 324: **Nomura, Tatsuji; Watanabe,**
Takeshi; Habu, Sonoko (Eds.):
Humanized Mice. 2008.
ISBN 978-3-540-75646-0

Vol. 325: **Shenk, Thomas E.; Stinski, Mark F. (Eds.):**
Human Cytomegalovirus. 2008.
ISBN 978-3-540-77348-1

Vol. 326: **Reddy, Anireddy S.N; Golovkin,**
Maxim (Eds.):
Nuclear pre-mRNA processing in plants. 2008.
ISBN 978-3-540-76775-6

Vol. 327: **Manchester, Marianne; Steinmetz,**
Nicole F. (Eds.):
Viruses and Nanotechnology. 2008.
ISBN 978-3-540-69376-5

Vol. 328: **van Etten, (Ed.):**
Lesser Known Large dsDNA Viruses. 2008.
ISBN 978-3-540-68617-0

Vol. 329: **Diane E. Griffin; Michael B.A.**
Oldstone (Eds.): Measles 2009.
ISBN 978-3-540-70522-2

Vol. 330: **Diane E. Griffin; Michael B.A.**
Oldstone (Eds.): Measles 2009.
ISBN 978-3-540-70616-8

Vol. 331: **Villiers, E. M. de (Eds):**
TT Viruses. 2009. ISBN 978-3-540-70917-8

Vol. 332: **Karasev A. (Ed.):**
Plant produced Microbial Vaccines. 2009.
ISBN 978-3-540-70857-5

Vol. 333: **Richard W. Compans;**
Walter A. Orenstein (Eds):
Vaccines for Pandemic Influenza. 2009.
ISBN 978-3-540-92164-6

Vol. 334: **Dorian McGavern;**
Micheal Dustin (Eds.):
Visualizing Immunity. 2009.
ISBN 978-3-540-93862-0

Vol. 335: **Beth Levine; Tamotsu Yoshimori;**
Vojo Deretic (Eds.):
Autophagy in Infection and Immunity, 2009.
ISBN 978-3-642-00301-1

Vol. 336: **Tammy Kielian (Ed.):**
Toll-like Receptors: Roles in Infection and
Neuropathology, 2009.
ISBN 978-3-642-00548-0

Vol. 337: **Chihiro Sasakawa (Ed.):**
Molecular Mechanisms of Bacterial Infection
via the Gut, 2009.
ISBN 978-3-642-01845-9

Vol. 338: **Alan L. Rothman (Ed.):**
Dengue Virus, 2009.
ISBN 978-3-642-02214-2

Vol. 339: **Paul Spearman; Eric O. Freed (Eds.):**
HIV Interactions with Host Cell Proteins, 2009.
ISBN 978-3-642-02174-9

Vol. 340: **Takashi Saito; Facundo D. Batista (Eds.):**
Immunological Synapse, 2010.
ISBN 978-3-642-12727-4

Vol. 341: **Øystein Bruserud (Ed.):**
The Chemokine System in Clinical
and Experimental Hematology, 2010.
ISBN 978-3-642-12638-3

Allison Abendroth • Ann M. Arvin •
Jennifer F. Moffat
Editors

Varicella-zoster Virus

Springer

Editors
Allison Abendroth, Ph.D
University of Sydney
Department of Infectious
Diseases and Immunology
Senior Lecturer and Head
VZV Research Laboratory
Room 601 Blackburn Building
D06
2006 Sydney New South Wales
Australia
allison.abendroth@sydney.edu.au

Prof. Dr. Ann M. Arvin, M.D
Stanford University
Professor of Pediatrics and Microbiology
and Immunology
School of Medicine
Dept. Microbiology & Immunology
300 Pasteur Drive
94305-5124 Stanford California
USA
aarvin@stanford.edu

Jennifer F. Moffat, Ph.D
Associate Professor
Department of Microbiology
and Immunology
Upstate Medical University
2215 Weiskotten Hall
750 East Adams Street
13210 Syracuse New York
USA
moffatj@upstate.edu

ISSN 0070-217X
ISBN: 978-3-642-12727-4 e-ISBN: 978-3-642-12728-1
DOI 10.1007/978-3-642-12728-1
Springer Heidelberg Dordrecht London New York

Library of Congress Control Number: 2010933953

Cover design: WMXDesign GmbH, Heidelberg, Germany

Printed on acid-free paper

Springer is part of Springer Science+Business Media (www.springer.com)

Preface

Varicella-zoster virus (VZV) is a medically important human herpesvirus, belonging to the subfamily *Alphaherpesviridae*. The capacity to persist in sensory neurons is a defining characteristic of the *Alphaherpesviridae* subgroup which also includes herpes simplex virus 1 and 2; like VZV, simian varicella virus (SVV), pseudorabies virus-1 (PRV-1), and equine herpesvirus-1 (EHV-1) belong to the *Varicellovirus* genus. The basic elements of the infectious cycle of VZV in the human host are that infection of the naïve host results in varicella, commonly known as chickenpox, latency is established in sensory ganglia, and reactivation causes zoster or "shingles." The relationship between the causative agent of varicella and zoster was demonstrated more than 100 years ago when children inoculated with material from zoster lesions were shown to develop varicella. The localized distribution of the zoster rash was also recognized as demarcating the dematome innervated by axons from neurons in each of the sensory ganglia. Early electron microscopy studies showed that virus particles were present in high concentrations in the vesicular fluid from both varicella and zoster lesions, and VZV was among the first viruses propagated in vitro by John Enders and Thomas Weller. The introduction of immunosuppressive therapies for malignancy led to observations suggesting the need for cell-mediated immunity in the host response to varicella and its role in maintaining VZV latency. Fortunately, early studies of the molecular virology of VZV revealed that it was inhibited by interference with the thymidine kinase gene, and the life-threatening and often fatal VZV infections experienced by these patients became treatable with antiviral drugs. Subsequently, the capacity to grow VZV in tissue culture was exploited to create a live attenuated VZV vaccine by Michiaki Tashihaki. While now taken for granted, these early insights about VZV and its characteristics as a human pathogen as well as the development of effective antiviral drugs and vaccines occurred over many decades. Importantly, these early observations set the stage for the remarkable progress that has been made in our understanding of the molecular biology of VZV, the subtleties of its tropism for differentiated human cells, including lymphocytes as well as skin and neurons, and the mechanisms by which the virus achieves an equilibrium with the host so that it persists not just in the individual but in the human population.

The purpose of this volume is to review key areas of progress in the field of VZV research, as well as work on the related SVV, written by those who have contributed many of the new findings that have enriched our knowledge of the unique characteristics of this ubiquitous human pathogen. Although the VZV genome is the smallest among the human herpesviruses, the rapidly accelerating pace of discovery about VZV and VZV–host interactions reflected in these reviews promises to continue as new tools are available and new hypotheses are generated to explain how VZV has created and maintained its niche in the human "virome" so successfully. Further improvements in the clinical management of VZV infection should emerge in parallel with better insights into VZV molecular virology and pathogenesis.

Stanford, CA, June, 2010

Allison Abendroth
Ann M. Arvin
Jennifer F. Moffat

Contents

Contributors

Allison Abendroth Department of Infectious Diseases and Immunology, University of Sydney, Blackburn Building, Room 601, Camperdown, NSW 2006, Australia and Centre for Virus Research, Westmead Millennium Institute, Westmead, NSW 2145, Australia, allison.abendroth@sydney.edu.au

Ann M. Arvin Stanford University School of Medicine, G311, Stanford, CA 94305, USA, aarvin@stanford.edu

Yevgeniy Azarkh Department of Neurology, University of Colorado Denver School of Medicine, 12700 E. 19th Avenue, Mail Stop B182, Aurora, CO 80045, USA

Barbara Berarducci Départment de Virologie, Institut Pasteur, 25 Rue du Docteur Roux, 75015, Paris, France, barbara.berarducci@pasteur.fr

Judith Breuer University College London, Windeyer Building, 46 Cleveland St, W1 2BE, Bloomsbury, London, j.breuer@ucl.ac.uk

Xibing Che Departments of Pediatrics and Microbiology and Immunology, Stanford University School of Medicine, Stanford, CA, USA

Jeffrey I Cohen Laboratory of Clinical Infectious Diseases, National Institutes of Health, Bldg. 10, Room 11N234, 10 Center Drive, Bethesda, MD 20892, USA, jcohen@niaid.nih.gov

Randall J. Cohrs Department of Neurology, University of Colorado Denver School of Medicine, 12700 E. 19th Avenue, Mail Stop B182, Aurora, CO 80045, USA, randall.cohrs@ucdenver.edu

Angela Erazo Graduate Program in Molecular Virology and Microbiology, School of Medicine, University of Pittsburgh, Pittsburgh, PA, USA, and Department of Ophthalmology, University of Pittsburgh, 1020 EEI building, 203 Lothrop Street, Pittsburgh, PA, 15213, USA

Anne A. Gershon Department of Pathology and Cell Biology, College of Physicians and Surgeons, Columbia University, 630 West 168th Street, New York, NY 10032, USA, and Department of Pediatrics, College of Physicians and Surgeons, Columbia University, 630 West, 168th Street, New York, NY 10032, USA, aag1@columbia.edu

Michael D. Gershon Department of Pathology and Cell Biology, College of Physicians and Surgeons, Columbia University, 630 West 168th Street, New York, NY 10032, USA, and Department of Pediatrics, College of Physicians and Surgeons, Columbia University, 630 West, 168th Street, New York, NY 10032, USA, mdg4@columbia.edu

Don Gilden Department of Neurology, University of Colorado Denver School of Medicine, 12700 E. 19th Avenue, Mail Stop B182, Aurora, CO 80045, USA, and Department of Microbiology, University of Colorado Denver School of Medicine, 12700 E. 19th Avenue, Mail Stop B182, Aurora, CO 80045, USA, don.gilden@ucdenver.edu

Wayne L. Gray Department of Microbiology and Immunology, University of Arkansas for Medical Sciences, 4301 West Markham Street, Little Rock, AR 72205, USA, graywaynel@uams.edu

Rebecca J. Greenblatt Department of Microbiology and Immunology, SUNY Upstate Medical University, 750 E. Adams Street, Syracuse, NY 13210, USA

Charles Grose Department of Pediatrics/2501 JCP, University of Iowa Hospital, Iowa City, IA 52242, USA, charles-grose@uiowa.edu

Teri K. Kenyon Department of Pediatrics/2501 JCP, University of Iowa Hospital, Iowa City, IA 52242, USA

Paul R. Kinchington Department of Ophthalmology, University of Pittsburgh, 1020 EEI building, 203 Lothrop Street, Pittsburgh, PA 15213, USA, and Department of Microbiology and Molecular Genetics, School of Medicine, University of Pittsburgh, Pittsburgh, PA, USA, kinchingtonp@upmc.edu

Chia-Chi Ku The Graduate Institute of Immunology, College of Medicine, National Taiwan University, Taipei, Taiwan

Myron J. Levin University of Colorado Denver, Bldg. 401, Mail Stop C227, 1784 Racine St., Room R09-108, Aurora, CO 80045, USA, Myron.Levin@ucdenver.edu

Ravi Mahalingam Departments of Neurology, University of Colorado Denver School of Medicine, 12700 E. 19th Avenue, Mailstop B182, Aurora, CO 80045, USA, ravi.mahalingam@ucdenver.edu

Ilhem Messaoudi Vaccine and Gene Therapy Institute and Division of Pathobiology and Immunology, Oregon National Primate Research Center, Oregon Health and Science, 505 NW 185th Avenue, Beaverton, OR 97006, USA

Jennifer F. Moffat Department of Microbiology and Immunology, SUNY Upstate Medical University, 750 E. Adams Street, Syracuse, NY 13210, USA, moffatj@upstate.edu

Yasuko Mori Division of Clinical Virology, Kobe University Graduate School of Medicine, 7-5-1, Kusunoki-cho, Chuo-ku, Kobe 650-0017, Japan and Laboratoy of Virology and Vaccinology, Division of Biomedical Research, National Institute of Biomedical Innovation, 7-6-8, Saito-Asagi, Ibaraki, Osaka 567-0085, Japan, ymori@med.kobe-u.ac.jp

Maria A. Nagel Department of Neurology, University of Colorado Denver School of Medicine, 12700 E. 19th Avenue, Mail Stop B182, Aurora, CO 80045, USA

Stefan Oliver Stanford University School of Medicine, S356, Stanford, CA 94305, USA, sloliver@stanford.edu

Mike Reichelt Stanford University School of Medicine, S356, Stanford, CA 94305, USA, reichelt@stanford.edu

William T. Ruyechan Department of Microbiology and Immunology, Witebsky Center for Microbial Pathogenesis and Immunology, University at Buffalo SUNY, 138 Farber Hall, Buffalo, NY 14214, USA, ruyechan@buffalo.edu

Tomohiko Sadaoka Division of Clinical Virology, Kobe University Graduate School of Medicine, 7-5-1, Kusunoki-cho, Chuo-ku, Kobe 650-0017, Japan and Laboratoy of Virology and Vaccinology, Division of Biomedical Research, National Institute of Biomedical Innovation, 7-6-8, Saito-Asagi, Ibaraki, Osaka 567-0085, Japan

D. Scott Schmid Herpesvirus Team and National VZV Laboratory, MMRHLB, Centers for Disease Control and Prevention, 1600 Clifton Rd/ Bldg 18/Rm 6-134/ MS G-18, Atlanta, GA 30333, USA, SSchmid@cdc.gov

Barry Slobedman Centre for Virus Research, Westmead Millennium Institute, Westmead, NSW 2145, Australia, barry.slobedman@sydney.edu.au

Marvin Sommer Stanford University School of Medicine, S356, Stanford, CA 94305, USA, msommer@stanford.edu

Megan Steain Department of Infectious Diseases and Immunology, University of Sydney, Blackburn Building, D06, Camperdown, NSW, 2006, Australia

Susan Vleck Departments of Pediatrics and Microbiology and Immunology, Stanford University School of Medicine, Stanford, CA, USA

Adriana Weinberg University of Colorado Denver, Research Complex 2, Mail Stop 8604, 12700 E. 19th Ave., Aurora, CO 80045, USA, Adriana.Weinberg@ucdenver.edu

Leigh Zerboni Stanford University School of Medicine, S366, Stanford, CA 94305, USA, zerboni@stanford.edu

The Varicella-Zoster Virus Genome

Jeffrey I. Cohen

Contents

Abstract The varicella-zoster virus (VZV) genome contains at least 70 genes, and all but six have homologs in herpes simplex virus (HSV). Cosmids and BACs corresponding to the VZV parental Oka and vaccine Oka viruses have been used to "knockout" 34 VZV genes. Seven VZV genes (ORF4, 5, 9, 21, 29, 62, and 68) have been shown to be required for growth in vitro. Recombinant viruses expressing several markers (e.g., beta-galactosidase, green fluorescence protein, luciferase) and several foreign viral genes (from herpes simplex, Epstein–Barr virus, hepatitis B, mumps, HIV, and simian immunodeficiency virus) have been constructed.

J.I. Cohen
Laboratory of Clinical Infectious Diseases, National Institutes of Health, Bldg. 10, Room 11N234, 10 Center Drive, Bethesda, MD 20892, USA
e-mail: jcohen@niaid.nih.gov

A.M. Arvin et al. (eds.), *Varicella-zoster Virus*,
Current Topics in Microbiology and Immunology 342, DOI 10.1007/82_2010_10

Further studies of the VZV genome, using recombinant viruses, may facilitate the development of safer and more effective VZV vaccines. Furthermore, VZV might be useful as a vaccine vector to immunize against both VZV and other viruses.

Abbreviations

HSV	Herpes simplex virus
IE	Immediate-early
IRL	Internal repeat long
IRS	Internal repeat short
TRL	Terminal repeat long
TRS	Terminal repeat short
UL	Unique long
US	Unique short
VZV	Varicella-zoster virus

1 Genome Structure and Organization

Varicella-zoster virus (VZV) is an alphaherpesvirus that is in the same subfamily as herpes simplex virus (HSV) 1 and 2. VZV is a member of varicellovirus genus, along with equine herpesvirus 1 and 4, pseudorabies virus, and bovine herpesvirus 1 and 5. Cercopithecine herpesvirus 9 (simian varicella virus) is the virus most homologous to VZV.

1.1 VZV Genome

The complete sequence of the VZV genome was determined by Davison and Scott (1986). The prototype strain VZV Dumas is 124,884 base pairs in length. The genome consists of a unique long (UL) region bounded by terminal long (TRL) and internal long (IRL) repeats, and a unique short (US) region bounded by internal short (IRS) and terminal short (TRS) repeats (Fig. 1). The US region can orientate in either of the two directions, while the UL region rarely changes its orientation; thus, there are usually two isomers of the genome in infected cells. The genome is linear in virions with an unpaired nucleotide at each end. In VZV-infected cells the ends pair and the genome circularizes.

The genome has five repeat regions. Repeat region 1 (R1) is located in open reading frame (ORF) 11, R2 is located in ORF14 (glycoprotein C), R3 in ORF22, R4 between ORF62 and the origin of viral replication, and R5 between ORF 60 and 61.

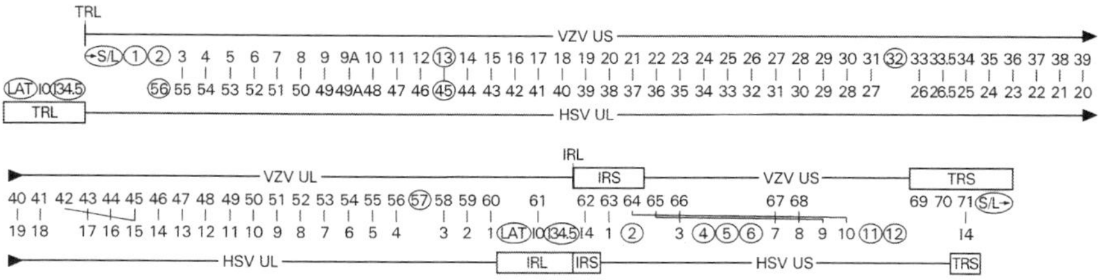

Fig. 1 Comparison of the VZV and HSV genomes. The VZV genome (*first rows*) contains unique long (UL) and unique short (US) regions flanked by terminal long (TRL), terminal short (TRS), internal long (IRL), and internal short (IRS) repeats. VZV genes (*second rows*) and HSV genes (*third rows*) are shown by numbers. VZV or HSV genes that do not have homologs in the corresponding virus are circled. The HSV genome contains UL and US regions, and TRL, TRS, IRL, and IRS repeats (*fourth rows*). Modified from Cohen 1999, with permission

The length of the repeat regions varies among different VZV strains and has been used to distinguish the strains.

1.2 VZV Genes

1.2.1 VZV Immediate-Early Genes

VZV encodes at least 70 genes, three (ORF62, 63, 64) of which are present in both of the short repeat regions (Cohen et al. 2007b). VZV encodes at least three immediate-early (IE) proteins that are located in the tegument of virions and regulate virus transcription (Table 1). IE4 and IE62 transactivate IE, early, and late promoters. IE63 represses several VZV promoters and inhibits the activity of interferon-alpha (Ambagala and Cohen 2007) and binds to antisilencing protein 1 (Ambagala et al. 2009). ORF61 protein, which is not present in the tegument of virions and has not been shown to be an IE gene, activates IE, early, and late viral promoters.

1.2.2 VZV Genes Encoding Replication Proteins

VZV encodes a viral DNA polymerase, likely composed of two subunits (ORF28 and ORF16) that is inhibited by acyclovir. The viral thymidine kinase (ORF36) phosphorylates deoxycytidine, thymidine, and acyclovir. VZV ORF18 and ORF19 encode the small and large subunits of ribonucleotide reductase that convert ribonucleotides to deoxyribonucleotides. VZV encodes at least two DNA binding proteins – ORF29 protein is a single-stranded DNA binding protein, and ORF 51 protein binds to the origin of DNA replication. VZV encodes two protein kinases. ORF47 protein phosphorylates VZV ORF32 protein, IE62,

Table 1 VZV gene products

VZV gene	HSV-1 homolog	VZV protein function
1	None	Membrane protein
2	None	
3	UL55	
4	UL54 (ICP 27)	Transactivator, tegument protein, expressed in latency
5	UL53 (gK)	gK
6	UL52 (HPC)	
7	UL51	
8	UL50	Deoxyuridine triphosphatase
9	UL49 (VP22)	Tegument protein
9A	UL49A	Syncytia formation, putative gN
10	UL48 (VP16)	Transactivator, tegument protein
11	UL47 (VP13/14)	
12	UL46 (VP11/12)	
13	None	Thymidylate synthetase
14	UL44 (gC)	gC
15	UL43	
16	UL42 (PPF)	Putative small subunit of viral DNA polymerase
17	UL41 (vhs)	Induces RNA cleavage
18	UL40	Ribonucleotide reductase, small subunit
19	UL39	Ribonucleotide reductase, large subunit
20	UL38 (VP19C)	
21	UL37	Nucleocapsid protein, expressed in latency
22	UL36	
23	UL35 (VP26)	Capsid assembly
24	UL34	
25	UL33	
26	UL32	
27	UL31	
28	UL30	DNA polymerase
29	UL29 (ICP8)	ssDNA binding protein, expressed in latency
30	UL28	
31	UL27 (gB)	gB
32	None	Probable substrate for ORF47 kinase
33	UL26 (VP24)	Protease
33.5	UL26.5 (VP22)	Assembly protein
34	UL25	
35	UL24	Cell-to-cell fusion
36	UL23	Thymidine kinase
37	UL22 (gH)	gH
38	UL21	
39	UL20	
40	UL19 (VP5)	Major nucleocapsid protein
41	UL18 (VP23)	
42/45	UL15	
43	UL17	
44	UL16	
46	UL14	
47	UL13	Protein kinase, tegument protein
48	UL12	Putative DNase
49	UL11	Virion protein

(*continued*)

Table 1 (continued)

VZV gene	HSV-1 homolog	VZV protein function
50	UL10 (gM)	gM
51	UL9	Origin binding protein
52	UL8 (HPC)	
53	UL7	
54	UL6	Putative portal protein
55	UL5 (HPC)	
56	UL4	
57	None	Cytoplasmic protein
58	UL3	
59	UL2	Uracil-DNA glycosylase
60	UL1 (gL)	gL, chaperone for gH
61	ICP0	Transactivator, transrepressor
62, 71	ICP4	Transactivator, tegument protein, expressed in latency
63, 70	US1 (ICP22)	Tegument protein, transrepressor, inhibits interferon-alpha, expressed in latency
64, 69	US10	
65	US9	Virion protein
66	US3	Protein kinase, expressed in latency
67	US7 (gI)	gI
68	US8 (gE)	gE
S/L	None	Cytoplasmic protein (also referred to as ORF0)

HPC helicase-primase complex; *PPF* polymerase processivity factor

IE63, and glycoprotein I. ORF66 protein phosphorylates IE62, which results in inclusion of IE62 into the virion tegument. VZV encodes other enzymes including a dUTPase (ORF8), thymidylate synthetase (ORF13), protease (ORF33), DNase (ORF48), and uracil-DNA glycosylase (ORF59).

1.2.3 VZV Genes Encoding Putative Late Proteins

VZV ORF10 encodes a tegument protein that forms a complex with transcription factors at the ORF62 promoter to activate transcription of ORF62. ORF17 protein induces cleavage of RNA. ORF33.5 encodes the assembly protein, which forms a scaffold thought to be involved in the construction of nucleocapsids. ORF40 encodes the major nucleocapsid protein, while ORF21 also encodes a nucleocapsid protein. ORF54 encodes the putative portal protein, which allows viral DNA to enter nucleocapsids.

1.2.4 VZV Genes Encoding Glycoproteins

VZV encodes seven viral glycoproteins – gB (ORF31), gC (ORF14), gE (ORF68), gH (ORF 37), gI (ORF67), gK (ORF5), gL (ORF60), gM (ORF50), and presumably gN (ORF9A). VZV gB, based on homology with HSV gB, is likely critical for entry of virus into cells. gE binds to a cellular receptor [insulin degrading enzyme (Li et al. 2006)] and gH and gM are important for cell-to-cell spread of virus (Yamagishi et al. 2008). gI facilitates maturation of gE, and gL is a chaperone for gH. gK may be important for syncytia formation.

2 Comparative Genomics of VZV and HSV

VZV and HSV are largely collinear, although the UL region of HSV is orientated in the opposite direction to that of VZV using the standard nomenclature. The VZV US region is much shorter (5.2 kb) than the HSV-1 US (13.0 kb), and the VZV, TRL and IRL regions are also shorter (0.9 kb) than their HSV-1 counterparts (9.2 kb).

2.1 Core Proteins Conserved with Herpesviruses in Other Subfamilies

The VZV genome contains about 41 "core genes" that are conserved with each of the three subfamilies of herpesviruses, alphaherpesvirus, betaherpesvirus, and gammaherpesvirus (Davison 1993). Core genes include IE4, the VZV DNA polymerase, helicase-primase components, single-stranded DNA binding protein, ribonucleotide reductase, uracil-DNA glycosylase, dUTPase, DNase, ORF47 protein kinase, major capsid protein, protease, assembly protein, several tegument proteins, gB, gH , gL, gM, and gN.

2.2 VZV Functional and Nonfunctional Homologs of HSV Genes

Several VZV genes can complement their HSV homologs. VZV ORF61 can substitute for HSV ICP0 (Moriuchi et al. 1992) and VZV ORF62 can complement HSV ICP0 (Felser et al. 1988). Although VZV ORF10 can complement the transactivating function of HSV VP16, HSV-1 VP16 is essential for replication of HSV while VZV ORF10 is dispensable (Cohen and Seidel 1994b; Moriuchi et al. 1993). VZV ORF 51 can complement HSV UL9 (Chen et al. 1995). In contrast, VZV ORF4 cannot complement HSV ICP27 (Moriuchi et al. 1994a).

2.3 VZV Genes Not Conserved with HSV

VZV encodes six genes (ORF1, 2, 13, 32, 57, and S/L) that are absent in HSV (Fig. 1). ORF 13 encodes the viral thymidylate synthetase, which has a homolog in herpesvirus saimiri and Kaposi's sarcoma-associated herpesvirus.

2.4 *HSV Genes Not Conserved with VZV*

HSV encodes nine genes (UL45, UL56, US2, US5, US6, US11, US12, and LAT) that are absent in VZV (Fig. 1). HSV US 12 encodes ICP47, which blocks presentation of MHC class I. HSV US6, US4, and US5 encode glycoproteins D, G, and J, respectively. VZV gE is the most abundant viral glycoprotein and shares many features of HSV gD, including binding to a cellular receptor that contributes to VZV entry (Li et al. 2006).

3 Mutagenesis with Cosmids and BACs

3.1 *Mutagenesis Using Marker Rescue*

The first genetically engineered mutant of VZV was constructed in 1987, in which the Epstein–Barr virus gp350 gene was inserted into the VZV genome (Lowe et al. 1987). Fibroblasts were cotransfected with VZV virion DNA and a plasmid with EBV gp350 flanked by VZV thymidine kinase sequences, and plaques were purified by limiting dilution. While the procedure was successful, the process of plaque purification that requires sequential rounds of sonication is very labor intensive. The same strategy in which VZV virion DNA is cotransfected with plasmids with flanking sequences that are homologous to VZV DNA was used to reinsert and thereby "rescue" essential genes into the genome of VZV lacking ORF4 (Cohen et al. 2005), ORF21 (Xia et al. 2003), and ORF68 (Ali et al. 2009).

3.2 *Mutagenesis Using Cosmids*

A cosmid system was first used to generate mutations in the VZV genome in 1993 (Cohen and Seidel 1993). Virion DNA was isolated from the Oka vaccine strain of VZV, the linear DNA was blunted with T4 DNA polymerase, and oligonucleotides containing Not I or Mst II restriction sites were ligated to the DNA. The resulting DNA was cut with Mst II or Not I and four large DNA fragments, which overlap the entire VZV genome, were ligated into a cosmid vector that was linearized with Mst II or Not I. Human melanoma cells were transfected with the four cosmids along with a plasmid encoding VZV IE62, which increases the infectivity of viral DNA (Moriuchi et al. 1994b). While the cosmids are able to produce infectious virus in the absence of the plasmid, the latter increases the reliability and enhances the efficiency of virus production. Recombinant VZV derived from the cosmids grew to similar titers as the nonrecombinant virus used to generate the cosmids.

Other cosmid systems have also developed to perform VZV mutagenesis. Mallory et al. (1997) used five cosmids derived from the Oka vaccine strain to produce recombinant virus, and Niizuma et al. (2003) described a cosmid library from the parental Oka vaccine. These cosmids have been used to "knockout" a large number of VZV genes and to study the phenotype of the resulting mutants in cell culture and rodents.

3.3 Mutagenesis Using BACs

Four separate BACs have been constructed for VZV mutagenesis. Three research groups have inserted the parental Oka virus into BACs (Nagaike et al. 2004; Zhang et al. 2007; Tischer et al. 2007) and one group inserted the vaccine Oka virus into a BAC (Yoshii et al. 2007). While each of the recombinant parental Oka viruses derived from BACs grew to titers similar to nonrecombinant parental Oka virus, recombinant virus derived from the vaccine Oka BAC grew to slightly lower titers than the nonrecombinant vaccine Oka virus.

3.4 Results of Mutagenesis Studies

VZV cosmids and BACs have been used to mutate multiple VZV genes (Table 2). Several VZV genes (ORF4, 5, 9, 21, 29, 62, and 68) have been shown to be essential for replication in cells in vitro. Other VZV genes that have been tested are not required for growth in cell culture, but some are required for growth in certain types of cells in vitro (Cohen and Nguyen 1997), in lymphocytes (Moffat et al. 1999; Soong et al. 2000), or in human skin (Moffat et al. 1999). Each of the six VZV genes that do not have HSV homologs (ORF1, 2, 13, 32, 57, and S/L) are not required for growth in cell culture (Cohen and Seidel 1993, 1995; Cox et al. 1998; Reddy et al. 1998a; Sato et al. 2002b; Zhang et al. 2007).

Chimeras have been constructed containing various portions of the parental and vaccine Oka genomes (Zerboni et al. 2005). These chimeras have demonstrated that attenuation of VZV is a multigenic trait due to mutations throughout the genome.

VZV cosmids have been used to prove that mutations in individual genes correspond with resistance to antiviral compounds. Mutagenesis of VZV ORF54 (homologous to the portal protein of HSV) conferred resistance to a thiourea inhibitor compound (Visalli et al. 2003). VZV cosmids have been used to produce virus expressing beta-galactosidase (Cohen et al. 1998), green fluorescence protein (Zerboni et al. 2000; Li et al. 2006), or luciferase (Oliver et al. 2008; Zhang et al. 2007), which have been useful for studies in animals and virus entry in vitro.

Most recombinant viruses have stable genomes with a few exceptions. Attempts to delete or mutate one copy of a gene that is normally present in both of the short repeat regions of the genome often result in recombination with wild-type sequences in both short repeat regions after several rounds of replication (Sommer et al. 2001; Sato et al. 2003a; Oliver et al. 2008). In addition, point mutations that

Table 2 Deletion and stop codon mutants constructed using cosmids or BACs in VZV and their effect on growth in vitro

VZV gene	Mutation	Growth in vitro	Reference
ORFS/L (ORF0)	Del	Impaired	Zhang et al. (2007)
ORF1	Stop	No change	Cohen and Seidel (1995)
	Del	Slight reduced	Zhang et al. (2007)
ORF2	Del	No change	Sato et al. (2002b)
	Del	No change	Zhang et al. (2007)
ORF3	Del	Slight reduced	Zhang et al. (2007)
ORF4	Del	Essential	Cohen et al. (2005)
	Del	Essential	Sato et al. (2003b)
	Del	Essential	Zhang et al. (2007)
ORF5 (gK)	Del	Essential	Mo et al. (1999)
ORF8	Stop	No change	Ross et al. (1997)
	Delete[a]	Reduced	Ross et al. (1997)
ORF9	Start	Essential	Tischer et al. (2007)
	Del	Essential	Che et al. (2008)
ORF9A	Stop	No change	Ross et al. (1997)
ORF10	Del	No change	Cohen and Seidel (1994b)
	Del	No change	Che et al. (2008)
ORF11	Del	No change	Che et al. (2008)
ORF12	Del	No change	Che et al. (2008)
ORF13	Stop	No change	Cohen and Seidel (1993)
ORF14 (gC)	Stop	No change	Cohen and Seidel (1994a)
ORF17	Del	Reduced	Sato et al. (2002a)
ORF19	Del	Reduced	Heineman and Cohen (1994)
ORF21	Del	Essential	Xia et al. (2003)
ORF23	Del	Reduced	Chaudhuri et al. (2008)
ORF29	Del	Essential	Cohen et al. (2007a)
ORF32	Del	No change	Reddy et al. (1998a)
ORF35	Del	Reduced	Ito et al. (2005)
ORF47	Stop	No change	Heineman and Cohen (1995)
ORF49	Del	Reduced	Sadaoka et al. (2007)
ORF50	Del	Reduced	Yamagishi et al. (2008)
ORF57	Del	No change	Cox et al. (1998)
ORF58	Del	No change	Yoshii et al. (2008)
ORF59	Del	No change	Reddy et al. (1998b)
ORF61	Del	Reduced	Cohen and Nguyen (1998)
ORF62	Del	Essential	Sato et al. (2003a)
ORF63	Del	Essential	Sommer et al. (2001)
	Del	Reduced	Cohen et al. (2004)
ORF65	Del	No change	Cohen et al. (2001)
	Del	No change	Niizuma et al. (2003)
ORF66	Stop	No change	Heineman et al. (1996)
ORF67 (gI)	Del	Reduced[b] or essential	Cohen and Nguyen (1997)
	Del	Reduced	Mallory et al. (1997)
ORF68 (gE)	Del	Essential	Mallory et al. (1997)
	Del	Essential	Ali et al. (2009)

Stop stop codons; *del* deletion; *start* mutations in start codon

[a]Interrupts expression of ORF8 and ORF9A

[b]Reduced in melanoma cells, essential in Vero cells

severely impair growth of the virus can sometimes undergo back mutation with reversion to wild-type virus over time (Cohen et al. 1998).

VZV genes thought to be required for cell growth have been proven to be essential by growing a virus mutant unable to express the protein in a complementing cell line (Xia et al. 2003) or in cells infected with baculovirus expressing the VZV protein (Ali et al. 2009; Cohen et al. 2005, 2007a) and then showing that the virus cannot be grown on noncomplementing cells. An alternative approach has been to insert the essential gene elsewhere in the viral genome. This latter method is not ideal, since the gene will not be formally proven to be required for growth. Inefficiency in cosmid transfections or impaired virus growth may be misinterpreted as showing that a virus gene is essential. One VZV gene, ORF63, has been reported to be essential in the absence of a complementation system (Sommer et al. 2001), but was actually shown to be nonessential for growth in cell culture (Cohen et al. 2004). The failure to obtain virus in the former study may have been due to the lack of a cotransfected plasmid expressing the VZV IE62 gene to transiently enhance virus growth after cosmid DNA transfection (Baiker et al. 2004).

3.5 VZV as an Expression Vector

VZV has been used to express a number of foreign viral proteins (Table 3). While these recombinant VZV mutants have been shown to induce immunity to the foreign virus and to protect animals from disease after challenging with the foreign virus in some cases (Heineman et al. 1995, 2004), in one case the recombinant VZV actually enhanced replication of the foreign virus (Strapans et al. 2004). Thus, it is critical to test such recombinant viruses in animal models whenever possible. Since children are vaccinated with the varicella vaccine to prevent chickenpox, expression of additional viral proteins might allow immunization against other viruses and ultimately reduce the number of vaccines required during childhood.

Table 3 Recombinant VZV expressing other viral proteins

Viral protein	Immunogenicity	Reference
EBV gp350	Not reported	Lowe et al. (1987)
Hepatitis B SAg	Induced antibody to hepatitis B SAg	Shiraki et al. (1991)
HIV env	Induced humoral and cellular immunity to HIV	Shiraki et al. (2001)
SIV gp160	Induced neutralizing antibody Enhanced SIV infection in monkeys	Strapans et al. (2004)
HSV gD	Induced neutralizing antibody to HSV-2 Reduced severity of HSV-2 in guinea pigs	Heineman et al. (1995)
Mumps HA-N	Induced neutralizing antibody to mumps	Somboonthum et al. (2007)
HSV gD and gB	Induced neutralizing antibody to HSV-2 Reduced severity of HSV-2 in guinea pigs	Heineman et al. (2004)

gp350 Glycoprotein 350; *SAg* surface antigen; *gD* glycoprotein D; *HA-N* hemagglutinin-neuraminidase; *gB* glycoprotein B

3.6 *Use of Genetics to Develop Safer VZV Vaccines*

The current vaccine is very safe in immunocompetant persons, but it has caused disease in severely immunocompromised persons. Replication-defective vaccines might be developed by growing viruses in complementing cells; however, attempts to produce high titers of cell-free virus using this approach has been unsuccessful (Cohen, unpublished data). An alternative approach is to express an essential gene, such as one expressed during latency, under a different promoter, which may allow high titers of virus during replication, but may impair latency (Cohen et al. 2007a).

Acknowledgments I thank the intramural research program of the National Institute of Allergy and Infectious Diseases for support.

References

Ali MA, Li Q, Fischer ER, Cohen JI (2009) The insulin degrading enzyme binding domain of varicella-zoster virus (VZV) glycoprotein E is important for cell-to-cell spread and VZV infectivity, while a glycoprotein I binding domain is essential for infection. Virology 386:270–279

Ambagala AP, Cohen JI (2007) Varicella-zoster virus IE63, a major viral latency protein, is required to inhibit the alpha interferon-induced antiviral response. J Virol 81:7844–7851

Ambagala AP, Bosma T, Ali MA, Poustovoitov M, Chen JJ, Gershon MD, Adams PD, Cohen JI (2009) Varicella-zoster virus immediate-early 63 protein interacts with human antisilencing function 1 protein and alters its ability to bind histones H3.1 and H3.3. J Virol 83:200–209

Baiker A, Bagowski C, Ito H, Sommer M, Zerboni L, Fabel K, Hay J, Ruyechan W, Arvin AM (2004) The immediate-early 63 protein of varicella-zoster virus: analysis of functional domains required for replication in vitro and for T-cell and skin tropism in the SCIDhu model in vivo. J Virol 78:1181–1194

Chaudhuri V, Sommer M, Rajamani J, Zerboni L, Arvin AM (2008) Functions of varicella-zoster virus ORF23 capsid protein in viral replication and the pathogenesis of skin infection. J Virol 82:10231–10246

Che X, Reichelt M, Sommer MH, Rajamani J, Zerboni L, Arvin AM (2008) Functions of the ORF9-to-ORF12 gene cluster in varicella-zoster virus replication and in the pathogenesis of skin infection. J Virol 82:5825–5834

Chen D, Stabell EC, Olivio PD (1995) Varicella-zoster virus gene 51 complements a herpes simplex virus type I UL9 mutant. J Virol 69:4515–4518

Cohen JI (1999) Genomic structure and organization of varicella-zoster virus. Contrib Microbiol 3:10–20

Cohen JI, Nguyen H (1997) Varicella-zoster virus glycoprotein I (gI) is essential for growth of virus in vero cells. J Virol 71:6913–6920

Cohen JI, Nguyen H (1998) Varicella-zoster virus ORF61 deletion mutants replicate in cell culture, but a mutant with stop codons in ORF61 reverts to wild-type virus. Virology 246:306–316

Cohen JI, Seidel KE (1993) Generation of varicella-zoster virus (VZV) and viral mutants from cosmid DNAs: VZV thymidylate synthetase is not essential for replication in vitro. Proc Natl Acad Sci USA 90:7376–7380

Cohen JI, Seidel KE (1994a) Absence of varicella-zoster virus (VZV) glycoprotein V does not alter growth of VZV in vitro or sensitivity to heparin. J Gen Virol 75:3087–3093

Cohen JI, Seidel KE (1994b) Varicella-zoster virus (VZV) open reading frame 10 protein, the homolog of the essential herpes simplex virus protein VP16, is dispensable for VZV replication in vitro. J Virol 68:7850–7858

Cohen JI, Seidel KE (1995) Varicella-zoster virus open reading frame 1 encodes a membrane protein that is dispensable for growth of VZV in vitro. Virology 206:835–842

Cohen JI, Wang Y, Nussenblatt R, Straus SE, Hooks JJ (1998) Chronic uveitis in guinea pigs infected with varicella-zoster virus expressing *Escherichia coli* beta-galactosidase. J Infect Dis 177:293–300

Cohen JI, Sato H, Srinivas S, Lekstrom K (2001) The varicella-zoster virus (VZV) ORF65 virion protein is dispensable for replication in cell culture and is phosphorylated by casein kinase II, but not by the VZV protein kinases. Virology 280:62–71

Cohen JI, Cox E, Pesnicak L, Srinivas S, Krogmann T (2004) The varicella-zoster virus ORF63 latency-associated protein is critical for establishment of latency. J Virol 78:11833–11840

Cohen JI, Krogmann T, Ross JP, Pesnicak LP, Prikhod'ko EA (2005) The varicella-zoster virus ORF4 latency associated protein is important for establishment of latency. J Virol 79:6969–6975

Cohen JI, Krogmann T, Pesnicak L, Ali MA (2007a) Absence or overexpression of the varicella-zoster virus (VZV) ORF29 latency-associated protein impairs late gene expression and reduces latency in a rodent model. J Virol 81:1586–1591

Cohen JI, Straus SE, Arvin AM (2007b) Varicella-zoster virus: replication, pathogenesis, and management. In: Knipe DM, Howley PM (eds) Fields virology, 5th edn. Lippincott-Williams & Wilkins, Philadelphia

Cox E, Reddy S, Iofin I, Cohen J (1998) Varicella-zoster virus ORF57, unlike its pseudorabies virus UL3.5 homolog, is dispensable for replication in cell culture. Virology 250:205–209

Davison AJ (1993) Herpesvirus genes. Rev Med Virol 3:237–244

Davison AJ, Scott J (1986) The complete DNA sequence of varicella-zoster virus. J Gen Virol 67:1759–1816

Felser JM, Kinchington PR, Inchauspe G, Straus SE, Ostrove JM (1988) Cell lines containing varicella-zoster virus open reading frame 62 and expressing the "IE" 175 protein complement ICP4 mutants of herpes simplex virus type 1. J Virol 62:2076–2082

Heineman TC, Cohen JI (1994) Deletion of the varicella-zoster virus large subunit of ribonucleotide reductase impairs the growth of virus in vitro. J Virol 68:3317–3323

Heineman TC, Cohen JI (1995) The varicella-zoster virus (VZV) open reading frame 47 (ORF47) protein kinase is dispensable for viral replication and is not required for phosphorylation of ORF63 protein, the VZV homolog of herpes simplex virus ICP22. J Virol 69:7367–7370

Heineman TC, Connelly BL, Bourne N, Stanberry LR, Cohen JI (1995) Immunization with recombinant varicella-zoster virus expressing herpes simplex virus type 2 glycoprotein D reduces the severity of genital herpes in guinea pigs. J Virol 69:8109–8113

Heineman TC, Seidel K, Cohen JI (1996) The varicella-zoster virus ORF66 protein induces kinase activity and is dispensable for viral replication. J Virol 70:7312–7317

Heineman T, Pesnicak L, Ali M, Krogmann T, Krudwig N, Cohen JI (2004) Varicella-zoster virus expressing HSV-2 glycoproteins B and D induces protection against HSV-2 challenge. Vaccine 22:2558–2565

Ito H, Sommer MH, Zerboni L, Baiker A, Sato B, Liang R, Hay J, Ruyechan W, Arvin AM (2005) Role of the varicella-zoster virus gene product encoded by open reading frame 35 in viral replication in vitro and in differentiated human skin and T cells in vivo. J Virol 79:4819–4827

Li Q, Ali MA, Cohen JI (2006) Insulin degrading enzyme is a cellular receptor for varicella-zoster virus infection and for cell-to-cell spread of virus. Cell 127:305–316

Lowe RS, Keller PM, Keech BJ, Davison AJ, Whang Y, Morgan AJ, Kieff E, Ellis RW (1987) Varicella-zoster virus as a live vector for the expression of foreign genes. Proc Natl Acad Sci USA 84:3896–3900

Mallory S, Sommer M, Arvin AM (1997) Mutational analysis of the role of glycoprotein I in varicella-zoster virus replication and its effects on glycoprotein E conformation and trafficking. J Virol 71:8279–8288

Mo C, Suen J, Sommer M, Arvin A (1999) Characterization of varicella-zoster virus glycoprotein K (open reading frame 5) and its role in virus growth. J Virol 73:4197–4207

Moffat JF, Zerboni L, Sommer MH, Heineman TC, Cohen JI, Kaneshima H, Arvin AM (1999) The ORF47 and ORF66 putative protein kinases of varicella-zoster virus determine tropism for human T cells and skin in the SCID-hu mouse. Proc Natl Acad Sci USA 95:11969–11974

Moriuchi H, Moriuchi M, Smith HA, Straus SE, Cohen JI (1992) Varicella-zoster virus open reading frame 61 protein is functionally homologous to herpes simplex virus type 1 ICP0. J Virol 66:7303–7308

Moriuchi H, Moriuchi M, Straus SE, Cohen JI (1993) Varicella-zoster virus open reading frame 10 protein, the herpes simplex virus VP16 homolog, transactivates herpesvirus immediate-early gene promoters. J Virol 67:2739–2746

Moriuchi H, Moriuchi M, Smith HA, Cohen JI (1994a) Varicella-zoster virus open reading frame 4 protein is functionally distinct from and does not complement its herpes simplex virus type 1 homolog ICP27. J Virol 68:1987–1992

Moriuchi M, Moriuchi H, Straus SE, Cohen JI (1994b) Varicella-zoster virus (VZV) virion-associated transactivator open reading frame 62 protein enhances the infectivity of VZV DNA. Virology 200:297–300

Nagaike K, Mori Y, Gomi Y, Yoshii H, Takahashi M, Wagner M, Koszinowski U, Yamanishi K (2004) Cloning of the varicella-zoster virus genome as an infectious bacterial artificial chromosome in *Escherichia coli*. Vaccine 22:4069–4074

Niizuma T, Zerboni L, Sommer MH, Ito H, Hinchliffe S, Arvin AM (2003) Construction of varicella-zoster virus recombinants from parent Oka cosmids and demonstration that ORF65 protein is dispensable for infection of human skin and T cells in the SCID-hu mouse model. J Virol 77:6062–6065

Oliver SL, Zerboni L, Sommer M, Rajamani J, Arvin AM (2008) Development of recombinant varicella-zoster viruses expressing luciferase fusion proteins for live in vivo imaging in human skin and dorsal root ganglia xenografts. J Virol Methods 154:182–193

Reddy SM, Cox E, Iofin I, Soong W, Cohen JI (1998a) Varicella-zoster virus (VZV) ORF32 encodes a phosphoprotein that is posttranscriptionally modified by the VZV ORF47 protein kinase. J Virol 72:8083–8088

Reddy SM, Williams M, Cohen JI (1998b) Expression of a uracil DNA glycosylase (UNG) inhibitor in mammalian cells: varicella-zoster virus can replicate in vitro in the absence of detectable UNG activity. Virology 251:393–401

Ross J, Williams M, Cohen JI (1997) Disruption of the varicella-zoster virus dUTPase and the adjacent ORF9A gene results in impaired growth and reduced syncytia formation in vitro. Virology 234:186–195

Sadaoka T, Yoshii H, Imazawa T, Yamanishi K, Mori Y (2007) Deletion in open reading frame 49 of varicella-zoster virus reduces virus growth in human malignant melanoma cells but not in human embryonic fibroblasts. J Virol 81:12654–12665

Sato H, Callanan LD, Pesnicak L, Krogmann T, Cohen JI (2002a) Varicella-zoster virus (VZV) ORF17 protein induces RNA cleavage and is critical for replication of VZV at 37°C, but not 33°C. J Virol 76:11012–11023

Sato H, Pesnicak L, Cohen JI (2002b) Varicella-zoster virus ORF2 encodes a membrane phosphoprotein that is dispensable for viral replication and for establishment of latency. J Virol 76:3575–3578

Sato B, Ito H, Hinchliffe S, Sommer MH, Zerboni L, Arvin AM (2003a) Mutational analysis of open reading frames 62 and 71, encoding the varicella-zoster virus immediate-early transactivating protein, IE62, and effects on replication in vitro and in skin xenografts in the SCID-hu mouse in vivo. J Virol 77:5607–5620

Sato B, Sommer M, Ito H, Arvin AM (2003b) Requirement of varicella-zoster virus immediate-early 4 protein for viral replication. J Virol 7:12369–12372
Shiraki K, Hayakawa Y, Mori H, Namazue J, Takamizawa A, Yoshida I, Yamanishi K, Takahashi M (1991) Development of immunogenic recombinant Oka varicella vaccine expressing hepatitis B virus surface antigen. J Gen Virol 72:1393–1399
Shiraki K, Sato H, Yoshida Y, Yamamura JI, Tsurita M, Kurokawa M, Kageyama S (2001) Construction of Oka varicella vaccine expressing human immunodeficiency virus env antigen. J Med Virol 64:89–95
Somboonthum P, Yoshii H, Okamoto S, Koike M, Gomi Y, Uchiyama Y, Takahashi M, Yamanishi K, Mori Y (2007) Generation of a recombinant Oka varicella vaccine expressing mumps virus hemagglutinin-neuraminidase protein as a polyvalent live vaccine. Vaccine 25:8741–8755
Sommer MH, Zagha E, Serrano OK, Ku CC, Zerboni L, Baiker A, Santos R, Spengler M, Lynch J, Grose C, Ruyechan W, Hay J, Arvin AM (2001) Mutational analysis of the repeated open reading frames, ORFs 63 and 70 and ORFs 64 and 69, of varicella-zoster virus. J Virol 75:8224–8239
Soong W, Schultz JC, Patera AC, Sommer MH, Cohen JI (2000) Infection of human T lymphocytes with varicella-zoster virus: an analysis with viral mutants and clinical isolates. J Virol 74:1864–1870
Strapans SI, Barry AP, Silvestri G, Safrit JT, Kozyr N, Sumpter B, Nguygen H, McClure H, Montefiori D, Cohen JI, Feinberg M (2004) Enhance simian immunodeficiency virus replication and accelerated AIDS in macaques primed to mount a CD4 T cell response to SIV Env. Proc Natl Acad Sci USA 101:13026–13031
Tischer BK, Kaufer BB, Sommer M, Wussow F, Arvin AM, Osterrieder N (2007) A self-excisable infectious bacterial artificial chromosome clone of varicella-zoster virus allows analysis of the essential tegument protein encoded by ORF9. J Virol 81:13200–13208
Visalli RJ, Fairhurst J, Srinivas S, Hu W, Feld B, DiGrandi M, Curran K, Ross A, Bloom JD, van Zeijl M, Jones TR, O'Connell J, Cohen JI (2003) Identification of small molecule compounds that selectively inhibit varicella-zoster virus replication. J Virol 77:2349–2358
Xia D, Srinivas S, Sato H, Pesnicak L, Straus SE, Cohen JI (2003) Varicella-zoster virus ORF21, which is expressed during latency, is essential for virus replication but dispensable for establishment of latency. J Virol 77:1211–1218
Yamagishi Y, Sadaoka T, Yoshii H, Somboonthum P, Imazawa T, Nagaike K, Ozono K, Yamanishi K, Mori Y (2008) Varicella-zoster virus glycoprotein M homolog is glycosylated, is expressed on the viral envelope, and functions in virus cell-to-cell spread. J Virol 82: 795–804
Yoshii H, Somboonthum P, Takahashi M, Yamanishi K, Mori Y (2007) Cloning of full length genome of varicella-zoster virus vaccine strain into a bacterial artificial chromosome and reconstitution of infectious virus. Vaccine 25:5006–5012
Yoshii H, Sadaoka K, Matsuura M, Nagaike K, Takahashi M, Yamanishi K, Mori Y (2008) Varicella-zoster virus ORF 58 gene is dispensable for viral replication in cell culture. Virol J 30(5):54
Zerboni L, Hinchliffe S, Sommer MH, Ito H, Besser J, Stamatis S, Cheng J, Distefano D, Kraiouchkine N, Shaw A, Arvin AM (2005) Analysis of varicella zoster virus attenuation by evaluation of chimeric parent Oka/vaccine Oka recombinant viruses in skin xenografts in the SCIDhu mouse model. Virology 332:337–346
Zerboni L, Sommer M, Ware CF, Arvin AM (2000) Varicella-zoster virus infection of a human CD4-positive T-cell line. Virology 270:278–285
Zhang Z, Rowe J, Wang W, Sommer M, Arvin A, Moffat J, Zhu H (2007) Genetic analysis of varicella-zoster virus ORF0 to ORF4 by use of a novel luciferase bacterial artificial chromosome system. J Virol 81:9024–9033

VZV Molecular Epidemiology

Judith Breuer

Contents

Abstract The molecular epidemiology of varicella zoster virus (VZV) has led to an understanding of virus evolution, spread, and pathogenesis. The availability of over 20 full length genomes has confirmed the existence of at least five virus clades and generated estimates of VZV evolution, with evidence of recombination both past and ongoing. Genotyping by restriction enzyme analysis (REA) and single nucleotide polymorphisms (SNP) has proven that the virus causing varicella is identical to that which later reactivates as zoster in an individual. Moreover, these methods have shown that reinfection, which is mostly asymptomatic, may also occur and the second virus may establish latency and reactivate. VZV is the only

J. Breuer
University College London, Windeyer Building, 46 Cleveland St, W1 2BE, Bloomsbury, London
e-mail: j.breuer@ucl.ac.uk

A.M. Arvin et al. (eds.), *Varicella-zoster Virus*,
Current Topics in Microbiology and Immunology 342, DOI 10.1007/82_2010_9

human herpesvirus that is spread by the respiratory route. Genotyping methods, together with epidemiological data and modeling, have provided insights into global differences in the transmission patterns of this ubiquitous virus.

Abbreviations

Bp	Base pair
DNA	Deoxyribonucleic acid
GC	Guanisine cytosine
HSV	Herpes simplex virus
ORF	Open reading frame
P72	Passage 72
PCR	Polymerase chain reaction
R4	Repeat 4
REA	Restriction enzyme analysis
RFLP	Restriction fragment length polymorphism
SNP	Single nucleotide polymorphism
US	United States
VZV	Varicella zoster virus

1 Introduction

1.1 What is Molecular Epidemiology?

The molecular epidemiology of infections uses information about pathogen genetic variation to investigate the evolution, transmission, and pathogenesis of infectious agents. For varicella-zoster virus (VZV), a pathogen that naturally infects only humans and for which there are few tractable models of disease, molecular epidemiological tools have provided important insights, which have changed our assumptions about the natural history of infection, have improved our understanding of virus evolution, spread, and transmission, and have informed public health policy.

1.2 VZV Molecular Epidemiology: The Historical Context

VZV was first isolated in 1953 by Thomas Weller (Weller et al. 1958). Although it was already known that the same virus was responsible for two diseases, varicella (chickenpox) and zoster (shingles) (Bokay 1909), and that susceptible individuals in contact with either could develop varicella, it was thought that varicella caught from zoster or vZoster differed in its characteristics from vVaricella or varicella transmitted from a case of varicella (Bokay 1909; Seiler 1949). In a classical epidemiological study, Edgar Hope Simpson showed that the clinical appearance and incubation period of vZoster was identical to that of vVaricella and therefore the two forms of the virus were identical (Simpson 1954). Hope-Simpson also observed that zoster always followed varicella and confirmed previous observations that immunity to varicella was lifelong (Hope-Simpson 1965). The observations that immunity was long-lived provided the rationale for the development of a vaccine to prevent varicella in the 1970s (Takahashi et al. 1974). Two subsequent developments fueled the beginnings of molecular epidemiological studies in VZV; the first was the availability of molecular biological tools such as restriction enzymes and the second was the realization that new tools were needed to distinguish the live vaccine from naturally circulating virus.

2 Genetic Variation in VZV: Tools for Molecular Epidemiology

Five main methods have been used to describe genetic variation of VZV: restriction fragment length polymorphisms (RFLP), length polymorphisms of repeat regions, restriction site SNP, SNP, and whole genome sequencing.

2.1 Restriction Fragment Length Polymorphisms

Restriction endonucleases are naturally occurring enzymes of bacteria and archea, which cleave foreign deoxyribonucleic acid (DNA) at signature nucleotide sequences. Evolving as host defenses against viruses, restriction enzymes were first used to analyze VZV DNA in 1977 (Oakes et al. 1977). Following digestion of DNA with a restriction enzyme, the fragments are separated on gels. The separation of fragments by size reflects the location of restriction sites and is altered by genetic variation. Initial studies using the enzymes EcoRI and HindI failed to distinguish between five US isolates of VZV, however, a later study of seven VZV isolates, two Japanese and five US, was more successful (Richards et al. 1979). In this study, cleavage patterns were generated using five enzymes: EcoRI, HindI, SmaI, BamHI, and AvaI. AvaI maps were identical for all seven isolates, while between 0 and 3 restriction fragment variants were generated by the other four enzymes. However, the authors still did not find any obvious groupings of isolates, and notably, there was no clear distinction between the Japanese and US isolates (Richards et al. 1979). The authors also observed that VZV was less variable than the herpes simplex virus (HSV) control and that no differences between low (10) and high (36) passage viruses were noted. The genomic stability of VZV and the applicability of restriction enzyme analysis (REA) as an epidemiological tool were confirmed by Zweerink and colleagues (Zweerink et al. 1981). They analysed two strains AW and KmcC, which had been passaged up to 30 and 71 times, respectively, in the human fibroblast line WI-38. There was no difference using six restriction enzymes between AW at p6 and p30. Minor differences with one enzyme were found in KMcC at passage 72 (p72) compared with p6 and p42. The data suggested that VZV was, for the most part, stable on tissue culture growth, and strains analyzed at up to $\sim$50 passages were likely to be closely representative of the original virus. In fact, later data on whole genome sequencing of passaged virus showed that substitutions begin to accumulate between p5 and p20, and that by p72, the virus has diverged by 2.3×10^{-4} substitutions per base (Tyler et al. 2007). In 1981, Dumas and colleagues used XbaI, BglI, and PstI restriction enzyme maps, southern blotting, and fragment hybridization to describe the organization of the VZV genome, including the existence of two unique sequences, one long and one short, and two internally repeated sequences, which were at opposite orientations with respect to one another (Dumas et al. 1981). These findings were confirmed by Straus and colleagues in 1981, who used 11 restriction enzymes (Straus et al. 1981). In addition, Straus observed that the results were consistent with that of the genome existing in a circular state (Straus et al. 1981). This physical mapping of the VZV genome was followed by confirmation, using REA, that VZV exists as four isomers, with inversion of the unique short regions (Ecker and Hyman 1982; Straus et al. 1982). These authors also confirmed earlier observations by Dumas and Straus that VZV resembled pseudorabies virus in this as in other aspects of its genome organization (Ecker and Hyman 1982).

Much of the impetus for molecular genotyping and molecular epidemiology of VZV followed the early success by Takahashi in developing a live attenuated vaccine strain vOka, which was quickly shown to be safe and immunogenic in children (Takahashi et al. 1974). Genotyping of parental and vaccine Oka strains revealed several differences using the HpaI enzyme, an extra band in vOka when digested by KpnI or SmaI, but no differences in Hind III, EcoRI, or BglI digests. (Ecker and Hyman 1981). Not long afterwards, Martin and colleagues published unique differences between BamHI, BglI, and HpaI digests of the vOka vaccine strain and six wild-type US strains (Martin et al. 1982). Small differences between the wild-type strains were noted and led Straus to examine a larger set of 17 viruses taken from patients who contracted with varicella and zoster, including four high passage viruses (Martin et al. 1982; Straus et al. 1983). Using a combination of seven enzymes, HindIII, XblI, EcoRI, BglI, PstI, BamHI, and SmaI, Straus and colleagues noted that all 17 isolates differed from each other, but that around five were more similar and that the enzymes EcoRI, HindIII, and SmaI appeared to be the most informative for distinguishing between strains (Straus et al. 1983). Importantly, this paper showed for the first time that two strains of VZV from members of the same family who contracted varicella from the same source were identical by BamHI, EcoRI, HindIII, and SmaI digests and different from other viruses in the study. Also viruses from three different vesicles in the same person were identical to each other and different from viruses isolated from other unrelated individuals.

Straus was also able to identify that most differences between viruses arose from length variation of certain key fragments. Making use of the restriction map generated by Dumas, Straus mapped the variable fragments back to four areas, which we now know to correspond to the variable R2 repeat region in open reading frame (ORF) 14, the R3 repeat region in ORF 22, and the two copies of the noncoding repeat 4 (R4) repeat region located between ORFs 62 and 63 and 70 and 71(Straus et al. 1983). The accumulating molecular epidemiological data led to a seminal publication, in which Straus used REA to show that the virus that had caused primary varicella in a leukaemic child was identical to the virus recovered from the lesions of herpes zoster occurring in the same child 3 years later (Straus et al. 1984). This finding provided proof of the original hypothesis outlined by Hope Simpson, namely that zoster is caused by reactivation of the virus that has persisted latently within the host following primary varicella infection (Hope-Simpson 1965).

2.2 Length Polymorphisms of Variable Regions

It was soon apparent that RFLP differences between strains largely arose from the variation in the repeat regions within the genome. Casey and colleagues showed that the guanisine cytosine (GC) rich R4 region, which occurs in both the internal and the terminal repeat regions, comprised variable numbers of 27 bp repeat units (Casey et al. 1985). The same group characterized the R2 repeat region in ORF 14 (glycoprotein C) and showed it to comprise 42 bp repeat units, which again varied in number between different strains (Kinchington et al. 1985). The sequencing of the first VZV genome, the Dumas strain, by Davison revealed the R1 and R3 repeat sequences located in ORFs 11 and 22, while R5, which lies in the noncoding regions between ORFs 60 and 61, was characterized by Hondo in 1988 (Hondo and Yogo 1988). Also in 1988, Kinoshita and colleagues cloned and sequenced fragments containing R1 and demonstrated that the combinations of 18 and 15 bp sequences which made up the repeat element differed between unrelated isolates (Kinoshita et al. 1988). Some length variation following tissue culture passage was observed, suggesting instability. Interestingly, R1 length variation was found between viruses isolated from vesicular fluid and ganglionic tissue in the same patient (Hondo et al. 1987). Other groups who have genotyped two or more viruses from different vesicles in the same patient have, however, not found differences by REA (Shibuta et al. 1974; Pichini et al. 1983; Hondo et al. 1987; Kinoshita et al. 1988; Takayama et al. 1988). The instability of R1 mirrored findings for R3 (Zweerink et al. 1981; Hayakawa et al. 1986). Hawakaya found that the HpaI fragment K, which contains the R3 site, showed length variation on passage while fragments F and G remained stable up to passage 85 (Hayakawa et al. 1986). The HpaI K fragment also varied among viruses from patients who had shared a room and infected one another (Hayakawa et al. 1986). The definitive analysis, however, came from whole genome sequencing by US/Canadian scientists

(Grose et al. 2004; Tyler et al. 2007; Tyler and Severini, 2006). Repeated passage of one virus revealed that R1, the origin of replication (OriS), and R4 elongated by 1–2 repeat units after 72 passages, while R2 and R5 remained stable and R3 was completely unstable (Grose et al. 2004; Tyler et al. 2007). The length variation of these repeat regions led a number of groups to explore their use for genotyping. Initial evaluation by Takada showed mispriming when the R2 region was amplified by polymerase chain reaction (PCR), resulting in band ladders that were hard to interpret, while the R4 was unstable on low passage (Takada et al. 1995). R5 was found to be stable and to have three alleles, comprising different combinations of the 88 and 22 bp units (Takada et al. 1995). Further studies confirmed the difficulties of amplifying R2, R3, and R4 and their instability in in vitro culture. Unfortunately, the stability of R5 was matched by its noninformative qualities, with only three alleles, one of which was present in over 90% of European strains (Hawrami et al. 1996; Na 1997; Yoshida and Tamura 1999; Sauerbrei et al. 2003; Yoshida et al. 2003). R1 sequencing revealed a large number of variants with certain genotypes predominating among UK and Japanese strains (Hawrami et al. 1996; Yoshida et al. 1999). Multiple R1 alleles were found in one patient who died from Herpes Zoster (Kinoshita et al. 1988) and in the ocular fluid from patients with acute retinal necrosis (Abe et al. 2000). Since all the patients were immunosuppressed, this finding may reflect slippage of the unstable repeat due to high rates of replication. Alternatively, infection with multiple strains may have occurred.

2.3 *Restriction Site Polymorphisms: BglI and PstI Digests*

Several groups (Martin et al. 1982; Gelb et al. 1987) observed that the restriction enzyme digestion patterns of BglI and PstI distinguised US wild type clinical strains from the Oka vaccine and vaccine-related strains that caused clinical symptoms. Following on from this, Adams in 1989 identified a BglI site in a BamHI fragment, D, which was stably present in the vaccine, parental Oka, and three wild type Japanese strains, but was absent in wild type American strains (Adams et al. 1989). Unlike other informative sites, none of the cleaved products of BamHI fragment D showed length variation in different viruses. Moreover, a 1940 bp HindIII subfragment of the BamHI D fragment could not be further cleaved by other enzymes. Together this suggested that the BglI site, which distinguished the Japanese vOka and other Japanese strains from wild-type US strains, was present as a SNP (Single Nucleotide Polymorphism) within a stable region. Exploiting this, La Russa and colleagues developed a PCR to detect this BglI site, which is located in ORF 54, for use in distinguishing Japanese vaccine from US wild-type rashes (LaRussa et al. 1992). Finding that the BglI site was also positive in 3 of 20 clinical varicella and zoster strains from the US, La Russa, using data from the restriction maps, identified a PstI site in ORF38, which had previously been found to be positive in all US strains tested and negative in the vaccine (Adams et al. 1989; LaRussa et al. 1992). Another SNP that distinguishes Japanese from European strains occurs in ORF 10 and results in the substitution of a histidine in Japanese strains instead of a proline in European strains at position 10 in the protein (Kinchington and Turse 1995). This amino acid substitution abrogates binding of an antibody directed against the N terminus of the ORF 10 peptide (Kinchington and Turse 1995) and theoretically could provide a serological test to differentiate Japanese and European strains.

2.4 Single Nucleotide Polymorphisms

VZV was the first full length human herpesvirus genome to be sequenced (Davison and Scott 1986). Following publication of the sequence (Dumas strain) and the advent of PCR, detection of SNPs became the most widely used method for genotyping of viruses as well as other organisms. The development of SNPs typing tools for molecular epidemiological analysis of VZV was pioneered simultaneously by two US groups as well as our group in the UK. As described by Grose in 2001, "SNPs are sites that contain single base pair (Bp) variations (Collins et al. 1997, 1998). These sites generally are considered biallelic due to the limited number of transversions that have been found (Kruglyak 1997; Brookes 1999). Each allele must occur in more than 1% of the population to be considered an SNP (Kruglyak 1997; Wang et al. 1998). Individuals who share many of the same SNPs are likely to have arisen from a common ancestor and can therefore be grouped by inheritance (Collins et al. 1997)" (Faga et al. 2001).

In 2001 we published data on SNPs occurring in 10 UK isolates of VZV: seven cases of varicella and three of zoster. To identify SNPs we designed 37 primer pairs, each amplifying a 500 bp fragment, which were spaced across the genome at 3 kb intervals. Denaturation and reannealing of PCR products produced homo and heteroduplexes. Running these fragments on a denaturing gel then allowed us to identify fragments for which the heteroduplex mobility was shifted compared with that of the homoduplex. Fifteen fragments within the 10 genomes showed heteroduplex mobility shifts and sequencing of these revealed 92 polymorphisms. From our data we identified three distinct groups of viruses circulating in the UK, labeled A, B, and C (Barrett-Muir et al. 2001). Further analysis of viruses from around the world identified a second variant of the A group, labeled A2, and a Japanese genogroup to which the Oka vaccine strain belonged (Barrett-Muir 2001). A subset of SNPs, in ORFs 1, 21, 50, and 54, discriminated between the three UK strains as well as the Japanese vaccine strain (Parker et al. 2006). The findings were confirmed and extended following a survey of over 65 samples from Africa (Zambia, Guinea Bissau), India, Bangladesh, Singapore, Japan, and the US. From this, additional genotypes A1 and A2 and J1 and J2 genotype were identified (Muir et al. 2002) (Fig. 1). The last (J2) is probably the same as Clade 4/M2/C as shown by the CDC group (Loparev et al. 2009). Grose took a slightly different approach. By sequencing five glycoprotein genes and ORF 62 in 10 samples, including the Oka vaccine strain, his group identified a total of 61 polymorphisms, 21 in the glycoproteins and 40 in ORF 62

(Faga et al. 2001; Wagenaar et al. 2003a). From this, a phylogenetic tree also identified four groupings, which were confirmed by analysis of a further seven viruses (Fig. 1). The Grose group also used letters A, B, C, and D to label the groups, but the labels do not correspond to the Breuer/Muir A, B, C, J lettering system, for example, Grose A corresponds to Breuer.Muir C.

In a third approach, Schmid's group at the Center for Disease Control located a 447 bp fragment in ORF 22 upstream of the C terminal coding region, which was sufficiently polymorphic to allow identification of three phylogenetic clades: European (E), Mosaic (M), and Japanese (J) (Table 1) (Loparev et al. 2004). The CDC group proposed the existence of additional variants of the M group M3 and M4 based on additional SNPs in ORF 22. However, the ORF 22 approach also but initially failed to identify one clade, group 3/B/D (Loparev et al. 2004). Since then, by including analysis of SNPs from ORFs 21 and 50, Loparev has shown good discrimination of all five groups and has mooted the existence of two further stable Clades M3 and M4 (Sergeev et al. 2006; Loparev et al. 2007b, 2009; Sauerbrei et al. 2008). M4 is BglI negative and has been shown by the CDC group to resemble the European Clades 1 and 3 (Loparev et al. 2009). As such they may be the same as the viruses labeled as Clade 1/C/E1/A–Clade 3/B/E2/D recombinants by the UK group (Muir 2002). Extensive SNP typing has shown M4 to have unique SNPS, which are not shared with any other Clade, suggesting that once full length sequence has been obtained, its status as an independent Clade will be confirmed (Loparev et al. 2009). The status of M3 is less certain. This BglI positive virus has been described only a few times and more data are needed to confirm its existence as an independent Clade (Sergeev et al. 2006; Loparev et al. 2007a)

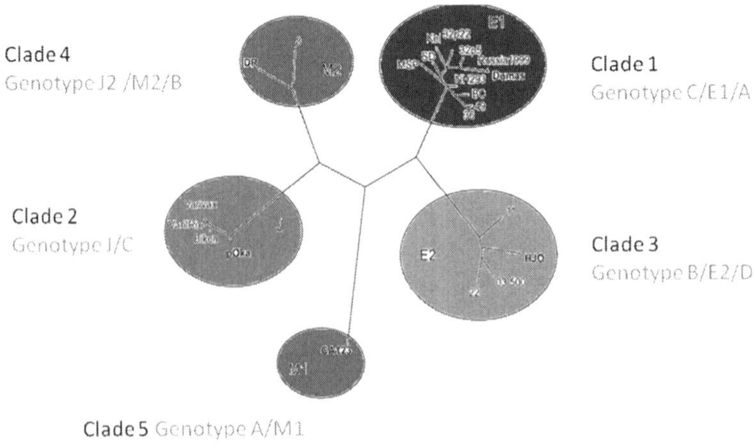

Fig. 1 Phylogenetic Tree to show VZV genogroups and their nomenclatures (clade/UK scattered SNP/CDC ORF 22/Iowa glycoprotein). (Reproduced with permission from ASM (Loparev et al. 2007b))

Table 9.1

				Universal Nomenclature for genotyping							
				1	3	2	5	4	VII	VI	
Scattered SNP Analysis (London Group)											
Study	Period	Region/Country	N	Genotype							
				C	B	J	A				Recomb.
Barrett-Muir 2002 and Muir 2003	NA	UK	35	62.8	14.3	0	17.1				5.7
Sengupta 2007	1998–2003	UK	298	58.1	20.7	8.7	8.7				0
Carr 2004	2002–2003	Ireland	16	37.5	12.5	12.5	25				12.5
Barrett-Muir 2002 and Muir 2003	NA	Japan	9	0	0	100	0				0
		India	5	0	0	0	100				0
		Bangladesh	9	0	0	0	100				0
		Zambia	5	0	0	0	100				0
		Guinea-Bissau	9	0	0	0	100				0
		Brazil	5	40	40	0	0				10
Glycoprotein/IE62 typing scheme (Iowa Group)											
Study	**Study Period**	**Country**	**N**	**Genotype**							
				A	**D**	**C**		**B**			0
Wagenaar 2003	Early 1990s	Singapore	6	0	0	50		50			0
	Late 1980s	Thailand	2	0	100	0		0			0
	N/A	USA	8	62.5	25	0		12.5			0
	N/A	Iceland	1	0	100	0		0			0
Peters 2006 and Tyler 2007	1996–2003	Canada	7	42.9	42.9	0		5.3			0
ORF 22 typing scheme											
Study	**Study Period**	**Country**	**N**	**Genotype**							
				E1	**E2**	**J**	**M1**	**M2**	**M3**	**M4**	**Recomb.**
Loparev 2004	1976–2002	USA	52	73.1		5.8	9.6	11.6	0	0	0
		USA (Hawaii)	3	0		100	0	0	0	0	0
		Canada	50	76		6	0	20	0	0	0
		Mexico	5	0		0	0	100	0	0	0
		Chile	2	50		0	0	50	0	0	0
		Nicaragua	1	0		0	100	0	0	0	0

(*continued*)

Table 9.1 (continued)

				Universal Nomenclature for genotyping							
				1	3	2	5	4	VII	VI	
Scattered SNP Analysis (London Group)											
Study	Period	Region/Country	N	Genotype							
				C	**B**	**J**	**A**				Recomb.
		Congo	19	0		0	100	0	0	0	0
		Chad	5	0		0	100	0	0	0	0
		Morocco	1	0		0	100	0	0	0	0
		Germany	21	100		0	0	0	0	0	0
		Iceland	17	94.1		0	0	5.9	0	0	0
		Czech Republic	1	100		0	0	0	0	0	0
		Poland	4	100		0	0	0	0	0	0
		Russia (Moscow)	20	100		0	0	0	0	0	0
		Belarus	5	100		0	0	0	0	0	0
		Ukraine	6	100		0	0	0	0	0	0
		Lithuania	5	100		0	0	0	0	0	0
		Latvia	9	89.9		0	0	11.1	0	0	0
		Moldavia	7	71.4		0	0	28.6	0	0	0
		Estonia	3	100		0	0	0	0	0	0
		Jordon	1	100		0	0	0	0	0	0
		Kazakhastan	1	100		0	0	0	0	0	0
		Russia (Novosibirisk)	12	100		0	0	0	0	0	0
		Japan	20	0		100	0	0	0	0	0
		Nepal	5	0		0	20	80	0	0	0
		India	16	0		0	37.5	62.5	0	0	0
		Bangladesh	7	0		0	100	0	0	0	0
		South China	3	0		0	0	100	0	0	0
		Western Australia	10	30		0	42.9	57.1	0	0	0
		Eastern Australia	10	100		0	0	0	0	0	0

Dayan 2004	2002	Argentina	13	100		0	0	0	0	0	0
Sergeev 2006	2001–2001	USA	130	81.5		3.1	12.3	3.1	0.8	0	0
Loparev 2007	1998–2004	France	19	73.7		0	10.5	5.3	0	10.5	0
	2004	Spain	12	75		0	0	0	16.6	8.3	0
Koskienemi 2007	1995–96	Finland	28	100		0	0	0	0	0	0
Loparev 2007	2003–2004**	Australia	127	52.8	23.6	16	7.9	3.1	0	0	0
	2004–2005**	New Zealand	38	65.8	21.1	0	13.1	0	0	0	0
Sauerbrel 2008	2002–2006**	Germany	77	37.7	39	0	22.1	0	0	0	1.3
Lopez 2008	2004	USA	4	0	-	0	100	0	0	0	0

2.5 Genome Sequencing

The phylogeny of VZV was finally resolved by whole genome sequencing carried out independently by two groups, one a US/Swedish collaboration and the other a US/Canadian collaboration (Norberg et al. 2006; Peters et al. 2006). Until this time, seven full length sequences were available: the Dumas strain, the Biken, Merck, and GSK strains of Oka vaccine strain, parental Oka, and MSP and BC in which the identical amino acid substitution D150N had been reported in glycoprotein E (Davison and Scott 1986; Gomi et al. 2002; Grose et al. 2004). The US/Canadian collaboration reported an additional 11 strains, while the US/Swedish collaboration reported sequence from an additional two strains. In the latter, repeat and tandem repeat regions including the R1–R5 and OriS regions were excluded (Norberg et al. 2006). In the first study, all 11 additional strains sequenced were collected from North America, while in the second, one strain originated from Morocco and the second from the USA (Norberg et al. 2006; Peters et al. 2006). At the same time, the US/Canadian group carried out the definitive investigation of virus stability. Repeated passage of a virus in tissue culture followed by whole genome sequencing showed that, excluding unstable repeat regions, one substitution, that is, 8×10^{-6} per base, occurred following 20 passages and 28 substitutions, that is, 2.2×10^{-4} per base, after 72 passages. This suggests that above 20 passages the virus becomes unrepresentative of the original (Tyler et al. 2007).

The three classifications continued to be confusing and, consequently, a meeting was organized in London in July, 2008, to propose a common nomenclature. Four principles were agreed upon: that the new nomenclature should be distinct from the three established nomenclatures, that it should reflect the phylogenetic structure, that it should reflect the order in which full genome sequences became available, and that it should describe the provenance of the strain. To that end, the proposed nomenclature for VZV and the prototypic viruses for each clade are shown in Table 1.

3 Molecular Epidemiology in Practice

3.1 Varicella-Zoster Virus Evolution and Clades

Early restriction enzyme analyses noted that variation between unrelated VZV strains was less than that between HSV strains (Richards et al. 1979). In 1995, Takayama and colleagues used long PCR to amplify fragments of DNA from 6.8 to 11.4 kb from 40 Japanese viruses (Takayama et al. 1996). Restriction digests of the fragments that covered ORFs 12–16, 38–43, and 54–60 were generated with ten enzymes yielding 12 variable restriction sites. Comparison with the only full length sequence then available, the Dumas strain, showed a total of 28 substitutions in 65,000 nucleotide residues anaylzed, giving an estimated interstrain variation of 0.043% or one base in 2,300 (Takayama et al. 1996). Our data from heteroduplex mobility assays of 37 fragments across 10 VZV genomes found 92 nucleotide substitutions compared with the Dumas in 15,059 nucleotides, giving a variation of 0.061% or 1 in every 1,637 nucleotides (Barrett-Muir et al. 2001). Both these estimates were considerably lower than those for herpes simplex 0.32–0.81%, cytomegalovirus 2.5%, human herpes virus 1.5–2%, and pseudorabies virus 2–3% (Barrett-Muir et al. 2001). Final confirmation came from whole genome studies that found, excluding variable regions, a maximum variation of 0.2% (1 in 1,000 bases) between Clades 5/A1/M1 and 2/J/B (Norberg et al. 2006) in one study, while in the other study Clades 1/C/E1/A and 2/J/B were the most divergent (Norberg et al. 2006; Peters et al. 2006). In addition, the US/Canadian group was able to calculate intra-clade diversity as 0.03–0.07% or 1 in 1,429 to 1 in 3,333 bases, using sequences from eight clade 1/C/E1/A viruses.

3.1.1 Mutation and Evolution of VZV

During the evolution of VZV, it is clear that mutation is occurring and that some have become fixed. In the first attempt to map SNPs along the full length of the genome, using heteroduplex analysis, we noted that genome variation, excluding the repeat regions, was not evenly distributed (Barrett-Muir et al. 2001) and this was confirmed from full genomes (Tyler et al. 2007). Per 100 bases, ORFs1 and 62 were the most variable while ORFs 3, 25, and 49 had no substitutions (Tyler et al. 2007). These data showed that hypervariable regions clustered towards the ends (ORFs 1–3) and (ORFs 57–68) of the genome as well as around ORFs 35–37. McGeoch in his phylogenetic analysis excluded the right hand end of the genome, observing that the number of substitutions per nucleotide was higher than that of the unique long region, suggesting a different evolutionary history (McGeoch 2009).

The finding by Grose's and later Tipple's group of the Clade 1 viruses MSP and BC, both of which carry an asparagine residue instead of an aspartic acid residue at position 150 in glycoprotein E, provided an opportunity to examine viral mutation in action (Santos et al. 2000). The effect of this amino acid substitution is to abrogate binding of the well characterized 3B3 antibody to its major B cell epitope. Although closely related, the MSP and BC strains did not form a distinct phylogenetic cluster, suggesting that the N150D the substitution is a homoplasy, that is, occurred independently as a result of parallel or convergent evolution probably as a result of antibody-driven viral escape (Tipples et al. 2002; Grose et al. 2004).

Among the eight Clade 1 full length viruses sequenced and analysed by Peters and colleagues, they identified two geographically clustered subgroupings (Peters et al. 2006). Viruses collected in Canada (BC, 36, and 49) formed one distinct subgroup within Clade 1, while strains from states located in the center of the USA (SD, Kel, and 32) formed another, suggesting each had evolved from their own common ancestor (Peters et al. 2006). Excluding the tandem repeat regions, which have a different evolutionary rate, the number of fixed substitutions between strains within the Canadian cluster was 18 and within the US cluster was 12 in a genome of 123453 bases. Making the assumption that the common ancestor for each cluster occurred at the time of the migration of Europeans to the US and Canada, approximately 400 years ago the substitution rate can be calculated as 1.8×10^{-7} per nucleotide per year for the Canadian strains and 1.2×10^{-7} per nucleotide per year for the US strains (Breuer unpublished data).

3.1.2 Recombination and VZV Evolution

The above estimates of the nucleotide substitution rate also assume that no recombination has occurred, which we know not to be the case. Both in vitro and in vivo recombinations between wild-type and vaccine Varicella were first described in the 1980s using the BglI site as a marker (positive in the vaccine and negative in US wild-type strains) (Gelb et al. 1987; Dohner et al. 1988). In 2002, using SNP typing we described three wild-type strains, which appeared to be recombinants (Muir et al. 2002). One was a clade 1/Clade 5 recombinant and the other two were deemed to be Clade 1/Clade 3 recombinants (Muir et al. 2002). More recent analysis suggests that the Clade1/3 recombinants may actually be the same as the M4 genotype described by the CDC group (Loparev et al. 2007a). The publication of full length genomes confirmed previously derived phylogenetic trees and allowed a more comprehensive analysis of which SNPS that are unique to each clade and which are shared. Both of the groups who generated the data identified ancestral recombination as having occurred. In particular, the US/Swedish group noted areas of homology between the Clade 4 and Clades 1 and 2 as well as between Clade 5 and Clades 1 and 2 (Norberg et al. 2006). Similarly, the US/Canadian group found homology between Clade 4 and Clades 1 and 2 (Peters et al. 2006). Interestingly the US/Canadian group identified their prototypic Clade 4/M2/J2/C virus,V-8, as being closest to Clade 2, with two areas of recombination with Clade 1, between ORFs 13 and 17 inclusive, and midway though ORF 22–26 inclusive (Peters et al. 2006). The same regions have been found to be recombinant in the simian alphaherpesvirus herpesvirus papio (Tyler and Severini 2006) and in recent recombinants between VZV Oka vaccine strain, which is Clade 2 and wild-type Clade 1 and 3 viruses (Schmid unpublished data). The authors make the case that this example and others of alphaherpesvirus recombination may be facilitated by reiterative subunits that are present in alphaherpesvirus genomes, notably for VZV, the R1 repeat in ORF11 and the R3 repeat in ORF 22 (Peters et al. 2006).

3.1.3 Reconstructing the Evolutionary History of VZV

Using full length genomes representing the three subfamilies within Herpesviridae, McGeoch and colleagues have presented compelling data for the co-evolution of herpesviruses with their hosts (McGeoch and Cook 1994; McGeoch et al. 1995). By their estimates the alphaherpesviruses, including VZV, diverged from the beta- and gamma-herpesvirinae over 400 million years ago. Divergence of VZV from HSV is estimated to have occurred around 80–100,000 years ago at the time of the human dispersals from Africa. We calculated, using SNP analysis, that if the rate of VZV mutation was the same as for HSV (i.e., 10^{-7}/substitutions/base/year), VZV clades would have separated between 3,000 and 19,000 years ago (Barrett-Muir et al. 2001). Alternatively, if the substitution rate were lower, either intrinsically, or because VZV is latent for more of its natural history and undergoes fewer replication cycles than other herpes viruses, divergence could be more ancient and spread could have coincided with human migrations (Muir et al. 2002; Wagenaar et al. 2003b).

The substitution rate estimated from the US and Canadian clusters is around 10^{-7}/base/year, suggesting that the substitution rate is the same as herpes simplex. Given that strains of HSV are generally more diverse than VZV, this means either that VZV spends more time latent or that divergence between VZV strains has occurred more recently. While both may well be true, several pieces of evidence suggest that the geographical distribution of VZV strains may be due to spread after the migration of humans out of Africa. In particular, the finding of closely related Clade 5 strains of VZV in Africa and the Indian subcontinent does not reflect the evolutionary history of the hosts; Human populations from these countries are not thought to have a particularly recent common ancestry (Muir et al. 2002). Based on the sequence data available to them, from six viruses, which did not include a representative Clade 3 virus, Norberg and colleagues suggested that Clades 1 and 2 were the oldest, and that Clades 4 and 5 were recombinants of 1 and 2 that have arisen more recently (Norberg et al. 2006). Since Clade 1 is a European strain and Clade 2 is a Japanese strain, this interpretation is clearly at odds with the notion that VZV clades arose at the same time as human speciation (McGeoch et al. 1995).

More recently, McGeoch has used 23 published full length viral sequences and Bayesian methods to estimate a new model of VZV evolution (McGeoch 2009). This model is based only on SNPs occurring on the unique long region because McGeoch has estimated that substitution rates to the right of the internal repeat sequence, including the unique short region, are sufficiently higher to suggest a distinct evolutionary history. From this analysis, McGeoch inferred that the main European genogroup, Clade 1, was a more recent recombinant between older Clade 3 (European) and Clade 4 (African/Asian) strains (McGeoch 2009). The paucity of sequences for the Clade 2 (Japanese) and Clade 5 (African) groups precluded anymore than vague timelines. The debate will only be resolved by sequencing of more full length genomes from all over the world.

3.2 Geographical Spread of VZV Strains

The application of molecular methods to VZV epidemiology really started with the development of tractable methods based on PCR of variable fragments and SNPs. Prior to that, the methods based on REA required cultured virus which, given the difficulties associated with VZV isolation, was not amenable to molecular epidemiology.

Initial observations from REA that VZV strains differed genetically were difficult to characterize. Many of the most informative restriction fragments, that is, those that discriminated best between different strains, mapped to one or more of three variable regions – R2, R3, and R4 (Straus et al. 1981). Early studies were able to establish identity between directly transmitted strains (Straus et al. 1983), differences between epidemiologically unlinked strains, and differences between UK and Japanese wild-type isolates (Martin et al. 1982; Straus et al. 1983; Adams et al. 1989), but no obvious virus groupings emerged.

Following the identification of BglI and PstI SNPs as a method for distinguishing the Japanese vaccine from US wild-type strains (Martin et al. 1982; Adams et al. 1989; LaRussa et al. 1992), we used these restriction sites to survey over 240 strains of varicella and zoster collected from our local population in East London (Hawrami et al. 1997). In common with the findings by La Russa that 3 of 20 wild-type US strains were BglI positive, we found 20 of 244 viruses (8%) circulating in London to be BglI positive (LaRussa et al. 1992; Hawrami et al. 1997). All were PstI positive. Two clear messages emerged from this. First that UK wild-type strains could be distinguished in the same way as US viruses from the Japanese vOka vaccine – a finding of public health utility, should the vaccine be introduced into the UK (Hawrami and Breuer 1997). Second, our data added weight to initial observations, made on small numbers of viruses, that different, stable VZV genotypes might exist. The finding that immigrants from African and Asian countries with zoster were 6.3 times ($p < 0.0005$) more likely to have BglI positive strains than the UK residents reinforced the notion that geographical differences in the distribution of these genotypes might exist (Hawrami et al. 1997). Further complexity was added by reports from Japan confirming that, while all viruses were expectedly BglI positive, 19 of 30 (63%) clinical isolates in one study and 30 of 40 (75%) in another were also PstI positive and thus indistinguishable from the vaccine (Takada et al. 1995; Takayama et al. 1996).

To investigate further the molecular epidemiology of BglI and PstI strains we undertook two studies. On the premise that the virus isolated from a patient with zoster is identical to the original infecting strain of varicella, we genotyped subjects living in an ethnically mixed area of East London, where up to 30% of the population are immigrants who presented with varicella (105) and zoster (144) (Hawrami et al. 1997). Among the cases of zoster, BglI positive viruses were significantly more common ($p < 0.05$) in subjects who had immigrated from the Africa, India, Asia, and the Far East than in subjects who had grown up in the UK (Hawrami et al. 1997). In contrast, we found no strong association of BglI positive varicella viruses with immigrants. Instead there was a significant rise ($p < 0.001$) from 5 to 40% over 25 years in the prevalence of BglI positive varicella cases, suggesting that BglI viruses had been introduced with immigration and were now spreading in the UK (Hawrami et al. 1997). Although limited in its scope, this study did provide the first indication that BglI and PstI genotyping, especially in view of its simplicity and low cost, were useful for investigating questions about VZV strain distribution, spread, and possibly even pathogenesis.

Further analysis of viruses from around the world confirmed the BglI positive genotype as being present in 100% of 100 viruses collected in the parts of Africa, the Far East, and the Indian subcontinent, but accounting for fewer than 20% of viruses collected in Europe, the USA, and other countries settled by Europeans (Table 2). A prospective study of over 400 patients presenting with clinical zoster allowed us to identify 200 white British-born subjects aged from 5 to 98 years (Quinlivan et al. 2002; Sengupta et al. 2007). Again, we assumed their zoster virus was identical to the virus that first caused varicella and that, having grown up in the UK, their varicella occurred at aged $\leq$10 years of age. We showed that BglI negative Clade 1 and 3 viruses accounted for 80–90% of viruses circulating in London over the past 100 years in London (Sengupta et al. 2007). This confirms that the data obtained from opportunistic sampling of varicella and zoster viruses, which is the basis of most molecular epidemiological studies carried out to date, provide a reasonably accurate picture of VZV strain prevalence across the world (Sengupta et al. 2007). Moreover, our data confirms that BglI genotyping can be used to distinguish strains of European (BglI negative Clade 1 and 3) and non-European origin (BglI positive Clades 2, 4 and 5) and thus provides an easy tool for molecular epidemiology (Sengupta et al. 2007; Loparev et al. 2009).

These simple tools have now been used by many to investigate strain distribution throughout the world. Europe (UK, Ireland, Germany, France, Italy, Iceland, Greece, Finland, Spain, Czech Republic, Bulgaria, Albania, Latvia, Lithuania,

Table 9.2 Updated VZV Nomenclature and Naming Convention, 2008

New Clade Designations	1	2	3	4	5	(VI)	VII)
Old CDC	E1	J	E2	M2	M1	M4	M3
Old Breuer/ Muir	C	J	B	J2	A		
Old Grose	A	B	D	C			
Reference Strains	Dumas	vOka	11	8	CA123	08-104p	
GenBank Number	NC 001348	DQ008355	DQ479955	DQ479960	DQ457052		
	MSP	vOka	HJO	DR			
	AY548170	DQ008354	AJ871403	DQ452050			

*Clades are numbered in order of whole genome publication date. Proposed new clades are indicated with roman numerals. p: partial genome sequenced

VZV Naming Convention

VZV (i or s) / city . country / week . year / sample number (v or z) [clade]

i cultured isolate

s sequenced directly from clinical sample

v varicella

z zoster

*use full city name, 3-letter country code, 2-digit year, clade in brackets.

Examples:

VZVi/Minneapolis.USA/32.98/2v[4]

A cultured VZV isolate collected in Minneapolis, USA, in the 32nd week of 1998, the second varicella sample sequenced of this description, phylogenetically clusters with VZV clade 4.

VZVs/London.UK/15.04/z[5]

A genome sequenced directly from the clinical sample (not cultured, but may be amplified), collected in London, UK, in the 15th week of 2004. The presentation was zoster, and the genome clusters with clade 5.

Russia, Poland, Estonia, Belorussia, Ukraine) (Faga et al. 2001; Sauerbrei et al. 2003, 2008; Wagenaar et al. 2003b; Carr et al. 2004; Dayan et al. 2004; Koskiniemi et al. 2007; Loparev et al. 2009), America (USA, Canada, Argentina and Brazil) (Faga et al. 2001; Muir et al. 2002; Wagenaar et al. 2003a; Dayan et al. 2004; Loparev et al. 2004; Peters et al. 2006), the Middle East (Morocco, Tunisia) (Loparev et al. 2004), Australia and New Zealand (Loparev et al. 2007b) have predominantly BglI negative and PstI positive European Clade 1 and 3 strain viruses. In contrast, Japan, Singapore, India, and China have BglI positive non-European Clade 2, 4, and 5 strains (Chow et al. 1993; Takada et al. 1995; Takayama et al. 1996; Faga et al. 2001; Wagenaar et al. 2003a; Loparev et al. 2004; Peters et al. 2006; Kaushik et al. 2008; Liu et al. 2009), while in Africa (Chad, Sudan, DRC, Guinea Bissau, Zambia) (Muir et al. 2002; Loparev et al. 2004) Clade 5 strains predominated (Table 2). Furthermore, genotyping has shown that African and Asian Clade 4 and 5 strains are spreading in countries with European populations (Hawrami et al. 1997; Sauerbrei et al. 2003; Carr et al. 2004; Sauerbrei and Wutzler 2007; Sengupta et al. 2007). In contrast, very few European Clade 1 and 3 viruses and no Japanese Clade 2 viruses have been found in surveys of over 100 strains from African countries including Guinea Bissau, Zambia, Sudan, Chad, and Democratic Republic of Congo (Wagenaar et al. 2003a; Loparev et al. 2004; Quinlivan et al. 2002). Two cases of Clade 1 viruses that were genotyped as part of an outbreak in Guinea Bissau were, in fact, imported by children returning from Europe (Poulsen pers. corresp.). However, European Clade 3 viruses have been found in the Congo, causing atypical palmar and plantar lesions that were initially confused clinically with monkeypox (MacNeil et al. 2009). One Clade 3 virus has been opportunistically identified in Sudan, again as part of genotyping of a monkeypox outbreak (MacNeil et al. 2009). Whether Clade 3 viruses are endogenous to Africa remains to be seen. However, were this to be the case, it would fit with McGeoch's hypothesis that Clade 3 is among the more ancient VZV Clades and one of the antecedents of European Clade 1 (McGeoch 2009).

3.3 Molecular Epidemiology and VZV Pathogenesis

3.3.1 Genotyping, Reinfection, and Reactivation

The use of genotyping to investigate questions of pathogenesis is well established for VZV. Straus's study in 1984 used REA to genotype varicella and zoster viruses from the same patient, thus proving Hope Simpson's hypothesis that zoster is due to reactivation of latent VZV originally acquired from chickenpox (Straus et al. 1984). This finding was confirmed by Hayakawa in three patients (Hayakawa et al. 1986). All three patients were immunosuppressed with primary varicella and zoster occurring close together.

In 2006, we made use of SNP genotyping to investigate a case of recurrent zoster in an immunocompetent man. This young man, who gave a history of one episode of varicella aged 5 years old, presented as an adult with two episodes of zoster 3 years apart, the first time in the ocular division of the left trigeminal nerve and the second in the left thoracic region (Taha et al. 2004). Genotyping showed the first strain to be a Clade 5/A/M1 and the second to be a Clade 4/B//E2/D strain. Microsatellite typing of the human DNA confirmed that both samples originated from the same patient. This result suggested that infection with two different viruses had occurred and that both had established latency. The data corroborated findings from earlier studies which, using BglI genotyping, had shown that zoster due to clade 1/C/E1/A viruses occurred in 30% of subjects who had immigrated as adults to the UK from areas where Clade 5/A/M1 viruses circulate (Quinlivan et al. 2002). In none of these adults was there a history of varicella in the UK. These results suggested that asymptomatic reinfection of latently infected individuals is occurring, with establishment of latency by the second strain. More recently, the reactivation of wild-type virus in Health Care Workers who seroconverted to vaccine reinforces silent reinfection as a component of VZV natural history (Hambleton et al. 2008).

While reinfection and boosting of both humoral and cellular immunity is well described, our findings raise questions about the pathogenesis of latency (Arvin et al. 1983). Much evidence supports the retrograde spread of virus from skin to sensory nerve root ganglia as a major route for establishment of VZV latency (Hardy et al. 1991; Chen et al. 2003). Other data also suggest that haematogenous transfer of virus to ganglia is possible (Zerboni et al. 2005) and this is supported by our data. Whether or not haematogenous transfer first requires asymptomatic infection of the skin with retrograde spread to the ganglia thereafter, or whether cell free virus can be transferred directly from lymphocytes to ganglia, remains to be established.

That infection and establishment of latency with more than one strain is possible is self evident both because recombination is occurring in wild-type and vaccine viruses and from the case reports and studies discussed (Hambleton et al. 2008; Quinlivan et al. 2002; Taha et al. 2004). Early reports in which REA was used to genotype multiple isolates from a single patient showed not only variation in the fragments contained in tandem repeats but also heterogeneity of a SNP restriction site, suggesting multiple infections (Shibuta et al. 1974; Hondo et al. 1987) while other reports have failed to confirm this (Takayama et al. 1989). Using SNP genotyping, we have demonstrated coinfection with two different viruses within the same individual (Quinlivan et al. 2009). In that study, limiting dilution of PCR product allowed genotyping of single molecules and the evidence that a vesicle fluid from a child with chickenpox contained Clade 1/C/E1/A and Clade 3/B/D/E2 viruses in a ration of 3:1. Whether or not each of the two clades inhabited different vesicles is not known.

3.3.2 Genotyping and Virulence

The notion that viral genetic variation is associated with differences in virulence is now well established for many viruses, including HIV, Hepatitis C, Hepatitis B, and

Influenza A. Until recently, this has been difficult to study in larger DNA viruses. The discovery of the MSP variant by Grose and colleagues provided the first evidence that more virulent VZV variants might exist (Santos et al. 1998, 2000). A mutation likely to cause the increased virulence was identified in the gE protein, N150D, and affected the binding of an antibody that is widely used for diagnosis of VZV. This mutation has now been detected in two patients, one in the US and one in Canada (Santos et al. 2000; Tipples et al. 2002). In addition two isolates from leukemic children in Sweden were found to have different mutation in gE, S152A, which also abrogated binding of the antibody (Wirgart et al. 2006). Finally, a fatal case of varicella in a 15-year-old Italian boy was caused by a virus with a mutation in the terminal amino acid of gE, G161R, though it is not known whether this mutation also affected antibody binding. Additionally in this case, polymorphisms in ORF 36, which codes for the VZV thymidine kinase, may have contributed to the pathology, although these particular mutations have not themselves been associated with acyclovir resistance (Natoli et al. 2006). Extensive analysis by the Grose group showed that the original gE mutated virus, MSP, replicated faster than a wild-type VZV virus both in monolayers and in SCID-hu epithelial implants, with extensive cell-to-cell spread (Santos et al. 2000). Full length sequencing of viruses from the US and Canadian cases, MSP and BC, found no other mutation to explain the alteration in biological phenotype of these viruses. This extremely important finding provides further support for the hypothesis that a single mutation can profoundly alter the pathogenesis of the virus, both in vivo and in vitro.

The detection of the glycoprotein E mutations was helped by the fact that they occurred in an epitope, which is the target of the commonly used diagnostic monoclonal antibody, 3B3. The only other viral genetic association with clinical phenotype has been the data suggesting that vaccine virus obtained from rashes occurring after VZV immunization are significantly more likely to carry one or more of four SNPs, three of which occur in ORF 62 and two of which result in a reversion to the wild-type amino acid (Quinlivan et al. 2007). Our efforts to investigate the possibility of viral genetic associations with virulence have included genotyping of viruses sampled from cases of rare clinical complications that clustered in time and space (Sengupta unpublished). Three unrelated cases of VZV encephalitis presented within 1 week to a hospital in Poland and varicella-associated arthritis associated with chickenpox presented within a 2 week period in three unrelated children to a pediatrician in Germany. These case clusters were analyzed using clade-specific genotyping. The finding of different clades within each cluster suggested that the occurrences were coincidental and a viral genetic cause was unlikely.

3.4 *Molecular Epidemiology Transmission and Infection Control*

A number of groups have reported confirmed transmissions of virus between family members or patients sharing a hospital room, by REA (Straus et al. 1983; Takayama et al. 1989), length polymorphism of variable tandem repeat regions (Hondo and Yogo

1988; Kinoshita et al. 1988; Hawrami et al. 1996), and SNP typing with repeat region length polymorphism (Muir et al. 2002; Molyneaux et al. 2006). However, at least one group found that isolates from six patients who had shared a hospital room and who developed varicella although broadly similar varied in the length of *Hpa-1* G fragment. This fragment contains the R4 repeat region (Hayakawa et al. 1986).

SNP typing with or without length polymorphism of variable regions has been used to link viruses isolated from staff and patients on a single hospital ward suggesting transmission (Molyneaux et al. 2006; Lopez et al. 2008). In particular, the finding of an unusual signature SNP profile among viruses sampled from a nursing home outbreak enabled the investigators to link potential cases and led to a number of interesting observations with implications for infection control (Lopez et al. 2008). In this outbreak, the virus strain WV was found to cluster with Clade 5 viruses, which are most prevalent in Africa and Asia and account for fewer than 10–20% of circulating strains in European countries (Quinlivan et al. 2002; Carr et al. 2004; Sauerbrei and Wutzler 2007; Loparev et al. 2009; Sengupta unpublished). Unusually, the virus carried a sequence change at position 107252, which resulted in a substitution of a serine for a glycine, a change that has hitherto only been seen in the Oka vaccine strain (Lopez et al. 2008). The unique identity of this strain allowed the authors to establish that, despite efforts to ensure that the lesions were covered, transmission occurred from a case of zoster to two patients and a member of staff. In at least one case there was no possible direct contact with the index case, and the additional finding of the causative virus in dust confirmed that infection was likely to have occurred by the airborne route (Lopez et al. 2008). This case, supported by the molecular epidemiology, highlighted a number of important principles for control of transmission from zoster: namely that lesions of zoster are infectious even if covered by clothing or loose dressings and that virus is shed into the environment from where it may be aerosolized to cause infection in susceptible individuals (Breuer 2008). Molecular epidemiological tools, in particular SNP typing, has also been used extensively to establish the origin, vaccine, or wild-type of post immunization rashes and to identify rare cases of vaccine strain transmission to third parties (Breuer and Schmid 2008).

3.5 Molecular Epidemiology of VZV Outbreaks

One of the most puzzling aspects of VZV epidemiology is the geographical variation in the average age of infection, in particular the late age of infection that is reported in many tropical regions of the world (Lee and Tan 1995). A number of reasons for the geographically related differences seen with varicella have been advanced, including climate, host genetics, population density, and social mixing patterns. To investigate these, we have used molecular epidemiological tools to estimate viral infectivity in a tropical country (Nichols unpublished). The slow evolutionary rate of most of the VZV genome – VZV has an estimated nucleotide substitution of less than 0.008% per 20 passages in tissue culture and probably less during infection – precludes a phylogenetic approach to analyze transmission in an outbreak as has successfully been used for some RNA viruses. However, apparent length polymorphism in the R4 region in an outbreak. Hayakawa et al. (1986) suggests that the rate of mutation in this and other variable regions (R1–R5 and OriS) might be sufficient to measure transmission in a single outbreak. Genome sequencing by the US/Canadian group showed that the R1, R4, and OriS regions had sufficient diversity between strains, and sufficient mutation rates during 20–72 passages, that they would likely be the most useful for measuring transmission in a single outbreak (Tyler et al. 2007). The OriS also contained flanking SNPs, which recapitulated the full genomic phylogenetic tree (Peters et al. 2006; Tyler et al. 2007). All three regions showed length polymorphism between different strains and elongation by 0–2 repeat units after tissue culture passage. The OriS repeat elements were TA/GA dinucleotides while the R1 and R4 repeat elements were 48 and 27 Bps, respectively. We chose to use OriS variation to investigate person to person transmissions within an outbreak as means of calculating infectivity. The OriS had also been shown to be shorter in the vaccine Oka strain compared with the parental Oka strain (Gomi et al. 2002), and early studies had indicated that loss of dinucleotide repeats in the VZV Ori was associated with loss of replicative activity (Stow and Davison 1986).

Our study involved the collection of samples and data from a population in Bissau, Guinea Bissau, during an outbreak of varicella in 2000–2001 (Poulsen et al. 2002). Genotyping of 400 viruses from an outbreak involving 1,485 individuals revealed that all except six were Clade 5 and that the OriS varied during the outbreak, with a mutation rate of 16% per transmission. By examining 49 households for which the genotype of the index case and all subsequent household cases were known and modeling the data, we were able to show that the probability that an infection occurring in a household had come from the index case increased above background between days 10 and 20, peaking at day 14 (Nichols unpublished). Since this corresponds to the observed incubation period of VZV, the model appeared to correctly reflect household transmission. The model showed that nearly 30% of household infections in this outbreak, which occurred in the incubation period, had come from outside. These results and other epidemiological data allowed us to calculate the household infectivity of VZV in Guinea Bissau, a tropical country, as 12.5%. This is considerably less than the 60–85% that has been estimated for household infectivity in temperate climates (Simpson 1952; Ross 1962). From these data we therefore surmise that the differences in epidemiology seen in tropical climates are due to lower viral infectivity. In crowded Bandim, the epidemic spread of VZV despite low infectivity is strongly influenced by high population density, school attendance by children, and large households that contain proportionately more susceptible newborns (Nichols unpublished data).

3.5.1 Molecular Epidemiology and Public Health

Genotyping of VZV from annual outbreaks shows co-circulation temporally and spacially of genetically distinct viruses, suggesting that VZV epidemics originate from multiple sources. Co-circulation of multiple VZV clades observed a varicella outbreak in a region of London, England (Sengupta unpublished). Multiple clades were also found among viruses circulating at low frequencies between epidemics. Co-circulation is also supported by findings in the Guinea Bissau outbreak, wherein many different genotypes appear early in the outbreak and persist throughout the epidemic (Breuer unpublished data). This ties in with the hypothesis first outlined in the 1950s (London and Yorke 1973; Yorke and London 1973) that in temperate climates varicella is continually circulating, with outbreaks occurring when seasonal and social factors, such as the beginning of the school year, increase the intimacy and frequency of contacts between infected and VZV naïve subjects. VZVs pattern of genotypic co-circulation contrasts with measles, which spreads very quickly by casual contact, resulting in one genotype sweeping through a community (London and Yorke 1973; Yorke and London 1973).

Another factor contributing to viral diversity in an outbreak is reactivation of latent VZV genotypes. Within the Guinea Bissau outbreak, six cases of zoster occurred, each with a distinct TA/GA genotype, and two of these transmitted to household contacts and beyond (Breuer unpublished). This confirms that even where the main force of infection is from varicella, there is measurable transmission from cases of zoster. Quantification of this will be important for public health planning in vaccinated populations, where exposure is likely to arise more commonly from zoster than varicella. The use of the zoster vaccine may also prove to be an important weapon in the armory of controlling breakthrough infections in vaccinated populations.

4 Summary

The molecular epidemiology of varicella-zoster virus has led to an understanding of the evolution and spread of this human herpesvirus. The development of rapid methods has facilitated the use of genotyping to investigate questions of pathogenesis including asymptomatic reinfection and co-infection. Full length genomes together with modeling of genetic variation are likely to provide increasing insights into the viral determinants of pathogenesis and disease.

Acknowledgements The author thank Richard Nichols, Ravinda Kanda, Nitu Sengupta, and Karin Averbeck for helpful discussions and critical reading of the manuscript. Work from the Breuer group was funded by The Barts and the London Special Trustees, the Medical Research Council, and The Welcome Trust.

References

Abe T, Sato M, Tamai M (2000) Variable R1 region in varicella zoster virus in fulminant type of acute retinal necrosis syndrome. Br J Ophthalmol 84:193–198

Adams SG, Dohner DE, Gelb LD (1989) Restriction fragment differences between the genomes of the Oka varicella vaccine virus and American wild-type varicella-zoster virus. J Med Virol 29:38–45

Arvin AM, Koropchak CM, Wittek AE (1983) Immunologic evidence of reinfection with varicella-zoster virus. J Infect Dis 148:200–205

Barrett-Muir W, Hawrami K, Clarke J, Breuer J (2001) Investigation of varicella-zoster virus variation by heteroduplex mobility assay. Arch Virol 17(Suppl 1):17–25

Bokay J (1909) Uber den ätiologischen Zusammenhang der Varizellen mit gewissen Fällen von Herpes Zoster. Wein Klin Wochenschr 22:1323–1326

Breuer J (2008) Herpes zoster: new insights provide an important wake-up call for management of nosocomial transmission. J Infect Dis 197:635–637

Breuer J, Schmid DS (2008) Vaccine Oka variants and sequence variability in vaccine-related skin lesions. J Infect Dis 197(Suppl 2):S54–57

Brookes AJ (1999) The essence of SNPs. Gene 234:177–186

Carr MJ, McCormack GP, Crowley B (2004) Genetic variation in clinical varicella-zoster virus isolates collected in Ireland between 2002 and 2003. J Med Virol 73:131–136

Casey TA, Ruyechan WT, Flora MN, Reinhold W, Straus SE, Hay J (1985) Fine mapping and sequencing of a variable segment in the inverted repeat region of varicella-zoster virus DNA. J Virol 54:639–642

Chen JJ, Gershon AA, Li ZS, Lungu O, Gershon MD (2003) Latent and lytic infection of isolated guinea pig enteric ganglia by varicella zoster virus. J Med Virol 70(Suppl 1):S71–78

Chow VT, Wan SS, Doraisingham S, Ling AE (1993) Comparative analysis of the restriction endonuclease profiles of the Dumas and Singapore strains of varicella-zoster virus. J Med Virol 40:339–342

Collins FS, Guyer MS, Charkravarti A (1997) Variations on a theme: cataloging human DNA sequence variation. Science 278:1580–1581

Collins FS, Brooks LD, Chakravarti A (1998) A DNA polymorphism discovery resource for research on human genetic variation. Genome Res 8:1229–1231

Davison AJ, Scott JE (1986) The complete DNA sequence of varicella-zoster virus. J Gen Virol 67 (Pt 9):1759–1816

Dayan GH, Panero MS, Debbag R, Urquiza A, Molina M, Prieto S, Del Carmen PM, Scagliotti G, Galimberti D, Carroli G, Wolff C, Schmid DS, Loparev V, Guris D, Seward J (2004) Varicella seroprevalence and molecular epidemiology of varicella-zoster virus in Argentina, 2002. J Clin Microbiol 42:5698–5704

Dohner DE, Adams SG, Gelb LD (1988) Recombination in tissue culture between varicella-zoster virus strains. J Med Virol 24:329–341

Dumas AM, Geelen JL, Weststrate MW, Wertheim P, van der Noordaa J (1981) XbaI, PstI, and BglII restriction enzyme maps of the two orientations of the varicella-zoster virus genome. J Virol 39:390–400

Ecker JR, Hyman RW (1981) Varicella-zoster virus vaccine DNA differs from the parental virus DNA. J Virol 40:314–318

Ecker JR, Hyman RW (1982) Varicella zoster virus DNA exists as two isomers. Proc Natl Acad Sci USA 79:156–160

Faga B, Maury W, Bruckner DA, Grose C (2001) Identification and mapping of single nucleotide polymorphisms in the varicella-zoster virus genome. Virology 280:1–6

Gelb LD, Dohner DE, Gershon AA, Steinberg SP, Waner JL, Takahashi M, Dennehy PH, Brown AE (1987) Molecular epidemiology of live, attenuated varicella virus vaccine in children with leukemia and in normal adults. J Infect Dis 155:633–640

Gomi Y, Sunamachi H, Mori Y, Nagaike K, Takahashi M, Yamanishi K (2002) Comparison of the complete DNA sequences of the Oka varicella vaccine and its parental virus. J Virol 76:11447–11459

Grose C, Tyler S, Peters G, Hiebert J, Stephens GM, Ruyechan WT, Jackson W, Storlie J, Tipples GA (2004) Complete DNA sequence analyses of the first two varicella-zoster virus

glycoprotein E (D150N) mutant viruses found in North America: evolution of genotypes with an accelerated cell spread phenotype. J Virol 78:6799–6807
Hambleton S, Steinberg SP, Larussa PS, Shapiro ED, Gershon AA (2008) Risk of herpes zoster in adults immunized with varicella vaccine. J Infect Dis 197(Suppl 2):S196–199
Hardy I, Gershon AA, Steinberg SP, LaRussa P (1991) The incidence of zoster after immunization with live attenuated varicella vaccine. A study in children with leukemia. Varicella vaccine collaborative study group. N Engl J Med 325:1545–1550
Hawrami K, Breuer J (1997) Analysis of United Kingdom wild-type strains of varicella-zoster virus: differentiation from the Oka vaccine strain. J Med Virol 53:60–62
Hawrami K, Harper D, Breuer J (1996) Typing of varicella zoster virus by amplification of DNA polymorphisms. J Virol Methods 57:169–174
Hawrami K, Hart IJ, Pereira F, Argent S, Bannister B, Bovill B, Carrington D, Ogilvie M, Rawstorne S, Tryhorn Y, Breuer J (1997) Molecular epidemiology of varicella-zoster virus in East London, England, between 1971 and 1995. J Clin Microbiol 35:2807–2809
Hayakawa Y, Yamamoto T, Yamanishi K, Takahashi M (1986) Analysis of varicella-zoster virus DNAs of clinical isolates by endonuclease HpaI. J Gen Virol 67(Pt 9):1817–1829
Hondo R, Yogo Y (1988) Strain variation of R5 direct repeats in the right-hand portion of the long unique segment of varicella-zoster virus DNA. J Virol 62:2916–2921
Hondo R, Yogo Y, Kurata T, Aoyama Y (1987) Genome variation among varicella-zoster virus isolates derived from different individuals and from the same individuals. Arch Virol 93:1–12
Hope-Simpson RE (1965) The nature of herpes zoster: a long-term study and a new hypothesis. Proc R Soc Med 58:9–20
Kaushik KS, Lahiri KK, Chumber SK, Gupta RM, Kumar S, Kapila K, Karade S (2008) Molecular characterization of clinical varicella-zoster strains from India and differentiation from the oka vaccine strain. Jpn J Infect Dis 61:65–67
Kinchington PR, Turse SE (1995) Molecular basis for a geographic variation of varicella-zoster virus recognized by a peptide antibody. Neurology 45:S13–14
Kinchington PR, Reinhold WC, Casey TA, Straus SE, Hay J, Ruyechan WT (1985) Inversion and circularization of the varicella-zoster virus genome. J Virol 56:194–200
Kinoshita H, Hondo R, Taguchi F, Yogo Y (1988) Variation of R1 repeated sequence present in open reading frame 11 of varicella-zoster virus strains. J Virol 62:1097–1100
Koskiniemi M, Lappalainen M, Schmid DS, Rubtcova E, Loparev VN (2007) Genotypic analysis of varicella-zoster virus and its seroprevalence in Finland. Clin Vaccine Immunol 14:1057–1061
Kruglyak L (1997) The use of a genetic map of biallelic markers in linkage studies. Nat Genet 17:21–24
LaRussa P, Lungu O, Hardy I, Gershon A, Steinberg SP, Silverstein S (1992) Restriction fragment length polymorphism of polymerase chain reaction products from vaccine and wild-type varicella-zoster virus isolates. J Virol 66:1016–1020
Lee BW, Tan AY (1995) Chickenpox in the tropics. BMJ 310:941
Liu J, Wang M, Gan L, Yang S, Chen J (2009) Genotyping of clinical varicella-zoster virus isolates collected in China. J Clin Microbiol 47:1418–1423
London WP, Yorke JA (1973) Recurrent outbreaks of measles, chickenpox and mumps I. Seasonal variation in contact rates. Am J Epidemiol 98:453–468
Loparev VN, Gonzalez A, Deleon-Carnes M, Tipples G, Fickenscher H, Torfason EG, Schmid DS (2004) Global identification of three major genotypes of varicella-zoster virus: longitudinal clustering and strategies for genotyping. J Virol 78:8349–8358
Loparev V, Martro E, Rubtcova E, Rodrigo C, Piette JC, Caumes E, Vernant JP, Schmid DS, Fillet AM (2007a) Toward universal varicella-zoster virus (VZV) genotyping: diversity of VZV strains from France and Spain. J Clin Microbiol 45:559–563
Loparev VN, Rubtcova EN, Bostik V, Govil D, Birch CJ, Druce JD, Schmid DS, Croxson MC (2007b) Identification of five major and two minor genotypes of varicella-zoster virus strains:

a practical two-amplicon approach used to genotype clinical isolates in Australia and New Zealand. J Virol 81:12758–12765

Loparev VN, Rubtcova EN, Bostik V, Tzaneva V, Sauerbrei A, Robo A, Sattler-Dornbacher E, Hanovcova I, Stepanova V, Splino M, Eremin V, Koskiniemi M, Vankova OE, Schmid DS (2009) Distribution of varicella-zoster virus (VZV) wild-type genotypes in northern and southern Europe: evidence for high conservation of circulating genotypes. Virology 383:216–225

Lopez AS, Burnett-Hartman A, Nambiar R, Ritz L, Owens P, Loparev VN, Guris D, Schmid DS (2008) Transmission of a newly characterized strain of varicella-zoster virus from a patient with herpes zoster in a long-term-care facility, West Virginia, 2004. J Infect Dis 197:646–653

Martin JH, Dohner DE, Wellinghoff WJ, Gelb LD (1982) Restriction endonuclease analysis of varicella-zoster vaccine virus and wild-type DNAs. J Med Virol 9:69–76

McGeoch DJ (2009) Lineages of varicella-zoster virus. J Gen Virol 90:963–969

McGeoch DJ, Cook S (1994) Molecular phylogeny of the alphaherpesvirinae subfamily and a proposed evolutionary timescale. J Mol Biol 238:9–22

McGeoch DJ, Cook S, Dolan A, Jamieson FE, Telford EA (1995) Molecular phylogeny and evolutionary timescale for the family of mammalian herpesviruses. J Mol Biol 247:443–458

Molyneaux PJ, Parker S, Khan IH, Millar CG, Breuer J (2006) Use of genomic analysis of varicella-zoster virus to investigate suspected varicella-zoster transmission within a renal unit. J Clin Virol 36:76–78

Muir WB, Nichols R, Breuer J (2002) Phylogenetic analysis of varicella-zoster virus: evidence of intercontinental spread of genotypes and recombination. J Virol 76:1971–1979

Na GY (1997) Herpes zoster in three healthy children immunized with varicella vaccine (Oka/Biken); the causative virus differed from vaccine strain on PCR analysis of the IV variable region (R5) and of a PstI-site region. Br J Dermatol 137:255–258

Natoli S, Ciotti M, Paba P, Testore GP, Palmieri G, Orlandi A, Sabato AF, Leonardis F (2006) A novel mutation of varicella-zoster virus associated to fatal hepatitis. J Clin Virol 37:72–74

Norberg P, Liljeqvist JA, Bergstrom T, Sammons S, Schmid DS, Loparev VN (2006) Complete-genome phylogenetic approach to varicella-zoster virus evolution: genetic divergence and evidence for recombination. J Virol 80:9569–9576

Oakes JE, Iltis JP, Hyman RW, Rapp F (1977) Analysis by restriction enzyme cleavage of human varicella-zoster virus DNAs. Virology 82:353–361

Parker SP, Quinlivan M, Taha Y, Breuer J (2006) Genotyping of varicella-zoster virus and the discrimination of Oka vaccine strains by TaqMan real-time PCR. J Clin Microbiol 44:3911–3914

Peters GA, Tyler SD, Grose C, Severini A, Gray MJ, Upton C, Tipples GA (2006) A full-genome phylogenetic analysis of varicella-zoster virus reveals a novel origin of replication-based genotyping scheme and evidence of recombination between major circulating clades. J Virol 80:9850–9860

Pichini B, Ecker JR, Grose C, Hyman RW (1983) DNA mapping of paired varicella-zoster virus isolates from patients with shingles. Lancet 2:1223–1225

Poulsen A, Qureshi K, Lisse I, Kofoed PE, Nielsen J, Vestergaard BF, Aaby P (2002) A household study of chickenpox in Guinea-Bissau: intensity of exposure is a determinant of severity. J Infect 45:237–242

Quinlivan M, Hawrami K, Barrett-Muir W, Aaby P, Arvin A, Chow VT, John TJ, Matondo P, Peiris M, Poulsen A, Siqueira M, Takahashi M, Talukder Y, Yamanishi K, Leedham-Green M, Scott FT, Thomas SL, Breuer J (2002) The molecular epidemiology of varicella-zoster virus: evidence for geographic segregation. J Infect Dis 186:888–894

Quinlivan ML, Gershon AA, Al Bassam MM, Steinberg SP, LaRussa P, Nichols RA, Breuer J (2007) Natural selection for rash-forming genotypes of the varicella-zoster vaccine virus detected within immunized human hosts. Proc Natl Acad Sci U S A 104:208–212

Quinlivan M, Sengupta N, Breuer J (2009) A case of varicella caused by co-infection with two different genotypes of varicella-zoster virus. J Clin Virol 44:66–69

Richards JC, Hyman RW, Rapp F (1979) Analysis of the DNAs from seven varicella-zoster virus isolates. J Virol 32:812–821
Ross AH (1962) Modification of chicken pox in family contacts by administration of gamma globulin. N Engl J Med 267:369–376
Santos RA, Padilla JA, Hatfield C, Grose C (1998) Antigenic variation of varicella zoster virus Fc receptor gE: loss of a major B cell epitope in the ectodomain. Virology 249:21–31
Santos RA, Hatfield CC, Cole NL, Padilla JA, Moffat JF, Arvin AM, Ruyechan WT, Hay J, Grose C (2000) Varicella-zoster virus gE escape mutant VZV-MSP exhibits an accelerated cell-to-cell spread phenotype in both infected cell cultures and SCID-hu mice. Virology 275:306–317
Sauerbrei A, Wutzler P (2007) Different genotype pattern of varicella-zoster virus obtained from patients with varicella and zoster in Germany. J Med Virol 79:1025–1031
Sauerbrei A, Eichhorn U, Gawellek S, Egerer R, Schacke M, Wutzler P (2003) Molecular characterisation of varicella-zoster virus strains in Germany and differentiation from the Oka vaccine strain. J Med Virol 71:313–319
Sauerbrei A, Zell R, Philipps A, Wutzler P (2008) Genotypes of varicella-zoster virus wild-type strains in Germany. J Med Virol 80:1123–1130
Seiler HE (1949) A study of herpes zoster particularly in its relationship to chickenpox. J Hyg (Lond) 47:253–262
Sengupta N, Taha Y, Scott FT, Leedham-Green ME, Quinlivan M, Breuer J (2007) Varicella-zoster-virus genotypes in East London: a prospective study in patients with herpes zoster. J Infect Dis 196:1014–1020
Sergeev N, Rubtcova E, Chizikov V, Schmid DS, Loparev VN (2006) New mosaic subgenotype of varicella-zoster virus in the USA: VZV detection and genotyping by oligonucleotide-microarray. J Virol Methods 136:8–16
Shibuta H, Ishikawa T, Hondo R, Aoyama Y, Kurata K, Matumoto M (1974) Varicella virus isolation from spinal ganglion. Arch Gesamte Virusforsch 45:382–385
Simpson RE (1952) Infectiousness of communicable diseases in the household (measles, chickenpox, and mumps). Lancet 2:549–554
Simpson RE (1954) Studies on shingles: is the virus ordinary chickenpox virus? Lancet 267:1299–1302
Stow ND, Davison AJ (1986) Identification of a varicella-zoster virus origin of DNA replication and its activation by herpes simplex virus type 1 gene products. J Gen Virol 67(Pt 8): 1613–1623
Straus SE, Aulakh HS, Ruyechan WT, Hay J, Casey TA, Vande Woude GF, Owens J, Smith HA (1981) Structure of varicella-zoster virus DNA. J Virol 40:516–525
Straus SE, Owens J, Ruyechan WT, Takiff HE, Casey TA, Vande Woude GF, Hay J (1982) Molecular cloning and physical mapping of varicella-zoster virus DNA. Proc Natl Acad Sci USA 79:993–997
Straus SE, Hay J, Smith H, Owens J (1983) Genome differences among varicella-zoster virus isolates. J Gen Virol 64:1031–1041
Straus SE, Reinhold W, Smith HA, Ruyechan WT, Henderson DK, Blaese RM, Hay J (1984) Endonuclease analysis of viral DNA from varicella and subsequent zoster infections in the same patient. N Engl J Med 311:1362–1364
Taha YA, Quinlivan M, Scott FT, Leedham-Green M, Hawrami K, Thomas JM, Breuer J (2004) Are false negative direct immnufluorescence assays caused by varicella zoster virus gE mutant strains? J Med Virol 73:631–635
Takada M, Suzutani T, Yoshida I, Matoba M, Azuma M (1995) Identification of varicella-zoster virus strains by PCR analysis of three repeat elements and a PstI-site-less region. J Clin Microbiol 33:658–660
Takahashi M, Otsuka T, Okuno Y, Asano Y, Yazaki T (1974) Live vaccine used to prevent the spread of varicella in children in hospital. Lancet 2:1288–1290
Takayama M, Takayama N, Hachimori K, Minamitani M (1988) Restriction endonuclease analysis of viral DNA from a patient with bilateral herpes zoster lesions. J Infect Dis 157:392–393

Takayama M, Takayama N, Kameoka Y, Hachimori K, Kaneda K, Minamitani M (1989) Comparative restriction endonuclease analysis of varicella-zoster virus clinical isolates. Med Microbiol Immunol 178:61–67

Takayama M, Takayama N, Inoue N, Kameoka Y (1996) Application of long PCR method of identification of variations in nucleotide sequences among varicella-zoster virus isolates. J Clin Microbiol 34:2869–2874

Tipples GA, Stephens GM, Sherlock C, Bowler M, Hoy B, Cook D, Grose C (2002) New variant of varicella-zoster virus. Emerg Infect Dis 8:1504–1505

Tyler SD, Severini A (2006) The complete genome sequence of herpesvirus papio 2 (Cercopithecine herpesvirus 16) shows evidence of recombination events among various progenitor herpesviruses. J Virol 80:1214–1221

Tyler SD, Peters GA, Grose C, Severini A, Gray MJ, Upton C, Tipples GA (2007) Genomic cartography of varicella-zoster virus: a complete genome-based analysis of strain variability with implications for attenuation and phenotypic differences. Virology 359:447–458

Wagenaar TR, Chow VT, Buranathai C, Thawatsupha P, Grose C (2003) The out of Africa model of varicella-zoster virus evolution: single nucleotide polymorphisms and private alleles distinguish Asian clades from European/North American clades. Vaccine 21:1072–1081

Wang DG, Fan JB, Siao CJ, Berno A, Young P, Sapolsky R, Ghandour G, Perkins N, Winchester E, Spencer J, Kruglyak L, Stein L, Hsie L, Topaloglou T, Hubbell E, Robinson E, Mittmann M, Morris MS, Shen N, Kilburn D, Rioux J, Nusbaum C, Rozen S, Hudson TJ, Lipshutz R, Chee M, Lander ES (1998) Large-scale identification, mapping, and genotyping of single-nucleotide polymorphisms in the human genome. Science 280:1077–1082

Weller TH, Witton HM, Bell EJ (1958) The etiologic agents of varicella and herpes zoster; isolation, propagation, and cultural characteristics in vitro. J Exp Med 108:843–868

Wirgart BZ, Estrada V, Jackson W, Linde A, Grose C (2006) A novel varicella-zoster virus gE mutation discovered in two Swedish isolates. J Clin Virol 37:134–136

Yorke JA, London WP (1973) Recurrent outbreaks of measles, chickenpox and mumps II. Systematic differences in contact rates and stochastic effects. Am J Epidemiol 98:469–482

Yoshida M, Tamura T (1999) An analytical method for R5 repeated structure in varicella-zoster virus DNA by polymerase chain reaction. J Virol Methods 80:213–215

Yoshida M, Tamura T, Hiruma M (1999) Analysis of strain variation of R1 repeated structure in varicella-zoster virus DNA by polymerase chain reaction. J Med Virol 58:76–78

Yoshida M, Tamura T, Miyasaka K, Shimizu A, Ohashi N, Itoh M (2003) Analysis of numbers of repeated units in R2 region among varicella-zoster virus strains. J Dermatol Sci 31:129–133

Zweerink HJ, Morton DH, Stanton LW, Neff BJ (1981) Restriction endonuclease analysis of the DNA from varicella-zoster virus: stability of the DNA after passage in vitro. J Gen Virol 55:207–211

Roles of Cellular Transcription Factors in VZV Replication

William T. Ruyechan

Contents

Abstract Varicella zoster virus (VZV) is the causative agent of chickenpox and shingles. During productive infection the complete VZV proteome consisting of some 68 unique gene products is expressed through interaction of a small number of viral transcriptional activators with the general transcription apparatus of the host

W.T. Ruyechan
Department of Microbiology and Immunology, Witebsky Center for Microbial Pathogenesis and Immunology, University at Buffalo SUNY, 138 Farber Hall, Buffalo, NY 14214, USA
e-mail: ruyechan@buffalo.edu

A.M. Arvin et al. (eds.), *Varicella-zoster Virus*,
Current Topics in Microbiology and Immunology 342, DOI 10.1007/82_2010_42

cell. Recent work has shown that the major viral transactivator, commonly designated the IE62 protein, interacts with the human Mediator of transcription. This interaction requires direct contact between the MED25 subunit of Mediator and the acidic N-terminal transactivation domain of IE62. A second cellular factor, host cell factor-1, has been shown to be the common element in two mechanisms of activation of the promoter driving expression of the gene encoding IE62. Finally, the ubiquitous cellular transcription factors Sp1, Sp3, and YY1 have been shown to interact with sequences near the VZV origin of DNA replication and in the case of Sp1/Sp3 to influence replication efficiency.

1 Introduction

Varicella zoster virus (VZV) is a member of the alphaherpesvirinae and the causative agent of varicella (chickenpox) during primary infection and zoster (shingles) upon reactivation from latency in sensory ganglia (Cohen et al. 2007; Mueller et al. 2008). The VZV genome is a linear double-stranded DNA molecule which encodes some 71 open reading frames (ORFs). It is 125-kbp nucleotides in size and consists of two covalently joined segments: the unique long (U_L) and unique short (U_S) regions, each bounded by a terminal repeat and an internal inverted repeat (TR_L/IR_L and TR_S/IR_S, respectively) (Davison and Scott 1986). The entire complement of VZV genes is believed to be expressed during lytic infection in the case of both varicella and zoster resulting in the production of infectious virions and subsequent spread of the virus. In contrast, numerous reports have indicated that only a small number of viral genes may be expressed during latency (Cohrs et al. 2003; Kennedy 2002; Mitchell et al. 2003; Mueller et al. 2008; Silverstein and Straus 2000). The reported levels, frequencies, and numbers of genes expressed during latent infection vary based on the nature of the experimental approach (analysis of human ganglia versus animal models) and could, at least in some cases, reflect initial or abortive steps in reactivation.

Expression of the VZV proteome during lytic infection involves both viral and cellular factors. Transcription of VZV genes is performed by the host cell RNA polymerase II (RNAPII) and the basal cellular transcription apparatus. VZV transcriptional regulatory sequences appear to be relatively typical RNAPII promoters with TATA-like elements, although these AT-rich sequences frequently do not conform to consensus TATA-sequences (Ruyechan 2006; Ruyechan et al. 2003; Smale and Kadonaga 2003). The majority of VZV promoters also contain predicted binding sites for cellular transcription factors including Sp1, USF, and ATF, and an increasing number of reports has verified that these proteins contribute to expression of specific genes both in situ and in vivo (Berarducci et al. 2007; Che et al. 2007; Ito et al. 2003; Meier et al. 1994; Narayanan et al. 2005; Peng et al. 2003; Rahaus and Wolff 2003; Rahaus et al. 2003; Wang et al. 2009; Yang et al. 2006). Efficient expression of the VZV genome, however, also requires a small group of VZV gene products including those encoded by ORFs 62, 4, 61, 63, and 10

(Baudoux et al. 1995, 2000; Bontems et al. 2002; Defechereux-Thibaut de Maisieres et al. 1998; Inchuaspe et al. 1989; Kost et al. 1995; Moriuchi et al. 1995).

The major viral transactivator is the product of ORF62 and its complement, ORF71, which lie within the inverted repeats bracketing the U_S region of VZV DNA (Davison and Scott 1986). This protein is commonly designated IE62 since its synthesis begins in the immediate early phase of lytic VZV gene expression. The first two sections of this review will describe new developments in our understanding of the interplay between IE62 and cellular factors in expression of VZV genes including activation of its own promoter. The third section will describe recent findings in the role of cellular transcription factors in VZV origin-dependent DNA replication.

2 Role of the Eukaryotic Mediator Complex in IE62-Directed Transcriptional Activation

IE62 is required for viral growth in tissue culture and increases the infectivity of VZV DNA (Moriuchi et al. 1994; Sato et al. 2003). The protein is 1310-amino acids (aa) in length and contains a potent N-terminal acidic activation domain and a centrally located DNA-binding domain (DBD). IE62 is believed to be a native homodimer through comparison of its sequence and hydrodynamic properties with those of HSV-1 ICP4 (Shepard et al. 1989; Wu and Wilcox 1991, Peng and Ruyechan unpublished data). Dimerization of a bacterially expressed fragment containing the DBD and sequences corresponding to the ICP4 dimerization domain has been experimentally established (Tyler and Everett 1994). IE62 can bind in a sequence specific manner to a short consensus site (5′-ATCGT-3′) (Tyler and Everett 1993; Wu and Wilcox 1991). However, since many IE62 responsive promoters lack this sequence its role in IE62-mediated transactivation remains obscure.

IE62 is divided into five regions (I–V) based on amino acid sequence alignments of it and its alphaherpesvirus orthologues (Cheung 1989; Ruyechan 2004, 2006; Xiao et al. 1997). Briefly, IE62 region I (aa 1–467) contains a potent acidic activation domain (aa 1–86), which has been shown to be critical for IE62-mediated transactivation. Region I also contains sequences (aa 226–299) necessary for direct interaction with the cellular transcription factors Sp1 and USF, which synergize with IE62 in transcriptional activation (Che et al. 2006; Peng et al. 2003; Rahaus et al. 2003; Wang et al. 2009; Yang et al. 2004b, 2006). The DBD of IE62 falls within region II of the protein. Region II is the most highly conserved region among the alphaherpesvirus major transactivators and also contains the domain responsible for dimerization (Tyler and Everett 1994). DNA binding is required for IE62 function, as mutations which diminished DNA binding cause a corresponding loss in transactivation (Tyler et al. 1994). Region II is additionally the portion of IE62 involved in interaction with the TATA binding protein (TBP) and the general transcription factor TFIIB (Perera 2000). Region III (aa 641–734) contains a nuclear localization signal and sites for phosphorylation by viral kinases

(Besser et al. 2003; Eisfeld et al. 2006; Kinchington et al. 2000). The function of region IV (aa 735–1147) is unknown but mutations in this region result in alteration of IE62 activity (Baudoux et al. 2000). Region V (aa 1148–1310) is the site of numerous mutations in VZV vaccine strains although the significance of this remains unknown (Gomi et al. 2002). Viral factors which physically interact with IE62 and directly influence IE62 activity at viral promoters include the VZV IE4 and IE63 proteins. The IE4 protein interacts within region I (aa 161–299) and the IE63 binding site has been broadly mapped to aa 406–733). Both of these viral proteins can also regulate viral and cellular gene expression in the absence of IE62 (Bontems et al. 2002; Defechereux-Thibaut de Maisieres et al. 1998; Inchuaspe et al. 1989; Kinchington and Cohen 2000; Lynch et al. 2002; Spengler et al. 2000).

The binding partner or partners of the N-terminal acidic transcriptional activation domain (TAD), however, have remained a mystery, despite the identification and mapping of this domain in the early 1990s (Cohen et al. 1993; Perera et al. 1993). The IE62 TAD shows compositional similarity (being primarily composed of acidic and aliphatic residues) but essentially no sequence homology to the herpes simplex virus type 1 (HSV-1) VP16 TAD or the TADs which have been identified at the N-termini of the major transcriptional activators of the animal alphaherpesviruses Pseudorabies (Pr) and equine herpesvirus type 1 (EHV-1) which infect swine and horses, respectively (Cheung 1989; Smith et al. 1994). It has long been assumed that a cellular transcription factor or factors would be the target of the IE62 TAD since IE62 is capable of transactivation in the absence of other viral proteins. Recently, Yang et al. (2008) have shown that the IE62 TAD interacts with the mammalian Mediator of transcription.

2.1 Mediator

The Mediator complex was first identified in yeast via biochemical and genetic screening as a multisubunit coactivator of RNA Pol II-directed transcription (Bjorklund and Gustafsson 2004; Kornberg 2005). Since that time Mediator complexes have been identified in all eukaryotes thus far investigated (Malik and Roeder 2005). Mediator complexes can contain over 22 polypeptides and some estimates of the total number of Mediator components ranges over 30 (Conaway et al. 2005; Sato et al. 2004). Mediator is believed to transduce signals between transcriptional activators such as Sp1, p53, the adenovirus E1A protein, the KSHV RTA protein, and the HSV-1 VP16 TAD and the general transcription apparatus (Boyer et al. 1999; Casamassimi and Napoli 2007; Conaway et al. 2005; Gwack et al. 2003; Malik and Roeder 2005). Each of these activators has been shown to have one (and in some cases two) Mediator subunit as binding targets (Casamassimi and Napoli 2007; Conaway et al. 2005; Malik and Roeder 2005).

Mediator complexes can be divided into three portions: an extended core, a module of loosely or variably associated factors that interact with activators, and a module which is not always associated with the larger Mediator complexes and contains kinase activities (Casamassimi and Napoli 2007). The presence of the kinase containing module has been associated with both repression and activation of transcription (Akoulitchev et al. 2000; Elmlund et al. 2006). Recent structural studies have shown that that the core subunits form distinct submodules designated the head middle and tail with the head submodule components being responsible for contact with RNA Pol II, the tail subunits being a platform for interactions with gene specific regulators and the middle involved in transfer of positive and negative regulatory signal transfer (Casamassimi and Napoli 2007). In this classification scheme, the kinase module remains separate and the Med24–Med27 subunits, some of which are known to interact with viral transcriptional activators (Boyer et al. 1999; Gold et al. 1996; Yang et al. 2004a), remain unassigned to any major structural submodule.

2.2 *Physical and Functional Interaction Between Mediator and IE62*

Yang et al. (2008) recently examined the physical and functional interactions of the IE62 TAD with the mammalian Mediator complex using a variety of biochemical and molecular biological techniques. The hypothesis driving this avenue of investigation was that IE62 targets Mediator via specific interactions between its TAD and one or more Mediator subunits. These interactions would then aid in stabilizing the presence of Mediator at viral promoters resulting in the increased activation observed in the presence of IE62. Confirmation of this hypothesis would be congruent with the fact that (1) the majority of VZV promoters show low levels of basal activity in the absence of IE62 (Che et al. 2007; Perera et al. 1992a, b; Yang et al. 2004b) and (2) the results of Fan et al. (2006) showing that Mediator is recruited in an activator-specific manner and does not associate with many highly active RNAPII promoters in yeast.

Two potential MED subunits were initially chosen as potential targets for the IE62 TAD. These were MED25 which interacts with the HSV-1 VP16 TAD and MED23 which interacts with the adenovirus E1A protein TAD (Boyer et al. 1999; Mittler et al. 2003; Stevens et al. 2002; Yang et al. 2004a). GST pulldown experiments indicated that the IE62 TAD, like the already well-characterized VP16 acidic activation domain is capable of capturing components of the Mediator complex. There were, however, significant differences between the results obtained with the two TADs. Less signal was observed for the Flag-tagged MED25 and endogenous MED23 subunits in pulldowns with the IE62 TAD as compared to the VP16 TAD. These results suggest that the interaction of the IE62 TAD with Mediator is weaker than that of the VP16 TAD. A second important difference consistently observed between the two TADs was that the CDK8 kinase subunit of Mediator was not present in pulldowns performed with the IE62 TAD but was present as previously reported (Gold et al. 1996; Yang et al. 2004a) in pulldowns with the VP16 TAD.

Thus, the IE62 TAD may only interact with a form of Mediator lacking the CDK8 kinase. Alternatively, the interaction of the IE62 TAD with Mediator may result in the dissociation of the CDK8 subunit and possibly other subunits within the kinase module. It is known that the kinase module can interact with the RNAPII CTD (Elmlund et al. 2006) and that HSV infection results in alteration of CTD phosphorylation (Jenkins and Spencer 2001; Rice et al. 1994). Thus, it is possible that VZV infection may result in specific alteration of the RNAPII CTD that does not require CDK8.

The activity of the IE62 TAD in the context of Gal4-fusion constructs was inhibited by the ectopic overexpression of MED25 but not MED23. Further, the inhibition of IE62 TAD activity was significantly greater than that observed in this study and by others for the VP16 TAD (Yang et al. 2004a, 2008). The more extensive inhibitory effect of the exogenous MED25 on the IE62 TAD activity as compared to the VP16 TAD activity is consistent with the apparently lower level of affinity of the IE62 TAD for Mediator observed in the protein pulldown assays. Due to this lower affinity the ectopically over-expressed MED25 would be expected to more efficiently displace the IE62 TAD interacting with MED25 in the endogenous Mediator complex. The requirement for MED25 in IE62 TAD-mediated activation was corroborated by siRNA experiments which showed that MED25 specific siRNA significantly reduced the activity of the IE62 TAD.

The involvement of Mediator in the context of full length IE62 activated transcription was confirmed by chromatin immune-precipitation (ChIP) assays. These experiments demonstrated increased recruitment of Mediator, as monitored by the increase in the presence of MED23, to a model promoter upon cotransfection of pCMV62 with the reporter plasmid. The ChIP assay data also supported the results from the protein pulldown experiments showing no increase in recruitment of CDK8 to the promoter. This suggests that the IE62 TAD not only alone but also in the context of the full length IE62 molecule may interact with or recruit a form of Mediator lacking the CDK8 subunit (Uhlmann et al. 2007).

Several additional findings indicated that Mediator is required for VZV replication during infection. First, ectopically expressed MED25 inhibited the IE62-mediated expression of the complex native VZV ORF28/29 bidirectional regulatory element but not basal level expression of that element in transient transfection assays. ORF28 and ORF29 encode the catalytic subunit of the viral DNA polymerase and the major DNA-binding protein, respectively. High levels of expression of both proteins are required during productive infection (Kinchington and Cohen 2000; Meier et al. 1994). Second, a remarkable redistribution of Mediator was observed during the early phase of viral infection resulting in virtually total overlap with the IE62 signal present in large foci reminiscent of viral replication compartments in herpes simplex virus infection (Lukonis and Weller 1997). During the late phase of viral infection, when IE62 redistributes from the nucleus to the cytoplasm, Mediator returned to its punctate, generalized nuclear distribution suggesting that interaction with IE62 is required for the alterations observed during the early phase of infection. This is the first published demonstration of Mediator redistribution in response to viral infection.

2.3 *Mapping of the Minimal IE62 TAD*

In a recent follow-up study, Yamamoto et al. (2009) showed that the minimal IE62 TAD consists of aa 19–67 of IE62, and that residues throughout this region are important for transactivation. Alternate subdivision of the originally mapped 86-aa TAD resulted in a loss of transactivating activity. Thus, unlike the VP16 TAD, that is divided into two subdomains (H1 and H2) which can transactivate independently (Ikeda et al. 2002; Regier et al. 1993), the minimal IE62 TAD appears to be a single domain. Protein pulldown assays using wild type and mutant TAD fusion proteins showed that the minimal TAD was capable of capturing components representing three different portions of Mediator suggesting that the entire complex is recruited by the TAD. Mutations throughout the minimal TAD that ablated or showed large decreases in transactivation also resulted in loss of binding of all of the Mediator components examined. These data again support the idea that the minimal IE62 TAD is a single discreet domain.

2.4 *The IE62 TAD Interacts Directly with MED25*

Yang et al. (2008) showed that ectopic expression or down-regulation of expression of MED25 had a significant effect on IE62 and IE62 TAD-mediated transactivation. This observation suggested that MED25 might serve as a direct contact site between the IE62 TAD and Mediator. However, the possibility of a bridging protein, perhaps another Mediator subunit, being involved in the interaction remained since, as described above, the interaction with Mediator subunits appeared weak relative to that seen with the VP16 TAD. In addition, it was not known that even if a direct interaction was involved, that the IE62 TAD utilized the same region of MED25 for interaction as did the VP16 TAD. Both of these questions were answered using ectopically expressed polypeptides encompassing aa 1–250 and aa 402–590 of MED25.

The N-terminal fragment of MED25 contains the von Willebrand Factor A domain that anchors MED25 to the Mediator complex while the fragment encompassing aa 402–590 contains the VBD region involved in interaction with the VP16 TAD (Mittler et al. 2003; Yang et al. 2004a). The data obtained by Yamamoto et al. (2009) show that it is the VBD that interacts with the IE62 TAD and that mutations which ablate capture of other Mediator subunits also result in loss of interaction with this region. Thus MED25 appears to be the major contact point for the IE62 TAD with Mediator. Indication of a direct interaction between The IE62 TAD and MED25 is bolstered by the fact that the fragment containing aa 402–590 lacked the WVA domain, and thus was incapable of interaction with the complete Mediator complex. To fine map the region of MED25 within the aa 402–590 VBD fragment, alanine block mutations were introduced at aa 447–450 and aa 484–488 both of which contain basic amino acids. These mutations resulted in loss of interaction with the IE62 TAD and suggest the region of MED25 encompassing aa 484–488 is particularly important for the interaction. These data also represent the finest mapping to date of the region of MED25 involved in interaction with an acidic TAD. A model depicting the interaction of IE62 with Mediator indicating its interaction with MED25 and cellular factors binding at upstream activation sites is presented in Fig. 1.

2.5 The IE62 TAD Is Unstructured in the Absence of a Binding Partner

The HSV VP16 TAD has previously been shown to lack preformed structures, but this TAD forms an induced α-helices upon interaction with its target proteins in both activation subdomains as determined by NMR analysis (Jonker et al. 2005; Uesugi et al. 1997). The IE62 TAD, in contrast, was predicted to contain an extended α-helix spanning aa 19–35. Mutation of the amino acids in this region resulted in both a dramatic loss of transactivating activity and loss of interaction

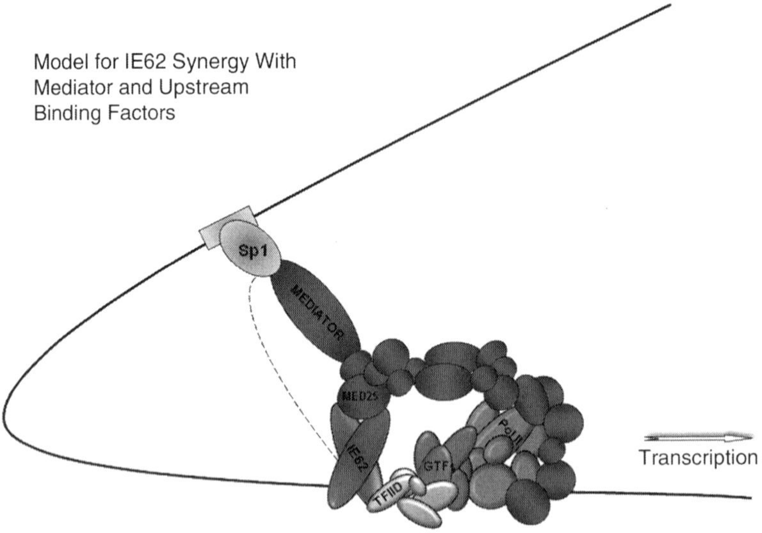

Fig. 1 Model for synergistic activation by IE62, Mediator, and cellular transcriptional activators. IE62 and a cellular factor that binds to a specific upstream site (illustrated here by Sp1) recruit Mediator to promoters. High levels of promoter activation result from the recruitment of Mediator and its stabilization at the promoter by both IE62 and Sp1. IE62 interacts with DNA (*solid line*) MED25, TBP/TFIID, and (not shown) the general transcription factor (GTF) TFIIB. The *dashed line* indicates the potential direct interaction between IE62 and Sp1. The fact that IE62 does not require a specific binding site would allow it to position itself along the DNA and interact with activators whose sites could be variably placed relative to the TATA box

with Mediator (Yamamoto et al. 2009). These results suggested that the IE62 TAD might indeed contain a preformed α-helix involved in interaction with target proteins and subsequent transcriptional activation. Therefore, this possibility was explored by 2D NMR analysis. The NMR data indicate that, contrary to the *in silico* prediction and in accord with accumulating evidence that the acidic TADs of transactivators generally are largely unstructured (Jonker et al. 2005; Schmitz et al. 1994), the IE62 TAD is also largely unstructured in the absence of binding partners. This situation may in part explain the promiscuous nature of IE62 transactivation. It has been suggested that the lack of specific structure in TADs until they interact with their binding partner is advantageous. In this scenario, the intermolecular interactions required for the initiation or propagation of transactivation can be formed with various target proteins, allowing a given TAD to bind to its target with high specificity and low affinity, properties considered to be critical for regulators of gene expression (Hermann et al. 2001). Based on this, the mechanism of transactivation of the IE62 TAD likely involves an initial electrostatic interaction with its target followed by a conformational change leading to the formation of a stable interaction via hydrophobic residues. This is supported by the fact that mutations of both acidic amino acids and the single phenyalanine present in the IE62 TAD resulted in loss of activity (Ferreira et al. 2005).

2.6 Future Directions

Our understanding of the interaction of the functional consequences is only in its initial stages. Possibilities for future work include the following. (1) MED25 has clearly been shown to be a direct physical target of the IE62 TAD this finding does not eliminate the possibility of other Mediator contacts with the IE62 molecule which are important for transactivation functions. Since no other activation domains have been identified using other fragments of the IE62 molecule such regions may be conformational in nature and require the full native IE62 structure for their existence. Such interactions may be difficult to identify but a clearly important to pursue in order to obtain a full understanding of IE62 function. (2) The structural changes that occur within both the TAD and MED25 should be examined by NMR spectroscopy and possibly X-ray crystallography. Such information will not only be important for analysis of the IE62-MED25 interaction but may also be generally applicable to the understanding of the molecular events involved in transactivation by other viral and cellular TADs. (3) It will also be important to determine if VZV infection alters specific Mediator components and/or changes the composition of the Mediator complex that is recruited to the viral replication compartments. (4) The effects of the TAD should be assessed in the context of the viral genome utilizing the SCID mouse–human xenograft model of VZV pathogenesis (Ku et al. 2005). It is possible that such mutations may reveal a tissue specific dependence on the IE62 TAD-MED25 interaction.

3 Mechanisms of Activation of the ORF62 Promoter

Examination of the putative promoter regions of the three VZV IE genes reveals that only the promoters for ORFs 62 and 71, which encode IE62, contains a regulatory element known as the enhancer core (EC) whereas the promoters driving expression of the VZV IE4 and IE63 genes do not (Moriuchi et al. 1995). This sequence (5′-ATGCTAATGARATTTCTTT-3′) has traditionally been assumed to be involved in the expression of alphaherpesvirus IE genes based on work with HSV-1(Kristie et al. 2009). It is bound by a tripartite complex made up of the cellular factors Oct1, host cell factor-1 (HCF-1) and the viral factor HSV-1 VP16 (and its homologues including the VZV ORF10 protein). The presence of the Oct-1 recognition sequence within the ORF62 promoter (Narayanan et al. 2005) and the existence of ORF10 (Che et al. 2006) indicate that IE62 can be expressed via the same mechanism as are the HSV IE genes. In contrast, the VZV IE4 and IE63 proteins may be expressed via mechanisms possibly novel to VZV (Kost et al. 1995; Michael et al. 1998).

By analogy with the HSV IE expression mechanism, VZV ORF10 would interact with Oct-1 bound to the promoter and the cellular transcription factor HCF-1 to activate IE62 expression. This is clearly one mechanism by which IE62 is expressed since the existence of Oct-1/ORF10/HCF-1 complexes interacting with the IE62 promoter has been demonstrated (Moriuchi et al. 1995) and work by Narayanan et al. (2005) has shown that depletion of HCF-1 by siRNA significantly reduces the ORF10-mediated transactivation of the IE62 core promoter.

IE62, however, has also been shown to regulate expression of its own promoter and, unlike HSV ICP4, IE62 is capable of positive as well as negative regulation (Baudoux et al. 1995; Perera et al. 1992b; Shepard et al. 1989). Thus, an alternate pathway for initial expression of the three VZV IE genes would involve their expression via the direct action of IE62 that enters the infected cell as part of the virion tegument (Kinchington et al. 1992) and/or is synthesized immediately after infection. Expression of IE4 and IE63 would then be dependent on IE62 whereas the ORF10 protein would be dispensable for viral growth. These predictions have proven true as demonstrated experimentally in terms of the ability of IE62 to activate the IE4 and IE63 promoters (Kost et al. 1995; Michael et al. 1998) and the fact that ORF10 is dispensable for viral growth in cultured cells (Che et al. 2006). This leads to the question as to what specific cellular factors are involved in IE62 mediated expression of all three IE genes. Surprisingly, the study by Narayanan et al. (2005) showed that IE62 activation of its own promoter was also abrogated by the depletion of HCF-1. Thus, HCF-1 appears to be a common element linking the two mechanisms of activation of the IE62 promoter.

3.1 HCF-1

HCF-1 or C1 is a unique cellular protein that consists of a group of polypeptides proteolytically derived from a common 230-kDa precursor. These cleavage products remain stably associated and are believed to function as a single entity. HCF-1 interacts with a number of other cellular transcription factors including Sp1 and GABP, chromatin modification factors, and coactivators (Gunther et al. 2000; Vogel and Kristie 2000; Wysocka et al. 2003). It has also been implicated in control of the cell cycle (Adams et al. 2000). As indicated above, the importance of HCF-1 with regard to alphaherpesvirus replication stems from the fact that HCF-1 is part of the multiprotein IE enhancer complex that also includes the cellular POU domain protein Oct-1 and the HSV VP16 protein and its homologues (Kristie et al. 2009). HCF-1 is required for the stability of this complex. HCF-1 does not exhibit any apparent site-specific DNA-binding activity but interacts with OCT-1 and VP16 through protein–protein interactions. Further, HCF-1 can activate herpesvirus IE genes whose regulatory elements contain Sp1 and GABP binding sites in the absence of Oct-1 based on work in Oct-1 negative cell lines (Noguiera et al. 2004). Narayanan et al. (2005) showed by siRNA knockdown that HCF-1 is also involved in the IE62-mediated expression of the IE62 gene using a reporter construct containing only a TATA element but not upstream Oct1, GABP, and Sp1 sites. Thus, the expression of IE62, which is critical for VZV replication, can proceed by two mechanisms, one via Oct1/HCF-1/ORF10 and the other via IE62/HCF-1.

Based on the above described findings, IE62 may both interact directly with HCF-1 and/or the interaction may be bridged by another cellular factor. Inspection of the IE62 sequence indicates that it lacks the four amino acid consensus motif (D/E–H–X–Y) required for interaction with HCF-1 which is present in both VP16 and the VZV ORF10 protein (Freiman and Herr 1997). Therefore, there may be a novel sequence within IE62 involved direct interaction with HCF-1. Both IE62 and HCF-1, however, have been shown to be able to interact directly with and utilize Sp1 as a partner in their activation mechanisms (Kristie et al. 2009; Peng et al. 2003). Thus, Sp1 could act to bridge and/or stabilize the IE62/HCF-1 interaction. Since the promoters for all three VZV IE genes contain predicted Sp1 binding sites (Ruyechan et al. 2003), Sp1 and HCF-1 would be leading candidates as mechanistic partners of IE62 in alternative mechanism for expression of the VZV IE genes.

3.2 HCF-1 and Chromatin Remodeling

If HCF-1 is the common factor in both mechanisms of activation of the IE62 promoter, the question arises as to what critical function this protein performs. This question was answered by the work of Narayanan et al. (2007) who showed that HCF-1 is a component of the Set1 and MLL histone H3K4 methyltransferase complexes. Histone 3-lysine 4 (H3K4) methylation is a histone modification which is enriched in transcriptionally active euchromation (Wysocka et al. 2006; Yokoyama et al. 2004). ChIP assays and siRNA depletion of HCF-1 in experiments utilizing full length and truncated forms of the ORF62 promoter showed that (1) HCF-1 and Sp1 were recruited directly to the promoter in an IE62-dependent manner; (2) Sp1 was required to bridge or stabilize the interaction of HCF-1 and

IE62; (3) Set1 and MLL1 recruitment was concomitant with and dependent on the presence of HCF-1; and (4) recruitment of HCF-1 correlated with loss of repressive chromatin marks and gain of the activating H3K4 methylation mark.

Thus, the current model for the two mechanisms of activation of the ORF62 promoter involves the recruitment of methyl transferases by HCF-1 whose activities result in the enrichment of transcriptionally active euchromatin at the promoter site (Fig. 2). The model also has broader implications for VZV transcription overall. Sp1 sites have previously been identified in at least half of the 71 putative VZV promoters (Ruyechan et al. 2003). A recent analysis using broader search criteria and a more robust prediction program has identified potential Sp1 sites in all VZV promoters (Khalil and Ruyechan, unpublished data). IE62 and Sp1 recruit Mediator, HCF-1, TBP/TFIID, RNAPII, and the general transcription apparatus to promoters (Narayanan et al. 2007; Peng et al. 2003). This recruitment and the chromatin modifying enzymes associated with HCF-1 would result in the assembly of the transcription apparatus at promoters enriched for markers of transcriptionally active chromatin. Such a mechanism is likely to be a central pathway for transcription of the entire VZV genome.

3.3 Future Directions

Questions which should be addressed in the future regarding the roles of IE62, HCF-1, and Sp1 in viral gene regulation would include the following topics. (1) The presence

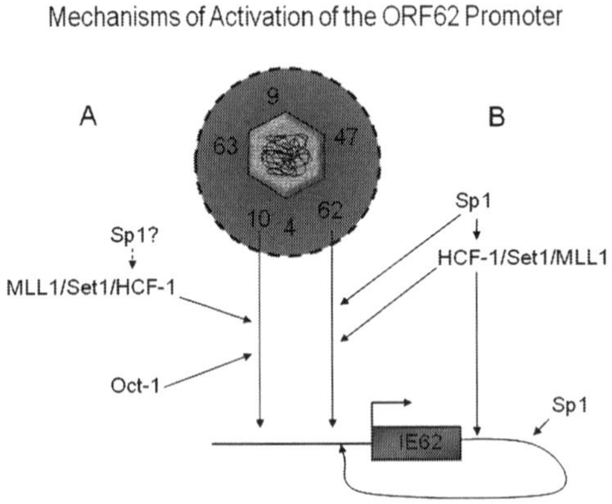

Fig. 2 Model of the mechanisms of activation of the ORF62 promoter. (**a**) The "classical" HSV-1 based model where Oct-1, HCF-1 with associated methyltransferases, and ORF10 delivered by the infecting virion bind to the enhancer core sequence and activate the promoter. It is assumed that ORF10 is capable of recruitment of Mediator. The role, if any that Sp1 plays in this mechanism is unknown. (**b**) The IE62-dependent mechanism where IE62 delivered by the infecting virion interacts with Sp1 and HCF-1 and associated methylases, recruits Mediator (not shown), and activates transcription. In both mechanisms, the newly synthesized IE62 can interact with Sp1 and HCF-1 in a positive feedback loop for increased IE62 production

of HCF-1 at viral promoters known to be dependent on IE62 and Sp1 should be evaluated in order to validate the hypothesis that HCF-1 is required for the expression of the majority of VZV genes rather than only those encoding IE62. (2) The ability of the ubiquitous cellular factor USF to recruit HCF-1 should be evaluated since some VZV genes require USF but not Sp1 sites for activation (Che et al. 2007; Yang et al. 2004b, 2006). (3) The potential interplay between IE62-dependent recruitment of HCF-1/Set1/MLL1 and the recently identified interaction between the VZV IE63 protein and the human antisilencing function 1a protein (ASF1a) (Ambagala et al. 2009) should be examined in the context of chromatin remodeling during viral infection. (4) The intracellular distribution of HCF-1 in VZV-infected cells should be investigated. HCF-1 is sequestered in the cytoplasm of sensory neurons, the sites of both HSV and VZV latency (Kristie et al. 1999). Redistribution of HCF-1 into the nucleus and its recruitment to HSV-1 IE promoters during reactivation as has been recently demonstrated by Whitlow and Kristie (Whitlow and Kristie 2009). Based on the model proposed above, such a redistribution could be a critical step in the induction of expression of the majority of VZV genes upon reactivation in latently infected ganglia. This may, in part, help to account for the highly destructive nature of VZV reactivation within the sensory nervous system.

4 Cellular Transcription Factors and Origin-Dependent DNA Replication

The VZV genome contains two origins of DNA replication (oriS) within the IRs/TRs repeats bounding the U_S segment at sites analogous to those within the HSV-1 genome (Davison and Scott 1986; Stow and McMonagle 1983). Both the VZV and HSV origins contain a central AT-rich palindrome. The overall architecture of the VZV and HSV-1 oriS regions, however, differs significantly (Fig. 3). In the HSV-1 oriS, binding sites for the UL9 origin binding protein (OBP) occur both upstream and downstream of the AT-rich region. These binding sites are designated Boxes I, II, and III. Boxes I and II are located upstream and downstream, on opposite strands of the DNA and are oriented in opposite directions. Box III is located upstream of Box I but is oriented in the same direction and occurs on the same DNA strand as Box II. Mutational analysis has shown that the minimal HSV-1 requires the presence of both Boxes I and II as well as the central AT-rich region (Nguyen-Huynh and Schaffer 1998; Weir and Stow 1990).

In contrast, the two VZV oriS contain an AT rich palindrome and three consensus 10-bp binding sites (5′-C(G/A)TTCGCACT-3′) for the VZV OBP encoded by VZV ORF51. The OBP binding sites, however, are all located upstream of the AT-rich palindrome (Stow and Davison 1986; Stow et al. 1990). They are designated Boxes A, B, and C and are identical or nearly identical to the consensus site for the HSV-1 U_L9 OPB with which the VZV ORF51 OBP shares 54.8% similarity and 46.5% identity (Chen and Olivo 1994). All three OPB-binding sites in the VZV origin are

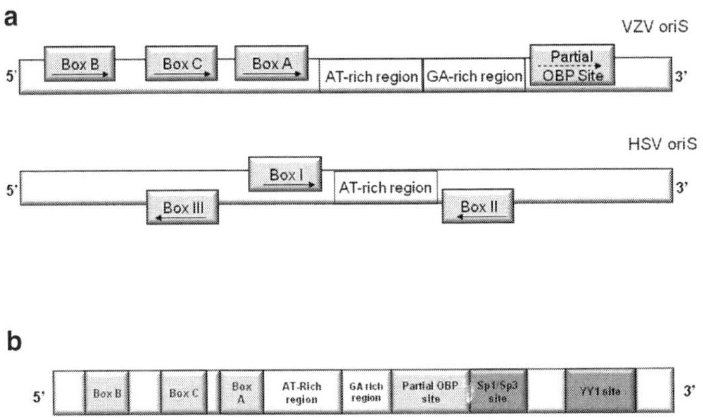

Fig. 3 Architecture of the VZV oriS. (**a**) Comparison of the architecture of the VZV and HSV oriS showing the relative positions and orientations of the OBP binding sites. (**b**) Schematic of the VZV oriS and downstream sequences indication the positions of the Sp1/Sp3 and YY1 sites relative to the AT-rich palindrome and OBP binding sites

oriented in the same direction and are present on the same strand of the viral DNA. Origin-dependent DNA replication in VZV requires the AT-rich region and Box A. Box C is not essential but its presence increases replication efficiency while Box B appears to be dispensable (Stow and Davison 1986; Stow et al. 1990). Nothing is currently known regarding the role played by sequences downstream of the VZV minimal origin in viral DNA replication. A partial (7 of 10 bp) OBP binding site is present downstream of the AT-rich palindrome and a GA-rich region absent in the HSV-1 oriS. This partial OBP site is in the same direction and is on the same strand as the upstream OBP boxes. However, gel shift and DNAse I protection assays failed to demonstrate VZV OBP binding to this sequence (Chen and Olivo 1994; Stow et al. 1990).

The structural differences between the VZV and HSV-1 oriS in terms of the positioning and orientation of the OBP binding sites suggests that the mechanism and/or transacting factors involved in initiation and/or regulation of origin-dependent DNA replication differs between HSV-1 and VZV. Stow et al. (1990) raised the possibility that cellular factors or as yet unidentified viral proteins bind downstream of the AT-rich region in the VZV oriS to facilitate unwinding of the origin and assembly of the viral DNA replication machinery.

There is precedence for this possibility based on work with Epstein-Barr virus (EBV), HSV-1, SV40, and human papillomaviruses (HPV). Sp1 binds to the downstream region of EBV ORI_{Lyt} and physically interacts with the EBV DNA polymerase catalytic and accessory subunits, increasing their recruitment and resulting in a positive effect on DNA replication (Baumann et al. 1999; Gruffat et al. 1995). Sp1 and the related factor Sp3 bind to the upstream region of the HSV-1 oriS and also have a positive effect on origin-dependent DNA replication although the specific mechanism remains unknown (Nguyen-Huynh and Schaffer 1998). Several transcription factors have been found to stimulate SV40 DNA replication including NF1, AP1, and Sp1 (Guo and DePamphilis 1992). Sp1 has also been reported to enhance replication of human papillomavirus 18 (HPV18) (Demeret et al. 1995). In contrast, the transcription factor YY1, has been shown to suppress HPV DNA replication through interference with the function of the HPV E2 protein (Lee et al. 1998).

4.1 Interaction of Transcription Factors with the Downstream Region of oriS

Computer analysis of the sequence downstream of the AT rich palindrome within VZV oriS by Khalil et al. (2008) indicated the presence of two consensus binding sites for cellular factors. One was a GC-rich element (5′-GGGGTGTGGGCGGGC-3′) predicted to bind Sp1 and other members of the Sp/XKLF family of transcription factors of which Sp1 is the prototype. Some 23 members of this family have been identified in the human genome based on the presence of a combination of three conserved Cys2His2 zinc fingers that form a DNA recognition motif (Bouwman and Philipsen 2002). However, only Sp1, Sp2, Sp3, and Sp4 have been extensively characterized as to their tissue specificity and role(s) in transcription (Li et al. 2004; Philipsen and Suske 1999). These four Sp family members all show a modular structure with activation and inhibitory domains present in their N-terminal regions and the characteristic set of three zinc fingers present near their C-termini. Sp1 and Sp3 are believed to be ubiquitously expressed in all tissues and both are capable of activating expression of mammalian genes via recognition of GC and GT-rich motifs (Bouwman and Philipsen 2002).

The second site identified (5′-CCAAATGGAG-3′) further downstream was predicted to bind the cellular factor Yin Yang 1 (YY1). YY1 is a 414-aa zinc finger protein that is capable of both repression and activation of cellular genes and influences transcription in several virus systems including AAV, HPV, EBV, parvovirus B19, HSV-1, and MuLV (Bushmeyer et al. 1995; Demeret et al. 1995; Lee et al. 1992; Pajunk et al. 1997; Zalani et al. 1997). YY1 sites are highly degenerate requiring verification of YY1 interaction via binding assays. Prior to the study of Khalil et al. (2008) no YY1 sites, either predicted or verified, had been reported to be present in the VZV genome.

Two complexes formed with GC-rich element were identified by EMSA analysis and antibody supershifts, one being specific for Sp1 and the other specific for Sp3. There was no evidence for the binding of Sp2 and Sp4 despite the fact that all four Sp family members are expressed in readily detectable amounts in the melanoma cell line used in these experiments. Partial deletion and site-specific mutation of the Sp1/Sp3 site abolished formation of both complexes providing further proof of specific binding. No difference in complex formation was observed using infected and uninfected cell extracts suggesting that upon infection, no additional viral or cellular proteins were recruited to these complexes. YY1 was found to bind to the predicted downstream site using the same EMSA, supershift, and mutational analysis. Site-specific mutation and deletion narrowed the YY1 site to 5′-CCAAATGG-3′ and indicated that the two 3′-terminal guanines are critical for YY1 binding. As was the case with the Sp1/Sp3 site, no difference in the formation of the YY1 specific complex was observed using uninfected and infected nuclear extracts. These results represent the first experimentally verified occurrence of YY1 binding to a VZV DNA sequence.

4.2 The Spl/Sp3 Site Is Involved in Origin-Dependent DNA Replication

The roles of the two transcription factor binding sites were assessed in Dpn I replication assays. These assays involve transfection of cells with a bacterially expressed plasmid containing the VZV OriS followed by superinfection with VZV-infected cells (Stow and Davison 1986; Stow et al. 1990). Replication efficiency is then determined via the differential susceptibility of the replicated and unreplicated plasmid DNA to digestion with the restriction enzyme DpnI which cleaves only bacterially synthesized and modified DNA. The results of these experiments showed that point mutation of the GC-rich site which ablated the formation of the Sp1 and Sp3 complexes observed in EMSA assays resulted in a 2–3 fold increase in replication efficiency as compared to that observed with the wild type virus. These findings concerning the function of the Sp1/Sp3 binding site within the VZV oriS indicate that interaction of Sp1 and Sp3 with the wild type sequence results in a suppression of origin-dependent replication. This is in direct contrast to the function of such sites in the EBV and HSV-1 origins and represents a novel finding for VZV.

The role or roles played by YY1 in VZV infection are unknown, thus the identification of a bonafide YY1 binding within the oriS downstream region site afforded the potential to identify a VZV-specific function for this ubiquitous cellular factor. Further, YY1 has been shown to be capable of direct interaction with Spl (Lee et al. 1993; Seto et al. 1993) raising the possibility that bridging could occur between Spl and YY1 bound at their respective sites in the downstream region of the VZV oriS. Mutation of the YY1 site, however, resulted in no statistically significant difference in DNA replication efficiency as compared to the wild type suggesting that YY1 bound to this site does not play a role in this fundamental process and eliminated the possibility that a potential Spl/YY1 bridging interaction might influence replication efficiency. They do not, however, rule out a role for YY1 in regulation of the ORF63 and, possibly the ORF62 genes which flank the VZV origin (Davison and Scott 1986).

4.3 Future Directions

The work described above represents just the beginning of our understanding of the nature and function of elements within or adjacent to the VZV oriS. The VZV oriS is clearly distinct from the HSV-1 oriS based on (1) its architecture, (2) the function of the downstream Sp1/Sp3 site described here, and (3) the fact that also unlike the HSV origin, the VZV DNA replication origin is part of a bidirectional promoter regulating expression of the ORF62 and ORF63 genes which flank it (Jones et al. 2006). Future avenues of investigation would include the following. (1) The Sp1/Sp3 and YY1 binding sites should be evaluated by site-specific mutagenesis regarding their roles in expression of the ORF62 and ORF63 genes flanking the oriS. (2) The relative occupancy of the GC-rich downstream element by Sp1 versus Sp3 should be evaluated by ChIP assays. Nothing is currently known concerning the role of Sp3 during VZV replication and it may be that Sp3 rather than Sp1 is involved in regulation of oriS function. (3) Since HCF-1 interacts with Sp1, the presence of HCF-1 and its associated chromatin modifying activities at oriS during viral replication should be determined.

5 Summary

The new information described here regarding the role of cellular transcription factors in VZV replication underscores the symbiotic relationship between this complex virus and its human host. Our knowledge concerning the molecular details of such interactions during lytic infection in permissive cells will allow rational experimental approaches to be developed for examination of viral growth and gene expression in the varied primary tissue types that are infected by VZV. It will be of particular interest to know if different subunits of Mediator are involved in tissue specific aspects of VZV pathogenesis and if HCF-1 proves to be a factor involved in a general mechanism for VZV gene expression. Finally, additional downstream elements in oriS, including the GA-rich region as well as the Sp1/Sp3 and YY1 sites may prove to be important in the tissue tropism and pathogenesis of VZV.

Acknowledgment The work from the author's laboratory described in this chapter was supported by Grants AI18449 and AI36884 from the National Institutes of Health.

References

Adams I, Kelso R, Cooley L (2000) The kelch repeat superfamily of proteins: propellers of cell function. Trends Cell Biol 10:17–24

Akoulitchev S, Chuikov S, Reinberg D (2000) TFIIH is negatively regulated by cdk8-containing mediator complexes. Nature 407:102–106

Ambagala AP, Bosma T, Ali MA, Poustovoitov M, Chen JJ, Gershon MD, Adams PD, Cohen JI (2009) Varicella-zoster virus immediate-early 63 protein interacts with human antisilencing function 1 protein and alters its ability to bind histones h3.1 and h3.3. J Virol 83:200–209

Baudoux L, Defechereux P, Schoonbroodt S, Merville MP, Rentier B, Piette J (1995) Mutational analysis of the varicella-zoster virus major immediate early-protein IE62. Nucleic Acids Res 23:1341–1349

Baudoux L, Defechereux P, Rentier B, Piette J (2000) Gene activation by varicella-zoster virus IE4 protein requires its dimerization and involves both the arginine-rich sequence, the central part, and the carboxyl-terminal cysteine-rich region. J Biol Chem 275:32822–32831
Baumann M, Feederle R, Kremmer E, Hammerschmidt W (1999) Cellular transcription factors recruit viral replication proteins to activate the Epstein-Barr virus origin of lytic DNA replication, oriLyt. EMBO J 18:6095–6105
Berarducci B, Sommer M, Zerboni L, Rajamani J, Arvin AM (2007) Cellular and viral factors regulate the varicella-zoster virus gE promoter during viral replication. J Virol 81:10258–10267
Besser J, Sommer MH, Zerboni L, Bagowski CP, Ito H, Moffat J, Ku C-C, Arvin AM (2003) Differentiation of varicella zoster virus ORF47 protein kinase and IE62 protein binding domains and their contributions to replication in human skin xenografts in the SCID-hu mouse. J Virol 77:5964–5974
Bjorklund S, Gustafsson CM (2004) The mediator complex. Adv Protein Chem 67:43–65
Bontems S, Valentin D, Baudoux EL, Rentier B, Sadzot-Delvaux C, Piette J (2002) Phosphorylation of varicella zoster virus IE63 protein by casein kinases influences its cellular localization and gene regulation activity. J Biol Chem 277:21050–21060
Bouwman P, Philipsen S (2002) Regulation of the activity of Sp1-related transcription factors. Mol Cell Endocrinol 195:27–38
Boyer TG, Martin ME, Lees E, Ricciardi RP, Berk AJ (1999) Mammalian Srb/Mediator complex is targeted by adenovirus E1A protein. Nature 399:276–279
Bushmeyer S, Park K, Atchison ML (1995) Characterization of functional domains within the multifunctional transcription factor, YY1. J Biol Chem 270:30213–30220
Casamassimi A, Napoli C (2007) Mediator complexes and eukaryotic transcription regulation: an overview. Biochimie 89:1439–1446
Che X, Zerboni L, Sommer MH, Arvin AM (2006) Varicella-zoster virus open reading frame 10 is a virulence determinant in skin cells but not in T cells in vivo. J Virol 80:3238–3248
Che X, Berarducci B, Sommer M, Ruyechan WT, Arvin AM (2007) The ubiquitous cellular transcription factor, USF, targets the varicella-zoster virus ORF10 promoter and determines virulence in human skin xenografts in SCIDhu mice in vivo. J Virol 81:3229–3239
Chen D, Olivo PD (1994) Expression of the varicella-zoster virus origin-binding protein and analysis of its site-specific DNA-binding properties. J Virol 68:3841–3849
Cheung AK (1989) DNA sequence analysis of the immediate early gene of pseudorabies virus. Nucleic Acids Res 17:4637–4646
Cohen JI, Heffel D, Seidel K (1993) The transcriptional activation domain of varicella-zoster virus open reading frame 62 protein is not conserved with its herpes simplex virus homolog. J Virol 67:4246–4251
Cohen JI, Straus SE, Arvin AM (2007) Varicella-zoster virus replication, pathogenesis and management. In: Knipe DM, Howley PM, Griffin DE, Lamb RA, Martin MA, Roizman B, Straus SE (eds) Fields virology, 5th edn. Lippincott Williams & Wilkins, Philadelphia, PA
Cohrs RJ, Gilden DH, Kinchington PR, Grinfeld E, Kennedy PG (2003) Varicella-zoster virus gene 66 transcription and translation in latently infected human ganglia. J Virol 77:6660–6665
Conaway RC, Sato S, Tomomori-Sato C, Yao T, Conaway JW (2005) The mammalian Mediator complex and its role in transcriptional regulation. Trends Biochem Sci 30:250–255
Davison AJ, Scott J (1986) The complete sequence of varicella-zoster virus. J Gen Virol 67:1759–1816
Defechereux-Thibaut de Maisieres P, Baudoux-Tebache L, Merville M-P, Rentier B, Bours V, Piette J (1998) Activation of the human immunodeficiency virus long terminal repeat by varicella-zoster virus IE4 protein requires nuclear factor-κB and involves both the amino-terminal and the carboxyl-terminal cysteine rich region. J Biol Chem 273:13636–13644
Demeret C, Le Moal M, Yaniv M, Thierry F (1995) Control of HPV 18 DNA replication by cellular and viral transcription factors. Nucleic Acids Res 23:4777–4784
Eisfeld AJ, Turse SE, Jackson SA, Lerner EC, Kinchington PR (2006) Phosphorylation of the varicella-zoster virus (VZV) major transcriptional regulatory protein IE62 by the VZV open reading frame 66 protein kinase. J Virol 80:1710–1723

Elmlund HV, Baraznenok M, Lindahl CO, Samuelsen PJ, Koeck S, Holmberg H, Hebert P, Gustafsson CM (2006) The cyclin-dependent kinase 8 module sterically blocks Mediator interactions with RNA polymerase II. Proc Natl Acad Sci USA 103:15788–15793
Fan X, Chou DM, Struhl K (2006) Activator-specific recruitment of Mediator in vivo. Nat Struct Mol Biol 13:117–120
Ferreira ME, Hermann S, Prochasson P, Workman JL, Berndt KD, Wright AP (2005) Mechanism of transcription factor recruitment by acidic activators. J Biol Chem 280:21779–21784
Freiman RN, Herr W (1997) Viral mimicry: common mode of association with HCF by VP16 and the cellular protein LZIP. Genes Dev 11:3122–3127
Gold MO, Tassan JP, Nigg EA, Rice AP, Herrmann CH (1996) Viral transactivators E1A and VP16 interact with a large complex that is associated with CTD kinase activity and contains CDK8. Nucleic Acids Res 24:3771–3777
Gomi Y, Sunamachi H, Mori Y, Nagaike K, Takahashi M, Yamanishi K (2002) Comparison of the complete DNA sequences of the varicella vaccine and its parental virus. J Virol 76:11447–11459
Gruffat HO, Renner D, Pich K, Hammerschmidt W (1995) Cellular proteins bind to the downstream component of the lytic origin of DNA replication of Epstein-Barr virus. J Virol 69:1878–1886
Gunther M, Laither M, Brison O (2000) A set of proteins interacting with transcription factor Sp1 identified in a two-hybrid screening. Mol Cell Biochem 210:131–142
Guo ZS, DePamphilis ML (1992) Specific transcription factors stimulate simian virus 40 and polyomavirus origins of DNA replication. Mol Cell Biol 12:2514–2524
Gwack YH, Baek J, Nakamura H, Lee SH, Meisterernst M, Roeder RG, Jung JU (2003) Principal role of TRAP/mediator and SWI/SNF complexes in Kaposi's sarcoma-associated herpesvirus RTA-mediated lytic reactivation. Mol Cell Biol 23:2055–2067
Hermann S, Berndt KD, Wright AP (2001) How transcriptional activators bind target proteins. J Biol Chem 276:40127–40132
Ikeda K, Stuehler T, Meisterernst M (2002) The H1 and H2 regions of the activation domain of herpes simplex virion protein 16 stimulate transcription through distinct molecular mechanisms. Genes Cells 7:49–58
Inchuaspe G, Nagpal S, Ostrove JM (1989) Mapping of two varicella-zoster virus encoded genes that activate the expression of viral early and late genes. Virology 173:700–709
Ito H, Sommer MH, Zerboni L, He H, Boucaud D, Hay J, Ruyechan W, Arvin AM (2003) Promoter sequences of varicella-zoster virus glycoprotein I targeted by cellular transcription factors Sp1 and USF determine virulence in skin and T cells in SCIDhu mice *in vivo*. J Virol 77:489–498
Jenkins HL, Spencer CA (2001) RNA polymerase II holoenzyme modifications accompany transcription reprogramming in herpes simplex virus type 1-infected cells. J Virol 75:9872–9884
Jones JO, Sommer M, Stamatis S, Arvin AM (2006) Mutational analysis of the varicella-zoster virus ORF62/63 intergenic region. J Virol 80:3116–3121
Jonker HR, Wechselberger RW, Boelens R, Folkers GE, Kaptein R (2005) Structural properties of the promiscuous VP16 activation domain. Biochemistry 44:827–839
Kennedy PGE (2002) Varicella zoster virus in human ganglia. Rev Med Virol 12:327–334
Khalil MI, Hay J, Ruyechan WT (2008) The cellular transcription factors Sp1 and Sp3 suppress varicella zoster virus origin-dependent DNA replication. J Virol 82:11723–11733
Kinchington PR, Cohen JI (2000) Viral proteins. In: Arvin AM, Gershon AA (eds) Varicella zoster virus: virology and clinical management. Cambridge University Press, Cambridge, UK
Kinchington PR, Hougland JK, Arvin AM, Ruyechan WT, Hay J (1992) The varicella-zoster virus immediate early protein IE62 is a major component of virus particles. J Virol 66:359–366
Kinchington PR, Fite K, Turse SE (2000) Nuclear accumulation of IE62, the varicella-zoster virus (VZV) major transcriptional regulatory protein, is inhibited by phosphorylation mediated by the VZV open reading frame 66 protein kinase. J Virol 74:2265–2277
Kornberg RD (2005) Mediator and the mechanism of transcriptional activation. Trends Biochem Sci 30:235–239

Kost RG, Kupinsky H, Straus SE (1995) Varicella-zoster virus gene 63: transcript mapping and regulatory activity. Virology 209:218–224

Kristie TM, Vogel JL, Sears AE (1999) Nuclear localization of the C1 factor in sensory neurons correlates with initiation of reactivation of HSV from latency. Proc Natl Acad Sci USA 96:1229–1233

Kristie TM, Liang Y, Vogel JL (2009) Control of alpha-herpesvirus IE gene expression by HCF-1 coupled chromatin modification activities. Biochim Biophys Acta. doi:10.1016/j.bbagrm.2009.08.003

Ku C-C, Besser J, Abendroth A, Grose C, Arvin AM (2005) Varicella zoster virus pathogenesis and immunobiology: new concepts emerging from investigations with the SCIDhu mouse model. J Virol 79:2651–2658

Lee T-C, Shi Y, Schwartz RJ (1992) Displacement of BrdUrd-induced YY1 by serum response factor activates skeletal α-actin transcription in embryonic myoblasts. Proc Natl Acad Sci USA 89:9814–9818

Lee JS, Galvin KM, Shi Y (1993) Evidence for physical interaction between the zinc-finger transcription factors YY1 and Sp1. Proc Natl Acad Sci USA 90:6145–6149

Lee K-Y, Broker TR, Chow LT (1998) Transcription factor YY1 represses cell-free replication from human papillomavirus origins. J Virol 72:4911–4917

Li L, He S, Sun JM, Davie JR (2004) Gene regulation by Sp1 and Sp3. Biochem Cell Biol 82:460–471

Lukonis CJ, Weller SK (1997) Formation of herpes simplex virus type 1 replication compartments by transfection: requirements and localization to nuclear domain 10. J Virol 71:2390–2399

Lynch JM, Kenyon TK, Grose C, Hay J, Ruyechan WT (2002) Physical and functional interaction between the varicella zoster virus IE63 and IE62 proteins. Virology 302:71–82

Malik S, Roeder RG (2005) Dynamic regulation of pol II transcription by the mammalian Mediator complex. Trends Biochem Sci 30:256–263

Meier JL, Luo X, Sawadogo M, Straus SE (1994) The cellular transcription factor USF cooperates with varicella-zoster virus immediate-early protein 62 to symmetrically activate a bi-directional viral promoter. Mol Cell Biol 14:6896–6906

Michael E, Kuck K, Kinchington PR (1998) Anatomy of the varicella zoster virus open reading frame 4 promoter. J Infect Dis 178:S27–S33

Mitchell BM, Bloom DC, Cohrs RJ, Gilden DH, Kennedy PGH (2003) Herpes simplex virus-1 and varicella-zoster virus latency in ganglia. J Neurovirol 9:194–204

Mittler G, Stuhler T, Santolin L, Uhlmann T, Kremmer E, Lottspeich F, Berti L, Meisterernst M (2003) A novel docking site on Mediator is critical for activation by VP16 in mammalian cells. EMBO J 22:6494–6504

Moriuchi M, Moriuchi H, Straus SE, Cohen JI (1994) Varicella zoster virus (VZV) virion associated transactivator open reading frame 62 protein enhances the infectivity of VZV DNA. Virology 200:297–300

Moriuchi H, Moriuchi M, Cohen JI (1995) Proteins and cis-acting elements associated with transactivation of the varicella zoster virus immediate-early gene 62 promoter by VZV open reading frame 10 protein. J Virol 69:4693–4701

Mueller NH, Gilden DH, Cohrs RJ, Mahalingam R, Nagel MA (2008) Varicella zoster virus infection: clinical features, molecular pathogenesis of disease, and latency. Neurol Clin 26:675–697

Narayanan A, Nogueira ML, Ruyechan WT, Kristie TM (2005) Combinatorial transcription of the HSV and VZV IE genes is strictly determined by the cellular coactivator HCF-1. J Biol Chem 280:1369–1375

Narayanan A, Ruyechan WT, Kristie TM (2007) The coactivator host cell factor-1 mediates Set1 and MLL H3K4 trimethylation at herpesvirus immediate early promoters for initiation of infection. Proc Natl Acad Sci USA 104:10835–10840

Nguyen-Huynh AT, Schaffer PA (1998) Cellular transcription factors enhance herpes simplex virus type 1 oriS-dependent DNA replication. J Virol 72:3635–3645

Noguiera ML, Wang VE, Tantin D, Sharp PA, Kristie TM (2004) Herpes simplex virus infections are arrested in OCT-1 deficient cells. Proc Natl Acad Sci USA 101:1473–1478

Pajunk HS, May C, Pfister H, Fuchs PG (1997) Regulatory interactions of transcription factor YY1 with control sequences of the E6 promoter of human papillomavirus type 8. J Gen Virol 78:3287–3295

Peng H, He H, Hay J, Ruyechan WT (2003) Interaction between the varicella zoster virus IE62 major transactivator and cellular transcription factor Sp1. J Biol Chem 278:38068–38075

Perera LP (2000) The TATA motif specifies the differential activation of minimal promoters by varicella zoster virus immediate-early regulatory protein IE62. J Biol Chem 275:487–496

Perera LP, Mosca J, Ruyechan WT, Hay J (1992a) Regulation of varicella zoster virus gene expression in human T-lymphocytes. J Virol 66:2468–2477

Perera LP, Mosca JD, Sadeghi-Zadeh M, Ruyechan WT, Hay J (1992b) The varicella-zoster virus immediate early protein, IE62, can positively regulate its cognate promoter. Virology 191:346–354

Perera LP, Mosca JD, Ruyechan WT, Hayward GS, Straus SE, Hay J (1993) A major transactivator of varicella-zoster virus, the immediate-early protein IE62, contains a potent N-terminal activation domain. J Virol 67:4474–4483

Philipsen S, Suske G (1999) A tale of three fingers: the family of mammalian Sp/XKLF transcription factors. Nucleic Acids Res 27:2991–3000

Rahaus M, Wolff MH (2003) Reciprocal effects of varicella zoster virus (VZV) and Ap1: activation of Jun, Fos, and ATF-2 after VZV infection and their importance for the regulation of viral genes. Virus Res 92:9–21

Rahaus M, Desloges N, Yang M, Ruyechan WT, Wolff MH (2003) Transcription factor USF, expressed during the entire phase of VZV infection, interacts physically with the major viral transactivator IE62 and plays a significant role in virus replication. J Gen Virol 84:2957–2967

Regier JL, Shen F, Triezenberg SJ (1993) Pattern of aromatic and hydrophobic amino acids critical for one of two subdomains of the VP16 transcriptional activator. Proc Natl Acad Sci USA 90:883–887

Rice SA, Long MC, Lam V, Spencer CA (1994) RNA polymerase II is aberrantly phosphorylated and localized to viral replication compartments following herpes simplex virus infection. J Virol 68:988–1001

Ruyechan WT (2004) Mechanism(s) of activation of varicella zoster virus promoters by the VZV IE62 protein. Rec Res Dev Virol 6:145–172

Ruyechan WT (2006) Varicella zoster virus transcriptional regulation and the roles of VZV IE proteins. In: Sandri-Goldin RM (ed) Alpha herpesviruses: molecular and cellular biology. Horizon Scientific, Norwich

Ruyechan WT, Peng H, Yang M, Hay J (2003) Cellular factors and IE62 activation of VZV promoters. J Med Virol 70:S90–S94

Sato B, Ito H, Hinchliffe S, Sommer MH, Zerboni L, Arvin AM (2003) Mutational analysis of open reading frames 62 and 71, encoding the varicella zoster virus immediate early transactivating protein, IE62, and effects on replication in vitro and in skin xenografts in the SCID-hu mouse in vivo. J Virol 77:5607–5620

Sato SC, Tomomori-Sato TJ, Parmely L, Florens B, Zybailov SK, Swanson CA, Banks J, Jin Y, Cai MP, Washburn JW, Conaway J, Conaway RC (2004) A set of consensus mammalian mediator subunits identified by multidimensional protein identification technology. Mol Cell 14:685–691

Schmitz ML, dos Santos Silva MA, Altmann H, Czisch M, Holak TA, Baeuerle PA (1994) Structural and functional analysis of the NF-kappa B p65 C terminus. An acidic and modular transactivation domain with the potential to adopt an alpha-helical conformation. J Biol Chem 269:25613–25620

Seto E, Lewis B, Shenk T (1993) Interaction between transcription factors Sp1 and YY1. Nature 365:462–464

Shepard AA, Imbalzano AN, DeLuca NA (1989) Separation of primary structural components conferring autoregulation, transactivation, and DNA-binding properties to the herpes simplex virus transcriptional regulatory protein ICP4. J Virol 63:3714–3728

Silverstein S, Straus SE (2000) Pathogenesis of latency and reactivation. In: Arvin AM, Gershon AA (eds) Varicella zoster virus: virology and clinical management. Cambridge University Press, Cambridge, UK
Smale ST, Kadonaga JT (2003) The RNA polymerase II core promoter. Annu Rev Biochem 72:449–479
Smith RH, Zhao Y, O'Callaghan DJ (1994) The equine herpesvirus type 1 immediate-early gene product contains an acidic transcriptional activation domain. Virology 202:760–770
Spengler ML, Ruyechan WT, Hay J (2000) Physical interaction between two varicella zoster gene regulatory proteins, IE4 and IE62. Virology 272:375–381
Stevens JL, Cantin GT, Wang G, Shevchenko A, Shevchenko A, Berk AJ (2002) Transcription control by E1A and MAP kinase pathway via the Sur2 mediator subunit. Science 296:755–758
Stow ND, Davison AJ (1986) Identification of a varicella-zoster virus origin of DNA replication and its activation by herpes simplex virus type 1 gene products. J Gen Virol 67:1613–1623
Stow ND, McMonagle EC (1983) Characterization of the TRS/IRS origin of DNA replication of herpes simplex virus type 1. Virology 130:427–438
Stow ND, Weir HM, Stow EC (1990) Analysis of the binding sites for the varicella-zoster virus gene 51 product within the viral origin of DNA replication. Virology 177:570–577
Tyler JK, Everett RD (1993) The DNA-binding domain of the varicella zoster virus gene 62 protein interacts with multiple sequences which are similar to the binding site of the related protein of herpes simplex virus type 1. Nucleic Acids Res 21:513–522
Tyler JK, Everett RD (1994) The DNA-binding domains of the varicella zoster virus gene 62 and herpes simplex virus type 1 ICP4 heterodimerize and bind to DNA. Nucleic Acids Res 22:711–721
Tyler JK, Allen KE, Everett RD (1994) Mutation of a single lysine residue severely impairs the DNA recognition and regulatory functions of the VZV gene 62 transactivator protein. Nucleic Acids Res 22:27–278
Uesugi M, Nyanguile O, Lu H, Levine AJ, Verdine GL (1997) Induced alpha helix in the VP16 activation domain upon binding to a human TAF. Science 277:1310–1313
Uhlmann T, Boeing S, Lehmbacher M, Meisterernst M (2007) The VP16 activation domain establishes an active mediator lacking CDK8 in vivo. J Biol Chem 282:2163–2173
Vogel JL, Kristie TM (2000) The novel coactivator C1 (HCF) coordinates multiprotein enhancer formation and mediates transcription activation by GABP. EMBO J 19:683–690
Wang L, Sommer M, Rajamani J, Arvin AM (2009) Regulation of the ORF61 promoter and ORF61 functions in varicella-zoster virus replication and pathogenesis. J Virol 83:7560–7752
Weir HM, Stow ND (1990) Two binding sites for the herpes simplex virus type 1 UL9 protein are required for efficient activity of the oriS replication origin. J Gen Virol 71:1379–1385
Whitlow Z, Kristie TM (2009) Recruitment of the transcriptional coactivator HCF-1 to viral immediate-early promoters during initiation of reactivation from latency of herpes simplex virus type 1. J Virol 83:9591–9595
Wu C-L, Wilcox KW (1991) The conserved DNA-binding domains encoded by the herpes simplex virus type 1 ICP4, Pseudorabies virus IE180, and varicella zoster virus IE62 genes recognize similar sites in the corresponding promoters. J Virol 65:1149–1159
Wysocka J, Myers MP, Laherty CD, Eisenman RN, Herr W (2003) Human Sin3 deacetylase and trithorax-related Set1/Ash2 histone H3-K4 methyltransferase are tethered together selectively by the cell-proliferation factor HCF-1. Genes Dev 17:896–911
Wysocka J, Swigut T, Xiao H, Milne TA, Kwon SY, Landry J, Kauer M, Tackett AJ, Chait BT, Badenhorst P, Wu C, Allis CD (2006) A PHD finger of NURF couples histone H3 lysine 4 trimethylation with chromatin remodelling. Nature 442:86–90
Xiao W, Pizer LI, Wilcox KW (1997) Identification of a promoter-specific transactivation domain in the herpes simplex virus regulatory protein ICP4. J Virol 71:1757–1765
Yamamoto S, Aletskii A, Szyperski T, Hay J, Ruyechan WT (2009) Analysis of the VZV IE62 N-terminal acidic transactivating domain and its interaction with the human mediator complex. J Virol 83:6300–6305

Yang F, DeBeaumont R, Zhou S, Naar AM (2004a) The activator-recruited cofactor/Mediator coactivator subunit ARC92 is a functionally important target of the VP16 transcriptional activator. Proc Natl Acad Sci USA 101:2339–2344
Yang M, Hay J, Ruyechan WT (2004b) The DNA element controlling expression of the varicella zoster virus ORF 28 and ORF 29 genes consists of two divergent unidirectional promoters which share a common USF site. J Virol 78:10939–10952
Yang M, Peng H, Hay J, Ruyechan WT (2006) Promoter activation by the varicella-zoster virus major transactivator IE62 and the cellular transcription factor USF. J Virol 80:7339–7353
Yang M, Hay J, Ruyechan WT (2008) The varicella zoster virus IE62 protein utilizes the human mediator complex in promoter activation. J Virol 82:12154–12163
Yokoyama A, Wang Z, Wysocka J, Sanyal M, Aufiero DJ, Kitabayashi I, Herr W, Cleary ML (2004) Leukemia proto-oncoprotein MLL forms a SET1-like histone methyltransferase complex with menin to regulate Hox gene expression. Mol Cell Biol 24:5639–5649
Zalani S, Coppage A, Holley-Guthrie E, Kenney S (1997) The cellular YY1 transcription factor binds a cis-acting, negatively regulating element in the Epstein-Barr virus BRLF1 promoter. J Virol 71:3268–3274

Effects of Varicella-Zoster Virus on Cell Cycle Regulatory Pathways

Jennifer F. Moffat and Rebecca J. Greenblatt

Contents

Abstract Varicella-zoster virus (VZV) grows efficiently in quiescent cells *in vivo* and in culture, and virus infection activates cell cycle and signaling pathways without cell division. VZV ORFs have been identified that determine the tissue tropism for nondividing skin, T cells, and neurons in SCID-Hu mouse models. The normal cell cycle status of human foreskin fibroblasts was characterized and was dysregulated upon infection by VZV. The expression of cyclins A, B1, and D3 was highly elevated but did not correspond with extensive cellular DNA synthesis. Cell cycle arrest may be due to activation of the DNA damage response during VZV DNA replication. Other host regulatory proteins were induced in infected cells, including p27, p53, and ATM kinase. A possible explanation for the increase in cell cycle regulatory proteins is activation of transcription factors during VZV infection. There is evidence that VZV infection activates transcription factors through the mitogen-activated protein kinase pathways extracellular-regulated kinase (ERK) and c-Jun N-terminal (transpose these parts of the compound noun) kinase (JNK), which could selectively increase cyclin levels. Some of these perturbed cell

J.F. Moffat (✉) and R.J. Greenblatt
Department of Microbiology and Immunology, SUNY Upstate Medical University, 750 E. Adams Street, Syracuse, NY 13210, USA
e-mail: moffatj@upstate.edu

A.M. Arvin et al. (eds.), *Varicella-zoster Virus*,
Current Topics in Microbiology and Immunology 342, DOI 10.1007/82_2010_28

functions are essential for VZV replication, such as cyclin-dependent kinase (CDK) activity, and reveal targets for interventions.

1 Introduction

Varicella-zoster virus (VZV) manipulates the intracellular environment to access functions it requires and to thwart intrinsic defenses that block virus replication. Activation of kinases involved in the cell cycle and signaling pathways are especially critical for VZV since the virus is tropic for memory T cells, neurons, and dermal fibroblasts that are normally not dividing (Arvin 2001). Using cultured human skin fibroblasts as a model, it has been shown that VZV infection results in an atypical intracellular state in which mitogen-activated and cell cycle pathways are induced without cell cycle progression. For a few of these cell kinases, their role in VZV replication is known and appears to be essential. Small molecule inhibitors against cyclin-dependent kinases (CDKs) and the c-Jun N-terminal kinase (JNK) prevent VZV replication in human foreskin fibroblasts (HFFs) (Taylor et al. 2004; Zapata et al. 2007). Uncovering these critical virus–cell interactions adds to our knowledge of VZV cell biology and exposes potential new targets for antiviral drugs.

1.1 VZV Tropism for Nondividing Cells

In humans infected with VZV, virus replication occurs in a variety of cell types that are dividing (basal keratinocytes), terminally differentiated (neurons), or quiescent (memory T cells, dermal fibroblasts) [reviewed in Gilden et al. (2006), Ku et al. (2005)]. In culture, VZV grows well in human tumor cell lines that are rapidly dividing, such as MeWo, and in primary cells that are either dividing or contact-inhibited, such as MRC-5, WI-38, and human fibroblasts from cornea, embryonic lung (HEL), and neonatal foreskin (HFF). In a direct comparison of VZV replication in MeWo cells at 60% confluence and HFFs at 100% confluence, there was no difference in the rate of spread or yield over 7 days (Leisenfelder and Moffat 2006). Lung and skin fibroblasts are normally quiescent *in vivo* and divide only when induced by extracellular signals that occur during development or wound healing. T cells are highly permissive for VZV, and virus spread occurs in thymocytes ($CD4^+$, $CD8^+$, and $CD4/8^+$) in the SCID-Hu thy/liv mouse model and in T cells isolated from cord blood or tonsils that are not dividing (Ku et al. 2002; Moffat et al. 1995; Soong et al. 2000). Although treating VZV-infected tonsil T cells with phorbol esters increases the percentage that are VZV-positive, it is not clear whether that is due to T cell division or increased virus replication and spread in dividing cells (Ku et al. 2002).

The phenotype of some VZV mutants has only been evident in models where most of the cells are not dividing, although undefined factors in addition to cell cycle status are clearly important. For example, VZV ORF10, ORF47, and ORF66 proteins are not essential in cultured cells but are required in skin and T cells (ORF47), only T cells (ORF66), or only skin (ORF10) (Besser et al. 2003; Che et al. 2006; Moffat et al. 1998; Schaap-Nutt et al. 2006). Recombinant VZV that lacks ORF47 kinase grows normally in fibroblasts and melanoma cells but is severely impaired for growth in skin and T cells in SCID-Hu mice. ORF47 is also required in immature but not mature dendritic cells (Hu and Cohen 2005), but since neither dendritic cell type is dividing in culture, this does not account for the mutant phenotype. Like ORF47, ORF66 kinase is required in primary human T cells isolated from cord blood and tonsils (Ku et al. 2002; Soong et al. 2000), and in thy/liv implants in SCID-Hu mice. Unlike ORF47, ORF66 is not essential in human skin (Moffat et al. 1998; Schaap-Nutt et al. 2006). However, mutations in ORF66 that delete the kinase or render the kinase domain inactive prevent VZV from growing in contact-inhibited primary corneal fibroblasts (PCFs) (Erazo et al. 2008). VZV ORF10 is a transcription factor in the virion tegument that is not essential in cultured cells (Cohen and Seidel 1994). Interestingly, deletion mutants lacking ORF10 grew only in the epidermis of skin xenografts, sparing the dermis where fibroblasts are the predominant cell type, although the mutant virus grew as well as the parental strain in human embryonic lung fibroblasts (HELFs) (Che et al. 2006, 2008). These studies on VZV virulence factors show that the cell type is important and the cellular milieu varies considerably, even among different types of fibroblasts.

1.2 Characterization of the Cell Cycle in VZV Host Cells

When VZV infects nondividing cells, the virus encounters a cellular environment that is not conducive to DNA synthesis. For a DNA virus, the relatively low abundance of DNA precursors in the nucleus of cells in G_1 or G_0 phase is an obstacle to efficient transcription and genome synthesis. Of the nondividing cells targeted by VZV, neurons are in G_0 phase, whereas skin and T cells are in G_1 phase because they have the potential for further cell divisions. Following proliferation and maturation in the thymus, peripheral T cells are resting and are maintained in G_1 phase by the cyclin-dependent kinase inhibitor proteins (CKIs) from the INK4 (p15, p16, p18, p19) and the Cip/Kip families (p21, p27, p57). CKIs bind to CDKs and prevent them from phosphorylating substrates that are involved in cell cycle progression, such as the retinoblastoma proteins (pRb, p107, p130) [reviewed in Sherr and Roberts (1999)]. Repression by CKIs must be lifted for T cells to proliferate, which occurs subsequent to activation from antigen presenting cells. Neurons are fully differentiated and do not divide in adults, and so they are considered in G_0 phase. The postmitotic state in neurons is maintained by p27 and p21 and sequestration of cyclin D in the cytoplasm (Politis et al. 2008). Less is known about the cell cycle status of fibroblasts in humans, although they are assumed to be nondividing unless triggered to proliferate by trauma or increase in body size. Complex interactions between dermal fibroblasts and epidermal keratinocytes create paracrine feedback loops of growth factors that can stimulate proliferation of both cell types (Werner et al. 2007). Fibroblasts vary widely in different organs, even depending on their depth in skin, and they are poorly characterized. VZV does not appear hindered by the $G_{0/1}$ environment, thus it was plausible to speculate that the virus could alter the cell cycle profile to obtain the nucleotide precursors it requires for replication.

The lack of information about the expression of CKIs, CDKs, and cyclins in dermal fibroblasts was a barrier to understanding the effects of VZV infection on the cell cycle. To study the interaction of VZV with the cell cycle apparatus in normal, human skin fibroblasts, it was first necessary to select a fibroblast culture and to describe the cell cycle protein profile. A primary culture of HFFs from a genetically normal neonate, deposited with ATCC (CCD-1137Sk), was chosen for its low passage, availability to other investigators, and exemption from internal review board approval for use (Leisenfelder and Moffat 2006). The cell cycle in this HFF culture is likely to be similar to other fibroblasts used to cultivate VZV, such as self-established HFFs, human embryonic lung (HEL or HELF), and MRC-5 cells, although differences may become apparent in direct comparisons, which would be quite interesting. To study the cell cycle in this HFF culture, several methods to synchronize the cells were evaluated. Biochemical inhibition of DNA synthesis using mimosine to arrest cells in S phase or nocodazole to arrest cells in G_2 phase by preventing microtubule polymerization was unsuccessful because the compounds were cytotoxic. Serum starvation and refeeding did not provide adequate synchrony. The most effective means of synchronizing the HFFs was to grow them to confluence, which arrests the cells by contact inhibition, and then subculture at a lower density. This resulted in approximately 25% of the cells entering the cell cycle in the first 24 h (Leisenfelder and Moffat 2006). The proportion of cells in S phase peaked at 24–28 h postplating, slightly preceding the peak in G_2 phase after 28 h. The cell cycle protein profile was then determined by immunoblot from samples collected at 4 h representing G_1 phase, 24 h for S, and 28 h for S–G_2. The results followed the known activation pattern of CDKs by sequential expression of cyclins (Fig. 1): (1) cyclin D3 activated CDK4/6 to force exit from G_1 (cyclins D1 and D2 were not expressed in this cell type), (2) cyclin E activated CDK2 briefly around the restriction point before S phase, (3) cyclin A activated first CDK2 then CDK1 during S phase, and (4) cyclin B1 activated CDK1 in G_2 (Sherr and Roberts 2004). In parallel with increased CDK activity, pRb appeared to become hyperphosphorylated (the changes in molecular weight were hard to distinguish by immunoblot), and p107 clearly accumulated in the hyperphosphorylated form. The cell cycle regulators also followed a predictable pattern, in that p27 levels decreased, p21 remained constant, phosphatase cdc25B peaked at 24 h, and cdc25A and cdc25C peaked at 28 h. The cdc25 family of phosphatases removes inhibitory phosphates on CDKs, adding a further level of regulation beyond cyclin binding (Boutros et al. 2006). Having information about normal cell cycle progression in this HFF culture made it possible to study the cell cycle status during VZV infection.

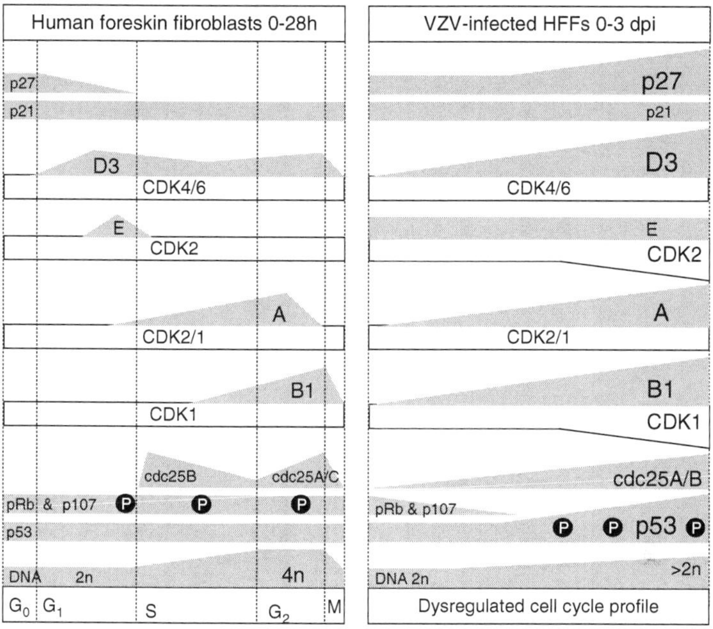

Fig. 1 Cell cycle profiles of normal and VZV-infected HFFs. Cell cycle regulatory proteins and DNA synthesis are shown as *shaded boxes*; CDKs are *open boxes*. Their relative abundance is indicated by the *area* and *font size*. Phosphorylation is denoted by P in a *black circle*. Quiescent HFFs are in G_0 phase and express basal levels of CKI proteins p27 and p21, which bind to cyclins and suppress their activity. When stimulated to enter the cell cycle, p27 levels drop and cyclin D3 levels rise, leading to increased CDK4 and CDK6 activity. CDK4/6 phosphorylate pRb and p107, which enables cyclin E expression that causes a burst of CDK2 activity to drive the cells into S phase. Increased levels of cdc25B and cyclin A in S phase activate first CDK2-cyclin A and then CDK1-cyclin A. Cyclin B accumulates in S phase and complexes with CDK1, but peak activity occurs in late G_2 phase when cdc25A and cdc25C remove inhibitory phosphates on CDK1. The pocket proteins pRb and p107 continue to be phosphorylated during the entire cell cycle, p53 is inactive unless activated by checkpoint control proteins, and DNA synthesis proceeds to reach 4n by the end of S phase. In contrast to normal HFFs, VZV-infected cells have a dysregulated cell cycle profile. The levels of p27, cyclin D3, cyclin A, cyclin B1, cdc25A, cdc25B, and phospho-p53 are above normal. The amounts of CDK1 and CDK2 are approximately threefold greater than uninfected cells, and this corresponds with high kinase activity. The levels of p21 remain constant while cyclin E levels persist longer than expected. The pocket proteins are undetectable and the amount of DNA is slightly greater than 2n. The dysregulated cell cycle profile in VZV-infected HFFs most closely resembles arrested G_2 phase, although the abundance of p27 and absence of cdc25C are atypical

1.3 VZV Dysregulates the Cell Cycle in Human Foreskin Fibroblasts

In confluent HFFs infected with VZV, the cell cycle could not be classified into a particular phase based on cyclin expression or CDK activity (Fig. 1). These findings refuted the initial working hypothesis that VZV would replicate optimally in S phase and contradicted the prediction that mainly CDK2–cyclin A/E would be active and not CDK1–cyclin B1, which is a marker for G_2 phase. The cell cycle profile of VZV-infected, confluent HFFs was paradoxical. Over 3 days of VZV infection, levels of most cell cycle regulatory proteins increased, cyclin E and p21 remained constant, and pRb and cdc25C disappeared (Leisenfelder and Moffat 2006). Most remarkable were the increases in cyclin A (3.3-fold), B1 (6.5-fold), and D3 (17.3-fold) in VZV-infected cells compared to confluent, uninfected cells. CDK1 and CDK2 also increased by roughly threefold and were active in kinase assays. The amount of p27 protein was not quantified on immunoblots, but it appeared approximately ten times greater in VZV-infected cells than in uninfected cells in G_1 phase, where its function is to prevent entry into S phase. The abundance of p27 was surprising, since CDK activity was high in infected cells. A possible explanation is that p27 and p21 can have both inhibitory and activating effects on CDK–cyclin complexes, depending on the cell type, intracellular localization, and whether cells are proliferating (Coqueret 2003; James et al. 2008). We noted that p27 was in the cytoplasm of VZV-infected cells, where it could function as an adaptor protein to facilitate CDK4/6–cyclin D3 binding, and was in the nuclei of uninfected cells where it might function as an inhibitor (unpublished observations). The role of p27 in VZV infection is a topic that deserves further study. The elevated, sustained, and simultaneous activity of CDK1 and CDK2 in VZV-infected cells, along with abnormally high levels of cyclin D3 and p27, was perplexing. Normally, inactive CDK1–cyclin B1 complexes accumulate during S phase as cyclin B1 is translated, and then kinase activity occurs in a burst at the G_2/M checkpoint. Cyclin B1 and all other cyclins in the cell are then rapidly degraded by the anaphase promoting complex (APC) in order for mitosis to finish and cytokinesis to mark the end of the cell cycle. Thus the prolonged CDK1–cyclin B1 activity in VZV-infected cells was unusual and suggested an arrest in late G_2 or early M phase. However, normal cells in G_2 phase have 4n DNA content, whereas VZV-infected cells did not appear to have duplicated chromosomes and exhibited an abnormal DNA content slightly greater than 2n.

1.4 Virus and Host DNA Synthesis in VZV-Infected Human Foreskin Fibroblasts

In the presence of such high CDK activity, usually associated with chromosome origin firing and DNA synthesis, it was obvious to ask whether this was occurring in VZV-infected cells. This important question turned out to be difficult to study. Cell cycle dysregulation is a hallmark of viral oncogenesis; gammaherpesviruses are associated with cell transformation and oncogenesis, such as lymphomas caused by EBV and KSHV, and even the poultry alphaherpesvirus Marek's disease virus causes T cell malignancies (Jarosinski et al. 2006). While no cancer has been attributed to VZV, epidermal hyperplasia has been observed around lesions in human skin biopsies and xenografts (Moffat et al. 1995; Santos et al. 2000). Keratinocyte proliferation could be a general response to localized infection or a pathogenic mechanism of the virus. Several approaches were used to measure cell DNA synthesis in VZV cultures, although the mixture of infected and uninfected cells precluded a purely biochemical analysis of nascent DNA. Flow cytometry of cultures costained with fluorescent antibody to VZV surface antigens (high titer human polyclonal antibody) and propidium iodide (PI) that binds stoichiometrically to DNA provided suggestive results that only virus DNA was synthesized in infected cells. Separate PI measurements on VZV-positive and VZV-negative cells indicated that the population of uninfected cells had a typical quiescent profile of 79% in G_1 phase, 3% in S, and 14% in G_2/M, and the infected cells' profile was 68% in G_2, 8% in S, and 16% in G_2/M (Leisenfelder and Moffat 2006). The apparent shift of ~5–10% of cells into S phase in the VZV+ population was deceptive because the mean fluorescence intensities (MFI) of the G_1 and G_2/M peaks were broader and shifted to the right, and lacked the gradual transition from S to G_2/M that is normally seen. The simplest explanation is that VZV-infected cells contained virus DNA, which would shift the peaks to the right, and in variable amounts, which would broaden the peaks. These results could not rule out cell DNA synthesis, and so further studies used confocal immunofluorescence microscopy to examine events in individual cell nuclei.

To visualize DNA synthesis in VZV-infected HFFs, the cultures were treated with the thymidine analog bromodeoxyuridine (BrDU), which is incorporated into nascent DNA strands before fixation for microscopy. VZV proteins and BrDU were detected with specific antibodies and fluorescent conjugates. In uninfected HFFs, BrDU was incorporated evenly throughout the nuclei of a small number of cells that were scattered randomly in the monolayer and were thus presumed to be those in S phase at the time of labeling. In contrast, BrDU was localized to replication compartments in the nuclei of VZV-infected cells. Their identification as replication compartments was supported by the colocalization of ORF29 single-stranded DNA binding protein and IE62 protein (Reichelt et al. 2009). Occasionally, a cell was observed at an early stage of VZV infection, categorized by IE62 expression only in the nucleus, and yet incorporated BrDU in a diffuse S phase pattern. This suggests that if a cell is committed to entering the cell cycle, or already in S phase, then VZV infection does not halt the process immediately. Despite the compelling observation of BrDU incorporation into replication compartments, it was still uncertain whether cell DNA synthesis was also occurring in the presence of VZV genome replication.

For more insight into this question, a BrDU pulse-chase experiment was performed based on the premise that newly synthesized virus DNA exits the nucleus in capsids while cell DNA remains in the nucleus. This approach was originally taken by Penfold and Mocarski to study replication compartments formed by human cytomegalovirus (Penfold and Mocarski 1997). Cultures of mock-infected or VZV-infected HFFs were pulsed with BrDU for 1 h and then analyzed by confocal IF microscopy after a 24-h chase. Mock-infected cells incorporated BrDU in S phase nuclei in which it remained. VZV-infected cells incorporated BrDU in virus replication compartments during the pulse. As predicted, this labeled viral DNA dissociated from ORF29 protein by 24 h and was detected throughout the nucleus and in cytoplasmic dots that were suggestive of virions. These results support the contention that VZV genome replication precludes cell DNA synthesis, but the remaining BrDU-labeled virus DNA in the nuclei at 24 h would have masked labeled cell DNA if it were there. On the other hand, no mitotic figures were seen in VZV-infected cells, implying that if cell DNA synthesis occurred, it was likely incomplete. So the question remains open and will require a different, and perhaps entirely novel, approach to answer.

If VZV infection suppresses cell DNA synthesis, whether completely or not, how might this occur? And how could this happen in the presence of high levels of CDK activity that normally drive the cell cycle forward? Our hypothesis was that virus genome replication triggered the host DNA damage checkpoint, which can arrest the cell cycle at any stage. Other herpesviruses are known to manipulate the DNA damage response, both activating and inhibiting the pathways in a highly selective manner that enhances virus replication [reviewed in (Lilley et al. 2007)]. Indeed, we found that VZV elicited an incomplete DNA damage response through the ATM-mediated homologous recombination repair pathway while the nonhomologous end-joining pathway was not activated (unpublished observations). Using a combination of confocal IF microscopy and immunoblot, active phospho-ATM and phospho-RPA32 were detected in VZV replication compartments, and the active forms of phospho-NBS1, phospho-p53, and Rad51 increased in VZV-infected cells after 2–3 days. The accumulation of p53 and its presence in the nuclei of infected cells was notable since it has many important functions in the cell, including transactivation of the p21 promoter and induction of apoptosis [reviewed in Sengupta and Harris (2005)]. The ATM kinase phosphorylates p53 in response to DNA double-strand breaks, which occur during herpesvirus DNA recombination and packaging. Phospho-p53 is stabilized and can potentially arrest the cell cycle by inducing p21 expression. However, p21 requires another protein, Mdm2 (also regulated by p53), for full inhibitory activity against CDK2 (Giono and Manfredi 2007). The status of Mdm2 in VZV-infected cells is not known, although one could speculate that it would be absent or inactive as we found abundant p21 and active CDK2 in infected cells despite the presence of phospho-p53. It would not be surprising if the DNA damage and p53 pathways were highly dysregulated in VZV-infected cells, similar to the situation with the CDK–cyclin complexes. These findings, albeit preliminary, suggest that activation of the homologous recombination repair pathway and p53 are likely to be involved in the cell cycle arrest observed in VZV-infected cells, while there is still insufficient evidence to explain why CDK activity is high and unscheduled.

CDK activity requires periodic transcription of cyclins that are regulated by transcription factors under the control of signaling pathways, particularly the mitogen-activated protein kinases (MAPKs). VZV infection affects MAPK pathways, including the main branches characterized by the extracellular-regulated kinases (ERK1 and −2) (Rahaus et al. 2006) and the c-Jun N-terminal kinases (JNK1 and −2) (Rahaus et al. 2005; Zapata et al. 2007). Activation of the ERK pathway by mitogens such as epidermal growth factor or serum results in the induction of key transcription factors that bind to the cyclin D1 promoter [reviewed in Meloche and Pouyssegur (2007)]. These transcription factors include AP-1, which is composed of c-Jun and c-Fos, SP-1, ATF-2, CREB, Oct-1, and NF-κB. The AP-1 family of transcription factors is also induced by activation of the JNK pathway. Less is known about the regulation of the cyclin D3 gene; it is induced by the E2F1 transcription factor that is under control of the Rb protein (Ma et al. 2003). Expression of cyclins A and B1 is activated by NF-Y and p300 in addition to SP-1, USF, and c-Myc transcription factors (Fung and Poon 2005; Porter and Donoghue 2003). VZV infection of MeWo cells results in a transient induction of AP-1 and ATF-2 (Rahaus et al. 2003). Thus VZV infection could possibly activate transcription factors that increase the expression of cyclins.

1.5 Concluding Remarks

The effects of VZV replication on the cell cycle and other major pathways cannot be predicted based on what is known about how cells function. Moreover, much of what is known about cells is learned from tumor cell lines that are not appropriate, or even permissive in the case of T cells, for VZV studies. Thus, it is necessary to fully characterize the pathway of interest in the relevant cell type, which may be quiescent primary neurons or T cells, before studying the impact of VZV infection. It may be worth considering a proteomic approach for questions surrounding VZV interactions with quiescent cells *in vivo*. A broader view of how VZV alters the cellular environment may come from a complete analysis of protein levels and their activation state in normal and infected tissues. The more difficult questions then follow, including whether the effects of the virus on the intracellular environment are incidental or essential for replication. Some of the critical issues still remaining are (1) whether drugs that inhibit essential cell functions can stop VZV replication without intolerable toxicity, (2) whether CDKs, ERK, JNK, and other cell kinases phosphorylate viral as well as host substrates, and (3) if there are differences in the way VZV alters cell cycle and signaling pathways in various cell types and tissues, both in culture and *in vivo*.

Acknowledgments We thank Dongmei Liu for her substantial technical assistance and Jenny Rowe for critically reading the manuscript. This work was supported by PHS AI052168 (JFM) and the Hendrick's Fund for Research Excellence (RJG).

References

Arvin AM (2001) Varicella-zoster virus, 4th edn. In: Knipe DM, Howley PM (eds) Fields virology, vol 2. Lippincott-Raven, Philadelphia, pp 2731–2768

Besser J, Sommer MH, Zerboni L, Bagowski CP, Ito H, Moffat J, Ku CC, Arvin AM (2003) Differentiation of varicella-zoster virus ORF47 protein kinase and IE62 protein binding

domains and their contributions to replication in human skin xenografts in the SCID-hu mouse. J Virol 77(10):5964–5974
Boutros R, Dozier C, Ducommun B (2006) The when and wheres of CDC25 phosphatases. Curr Opin Cell Biol 18(2):185–191
Che X, Zerboni L, Sommer MH, Arvin AM (2006) Varicella-zoster virus open reading frame 10 is a virulence determinant in skin cells but not in T cells *in vivo*. J Virol 80(7):3238–3248
Che X, Reichelt M, Sommer MH, Rajamani J, Zerboni L, Arvin AM (2008) Functions of the ORF9-to-ORF12 gene cluster in varicella-zoster virus replication and in the pathogenesis of skin infection. J Virol 82(12):5825–5834
Cohen JI, Seidel K (1994) Varicella-zoster virus (VZV) open reading frame 10 protein, the homolog of the essential herpes simplex virus protein VP16, is dispensable for VZV replication *in vitro*. J Virol 68(12):7850–7858
Coqueret O (2003) New roles for p21 and p27 cell-cycle inhibitors: a function for each cell compartment? Trends Cell Biol 13(2):65–70
Erazo A, Yee MB, Osterrieder N, Kinchington PR (2008) Varicella-zoster virus (VZV) open reading frame 66 protein kinase is required for efficient viral growth in primary human corneal stromal fibroblast cells. J Virol 82:7653–7665
Fung TK, Poon RY (2005) A roller coaster ride with the mitotic cyclins. Semin Cell Dev Biol 16(3):335–342
Gilden DH, Mahalingam R, Deitch S, Cohrs RJ (2006) Varicella-zoster virus neuropathogenesis and latency. In: Sandri-Goldin RM (ed) Alpha herpesviruses: molecular and cellular biology. Caister Academic, Norwich, UK, pp 305–324
Giono LE, Manfredi JJ (2007) Mdm2 is required for inhibition of Cdk2 activity by p21, thereby contributing to p53-dependent cell cycle arrest. Mol Cell Biol 27(11):4166–4178
Hu H, Cohen JI (2005) Varicella-zoster virus open reading frame 47 (ORF47) protein is critical for virus replication in dendritic cells and for spread to other cells. Virology 337(2):304–311
James MK, Ray A, Leznova D, Blain SW (2008) Differential modification of p27Kip1 controls its cyclin D-cdk4 inhibitory activity. Mol Cell Biol 28(1):498–510
Jarosinski KW, Tischer BK, Trapp S, Osterrieder N (2006) Marek's disease virus: lytic replication, oncogenesis and control. Expert Rev Vaccines 5(6):761–772
Ku CC, Padilla JA, Grose C, Butcher EC, Arvin AM (2002) Tropism of varicella-zoster virus for human tonsillar CD4(+) T lymphocytes that express activation, memory, and skin homing markers. J Virol 76(22):11425–11433
Ku CC, Besser J, Abendroth A, Grose C, Arvin AM (2005) Varicella-zoster virus pathogenesis and immunobiology: new concepts emerging from investigations with the SCIDhu mouse model. J Virol 79(5):2651–2658
Leisenfelder SA, Moffat JF (2006) Varicella-zoster virus infection of human foreskin fibroblast cells results in atypical cyclin expression and cyclin-dependent kinase activity. J Virol 80(11):5577–5587
Lilley CE, Schwartz RA, Weitzman MD (2007) Using or abusing: viruses and the cellular DNA damage response. Trends Microbiol 15(3):119–126
Ma Y, Yuan J, Huang M, Jove R, Cress WD (2003) Regulation of the cyclin D3 promoter by E2F1. J Biol Chem 278(19):16770–16776
Meloche S, Pouyssegur J (2007) The ERK1/2 mitogen-activated protein kinase pathway as a master regulator of the G1- to S-phase transition. Oncogene 26(22):3227–3239
Moffat JF, Stein MD, Kaneshima H, Arvin AM (1995) Tropism of varicella-zoster virus for human CD4+ and CD8+ T lymphocytes and epidermal cells in SCID-hu mice. J Virol 69(9):5236–5242
Moffat JF, Zerboni L, Sommer MH, Heineman TC, Cohen JI, Kaneshima H, Arvin AM (1998) The ORF47 and ORF66 putative protein kinases of varicella-zoster virus determine tropism for human T cells and skin in the SCID-hu mouse. Proc Natl Acad Sci USA 95(20):11969–11974
Penfold ME, Mocarski ES (1997) Formation of cytomegalovirus DNA replication compartments defined by localization of viral proteins and DNA synthesis. Virology 239(1):46–61

Politis PK, Thomaidou D, Matsas R (2008) Coordination of cell cycle exit and differentiation of neuronal progenitors. Cell Cycle 7(6):691–697

Porter LA, Donoghue DJ (2003) Cyclin B1 and CDK1: nuclear localization and upstream regulators. Prog Cell Cycle Res 5:335–347

Rahaus M, Desloges N, Yang M, Ruyechan WT, Wolff MH (2003) Transcription factor USF, expressed during the entire phase of varicella-zoster virus infection, interacts physically with the major viral transactivator IE62 and plays a significant role in virus replication. J Gen Virol 84(Pt 11):2957–2967

Rahaus M, Desloges N, Wolff MH (2005) ORF61 protein of varicella-zoster virus influences JNK/SAPK and p38/MAPK phosphorylation. J Med Virol 76(3):424–433

Rahaus M, Desloges N, Wolff MH (2006) Varicella-zoster virus influences the activities of components and targets of the ERK signalling pathway. J Gen Virol 87(Pt 4):749–758

Reichelt M, Brady J, Arvin AM (2009) The replication cycle of varicella-zoster virus: analysis of the kinetics of viral protein expression, genome synthesis, and virion assembly at the single-cell level. J Virol 83(8):3904–3918

Santos RA, Hatfield CC, Cole NL, Padilla JA, Moffat JF, Arvin AM, Ruyechan WT, Hay J, Grose C (2000) Varicella-zoster virus gE escape mutant VZV-MSP exhibits an accelerated cell-to-cell spread phenotype in both infected cell cultures and SCID-hu mice. Virology 275(2):306–317

Schaap-Nutt A, Sommer M, Che X, Zerboni L, Arvin AM (2006) ORF66 protein kinase function is required for T-cell tropism of varicella-zoster virus *in vivo*. J Virol 80(23):11806–11816

Sengupta S, Harris CC (2005) p53: Traffic cop at the crossroads of DNA repair and recombination. Nat Rev Mol Cell Biol 6(1):44–55

Sherr CJ, Roberts JM (1999) CDK inhibitors: positive and negative regulators of G1-phase progression. Genes Dev 13(12):1501–1512

Sherr CJ, Roberts JM (2004) Living with or without cyclins and cyclin-dependent kinases. Genes Dev 18(22):2699–2711

Soong W, Schultz JC, Patera AC, Sommer MH, Cohen JI (2000) Infection of human T lymphocytes with varicella-zoster virus: an analysis with viral mutants and clinical isolates. J Virol 74(4):1864–1870

Taylor SL, Kinchington PR, Brooks A, Moffat JF (2004) Roscovitine, a cyclin dependent kinase inhibitor, prevents replication of varicella-zoster virus. J Virol 78(6):2853–2862

Werner S, Krieg T, Smola H (2007) Keratinocyte–fibroblast interactions in wound healing. J Invest Dermatol 127(5):998–1008

Zapata HJ, Nakatsugawa M, Moffat JF (2007) Varicella-zoster virus infection of human fibroblast cells activates the c-Jun N-terminal kinase pathway. J Virol 81(2):977–990

Varicella-Zoster Virus Open Reading Frame 66 Protein Kinase and Its Relationship to Alphaherpesvirus US3 Kinases

Angela Erazo and Paul R. Kinchington

Contents

A. Erazo
Graduate Program in Molecular Virology and Microbiology, School of Medicine, University of Pittsbusrgh, Pittsburgh, PA, USA

Department of Ophthalmology, University of Pittsburgh, 1020 EEI building, 203 Lothrop Street, Pittsburgh, PA, 15213, USA

P.R. Kinchington (✉)
Department of Ophthalmology, University of Pittsburgh, 1020 EEI building, 203 Lothrop Street, Pittsburgh, PA, 15213, USA
Microbiology and Molecular Genetics, School of Medicine, University of Pittsburgh, Pittsburgh, PA, USA
e-mail: kinchingtonp@upmc.edu

A.M. Arvin et al. (eds.), *Varicella-zoster Virus*,
Current Topics in Microbiology and Immunology 342, DOI 10.1007/82_2009_7

Abstract The varicella-zoster virus (VZV) open reading frame (ORF) 66 encodes a basophilic kinase orthologous to the US3 protein kinases found in all alphaherpesviruses. This review summarizes current information on the ORF66 kinase, and outlines apparent differences from other US3 kinases, as well as some of the conserved functions. One critical difference is the VZV ORF66 kinase targeting of the major regulatory VZV IE62 protein to control its nuclear import and assembly into the VZV virion, which is so far unprecedented in the alphaherpesviruses. However, ORF66 targets some cellular targets which are also targeted by US3 kinases of other herpesviruses, including the histone deacetylase-1 and 2 proteins, pathways that lead to changes in actin dynamics, and the targeting of substrates of protein kinase A, including the nuclear matrix protein matrin 3.

1 Introduction

The open reading frame (ORF) 66 protein kinase is one of two varicella-zoster virus (VZV) protein kinases initially identified based on genomic position and homology to herpes simplex virus (HSV) kinases and the presence of classical structural motifs found common to all ser/thr kinases. Homologs of ORF66 are often termed the US3 kinases, since they are found in the unique short region of the genome of all sequenced neurotrophic alphaherpesviruses (and probably all alphaherpesviruses), but are absent in members of beta and gamma herpesviruses. By influencing phosphorylation states – the key means of reversible protein functional modulation – they affect many events in infection. Members of the family as a whole influence processes such as survival of the infected cell to apoptosis, the state of permissivity to gene expression, avoidance of immunity, modulating cellular pathways affecting host actin dynamics, and influencing the nuclear structure and nuclear membrane to enable assembly of virus components. The ORF66 kinase is clearly important for VZV growth in certain cell types relevant to human disease. Thus, interest in the ORF66 kinase and the search for its targets continues.

2 Genetics

VZV ORF66 lies in the unique short region of the VZV genome (nucleotides 113,037–114218 in VZV Dumas, 113142–114323 in POka). Its genetic disruption in VZV, first reported by Heineman et al. 1996, established it as not required for growth in cell cultures used for VZV propagation. In this regard, ORF66 mirrors similar US3 mutants of HSV, pseudorabiesvirus (PRV), and marek's disease virus (MDV). In the vaccine Oka background, ORF66 disruption had no effect on viral growth rates, but in the parent Oka VZV background, disruption caused 3–20 fold drop in peak growth levels compared to parental virus, depending on host cell type. As the gI gene lies immediately downstream of ORF66, the complete ORF66 gene cannot be deleted entirely without affecting gI expression, as ORF66 contains control elements in the gI promoter.

VZV mutants lacking ORF66 kinase activity do show more impaired growth in certain cell types or in organ culture models, suggesting that the host cell dictates the importance of the kinase to infection. VZV lacking ORF66 grows poorly in cultured T cells (Soong et al. 2000) and in human thymus/liver xenografts in severe combined immunodeficiency (SCID-hu) mice (Moffat et al. 1998; Schaap-Nutt et al. 2006; Schaap et al. 2005). This has relevance to human disease, since current models of VZV infection propose that T cells transport VZV from the tonsillar respiratory epithelium to the skin (Ku et al. 2005). In T cells, VZV lacking ORF66 kinase shows greater sensitivity to IFN-γ treatment, and increased levels of apoptosis. Electron microscopic examination of such cells reveals an apparent defect in the formation of nucleocapsids, but do not show the accumulation of abundant nucleocapsids at invaginations of the inner nuclear membrane, as seen for US3 kinase-deficient HSV-1 and PRV (see Sect. 5.3). The molecular basis for the VZV phenotypes is not yet understood.

VZV without functional ORF66 also replicates poorly in primary corneal fibroblasts obtained from human corneal stroma donor rims (Erazo et al. 2008). This is significant to human disease, as the cornea is often infected during zoster reactivating from the fifth cranial nerve. Corneal fibroblasts were initially evaluated to investigate possible roles of the ORF66 kinase on actin dynamics, as these cells develop prominent stress fibers when cultured on plastic support. Using recombinant VZV in which GFP was tagged N-terminally to ORF66 kinase, a truncated form or a kinase-inactive form, it was found that VZV without kinase was blocked for replication at a stage following the initial round of replication after infection with infected human MRC-5 cells. VZV lacking ORF66 formed microfoci of GFP positive cells that subsequently fail to expand over time. The basis for growth impairment is not yet known, but data suggested it was not a result of differential regulation of apoptosis or regulation of cellular localization of IE62 (Erazo et al. 2008).

3 ORF66 Structure and Characteristics

The ORF66 protein kinase, at 393 residues, has a predicted weight of 44 kDa. It is a phosphoprotein (Stevenson et al. 1994) that migrates in our hands at 55 kDa as two forms that appear to be differentially phosphorylated. There is no evidence to suggest alternative forms initiating at alternative ATG residues, as seen for US3 kinases of HSV and PRV. ORF66 has a conserved 285 residue kinase (catalytic) domain spanning amino acids 93–378 that has homology to all ser/thr kinases, which consists of 12 subdomains that fold into a characteristic 3-dimensional active core structure to transfer a γ-phosphate from ATP to the hydroxyl group of a specific S/T residue within its protein substrate (see Fig. 1). These subdomains are remarkably invariant within the eukaryotic protein kinase superfamily (Hanks

```
      * *  *                                    *     * ** *   *   *
  1 MNDVDATDTF VGQGKFRGAI STSPSHIMQT CGFIQQMFPV EMSPGIESED DPNYDVNMDI

               * * *        *                    I *        *    *
 61 QSFNIFDGVH ETEAEASVAL CAEARVGINK AGFVILKTFT PGAEGFAFAC MDSKTCEHVV

     II         *III                                  *
121 IKAGQRQGTA TEATVLRALT HPSVVQLKGT FTYNKMTCLI LPRYRTDLYC YLAAKRNLPI

     *                         *   * VIB           * VII
181 CDILAIQRSV LRALQYLHNN SIIHRDIKSE NIFINHPGDV CVGDFGAACF PVDINANRYY

                *VIII*       *IX        *          * *  *      *
241 GWAGTIATNS PELLARDPYG PAVDIWSAGI VLFEMATGQN SLFERDGLDG NCDSERQIKL

               *                                     *   *  *
301 IIRRSGTHPN EFPINPTSNL RRQYIGLAKR SSRKPGSRPL WTNLYELPID LEYLICKMLS

     *  XI                  *          *
361 FDARHRPSAE VLLNHSVFQT LPDPYPNPME VGD
```

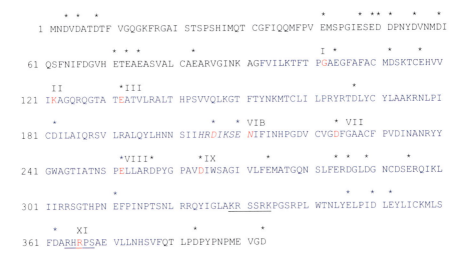

Fig. 1 ORF66 kinase protein sequence. Catalytic domain of ORF66 is highlighted in *blue*. Letters highlighted in *red* include the nonvariant residues found amongst kinase domains indicated by the *Roman numeral* above the residues, which are conserved for all US3 kinases. Not all 12 kinase domains are represented. The catalytic loop is represented in *italics* and potential autophosphorylation target sites are *underlined*. Acidic residues are *starred*

and Hunter 1995). Conservative mutations (D206E and/or K208R) made in the central catalytic domain spanning residues 203–211 disrupt kinase activity. Comparing ORF66 to the cellular ser/thr kinases predicts the ATP binding residue is likely K122, and a K122A mutation also abrogates kinase activity. Using the entire gene in blast searches, the closest cellular homologs are human serine/threonine kinase 9 (also known as cyclin dependent kinase – like 5) and the yeast cell cycle regulator cdc28 (McGeoch and Davison 1986; Schaap et al. 2005). However, the amino terminal region of the protein has a high ratio of acidic residues, as found in all US3 kinases and also in the p21-activated kinases upstream of Cdc42/Rac pathways (see starred residues in Fig. 1). The precise role of the acidic domain is not clear.

A significant fraction of ORF66 kinase is insoluble in most buffers designed to solubilize the protein without disturbing its kinase activity. ORF66 solubility is increased in higher pH buffers, as found for HSV-1 US3 kinase, and our optimal buffer used to solubilize GST-tagged ORF66 from baculovirus-infected cells contains 20 mM Tris-HCl pH8.5, 50 mM KCl, 1 mM EDTA, 1 mM DTT, 1%NP40, and 0.5% DOC. Kinase activity *in vitro* is optimal in 20 mM Hepes pH 7.5, 50 mM Mn^{2+}, and 50 mM KCl. ORF66 is not inhibited by 10 μg/ml heparin (which effectively blocks casein kinase II activity), so this is included in assays. Levels of 10 mM Mg^{2+} can also be used as the cation in the ORF66 *in vitro* kinase assay (Eisfeld et al. 2006).

The cellular localization of ORF66 protein has an unusual distribution. While initial studies using ORF66 specific antibodies first indicated ORF66 as a cytoplasmic protein (Stevenson et al. 1994), studies from our lab using epitope tagged or functional GFP-ORF66 fusions indicate both nuclear and cytoplasmic distribution in VZV-infected cells and in cells expressing the kinase autonomously, with nuclear forms predominating (Eisfeld et al. 2007; Kinchington et al. 2000; Schaap-Nutt et al. 2006). Nuclear ORF66 shows a discrete and distinct punctate nuclear accumulation, forming rings of puncta surrounding the infected cell nucleolus (Eisfeld et al. 2007; Kinchington et al. 2000). Characterization of these ORF66 speckles is in progress, and these appear dynamic (Eisfeld and Kinchington, manuscript in preparation). Functional ORF66 also associates with replication compartments early in infection in MRC-5 cells, whereas kinase dead (kd) forms (D206E, K208R) accumulate in both nuclear replication compartments and nuclear rim of late stage VZV-infected cells, colocalizing with major capsid protein (MCP) (Eisfeld and Kinchington, manuscript in preparation). This suggests that ORF66 kinase activity influences its own cellular distribution, and may be associated with capsid assembly and/or egress.

4 ORF66 Targets

Only one target of ORF66, the IE62 regulatory protein, has been extensively characterized at the time of this review (Table 1). The two sites targeted strongly suggest ORF66 is a basophilic kinase that phosphorylates ser/thr residues preceded by multiple arginine or lysine residues, particularly at −2 and −3 positions. This is consistent with target motifs of PRV and HSV US3 kinases determined by *in vitro* peptide substrates, with an optimal motif of $(R)_n X$–(S/T)–Y–Y (where n is >2, S/T is the target site where either serine or threonine is phosphorylated, X can be absent or any amino acid but preferably Arg, Ala, Val, Pro, or Ser, and Y is similar to X except that it cannot be an absent amino acid, proline, or an acidic residue) (Benetti and Roizman 2004). However, studies on the US3 kinase suggest the optimal motif is overly restrictive, and sites of phosphorylation with much lower matches to the consensus have been reported on lamin C (Mou et al. 2007). Of particular note is that both VZV ORF66 and HSV-1 US3 kinase target motifs overlap that targeted by Protein kinase A (PKA). Using antibodies to the phosphorylated serine in the PKA target motif, novel substrates are detected in extracts of VZV-infected cells that are not found in VZV kinase-deficient infected cells, suggesting the kinase targets multiple cellular proteins or induces activation of cellular kinases that target phospho-PKA motifs. Interestingly, the antibody identifies radically different protein profiles in the same cell type infected with VZV, HSV, and PRV (Erazo et al. manuscript in preparation).

4.1 Autophosphorylation

Protein kinase autophosphorylation is frequently employed to uphold the specificity of kinase functions (Wang and Wu 2002); thus it is not surprising that ORF66 autophosphorylates. Disruption of the kinase catalytic domain or the ATP binding

Table 1 *In vivo* and *in vitro* proteins substrates of the ORF66/US3 kinases

Alphaherpesvirus	ORF66/US3 phosphorylated protein substrate	*In vivo*/ *in vitro* target?	Function
VZV (Eisfeld et al. 2006; Erazo et al. 2008; Kinchington et al. 2000; Kinchington and Turse 1998)	IE62 (ICP4)	+/+	IE62 cytoplasmic accumulation/IE62 tegument inclusion
VZV, HSV, PRV	Matrin 3	+/ND	?
VZV (Walters et al. 2009), HSV-1 (Poon and Roizman 2007)	HDAC 1 and 2	+/ND	Block HDAC transcriptional repression
HSV-1 (Kato et al. 2009; Wisner et al. 2009)	gB	+/+	Downregulate gB surface expression, promote virion nuclear egress
HSV-1 (Kato et al. 2005; Mou et al. 2009)	UL31	+/+	Promote virion nuclear egress
HSV-1 (Kato et al. 2005; Purves et al. 1991; Ryckman and Roller 2004)	UL34	+/+	?
HSV-1 (Mou et al. 2007)	Lamin A/C	+/+	Disrupt nuclear lamina, promote virion nuclear egress
HSV-1 (Leach et al. 2007)	Emerin	+/ND	Disrupt nuclear lamina, promote virion nuclear egress
HSV1,2 (Daikoku et al. 1994; Kato et al. 2005)	US9	+?/+	?
HSV-1 (Kato et al. 2005; Purves et al. 1993; Smith-Donald and Roizman 2008)	ICP22	+?/+	?
HSV-2 (Daikoku et al. 1995)	UL12	ND/+	?
HSV-2 (Murata et al. 2002)	Cytokeratin 17	+?/+	Cell morphological changes
HSV-1 (Cartier et al. 2003b; Kato et al. 2005)	Bad	+/+	Block apoptosis
HSV-1 (Cartier et al. 2003a; Kato et al. 2005)	Bid	ND/+/−	Block apoptosis, mediate protection from granzyme B cleavage of Bid
HSV-1 (Liang and Roizman 2008)	IFNRα	+/ND	Inhibit activation of IFN-γ genes
HSV-1 (Benetti and Roizman 2004)	PKA	+/ND	PKA activation
PRV (Van den Broeke et al. 2009b)	PAK1 and PAK2	+/+	Actin projection formation and stress fiber dissasembly
HSV-1 (Benetti and Roizman 2007)	Procaspase 3	ND/+	Block activation of procaspase 3 and apoptosis
HSV-1 (Smith-Donald and Roizman 2008)	cdc25C Phosphatase	ND/+	Enhance interaction with ICP22, optimize viral gene expression

Specific US3 kinase and its protein target are noted. Evidence of US3 induced phosphorylation *in vivo* or *in vitro* is denoted by a plus (+) sign or not determined (ND). Functions associated with phosphorylation of each protein are also listed residue results in a poor ^{32}P-incorporation into the protein within VZV-infected cells and in *in vitro* reactions with purified kinase. The sites of phosphorylation remain to be determined (Eisfeld et al. 2006; Kinchington et al. 2000). Assuming kinase targeting of serines is preceded by basophilic residues, likely candidate sites are located at $KRS_{331}SRK$ and $RHRPS_{368}$. However, mutagenesis studies indicate that the S_{331} residue is not required for kinase activity (Schaap et al. 2005). VZV ORF66 has no obvious equivalent to the S147 autophosphorylation residue found to be the site of HSV-1 US3 autophosphorylation (Kato et al. 2008), and it is not yet known if ORF66 is phosphorylated by ORF47. The US3 kinase is phosphorylated by the UL13 kinase (Kato et al. 2006) in HSV-1 infected cells.

4.2 IE62

ORF66 kinase targets IE62, the major regulatory protein of VZV. IE62 is a nuclear transcriptional regulatory protein that drives VZV transcription by interacting with transcriptional activators, components of the mediator complex, and members of the general factors involved in recruitment of RNA pol II complex (Ruyechan et al. 2003; Yang et al. 2008). While the underlying mechanisms by which IE62 acts are not resolved, its ability to partly substitute for HSV ICP4 infers that both IE62 and ICP4 act in a similar manner. The targeting of IE62 by the VZV ORF66 kinase was serendipitously discovered in studies to examine the influence of ORF47 kinase on IE62 functions, as IE62 is an ORF47 kinase target (Ng et al. 1994). Cells transfected to express IE62 with or without the ORF47 kinase showed IE62 as a predominantly nuclear protein. The IE62 nuclear localization signal is a classical SV40-like signal high in arg/lys rich residues mapping to residues 677–85 (Kinchington and Turse 1998). However, IE62 coexpressed with the ORF66 kinase showed accumulation of abundant cytoplasmic forms of IE62, mirroring that seen in late stage VZV-infected cells. While IE62 is nuclear early in VZV infection before ORF66 is expressed, IE62 levels build in the cytoplasmic compartment as ORF66 accumulates, until some infected cell nuclei appear devoid of IE62. Cytoplasmic IE62 does not form in cells infected with VZV lacking functional ORF66 kinase, establishing that kinase activity is required. The sites of phosphorylation on IE62, mapped using plasmids expressing IE62 peptides in ORF66 transfected and VZV-infected cells, are predominantly restricted to IE62 residues S686 and S722. ORF66 directly phosphorylated IE62 *in vitro*, and bacterially expressed IE62 peptides with both or one serine intact remained a target for purified VZV ORF66 kinase *in vitro*, whereas loss of S686 and S722 abrogated the ability of IE62 to be an ORF66 target.

The ORF66 kinase-mediated regulation of IE62 has not been reported for corresponding proteins of other alphaherpesviruses, but it reflects the regulated nuclear import of many cellular proteins through phosphorylation (Harreman et al. 2004; Jans and Hubner 1996). As phosphorylation is reversible, it can enable multifunctional proteins to be controlled by their relocation to different cellular compartments. The nuclear exclusion of IE62 in VZV infection enables the packaging of abundant levels of IE62 into VZV virions, at about 50% of the level of the major capsid protein. Virions obtained from VZV-infected cells lacking kinase show virtually no structural forms of IE62 (Kinchington et al. 2001). It was concluded that nuclear exclusion of IE62 allows it to relocate to the *trans*-Golgi network, where VZV tegument is added to the egressing nucleocapsid (Kinchington et al. 2001). Virion packaging of IE62 may allow the introduction of preformed IE62 into the newly infected cell to promote the first events of infection, although this remains to be formally shown. The importance of the targeting of S686 in the ORF66 driven relocation of IE62 was shown using a VZV recombinant containing S686A changes in both copies of IE62 in the VZV genome. Such virus expressed IE62 protein which did not relocate to cytoplasm or become packaged during infection, despite the presence of a functional ORF66 kinase (Erazo et al. 2008).

We postulate that this interaction may come to play during VZV latency. VZV infects sensory nerve endings during varicella and establishes latency in neural nuclei in dorsal root ganglia. In contrast to HSV-1, where there is predominant silencing of protein expression and expression of non-coding latency associated RNA transcripts, VZV latency is characterized by expression of several lytic mRNAs and some regulatory proteins which show nuclear exclusion. Transcripts and proteins of ORF62 and the ORF66 kinase have been reported in human latently infected tissue (Cohrs and Gilden 2007; Cohrs et al. 2003), and IE62 shows predominantly cytoplasmic distribution (Cohrs et al. 2003; Lungu et al. 1998). It has been proposed that VZV latency is maintained by preventing nuclear functions of the regulatory proteins through nuclear exclusion. Our discovery may mechanistically explain IE62 nuclear exclusion during latency.

Interestingly S686 in IE62, which immediately follows the nuclear import signal, is highly conserved in virtually all the alphaherpesvirus ICP4/IE62 homologs. This suggests that their cellular localization may also be regulated by phosphorylation. Indeed, cotransfection studies suggest US3 kinases reduce nuclear import of the corresponding IE62/ICP4 homolog (Yee and Kinchington, unpublished data). However, HSV-1 ICP4 cytoplasmic forms are more reliant upon the functionality of the HSV ICP27 protein (Sedlackova and Rice 2008).

4.3 Matrin 3

The ORF66 kinase targeted sites on IE62 suggest the motif targeted by ORF66 overlaps that of PKA. Roizman and colleagues addressed novel targets of the US3 kinase by probing cell extracts with antibodies directed to the PKA phosphorylated substrates. While Desloges et al. suggested VZV modulated PKA activity (Desloges et al. 2008), we will shortly report (Erazo, Yee and Kinchington, manuscript in preparation) that VZV generates a profile of substrates that includes a prominent species of 125 kDa that was not detected in uninfected cells or in cells infected with VZV deficient in ORF66 kinase activity. Furthermore, this 125 kDa species was seen in cells autonomously expressing functional ORF66 kinase, but not the kinase-inactive form. Thus, this reagent identified a cellular protein significantly phosphorylated, directly or indirectly, by the ORF66 kinase. LC MS/MS analyses of immunoprecipitates with the PKA phospho-specific antibody show the 125 kDa species is matrin 3. Using antibodies to matrin 3 in conjunction with the PKA phospho-substrate antibodies, Matrin 3 phosphorylation only occurred in cells expressing ORF66 by adenovirus mediated transduction or by VZV infection, but not in the same cells if the expressed ORF66 kinase is disrupted or inactivated.

The consequences of matrin 3 phosphorylation to infection are not yet clear. Matrin 3 is one of the 12 major nuclear matrix proteins, but remains only scantily studied. One prior report detailing matrin 3 phosphorylation suggested rapid degradation of matrin 3 following PKA-mediated phosphorylation, induced by NMDA receptor activation of cultured neurons (Giordano et al. 2005). This degradation eventually led to cell death. VZV ORF66 may drive matrin 3 degradation to disrupt nuclear structure to promote nucleocapsid assembly, as suggested for nuclear matrix in HSV-1 capsid assembly (Bibor-Hardy et al. 1985). Matrin 3 also has structural features homologous to RNA binding motifs, and is involved in retaining hyper-edited RNAs and double-stranded RNAs in the nucleus that arise through errant processing (Zhang and Carmichael 2001). Matrin 3 acts as a gatekeeper of such RNAs to prevent their erroneous translation. Matrin 3 also interacts with hnRNP-L involved in regulating RNA splicing (Zeitz et al. 2009). Thus its phosphorylation by ORF66 may influence RNA processing in VZV-infected cells. Matrin 3 is also phosphorylated by the US3 kinases of PRV and HSV-1, despite the different profiles of proteins recognized by the anti-phospho-PKA-substrate antibody (Erazo, Yee and Kinchington, manuscript in preparation).

4.4 *Histone Deacetylases*

Transcription is strongly influenced by the chromatin state of the template DNA that, in turn, is under an elaborate control system that modulates histone binding and condensation. A key component is the reversible post translational modification of histones through the addition and subtraction of acetyl groups to their lysine tails. In general, permissive gene expression is promoted by histone acetyl transferases which acetylate histones to relax DNA binding and condensation. Silencing of expression is partly driven by their deacetylation, mediated by histone deacetylases (HDACs). HDACs are an ancient family of enzymes that have a major role in numerous biological processes. Eleven different HDAC isoforms have been identified in mammalian genomes and these are classified into four different families: class I (HDAC1, 2, 3, and 8), class II, (HDAC4, 5, 6, 7, 9, and 10), sirtuin class III and class IV (HDAC11) (Haberland et al. 2009;

Schwer and Verdin 2008). Because HDACs lack intrinsic DNA-binding activity, they are recruited to target genes through direct association with transcription regulatory proteins. HDAC activity is controlled by phosphorylation by numerous cellular kinases (Pflum et al. 2001). This blocks their deacetylase functions and promotes a cellular permissive state of transcription. The herpesviral infected cell is favored by a pro-active transcriptional state in which deacetylation is inhibited. In HSV-1 infected cells, multiple mechanisms are involved in the inactivation of HDAC activity, including HSV-1 ICP0, which dislodges the LSD1/CoREST/REST complex from HDAC1 and HDAC2, disrupting the silencing effects of this repressor complex on viral promoters. More recently, it was reported that HDAC-1 and 2 showed novel forms which were induced by the Us3 kinase, and that cells expressing the US3 kinase showed a more pro-active transcriptonal state.

Recent work suggests that HDAC-1 and 2 are also modulated in VZV-infected cells in an ORF66-dependent manner (Walters et al. 2009). HDAC-1 and 2 show novel slower mobility forms in SDS-PAGE gels of VZV-infected cell extracts that are not apparent if the ORF66 kinase is deleted. The slower form is differentially phosphorylated and is also seen in cells expressing the ORF66 protein kinase autonomously by transfection or by transduction with ORF66 expressing adenoviruses. Mapping of the sites of phosphorylation show it occurs at a specific residue in the C terminal domain of both proteins which are preceded by basic residues at –2 and –3 positions, consistent with PKA target motifs.

Functional consequences of ORF66 activity affecting HDACs has been suggested from studies using the HDAC inhibitors sodium butyrate. At 1 mM, this inhibitor relieves some of the attenuation of the ORF66 negative VZV as compared to the parental virus. Thus it seems that a prime function of the kinase is to regulate cell permissivity at the transcriptional level through interactions with HDAC-1 and 2 and possibly other HDACs.

5 Cellular and Viral Activities Modulated by the ORF66 Protein Kinase

5.1 MHC-I Surface Presentation

The ORF66 protein kinase mediates VZV-encoded immune evasion strategies. In the host, viral and cellular antigenic peptides are presented on the cell surface for CD8+ T cell recognition in conjunction with the major histocompatibility complex type I or MHC-I. Most herpesviruses have mechanisms to reduce surface presentation of MHC-I coupled viral antigens, presumably to allow prolonged survival of the cell in the presence of a developed immune system. In MHC biogenesis, antigenic peptides generated by the host 26S proteasome are actively transported to the ER lumen by the Transporter of Antigen Presentation (TAP), composed of a heterodimer of TAP-1 and TAP-2. TAP is inhibited by many viruses, because its inhibition affects MHC-I A and B, the main antigen presenters, but not MHC-I types that are needed to signal to natural killer cells. TAP is blocked in HSV-1 infected cells by the immediate early protein ICP47, of which there is no homolog in VZV. Varicelloviruses are reported to have a second gene that blocks TAP, of which the bovine herpesvirus UL49.5 is the most well characterized (Koppers-Lalic et al. 2008). However, we and others have not seen evidence that VZV ORF 9.5 has similar activities (Eisfeld et al. 2007). In the ER lumen, MHC-I heavy chain (Hc) bound to beta 2-microglobulin is stabilized by several chaperones (tapasin, ERp57 and calreticulin) until it couples with TAP and the antigenic peptide. The peptide loading complex can be disrupted or actively inhibited by some viral MHC-I modulators (e.g., human Cytomagalovirus US2). Once loaded, the antigenic peptide is processed to high affinity forms, which mature through the secretory pathway via the Golgi to the cell surface. *Cis- to medial*-Golgi transport is concurrent with conversion of high mannose glycan side chains to complex endoglycosidase-H (endo H) resistant forms.

It is not surprising that VZV downmodulates surface antigen presentation (Cohen 1998; Abendroth et al. 2001), as VZV has lymphotropic parameter in its human pathogenesis, and can sustain infection in multiple cell types over a prolonged period, including professional antigen presenting cells and chronic antigen-expressing neurons during latency. The lack of an ICP47 homolog suggests VZV uses novel mechanisms to mediate this block. Following an initial report by Abendroth et al. in which reduced MHC-I expression in ORF66 expressing cells was observed, we reported that surface MHC-I was reduced in ORF66 expressing cells mediated by transfection, adenovirus mediated transduction, or in recombinant viruses expressing GFP tagged forms of ORF66, but was not downregulated to the same extent in the corresponding conditions when the expressed kinase was disrupted or abrogated (Eisfeld et al. 2007). In both adenovirus transduced and VZV-infected cells, the ORF66 kinase induces the accumulation of endoglycosidase H sensitive MHC-1 forms, suggesting a block either at the assembly stage or the Golgi maturation step prior to *cis to medial* Golgi processing. In VZV infections without ORF66 kinase, MHC-1 processing is still partly blocked as compared to control cells, suggesting that additional mechanisms exist for VZV to block surface MHC-I (Eisfeld et al. 2007). In this respect, VZV is like many herpesviruses, and employs overlapping mechanisms. This is currently under further study.

5.2 IFN Signaling

Schaap et al. demonstrated that expression of ORF66 correlated with a differential level of signaling following IFNγ treatment of VZV-infected T cells. Specifically, the formation of phospho-Stat in T cells following IFNγ binding to its receptor was significantly diminished with ORF66 expression as compared to VZV infections lacking functional ORF66 (Schaap et al. 2005). It is not yet resolved as to how ORF66 blocks this activity, but it is notable that Roizman and colleagues have recently indicated that the HSV-1 US3 kinase may phosphorylate IFN-γRα to prevent its signaling (Liang and Roizman 2008).

5.3 Apoptosis

Arvin and colleagues also demonstrated that ORF66 protein kinase modulates the apoptosis of T cells that have been proposed to mediate dissemination of VZV from respiratory sites of infection to skin (Ku et al. 2005). VZV lacking ORF66 grew to levels 2 logs lower than parental virus in cultured human T cells but not in MeWo cells and the expression of the kinase conferred marginal growth advantage in skin xenografts (Schaap et al. 2005). T cells infected with a kinase-inactive G102A mutant showed increased levels of active caspase 3, the executioner protease in apoptosis, suggesting that loss of the ORF66 kinase correlated with VZV inability to check the development of apoptosis from VZV infection in this cell type (Schaap et al. 2005). These findings imply that ORF66 has an important function in extending the survival of infected T cells until they are able to home in on the skin (Schaap-Nutt et al. 2006). Therefore, inhibition of apoptosis may be the contributing function or the defining function of ORF66 needed for VZV propagation in T cells. It has also been reported that VZV modulates the PI3K/Akt pathway, involved in regulation of apoptosis. Expression of ORF66 transiently or by VZV-infected MeWo cells is involved in pro-survival signaling by activation of Akt, indicated by the increase of Akt phosphorylation at serine 473, that decreased when 66 was not expressed (Rahaus et al. 2007). This may also partly explain the growth deficit of VZV in this cell type (Moffat et al. 1998). However results from studies of ORF66 kinase deficient infections in human corneal fibroblasts, which are very restrictive for such mutants, indicated no significantly increased levels of apoptosis. Recent work has revealed that the role of apoptosis in HSV-1 infection is more important in highly replicating or transformed cells than in primary cell lines (Nguyen et al. 2007).

6 Alphaherpesvirus US3 Kinase Studies that Guide the Search for Roles of ORF66

Several roles of ORF66 may be speculated from identified roles of the HSV and PRV US3 kinases, since there are clear structural similarities. In addition to the kinase catalytic domain, all have a high proportion of acid residues in the amino terminal region, although this is functionally ill-defined. Thus we summarize the known features of the US3 kinases of other alphaherpesviruses.

6.1 US3 Kinases and Inhibition of Apoptosis

Viral perturbation of the host cell often triggers apoptosis, and herpesviruses have mechanisms to block programmed cell death and thus extend cell survival time to allow for viral replication (Aubert and Blaho 2001). The ability of the HSV-1 US3 kinase to block apoptosis has been the most extensively examined, although the US3 kinases of VZV, HSV-2, and PRV may have similar activities. US3 kinases block apoptosis induced by the virus as well as by a variety of ectopic treatments. HSV-1 US3 is one of at least four blockers of apoptosis (in addition to UL39, glycoproteins gD, gJ). Kinase activity is required in HSV and PRV, suggesting that cellular components are phosphorylated. The possible targets include the pro-apoptotic protein Bad, whose phosphorylation (and inactivation) was found to be US3-dependent (Cartier et al. 2003b). Recent studies indicate HSV-1 US3 kinase acts at a post-mitochondrial level, since US3, but not its inactive form US3 K220N, inhibited the cleavage and activation of procaspase 3, the zymogen form of caspase 3 – a major effector of the pro-apoptotic pathway. This was also phosphorylated *in vitro* by US3 (Benetti and Roizman 2007).

Two forms of HSV-1 US3 have been seen initiating at different ATGs and these appear to differ in their ability to block apoptosis. A US3 blocked apoptosis but US3.5 did not, even though both were able to localize in mitochondria (Poon et al. 2006a). PRV also encodes two forms, a long (US3a) and short isoform (US3b) differing by an additional N terminal 54 residues from US3a that encodes a mitochondrial localization signal (Calton et al. 2004). However both block apoptosis, suggesting that the mitochondrial signal and localization of the protein may only be partly responsible for the increase in its anti-apoptotic function (Geenen et al. 2005). There is evidence for the block in apoptosis to be downstream of cell signaling pathways activated by US3. HSV-1 US3 activates PKA, and PKA activation by forskolin inhibits apoptosis (Benetti and Roizman 2004). In sum, evidence points to the US3 kinases as one means to prevent programmed cell death, but several cellular targets may be involved.

6.2 US3 Modulation of HDAC

As just detailed, inhibition of HDACS is needed for efficient viral gene expression (Hobbs and DeLuca 1999; Poon et al. 2003). Although ICPO is thought to be the key player in blocking genomic silencing in HSV-1, US3 may contribute by post translational modification of HDAC1 and HDAC2 (Poon et al. 2003; Poon and Roizman 2007). HSV-1 US3.5 shares this ability (Poon et al. 2006a). US3 and 3.5 enable viral or host gene expression from restrictive cells and enhance expression from permissive cells transduced with baculovirus carrying CMV immediate early promoter-driven genes (Poon et al. 2006b). While phosphorylation regulates HDAC1 and 2 enzymatic activity, it is not yet clear how HSV US3 or VZV

ORF66 facilitates viral gene expression, though the suspicion is that HDAC1 and 2 are direct phosphorylation targets of US3 in infection (Poon and Roizman 2007).

6.3 Nucleocapsid Egress

Herpesvirus nucleocapsids assemble in the nucleus, and DNA packaged nucleocapsids pass through the inner (INM) and outer nuclear (ONM) membranes to the cytoplasm in an envelopment/de-envelopment fusion mechanism. Capsids then acquire most of the tegument and their final envelope as they bud through membranes of the *trans*-Golgi network (Gershon et al. 1994). In both HSV and PRV infection, deletion of the US3 kinase only moderately affects growth rates and infectious virus production in culture. However, EM analyses reveal that such mutants accumulate capsids in extended folds in between the INM and ONM, or perinuclear space (Klupp et al. 2001; Reynolds et al. 2002; Wagenaar et al. 1995). The US3 kinase thus facilitates egress of the capsid from the intranuclear space. US3 kinase mediated phosphorylation of several viral and cellular proteins have been implicated in this process, including UL34, a type 2 integral membrane protein that localizes to the INM, and UL31, Both are critical regulators of primary envelopment of nucleocapsids (Kato et al. 2005; Mou et al. 2009; Purves et al. 1991; Ryckman and Roller 2004). In HSV-1 infection, the UL34 forms a complex with UL31 and displays smooth localization along the nuclear envelope, while in the absence of US3 kinase activity, the complex forms aberrant punctate accumulations at the nuclear membrane (Reynolds et al. 2001; Reynolds et al. 2002). UL34 was the first established target of the US3 kinase, but studies now indicate that its phosphorylation by US3 at the C terminal end does not appear to be directly involved (Kato et al. 2006) in directing the normal localization of the UL34-UL31 complex. The US3 kinase does phosphorylate UL31 *in vitro* (Kato et al. 2005), and prevention of phosphorylation at the serine rich N terminus of UL31 leads to virions accumulating in the perinuclear space. Thus the US3 specific phosphorylation of UL31 may facilitate virion nuclear egress (Mou et al. 2009).

Additional roles of the US3 kinase in nucleocapsid egress stem from recent studies indicating the kinase phosphorylates nuclear membrane forms of the major glycoprotein gB at the cytoplasmic tail. HSV with gB altered at the site of US3 mediated phosphoryation shows the same phenotype to the US3 kinase deletions, in that nucleocapsids accumulate in perinuclear invaginations that protrude into the nucleoplasma. The US3 kinase may modulate gB-mediated fusion events at the ONM to allow de-envelopment during nuclear egress (Wisner et al. 2009). With regards to VZV ORF66, VZV gB does have predictable IE62-like potential motifs for phosphorylation in the cytoplasmic tail which bear similarities to that targeted by US3, but it is not yet known if ORF66 phosphorylates gB. Schaap-Nutt indicated no obvious accumulation of VZV nucleocapsids at the nuclear rim in T cells, but rather showed vastly decreased nucleocapsid formation.

The HSV US3 kinase also modulates host components of the nuclear envelope, including lamin A/C and emerin, an integral nuclear membrane protein which associates with lamin proteins (Leach et al. 2007; Mou et al. 2007). The US3 kinase leads to the redistribution of these proteins in HSV-1 infection, which normally lie just inside the nuclear membrane and act as a barrier for virions budding into the INM. In the presence of US3 kinase, emerin becomes hyperphosphorylated, increasing its mobility during infection. It was postulated that this causes dissociation of emerin and lamin A/C to facilitate the nuclear egress of nucleocapids (Leach et al. 2007). In the case of lamin A/C, it is theorized that US3 is involved in a careful balance in breaching the lamina network barrier to ease virion access to budding sites at the INM, yet maintaining laminar structure (Mou et al. 2007).

6.4 Alteration of the Host Cytoskeleton

Many viruses restructure the host cell cytoskeleton to promote viral inter and intracellular spread. The HSV-1 PRV and MDV US3 kinases have joined the group of increasing viral effector proteins reported to manipulate the host cell cytoskeleton (Smith and Enquist 2002). PRV US3 has been shown to disassemble the actin cytoskeleton and induce novel formation of actin and microtubule-containing cell projections in both the context of viral infection and in US3-transfected cells (Favoreel et al. 2005; Van Minnebruggen et al. 2003). Viral particles found within these dynamic projections were shown to move directionally towards the tip of projections as infection progressed (Favoreel et al. 2005) enabling more efficient spread to adjacent cells. This may allow more efficient infection in the presence of virus neutralizing antibodies. The US3-mediated actin disassembly also induced loss of cell-cell contacts and disassembly of focal adhesion which may be important to PRV spread (Van den Broeke et al. 2009a). Interestingly, US3 was found to induce activation of group A p21-activated kinases through a threonine residue in the activation loop, and phosphorylation of PAK1 and PAK2 *in vitro*. These host proteins are players in Rho GTPase signaling pathways involved in actin disassembly and lamellipodia or filopodia formation. PAK1 and PAK2 were found to be required for infection-induced actin cell projections and disassembly, respectively (Van den Broeke et al. 2009b).

In HSV-2, US3 kinase induced cell rounding and dissolution of actin stress fibers in transfected cells and US3 expressing cell lines. Dominant active forms of the RhoGTPases, Rac, and Cdc42, co-transfected with US3 kinase reduced cell rounding, suggesting that US3 affects the Cdc42/Rac signaling pathway. Interestingly, the Rho family proteins regulate various aspects of actin dynamics and can activate PAKs (Murata et al. 2000). Transient dissassembly of the actin cytoskeleton was also seen for MDV in infections and was dependent on expression of US3. Addition of cytochalasin D which inhibits G actin repolymerization, blocked MDV plaque formation. This taken together with the growth defects observed for US3-deficient MDV and the US3-dependent actin disassembly, support the idea that actin and not microtubule restructuring is important for virus intercellular spread (Schumacher et al. 2005). In contrast to PRV, MDV US3 kinase activity was not needed for actin remodeling in transfected cells (Schumacher et al. 2008), indicating US3-dependent actin disassembly may be dependent on its structure. With regard to VZV, we have found that cellular stress fibers are reduced in wild type VZV-infected cells but remain prominent and abundant in VZV-infected cells if the kinase is disrupted. Thus, it seems likely that a common target for ORF66 and the US3 kinases is the modulation of the actin-based cytoskeleton.

7 Concluding Remarks

VZV ORF66 clearly has important multifunctional roles in the infectious processes which are highly cell type dependent. Thus it is likely that the critical functions of the kinase induce the phosphorylation of cellular targets or of cellular signaling pathways to drive the altered phosphorylation states of cellular proteins. It is clear that VZV ORF66 has functions that are specific for VZV, as well as functions that may be conserved with other alphaherpesvirus US3 kinases. What those common pathways and cellular targets are remain to be resolved.

Acknowledgments The authors wish to acknowledge support from NIH grants NS064022, EY07897, EY08098, funds from the Research to Prevent Blindness, Inc and the Eye & Ear Foundation of Pittsburgh.

References

Abendroth A, Lin I, Slobedman B, Ploegh H, Arvin AM (2001) Varicella-zoster virus retains major histocompatibility complex class I proteins in the Golgi compartment of infected cells. J Virol 75:4878–4888

Aubert M, Blaho JA (2001) Modulation of apoptosis during herpes simplex virus infection in human cells. Microbes Infect 3:859–866

Benetti L, Roizman B (2004) Herpes simplex virus protein kinase US3 activates and functionally overlaps protein kinase A to block apoptosis. Proc Natl Acad Sci USA 101:9411–9416

Benetti L, Roizman B (2007) In transduced cells, the US3 protein kinase of herpes simplex virus 1 precludes activation and induction of apoptosis by transfected procaspase 3. J Virol 81:10242–10248

Bibor-Hardy V, Dagenais A, Simard R (1985) In situ localization of the major capsid protein during lytic infection by herpes simplex virus. J Gen Virol 66(Pt 4):897–901

Calton CM, Randall JA, Adkins MW, Banfield BW (2004) The pseudorabies virus serine/threonine kinase Us3 contains mitochondrial, nuclear and membrane localization signals. Virus Genes 29:131–145

Cartier A, Broberg E, Komai T, Henriksson M, Masucci MG (2003a) The herpes simplex virus-1 Us3 protein kinase blocks CD8T cell lysis by preventing the cleavage of Bid by granzyme B. Cell Death Differ 10:1320–1328

Cartier A, Komai T, Masucci MG (2003b) The Us3 protein kinase of herpes simplex virus 1 blocks apoptosis and induces phosporylation of the Bcl-2 family member Bad. Exp Cell Res 291:242–250

Cohen JI (1998) Infection of cells with varicella-zoster virus down-regulates surface expression of class 1 major histocompatibility complex antigens. J Infect Dis 177:1390–1393

Cohrs RJ, Gilden DH (2007) Prevalence and abundance of latently transcribed varicella-zoster virus genes in human ganglia. J Virol 81:2950–2956

Cohrs RJ, Gilden DH, Kinchington PR, Grinfeld E, Kennedy PG (2003) Varicella-zoster virus gene 66 transcription and translation in latently infected human Ganglia. J Virol 77:6660–6665

Daikoku T, Kurachi R, Tsurumi T, Nishiyama Y (1994) Identification of a target protein of US3 protein kinase of herpes simplex virus type 2. J Gen Virol 75(Pt 8):2065–2068

Daikoku T, Yamashita Y, Tsurumi T, Nishiyama Y (1995) The US3 protein kinase of herpes simplex virus type 2 is associated with phosphorylation of the UL12 alkaline nuclease in vitro. Arch Virol 140:1637–1644

Desloges N, Rahaus M, Wolff MH (2008) The phosphorylation profile of protein kinase A substrates is modulated during Varicella-zoster virus infection. Med Microbiol Immunol 197:353–360

Eisfeld AJ, Turse SE, Jackson SA, Lerner EC, Kinchington PR (2006) Phosphorylation of the varicella-zoster virus (VZV) major transcriptional regulatory protein IE62 by the VZV open reading frame 66 protein kinase. J Virol 80:1710–1723

Eisfeld AJ, Yee MB, Erazo A, Abendroth A, Kinchington PR (2007) Downregulation of class I major histocompatibility complex surface expression by varicella-zoster virus involves open reading frame 66 protein kinase-dependent and -independent mechanisms. J Virol 81:9034–9049

Erazo A, Yee MB, Osterrieder N, Kinchington PR (2008) Varicella-zoster virus open reading frame 66 protein kinase is required for efficient viral growth in primary human corneal stromal fibroblast cells. J Virol 82:7653–7665

Favoreel HW, Van Minnebruggen G, Adriaensen D, Nauwynck HJ (2005) Cytoskeletal rearrangements and cell extensions induced by the US3 kinase of an alphaherpesvirus are associated with enhanced spread. Proc Natl Acad Sci USA 102:8990–8995

Geenen K, Favoreel HW, Olsen L, Enquist LW, Nauwynck HJ (2005) The pseudorabies virus US3 protein kinase possesses anti-apoptotic activity that protects cells from apoptosis during infection and after treatment with sorbitol or staurosporine. Virology 331:144–150

Gershon AA, Sherman DL, Zhu Z, Gabel CA, Ambron RT, Gershon MD (1994) Intracellular transport of newly synthesized varicella-zoster virus: final envelopment in the trans-Golgi network. J Virol 68:6372–6390

Giordano G, Sanchez-Perez AM, Montoliu C, Berezney R, Malyavantham K, Costa LG, Calvete JJ, Felipo V (2005) Activation of NMDA receptors induces protein kinase A-mediated phosphorylation and degradation of matrin 3. Blocking these effects prevents NMDA-induced neuronal death. J Neurochem 94:808–818

Haberland M, Montgomery RL, Olson EN (2009) The many roles of histone deacetylases in development and physiology: implications for disease and therapy. Nat Rev Genet 10:32–42

Hanks SK, Hunter T (1995) Protein kinases 6. The eukaryotic protein kinase superfamily: kinase (catalytic) domain structure and classification. FASEB J 9:576–596

Harreman MT, Kline TM, Milford HG, Harben MB, Hodel AE, Corbett AH (2004) Regulation of nuclear import by phosphorylation adjacent to nuclear localization signals. J Biol Chem 279:20613–20621

Heineman TC, Seidel K, Cohen JI (1996) The varicella-zoster virus ORF66 protein induces kinase activity and is dispensable for viral replication. J Virol 70:7312–7317

Hobbs WE 2nd, DeLuca NA (1999) Perturbation of cell cycle progression and cellular gene expression as a function of herpes simplex virus ICP0. J Virol 73:8245–8255

Jans DA, Hubner S (1996) Regulation of protein transport to the nucleus: central role of phosphorylation. Physiol Rev 76:651–685

Kato A, Arii J, Shiratori I, Akashi H, Arase H, Kawaguchi Y (2009) Herpes simplex virus 1 protein kinase Us3 phosphorylates viral envelope glycoprotein B and regulates its expression on the cell surface. J Virol 83:250–261

Kato A, Tanaka M, Yamamoto M, Asai R, Sata T, Nishiyama Y, Kawaguchi Y (2008) Identification of a physiological phosphorylation site of the herpes simplex virus 1-encoded protein kinase Us3 which regulates its optimal catalytic activity in vitro and influences its function in infected cells. J Virol 82:6172–6189

Kato A, Yamamoto M, Ohno T, Kodaira H, Nishiyama Y, Kawaguchi Y (2005) Identification of proteins phosphorylated directly by the Us3 protein kinase encoded by herpes simplex virus 1. J Virol 79:9325–9331

Kato A, Yamamoto M, Ohno T, Tanaka M, Sata T, Nishiyama Y, Kawaguchi Y (2006) Herpes simplex virus 1-encoded protein kinase UL13 phosphorylates viral Us3 protein kinase and regulates nuclear localization of viral envelopment factors UL34 and UL31. J Virol 80:1476–1486

Kinchington PR, Fite K, Seman A, Turse SE (2001) Virion association of IE62, the varicella-zoster virus (VZV) major transcriptional regulatory protein, requires expression of the VZV open reading frame 66 protein kinase. J Virol 75:9106–9113

Kinchington PR, Fite K, Turse SE (2000) Nuclear accumulation of IE62, the varicella-zoster virus (VZV) major transcriptional regulatory protein, is inhibited by phosphorylation mediated by the VZV open reading frame 66 protein kinase. J Virol 74:2265–2277

Kinchington PR, Turse SE (1998) Regulated nuclear localization of the varicella-zoster virus major regulatory protein, IE62. J Infect Dis 178(Suppl 1):S16–S21

Klupp BG, Granzow H, Mettenleiter TC (2001) Effect of the pseudorabies virus US3 protein on nuclear membrane localization of the UL34 protein and virus egress from the nucleus. J Gen Virol 82:2363–2371

Koppers-Lalic D, Verweij MC, Lipinska AD, Wang Y, Quinten E, Reits EA, Koch J, Loch S, Marcondes Rezende M, Daus F, Bienkowska-Szewczyk K, Osterrieder N, Mettenleiter TC, Heemskerk MH, Tampe R, Neefjes JJ, Chowdhury SI, Ressing ME, Rijsewijk FA, Wiertz EJ (2008) Varicellovirus UL 49.5 proteins differentially affect the function of the transporter associated with antigen processing, TAP. PLoS Pathog 4(5):e1000080

Ku CC, Besser J, Abendroth A, Grose C, Arvin AM (2005) Varicella-Zoster virus pathogenesis and immunobiology: new concepts emerging from investigations with the SCIDhu mouse model. J Virol 79:2651–2658

Leach N, Bjerke SL, Christensen DK, Bouchard JM, Mou F, Park R, Baines J, Haraguchi T, Roller RJ (2007) Emerin is hyperphosphorylated and redistributed in herpes simplex virus type 1-infected cells in a manner dependent on both UL34 and US3. J Virol 81:10792–10803

Liang L, Roizman B (2008) Expression of gamma interferon-dependent genes is blocked independently by virion host shutoff RNase and by US3 protein kinase. J Virol 82:4688–4696

Lungu O, Panagiotidis CA, Annunziato PW, Gershon AA, Silverstein SJ (1998) Aberrant intracellular localization of Varicella-Zoster virus regulatory proteins during latency. Proc Natl Acad Sci USA 95:7080–7085

McGeoch DJ, Davison AJ (1986) Alphaherpesviruses possess a gene homologous to the protein kinase gene family of eukaryotes and retroviruses. Nucleic Acids Res 14:1765–1777

Moffat JF, Zerboni L, Sommer MH, Heineman TC, Cohen JI, Kaneshima H, Arvin AM (1998) The ORF47 and ORF66 putative protein kinases of varicella-zoster virus determine tropism for human T cells and skin in the SCID-hu mouse. Proc Natl Acad Sci USA 95:11969–11974

Mou F, Forest T, Baines JD (2007) US3 of herpes simplex virus type 1 encodes a promiscuous protein kinase that phosphorylates and alters localization of lamin A/C in infected cells. J Virol 81:6459–6470

Mou F, Wills E, Baines JD (2009) Phosphorylation of the U(L)31 protein of herpes simplex virus 1 by the U(S)3-encoded kinase regulates localization of the nuclear envelopment complex and egress of nucleocapsids. J Virol 83:5181–5191

Murata T, Goshima F, Daikoku T, Takakuwa H, Nishiyama Y (2000) Expression of herpes simplex virus type 2 US3 affects the Cdc42/Rac pathway and attenuates c-Jun N-terminal kinase activation. Genes Cells 5:1017–1027

Murata T, Goshima F, Nishizawa Y, Daikoku T, Takakuwa H, Ohtsuka K, Yoshikawa T, Nishiyama Y (2002) Phosphorylation of cytokeratin 17 by herpes simplex virus type 2 US3 protein kinase. Microbiol Immunol 46:707–719

Ng TI, Keenan L, Kinchington PR, Grose C (1994) Phosphorylation of varicella-zoster virus open reading frame (ORF) 62 regulatory product by viral ORF 47-associated protein kinase. J Virol 68:1350–1359

Nguyen ML, Kraft RM, Blaho JA (2007) Susceptibility of cancer cells to herpes simplex virus-dependent apoptosis. J Gen Virol 88:1866–1875

Pflum MK, Tong JK, Lane WS, Schreiber SL (2001) Histone deacetylase 1 phosphorylation promotes enzymatic activity and complex formation. J Biol Chem 276:47733–47741

Poon AP, Benetti L, Roizman B (2006a) U(S)3 and U(S)3.5 protein kinases of herpes simplex virus 1 differ with respect to their functions in blocking apoptosis and in virion maturation and egress. J Virol 80:3752–3764

Poon AP, Gu H, Roizman B (2006b) ICP0 and the US3 protein kinase of herpes simplex virus 1 independently block histone deacetylation to enable gene expression. Proc Natl Acad Sci USA 103:9993–9998

Poon AP, Liang Y, Roizman B (2003) Herpes simplex virus 1 gene expression is accelerated by inhibitors of histone deacetylases in rabbit skin cells infected with a mutant carrying a cDNA copy of the infected-cell protein no. 0. J Virol 77:12671–12678

Poon AP, Roizman B (2007) Mapping of key functions of the herpes simplex virus 1 U(S)3 protein kinase: the U(S)3 protein can form functional heteromultimeric structures derived from overlapping truncated polypeptides. J Virol 81:1980–1989

Purves FC, Ogle WO, Roizman B (1993) Processing of the herpes simplex virus regulatory protein alpha 22 mediated by the UL13 protein kinase determines the accumulation of a subset of alpha and gamma mRNAs and proteins in infected cells. Proc Natl Acad Sci USA 90:6701–6705

Purves FC, Spector D, Roizman B (1991) The herpes simplex virus 1 protein kinase encoded by the US3 gene mediates posttranslational modification of the phosphoprotein encoded by the UL34 gene. J Virol 65:5757–5764

Rahaus M, Desloges N, Wolff MH (2007) Varicella-zoster virus requires a functional PI3K/Akt/GSK-3alpha/beta signaling cascade for efficient replication. Cell Signal 19:312–320

Reynolds AE, Ryckman BJ, Baines JD, Zhou Y, Liang L, Roller RJ (2001) U(L)31 and U(L)34 proteins of herpes simplex virus type 1 form a complex that accumulates at the nuclear rim and is required for envelopment of nucleocapsids. J Virol 75:8803–8817

Reynolds AE, Wills EG, Roller RJ, Ryckman BJ, Baines JD (2002) Ultrastructural localization of the herpes simplex virus type 1 UL31, UL34, and US3 proteins suggests specific roles in primary envelopment and egress of nucleocapsids. J Virol 76:8939–8952

Ruyechan WT, Peng H, Yang M, Hay J (2003) Cellular factors and IE62 activation of VZV promoters. J Med Virol 70(Suppl 1):S90–S94

Ryckman BJ, Roller RJ (2004) Herpes simplex virus type 1 primary envelopment: UL34 protein modification and the US3-UL34 catalytic relationship. J Virol 78:399–412

Schaap-Nutt A, Sommer M, Che X, Zerboni L, Arvin AM (2006) ORF66 protein kinase function is required for T-cell tropism of varicella-zoster virus in vitro. J Virol 80:11806–11816

Schaap A, Fortin JF, Sommer M, Zerboni L, Stamatis S, Ku CC, Nolan GP, Arvin AM (2005) T-cell tropism and the role of ORF66 protein in pathogenesis of varicella-zoster virus infection. J Virol 79:12921–12933

Schumacher D, McKinney C, Kaufer BB, Osterrieder N (2008) Enzymatically inactive U(S)3 protein kinase of Marek's disease virus (MDV) is capable of depolymerizing F-actin but results in accumulation of virions in perinuclear invaginations and reduced virus growth. Virology 375:37–47

Schumacher D, Tischer BK, Trapp S, Osterrieder N (2005) The protein encoded by the US3 orthologue of Marek's disease virus is required for efficient de-envelopment of perinuclear virions and involved in actin stress fiber breakdown. J Virol 79:3987–3997

Schwer B, Verdin E (2008) Conserved metabolic regulatory functions of sirtuins. Cell Metab 7:104–112

Sedlackova L, Rice SA (2008) Herpes simplex virus type 1 immediate-early protein ICP27 is required for efficient incorporation of ICP0 and ICP4 into virions. J Virol 82:268–277

Smith-Donald BA, Roizman B (2008) The interaction of herpes simplex virus 1 regulatory protein ICP22 with the cdc25C phosphatase is enabled in vitro by viral protein kinases US3 and UL13. J Virol 82:4533–4543

Smith GA, Enquist LW (2002) Break ins and break outs: viral interactions with the cytoskeleton of Mammalian cells. Annu Rev Cell Dev Biol 18:135–161

Soong W, Schultz JC, Patera AC, Sommer MH, Cohen JI (2000) Infection of human T lymphocytes with varicella-zoster virus: an analysis with viral mutants and clinical isolates. J Virol 74:1864–1870

Stevenson D, Colman KL, Davison AJ (1994) Characterization of the putative protein kinases specified by varicella-zoster virus genes 47 and 66. J Gen Virol 75(Pt 2):317–326

Van den Broeke C, Deruelle M, Nauwynck HJ, Coller KE, Smith GA, Van Doorsselaere J, Favoreel HW (2009a) The kinase activity of pseudorabies virus US3 is required for modulation of the actin cytoskeleton. Virology 385:155–160

Van den Broeke C, Radu M, Deruelle M, Nauwynck H, Hofmann C, Jaffer ZM, Chernoff J, Favoreel HW (2009b) Alphaherpesvirus US3-mediated reorganization of the actin cytoskeleton is mediated by group A p21-activated kinases. Proc Natl Acad Sci USA 106:8707–8712

Van Minnebruggen G, Favoreel HW, Jacobs L, Nauwynck HJ (2003) Pseudorabies virus US3 protein kinase mediates actin stress fiber breakdown. J Virol 77:9074–9080

Wagenaar F, Pol JM, Peeters B, Gielkens AL, de Wind N, Kimman TG (1995) The US3-encoded protein kinase from pseudorabies virus affects egress of virions from the nucleus. J Gen Virol 76(Pt 7):1851–1859

Walters MS, Erazo A, Kinchington PR, Silverstein S (2009) Histone deacetylases 1 and 2 are phosphorylated at novel sites during varicella-zoster virus infection. J Virol 83:11502–11513

Wang ZX, Wu JW (2002) Autophosphorylation kinetics of protein kinases. Biochem J 368:947–952

Wisner TW, Wright CC, Kato A, Kawaguchi Y, Mou F, Baines JD, Roller RJ, Johnson DC (2009) Herpesvirus gB-induced fusion between the virion envelope and outer nuclear membrane during virus egress is regulated by the viral US3 kinase. J Virol 83:3115–3126

Yang M, Hay J, Ruyechan WT (2008) Varicella-zoster virus IE62 protein utilizes the human mediator complex in promoter activation. J Virol 82:12154–12163

Zeitz MJ, Malyavantham KS, Seifert B, Berezney R (2009) Matrin 3: chromosomal distribution and protein interactions. J Cell Biochem 108(1):125–133

Zhang Z, Carmichael GG (2001) The fate of dsRNA in the nucleus: a p54(nrb)-containing complex mediates the nuclear retention of promiscuously A-to-I edited RNAs. Cell 106:465–475

VZV ORF47 Serine Protein Kinase and Its Viral Substrates

Teri K. Kenyon and Charles Grose

Contents

Abstract ORF47, a serine protein kinase of varicella-zoster virus (VZV) and homolog of herpes simplex virus UL13, is an interesting modulator of VZV pathogenesis. This chapter summarizes research showing that ORF47 protein kinase activity, by virtue of phosphorylation of or binding to various viral substrates, regulates VZV proteins during all phases of viral infection and has a pronounced effect on the trafficking of gE, the predominant VZV glycoprotein, which in turn is critical for cell-to-cell spread of the virus. Casein kinase II, an ubiquitous cellular protein kinase, recognizes a similar but less stringent phosphorylation consensus sequence and can partially compensate for lack of ORF47 activity in VZV-infected cells. Differences between the phosphorylation consensus sites of the viral and cellular kinases are outlined in detail.

T.K. Kenyon and C. Grose (✉)
Department of Pediatrics/2501 JCP, University of Iowa Hospital, Iowa City, IA 52242, USA
e-mail: charles-grose@uiowa.edu

A.M. Arvin et al. (eds.), *Varicella-zoster Virus*,
Current Topics in Microbiology and Immunology 342, DOI 10.1007/82_2009_5

1 Introduction

The ancestral primordial herpesvirus is estimated to have emerged 180–220 million years ago (Wagenaar et al. 2003). The alpha herpesviruses include herpes simplex 1 and 2 and varicella-zoster virus (VZV). Varicelloviruses diverged from the ancestral simplex virus about 80 million years ago, near the widespread speciation of mammals (Davison and Scott 1986; McGeoch 1990). VZV, simian varicella virus (cercopithecine herpesvirus 9), pseudorabiesvirus (pig herpesvirus 1), and equine herpesvirus 1 are grouped together in the genus *Varicellovirus* (Davison 1991). VZV is the etiologic agent of both chicken pox in children (primary varicella) and shingles (or herpes zoster, after reactivation from latency) in older adults (Weller 1953).

VZV encodes at least 70 open reading frames (ORFs). Of these, two ORFs are known to encode active serine/threonine kinases. VZV ORF47 includes regions with extensive homology to serine/threonine kinase activity domains and ATP binding sites. VZV ORF47 is the homolog of herpes simplex UL13 kinase and human cytomegalovirus UL97 kinase (Davison and Scott 1986).

2 Molecular Evolution of VZV and ORF47

The ORF47 gene from eight different strains of VZV has been sequenced (Wagenaar et al. 2003). Of these, five are identical to VZV-Dumas, the prototype sequenced strain from Holland: VZV-32 (isolated in Texas in the 1970s), VZV-VIA (isolated in Iowa from a child with chicken pox in the 1990s), VZV-VSD (a wildtype virus collected in South Dakota in the 1980s), VZV-MSP (isolated from a child with chicken pox in 1995 in Minneapolis), and VZV-BC (isolated in 2000 from a 75-year-old adult with shingles) in Vancouver, BC. Three VZV strains encoded a silent SNP (no amino acid change) in ORF47: VZV-LAX1 (collected in Los Angeles in the early 1990s, before the introduction of the varicella vaccine), VZV-ICE1 (isolated in Iceland from the vesicle fluid of a child with chicken pox in the 1990s), and VZV-Oka (the vaccine strain originally isolated in Japan in the 1970s (Gomi et al. 2002; Tyler et al. 2007).

Only one VZV strain, VZV-Ellen, included a mutation that resulted in an amino acid change in the ORF47 gene. This substitution was within the serine/threonine kinase homology domain, though not within an ATP binding site. This change is of particular interest because VZV-Ellen was originally isolated from a child in Georgia in the 1960s and has been passaged in tissue culture at least a hundred times (Gershon et al. 1973).

The SCID-hu mouse system is an excellent model for comparing virulence among VZV strains in both skin and thymus implants. In this system, VZV-Ellen and other tissue culture-adapted strains were not fully virulent in skin when compared to low passage clinical VZV isolates (Moffat et al. 1998a). Many of the isolates mentioned above are low-passage isolates, purposefully retained at low passage numbers by propagating a few times in tissue culture (less than five) and freezing aliquots. Experiments are performed with fresh virus stocks to reduce the likelihood of interference from cell culture-acquired mutations. Thus, the only VZV strain with any non-silent mutation in ORF47 is a highly cell culture-adapted strain.

The notable exception to the hypothesis that tissue culture adaptation results in ORF47 mutation is VZV-Oka, the vaccine strain. VZV-Oka was isolated in Japan and passaged in guinea pig embryo fibroblasts and W138 cells (Takahashi et al. 1975). This propagation attenuated the virus such that a vaccine containing as much as 9,000 PFU induces immunity to VZV but does not produce clinical symptoms. Thus, it is interesting that in the vaccine strain, which must be virulent enough to induce immunity (perhaps by limited replication in T lymphocytes), ORF47 is identical in amino acid sequence to wildtype isolates. Other genes, including IE62 and gC, have acquired a large number of mutations in VZV-V-Oka (Peters et al. 2006).

3 Kinase Activity of ORF47 Protein

The story of the ORF47 protein kinase began with the publication that VZV gE (formerly called VZV gpI) was selectively phosphorylated by a virus-induced protein kinase (Montalvo and Grose 1986). In this chapter, gE (the predominant VZV glycoprotein) coprecipitated a virus-induced 50-kDa protein that appeared to be a serine protein kinase. As this chapter was published before the seminal Davison and Scott sequencing of the VZV-Dumas genome, there was no way to determine which gene of the VZV genome (125,000 base pairs of unknown sequence) was the likely candidate for the virus-induced kinase.

With the availability of VZV sequence data detailing likely kinase domains, it was possible to pinpoint the serine protein kinase, namely ORF47. With knowledge about the candidate gene, an immunogenic peptide was fused to beta-galactosidase, and the purified protein was used to immunize a rabbit. This polyclonal monospecific antibody precipitated the ORF47 protein, and parameters for an *in vitro* kinase assay were defined (Ng and Grose 1992). Subsequently, ORF47 derived from VZV-infected cells was found to phosphorylate an important viral protein: IE62 (also called ORF62) (Ng et al. 1994). IE62 is the major viral transactivator during primary VZV infection (Perera et al. 1992).

4 Expression of a Cloned VZV ORF47 Gene

All data regarding alphaherpesvirus kinases produced prior to 2001 were obtained using viral kinases immunoprecipitated from infected cell lysates or kinase-deletion mutant viruses. Previous attempts to clone, independently express, and purify any alphaherpesvirus kinase, including VZV ORF47, HSV-1 and HSV-2 UL13, VZV ORF66, and HSV-1 and HSV-2 Us3, while retaining their intrinsic kinase activity, had not been accomplished. The cellular kinase CKII was often utilized to assess the role of phosphorylation states of viral proteins, such as VZV gE, ORF65, and IE63, because no alpha herpesviral protein kinase had been active when purified from transfected cells. ORF47 has often been compared to CKII (Ng et al. 1994). We hypothesized that conditions that increase CKII phosphorylation may preserve ORF47.12 intrinsic kinase activity (Kenyon et al. 2001).

The ORF47 reading frame was epitope-tagged with the VZV gE 12-amino acid 3B3 epitope defined by Santos et al. 1998 (DQRQYGDVFKGD) near the amino-terminus using PCR mutagenesis (Hatfield et al. 1997). The epitope-tagged kinase protein was designated ORF47.12. An unexpected finding was that the VZV ORF47.12 protein kinase activity required polyamines *in vitro*. Polyamines enhance casein kinase II (CKII) activity but are not required in the protein kinase assay (Kenyon et al. 2003). Polyamines are basic cations synthesized in cells from ornithine. Ornithine decarboxylase catalyzes the production of the divalently positive putrescine (1,4-diamino butane) from ornithine; two putrescines are deaminated to form spermidine; which is trivalent; finally, a spermidine and a putrescine in a deamination reaction form spermine, also called cadaverine, which is tetravalently positive.

In the cellular environment, positively-charged polyamines stabilize protein-protein interactions and protein complex formation. Spermine or spermidine added to CKII *in vitro* kinase reactions stimulate CKII up to twenty times its basal activity level, but putrescine does not stimulate CKII. The regulatory beta-subunits of CKII bind spermine. The alpha subunits contain the catalytic domain and are not (when expressed without the beta-subunits) activated by the presence of polyamines. Polyamines thus regulate the activity of CKII by binding directly to the regulatory beta subunit of the kinase (Tuazon and Traugh 1991).

ORF47.12 autophosphorylation was stimulated by polyamines in that the polyamines were binding to and stabilizing either the tertiary structure of ORF47.12 or the formation of dimer complexes of ORF47.12, thus facilitating auto- or allophosphorylation and, in turn, substrate phosphorylation. All three polyamines increased the intrinsic kinase activity of ORF47.12. This stimulation was not merely due to the presence of positively-charged cations, as neither magnesium nor calcium recapitulated this effect.

Addition of poly-DL-lysine to ORF47.12 *in vitro* reactions decreased polyamine-stimulated ORF47.12 autophosphorylation in proportion to the amount of polyamine-induced increase in the kinase activity. For example, spermidine alone stimulated ORF47.12 autophosphorylation the most, and addition of poly-DL-lysine to a spermidine-stimulated reaction decreased autophosphorylation the most (Kenyon et al. 2003).

Poly-DL-lysine stimulates CKII kinase activity by binding to the substrate and increasing the stoichiometry of the substrate–kinase reaction. Polyamines do not stimulate CKII activity via this mechanism. Thus, while polyamines may stimulate ORF47.12 and CKII by a similar mechanism, the substrate-binding activity of poly-DL-lysine does not stimulate ORF47.12 kinase activity. This difference may be due to any of several reasons. Poly-DL-lysine is much larger than the polyamine compounds, the differing effect on kinase activity may be due to steric hindrance. Polyamines and poly-DL-lysine attach to different epitopes on CKII. It is suggested by these data that ORF47 possesses an epitope similar to the one by which CKII binds polyamines but lacks an epitope similar to the one by which CKII binds poly-DL-lysine.

Several other bands were observed in the ORF47.12 lanes in addition to the 54 kDa ORF47-dependent band. One of these (about 45 kDa) below the kinase band was reactive with an anti-mouse IgG antibody on an immunoblot, suggesting that this is the IgG heavy chain. A band that is immunoreactive with both Rab22 (anti-ORF47) and MAb 3B3 (anti-3B3 inserted epitope) exists at about 110 kDa and suggests the existence of an ORF47 dimer. An ORF47 homodimer has been detected in infected cells and is notably present in ORF47 mutants with kinase domain mutations (Besser et al. 2003). Radioactive bands that were also immunopositive for the two antibodies at higher molecular weights may suggest the existence of multimers or limited solubility of the kinase. These bands also appear on immunoblots of native ORF47 immunoprecipitated or coprecipitated from infected cells.

Kinases encode an invariant lysine upstream of the kinase activation domain that is required for kinase activity. To definitively attribute the phosphotransferase activity to ORF47, the invariant lysine (codon 169) near the kinase catalytic domain was replaced using PCR mutagenesis, and the mutant was designated ORF47.12d. PCR mutagenesis of the lysine at position 169 in the mutant ORF47.12d reduced autophosphorylation by 63% when compared to autophosphorylation of wildtype ORF47.12. This marked reduction in phosphorylation due to one amino acid substitution mutation supported our hypothesis that the phosphorylation observed was due solely to the immunoprecipitated ORF47.12. Radioimmunoprecipitation from lysates labeled with [S-35] Promix showed no difference in bands coprecipitated with ORF47.12d compared to bands seen in the radioimmunoprecipitation of ORF47.12. For all the above reasons, we concluded that autophosphorylation was dependent upon the ORF47.12 intrinsic kinase activity.

5 Autophosphorylation of ORF47 Protein

In the *in vitro* kinase assay using ORF47 kinase precipitated from VZV-infected cell lysates as described in our laboratory, ORF47 required manganese as a cofactor and phosphorylated itself and casein, a protein with a highly acidic domain. CKII can also phosphorylate casein, and thus comparing the two kinases was a natural extension of the work (Tuazon and Traugh 1991).

CK II is inhibited by 2 mM heparin, an acidic compound that binds an invariant lysine near the kinase domain (Tuazon and Traugh 1991). ORF47 is not inhibited by comparatively high concentrations of heparin in an *in vitro* kinase assay, an important biochemical quality that distinguishes it from CKII. ORF47 also utilizes GTP as a phosphate donor (Ng and Grose 1992). Another interesting finding is that ORF47 dimerization does not require autophosphorylation catalytic activity (Besser et al. 2003).

6 Phosphorylation of VZV IE62: The Major Transactivator

At the initiation of VZV infection, ORF47 enters an uninfected cell as a resident of the virion tegument and is thus present at the earliest stage of VZV infection. As it is produced during infection and packaged back into the virion, it is present throughout the viral lifecycle.

ORF47 phosphorylates the major VZV transactivator IE62. The VZV IE62 is encoded by both ORF62 and ORF71 within the inverted repeat sequences bounding the US region of the VZV genome. IE62 is the major VZV transcriptional regulatory protein and activates expression of all three kinetic classes of viral genes. It is homologous in large parts to HSV ICP4, but unlike the protein IE62, possesses a potent N-terminal acidic activation domain (Perera et al. 1992). VZV IE62 is expressed as an immediate early (IE) protein during the replication cycle and is localized to the nucleus during initial stages of infection. IE62 is phosphorylated by the viral ORF47 and ORF66 kinases and CKII (Lynch et al. 2002).

At late stages in infection, the IE62 protein is incorporated into the viral tegument in a process mediated by the VZV ORF66 kinase (Kinchington et al. 2001). Thus IE62 is present immediately upon infection and is believed to enter the nucleus along with the viral DNA where it can play a critical role in triggering and regulating the replicative cycle of VZV. IE62 has recently been shown to directly interact with the VZV IE4 and IE63 proteins (Sommer et al. 2001).

7 Phosphorylation of VZV IE63: A Latency-Associated Protein

In VZV, IE63 is the major transcript and protein found during VZV latency in both human ganglia and animal models (Cohrs and Gilden 2003). ORF47 also phosphorylates IE63, a viral protein that up-regulates transcription from a late VZV promoter – that of gI. The IE63 protein is encoded by VZV ORFs63 and 70 and is the putative homolog of HSV ICP22 (Sommer et al. 2001). Unlike that protein, however, IE63 has recently been shown to be required for growth in cell culture and is incorporated into the virion tegument. The predicted molecular weight of IE63 is 30.5 kD as compared to a predicted molecular weight of 46.5 kDa for HSV ICP22. The major region of similarity between IE63 and ICP22 lies within the carboxy-terminal portions of the two proteins and is quite limited (> 25%). Thus, the VZV IE63 protein appears to be both functionally and structurally distinct from HSV ICP22.

Using an *in vitro* system that preserved the biological activity of the cloned ORF47.12, we showed that VZV IE62 (homolog of HSV ICP4) and IE63 (homolog of HSV ICP22) were phosphorylated in a manner dependent upon the kinase activity of the cloned ORF47.12 protein kinase. Both VZV IE62 and IE63 are known residents of the VZ virion tegument. This is of great interest because HSV phosphorylation analyses have demonstrated that phosphorylation of the HSV-1 VP22 (homolog of VZV ORF10) and its release from the tegument occur concurrently and can be mediated by either HSV UL13 or CKII. Presumably, therefore, at least one functional site of VZV ORF47 kinase is also within the tegument.

In addition, the observation that ORF47 bound so tightly to IE63 that ORF47 precipitated IE63 from the kinase reaction supernatant was an unusual finding. Although kinases may often be identified by precipitation of substrate with concurrent coprecipitation of kinase, only rarely can phosphorylated substrates be identified by precipitation of the relevant kinase. Typically the substrate is released immediately after phosphorylation (Tuazon and Traugh 1991). Thus, the latter result suggests that ORF47 and its substrate IE63 exist as a complex. Subsequent samples with addition of exogenous heparin demonstrated that ORF47 and CKII bind IE63, perhaps each at the expense of the other. As the ORF47 consensus sequence is similar to that of CKII (see Sect. "Phosphorylation of VZV gE and gE Endocytosis") it is likely that the two kinases bind the same region.

Because these two kinases likely bind the same region of IE63, it is also likely that the addition of heparin, which is known to bind and inactivate CKII, reduces active CKII in the kinase reaction and thus allows ORF47 greater access to the substrate. Because heparin is anionic, it may also, of course, be changing the ionic strength or overall charge availability of the buffer, though the ORF47 buffer ionic strength was previously optimized for ORF47 activity. Heparin could also be reducing the activity of another, yet unidentified, contaminating kinase that phosphorylates ORF47 and thus represses its activity.

8 Phosphorylation of VZV gE and gE Endocytosis

As the viral kinase is present during viral infection in both the nuclei and cytoplasm of infected cells, the kinase can play multiple roles during infection. Late in VZV infection, ORF47 associates with the VZV predominant glycoprotein, gE. The cytoplasmic tail of gE contains an acidic cluster with four phosphorylatable residues: EDSESTDTEEE. VZV ORF47 competes with CKII, the cellular kinase, to differentially phosphorylate these four residues *in vivo* (Kenyon et al. 2002). Utilizing cloned, independently expressed gE with point mutations in the acidic cluster, the difference between the phosphorylation due to the two kinases is clear.

ORF47 equally phosphorylated both the initial serines and the trailing threonines in the acidic cluster. In the analyzed data, approximately equal amounts of radiophosphate were incorporated into the SSAA (EDSESADAEEE) gE and the AATT (EDAEATDTEEE) gE in the ORF47.12 kinase assay samples, and these amounts each approximated half the amount of phosphorylation observed in the ORF47/gE wildtype sample. No radiophosphate above background levels was incorporated into the AAAA gE (EDAEAADAEEE) by ORF47.12.

CKII, however, heavily phosphorylated the AATT gE mutant, but CKII-dependent phosphorylation of the SSAA gE mutant was reduced. Thus, CKII exhibited marked preference for the two threonine residues, especially when the lower affinity serines were replaced by alanines (Kenyon et al. 2002).

In immunoblots from cells infected with the ORF47-null VZV (rOka-47S, which has contiguous stop codons in the reading frame after codon 165), an additional band was detected at 150 kDa in the VZV gE sample above the usual 55–97 kDa glycoprotein band. This band had the same molecular weight and appearance as an underglycosylated gE dimer that is often found in gE samples purified from transfected cells and from insect cells (Olson and Grose 1997). Thus, a gE dimer exists in rOka-47S VZV-infected cells that was not found in wildtype VZV-infected cells. It is possible that phosphorylation of the two initial serines inhibits dimer formation.

A recombinant ROka-47S virus, in which stop codons interrupt the ORF47 transcript after codon 165, is the most extreme example of the ORF47 knockout phenotype and has been rightfully referred to as an ORF47-null mutant virus. When immunoblotted, no ORF47 protein has been found during infection by this mutant. In the SCID-hu mouse model, rOka-47S does not replicate in human skin or T cell implants, and no VZV proteins were detected at all (Moffat et al. 1998b). In ORF47-null VZV rOka-47S, gE does not internalize from the plasma membrane to the interiors of the syncytia. In contrast, during normal wildtype infection in confocal microscopy endocytosis assays, gE is endocytozed from the plasma membrane to the centers of syncytia, the ultimate destination of the *trans*-Golgi network (TGN) in VZV-associated syncytia, and colocalized with TGN markers (Olson et al. 1997).

9 ORF47 and the Role of VZV gE in Cell-to-Cell Spread

VZV gE plays a critical role in virus-induced cell-to-cell spread. As shown in Fig. 1, replicate monolayers were inoculated with synchronized cultures of either wildtype VZV or ORF47-null mutant VZV. At various stages after infection (0, 12, 24, and 36 hpi), the samples were processed for confocal microscopy with monoclonal antibodies to VZV gE and counterstained with TOTO-3 nuclear stain. The color channels were split on the resulting micrographs and each channel was produced separately in grayscale and in the photographic negative, as black-on-white imaging is easier to reproduce and interpret in monochrome than a white-on-black image.

The ORF47-null virus spreads faster in tissue culture than the comparable wildtype strain. For example, by 12 hpi, the wildtype strain has begun to infect other cells, and occasional small syncytia are forming. In the ORF47-null-infected

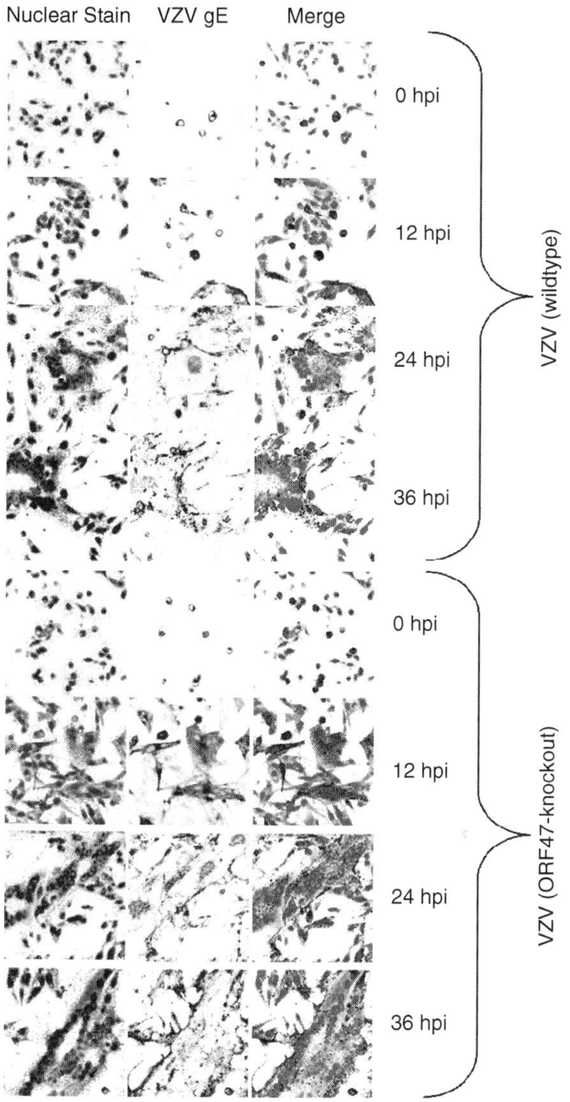

Fig. 1 Enhanced cell-to-cell spread in cells infected with a recombinant VZV genome lacking VZV ORF47 kinase activity. At increasing intervals, monolayers of infected cells were immunostained with an anti-VZV gE antibody plus a specific anti-nuclear reagent, and imaged by confocal microscopy. See the text for a further discussion of the figure

cells, however, plentiful amounts of gE are being synthesized. The plasma membranes and Golgi are rife with gE. At 36 hpi, most of the monolayer infected with wildtype VZV are involved in syncytia, though a few single infected cells persist. In the ORF47-null mutant-infected cells at 36 hpi, the monolayer is completely involved in syncytia, and the syncytia themselves are collapsing. The nuclear rings have condensed. The gE endocytosis defect has depleted the syncytial centers of gE, and the plasma membranes stain darkly for the membrane-trapped gE. Here, the promiscuity of CKII has directed large quantities of gE to the plasma membrane, where it acts to greatly accelerate cell-cell spread.

9.1 Deduction of the ORF47 Kinase Consensus Sequence

ORF47 protein kinase activity has been often compared to CKII protein kinase activity, as both kinases phosphorylate proteins with acidic domains surrounding the target residues. However, as few authentic substrates of ORF47 have been defined, determining the minimal phosphorylation consensus sequence for the viral kinase has been problematic. No specific substrates (substrates that only one of the two kinases phosphorylated) have been published.

In a detailed analysis, we surveyed numerous potential substrates of ORF47 (Kenyon et al. 2003). The most important substrates are included in Table 1. The impression was that ORF47 requires greater acidic component in the phosphorylation consensus sequence than does CKII, and thus the viral kinase is less promiscuous and more specifically targeted than is the cellular kinase, CKII. As shown above, two substrates of the ORF47 kinase were the regulatory IE62 and IE63 proteins. Both these proteins include extensive acidic domains around serine or threonine phosphorylation targets. It must again be emphasized that IE63 was so tightly bound by ORF47.12 that precipitation of the viral kinase coprecipitated the IE63 substrate – an unusual result.

All substrates phosphorylated by ORF47.12 in the highly specific *in vitro* kinase assays included extensive acidic domains around a serine or threonine. IE62 has four exaggerated acidic domains: 293-ETDDT, 363-SSSEDEDDE, 397-SDDSDS,

Table 1 Potential consensus sequences recognized by ORF47 kinase

Substrate	Serines	Threonines
Casein	17-SSSEES	40-QTEDE
	34-QSEEQ	
VZV IE62	363-SSEDE	77-RTEDV
	365-SEDEDDE	85-LTQDD
	397-SDDSDS	293-ETDDT
	1611-SADE	1218-TADD
	1290-SEDEDD	
VZV IE63	11-DSSESK	239-ETAE
	156-SDDGGED	
	163-DDSDDD	
	180-DSDAE	
	184-ESSDGED	
	194-EEESEES	
	199-ESTDSCE	
VZV gE	588-DDFEDSES	596-TDTEEE

and 1290-SEDEDD. Casein, which competes with viral substrates for ORF47.12 phosphorylation, includes the sequence ESLSSSEES at codon 14. VZV IE63, which ORF47 phosphorylates and precipitates, includes three closely-grouped acidic domains: 163-DDSDDD, 184-ESSDGED, and 194-EEESEESTDSCE.

As examined above and in Table 1, ORF47 phosphorylation substrates include an extensive acidic motif, and many of the proteins encode multiple serines or threonines within the acidic domain, such as the large domain within IE62 at amino acid 1285 (SSSSSSEDEDD). If the acidic cluster (EDEDD) attracts the ORF47 kinase and the N-proximal serine is phosphorylated, the phosphate added to that serine then increases the overall regional negative charge, and thus the kinase may bind with more affinity and thus increase the likelihood of the phosphorylation of each of the preceding serines, like a string of firecrackers sequentially lighting each succeeding fuse. This increase in negative change may increase the avidity of the kinase for the protein and increase interactions necessary for viral replication.

In addition to deducing the consensus sequence, the marked preference of ORF47 kinase for acidic amino acids elucidates some earlier results. Previously, we showed that ORF47.12 not only phosphorylated but bound and precipitated up to 80% of the exogenous VZV IE63 protein from *in vitro* kinase reactions, even after extensive washing with lysis and kinase buffers (Kenyon et al. 2001). One would postulate that VZV IE63 should have a very acidic region that ORF47 tightly binds. Indeed, VZV IE63 protein includes three closely-grouped regions with extensive serine and acidic sequences (S-157: DVSDDG GEDDSDDD, S-181: D SDAESSDGED, and S-197, EEESEES TDSCE). In those experiments, CKII did not bind and precipitate VZV IE63 from the *in vitro* kinase reaction. Thus, though VZV ORF47 recognizes a more stringent phosphorylation consensus sequence than CKII, the viral kinase binds extensive acidic clusters with greater avidity than the cellular kinase.

In summary, based on extensive data in this chapter, the most reasonable serine/threonine phosphorylation consensus sequence recognized by the VZV ORF47 kinase is the following motif: S/T – X – D/E – D/E, with a marked preference for additional acidic amino acids in the −1 and +1 position, and an exclusion for a basic, positively-charged residue in the +1 position.

References

Besser J, Sommer MH, Zerboni L, Bagowski CP, Ito H, Moffat J, Ku CC, Arvin AM (2003) Differentiation of varicella-zoster virus ORF47 protein kinase and IE62 protein binding domains and their contributions to replication in human skin xenografts in the SCID-hu mouse. J Virol 77:5964–5974

Cohrs RJ, Gilden DH (2003) Varicella zoster virus transcription in latently-infected human ganglia. Anticancer Res 23:2063–2069

Davison AJ, Scott JE (1986) The complete DNA sequence of varicella-zoster virus. J Gen Virol 67:1759–1816

Davison AJ (1991) Varicella-zoster virus; the fourteenth Fleming lecture. J Gen Virol 72:475–486

Gershon A, Cosio L, Brunell PA (1973) Observations on the growth of varicella-zoster virus in human diploid cells. J Gen Virol 18:21–31

Gomi Y, Sunamachi H, Mori Y, Nagaike K, Takahashi M, Yamanishi K (2002) Comparison of the complete DNA sequences of the Oka varicella vaccine and its parental virus. J Virol 76: 11447–11459

Hatfield C, Duus KM, Jones DH, Grose C (1997) Epitope mapping and tagging by recombination PCR mutagenesis. Biotechniques 22:332–337

Kenyon TK, Lynch J, Hay J, Ruyechan W, Grose C (2001) Varicella-zoster virus ORF47 protein serine kinase: characterization of a cloned, biologically active phosphotransferase and two viral substrates, ORF62 and ORF63. J Virol 75:8854–8858

Kinchington PR, Fite K, Seman A, Turse SE (2001) Virion association of IE62, the varicella-zoster virus major transcriptional regulatory protein, requires expression of the VZV open reading frame 66 protein kinase. J Virol 75:9106–9113

Kenyon TK, Cohen JI, Grose C (2002) Phosphorylation by the varicella-zoster virus ORF47 protein serine kinase determines whether endocytosed viral gE traffics to the trans-Golgi network or recycles to the cell membrane. J Virol 76:10980–10993

Kenyon TK, Homan E, Storlie J, Ikoma M, Grose C (2003) Comparison of varicella-zoster virus ORF47 protein kinase and casein kinase II and their substrates. J Med Virol 70(Suppl 1): S95–S102

Lynch JM, Kenyon TK, Grose C, Hay J, Ruyechan WT (2002) Physical and functional interaction between the varicella zoster virus IE63 and IE62 proteins. Virology 302:71–82

McGeoch DJ (1990) Evolutionary relationships of virion glycoprotein genes in the S regions of alphaherpesvirus genomes. J Gen Virol 71(Pt 10):2361–2367

Moffat JF, Zerboni L, Kinchington PR, Grose C, Kaneshima H, Arvin AM (1998a) Attenuation of the vaccine Oka strain of varicella-zoster virus and role of glycoprotein C in alphaherpesvirus virulence demonstrated in the SCID-hu mouse. J Virol 72:965–974

Moffat JF, Zerboni L, Sommer MH, Heineman TC, Cohen JI, Kaneshima H, Arvin AM (1998b) The ORF47 and ORF66 putative protein kinases of varicella-zoster virus determine tropism for human T cells and skin in the SCID-hu mouse. Proc Natl Acad Sci USA 95:11969–11974

Montalvo EA, Grose C (1986) Varicella zoster virus glycoprotein gpI is selectively phosphorylated by a virus-induced protein kinase. Proc Natl Acad Sci USA 83:8967–8971

Ng TI, Grose C (1992) Serine protein kinase associated with varicella-zoster virus ORF47. Virology 191:9–18

Ng TI, Keenan L, Kinchington PR, Grose C (1994) Phosphorylation of varicella-zoster virus open reading frame (ORF) 62 regulatory product by viral ORF47-associated protein kinase. J Virol 68:1350–1359

Olson JK, Grose C (1997) Endocytosis and recycling of varicella-zoster virus Fc receptor glycoprotein gE: internalization mediated by a YXXL motif in the cytoplasmic tail. J Virol 71:4042–4054

Olson JK, Bishop GA, Grose C (1997) Varicella-zoster virus Fc receptor gE glycoprotein: serine/threonine and tyrosine phosphorylation of monomeric and dimeric forms. J Virol 71:110–119

Perera LP, Mosca JD, Sadeghi-Zadeh M, Ruyechan WT, Hay J (1992) The varicella-zoster virus immediate early protein, IE62, can positively regulate its cognate promoter. Virology 191:346–354

Peters G, Tyler S, Grose C, Severini A, Gray M, Upton C, Tipples G (2006) A full genome phylogenetic analysis of varicella-zoster virus reveals a novel origin of replication-based genotyping scheme and evidence of recombination between major circulating clades. J Virol 80:9850–9860

Sommer MH, Zagha E, Serrano OK, Ku CC, Zerboni L, Baiker A, Santos R, Spengler M, Lynch J, Grose C, Ruyechan W, Hay J, Arvin AM (2001) Mutational analysis of the repeated open reading frames, ORFs63 and 70 and ORFs64 and 69, of varicella-zoster virus. J Virol 75:8224–8239

Santos RA, Padilla J, Hatfield C, Grose C (1998) Antigenic variation of varicella-zoster virus Fc receptor gE: loss of a major B cell epitope in the ectodomain. Virology 249:21–31

Takahashi M, Okuno Y, Otsuka T, Osame J, Takamizawa A (1975) Development of a live attenuated varicella vaccine. Biken J 18:25–33
Tuazon PT, Traugh JA (1991) Casein kinase I and II – multipotential serine protein kinases: structure, function, and regulation. Adv Second Messenger Phosphoprotein Res 23:123–164
Tyler SD, Peters GA, Grose C, Severini A, Gray MJ, Upton C, Tipples GA (2007) Genomic cartography of varicella-zoster virus: a complete genome-based analysis of strain variability with implications for attenuation and phenotypic differences. Virology 359:447–458
Wagenaar TR, Chow VT, Buranathai C, Thawatsupha P, Grose C (2003) The out of Africa model of varicella-zoster virus evolution: single nucleotide polymorphisms and private alleles distinguish Asian clades from European/North American clades. Vaccine 21:1072–1081
Weller TH (1953) Serial propagation *in vitro* of agents producing inclusion bodies derived from varicella and herpes zoster. Proc Soc Exp Biol Med 83:340–346

Overview of Varicella-Zoster Virus Glycoproteins gC, gH and gL

Charles Grose, John E. Carpenter, Wallen Jackson, and Karen M. Duus

Contents

Abstract The VZV genome is smaller than the HSV genome and only encodes nine glycoproteins. This chapter provides an overview of three VZV glycoproteins: gH (ORF37), gL (ORF60), and gC (ORF14). All three glycoproteins are highly conserved among the alpha herpesviruses. However, VZV gC exhibits unexpected differences from its HSV counterpart gC. In particular, both VZV gC transcription and protein expression are markedly delayed in cultured cells. These delays occur regardless of the virus strain or the cell type, and may account in part for the aberrant assembly of VZV particles. In contrast to VZV gC, the general properties of gH and gL more closely resemble their HSV homologs. VZV gL behaves as a

C. Grose (✉), J.E. Carpenter, and W. Jackson
Department of Pediatrics/2501 JCP, University of Iowa Hospital, Iowa City, IA 52242, USA
e-mail: charles-grose@uiowa.edu

K.M. Duus
Center for Immunology, Albany Medical College, Albany, NY, USA

A.M. Arvin et al. (eds.), *Varicella-zoster Virus*,
Current Topics in Microbiology and Immunology 342, DOI 10.1007/82_2009_4

chaperone protein to facilitate the maturation of the gH protein. The mature gH protein in turn is a potent fusogen. Its fusogenic activity can be abrogated when infected cultures are treated with monoclonal anti-gH antibodies.

1 Introduction

The VZV genome codes for at least nine glycoproteins. The glycoproteins are critical inducers of cell-to-cell spread and fusion, as well as required components of the mature virion. In this chapter, we provide an overview of three viral glycoproteins: gC, gH, and gL. Similarities and differences from their homologous glycoproteins among the other alpha herpesviruses are reviewed. In particular, VZV gC exhibits remarkably different properties from its counterpart HSV gC, as discussed below, and these differences may account in part for the invariably low titer of VZV grown in cultured cells. A genomic survey of the VZV glycoprotein sequences has been recently published (Storlie et al. 2008b).

2 VZV gC Protein Structure

Shortly after the publication of the sequence of the complete VZV genome (Davison and Scott 1986), investigations established the basic characteristics of VZV gC or gpV as it was then known. The gC protein is encoded by ORF14 and is a glycosylated polypeptide around 105 kD in mass that traffics to the plasma membrane (Kinchington et al. 1990b). ORF14 contains the variable repeat region R2 that differs in length in strains of VZV (Kinchington et al. 1986). Examination of its amino sequence shows 560 amino acids, including the repeated region from amino acids 33 to 140 that is expressed as approximately eight repeated sequences of TSAATRKPDPAVAP in the Dumas strain (Fig. 1). The unglycosylated protein has a mass of 61 kD while the observed mass is approximately 105 kD – a difference that indicates over 40 kD of post-translational additions to the protein (Fig. 2). Using prediction algorithms, four N-linked glycosylation sites and 25 O-linked mucin type sites exclusively on the repeated pattern were identified (Fig. 1). The observed molecular weight and glycosylation prediction indicates that more than one third of VZV gC molecular weight is due to oligosaccharide chains. The membrane spanning helical domain is predicated to be near the C terminus at amino acids 531–554, leaving only six amino acids extending into the cytosol. Genomic analysis showed that VZV gC is approximately 30% similar to HSV gC, indicating that both descended from a common ancestor gene (Grose 1990a).

```
MKRIQINLILTIACIQLSTESQPTPVSITELYTSAATRKP   1 -  40
---------------------       o   oo  o
DPAVAPTSAASRKPDPAVAPTSAASRKPDPAVAPTSAASR  41 -  80
      oo  o         oo  o         oo  o
KPDPAVAPTSAATRKPDPAVAPTSAASRKPDPAVAPTSAA  81 - 120
        oo  o         oo  o         oo
TRKPDPAVAPTSAASRKPDPAANTQHSQPPFLYENIQCVH 121 - 160
o         oo  o
GGIQSIPYFHTFIMPCYMRLTTGQQAAFKQQQKTYEQYSL 161 - 200

DPEGSNITRWKSLIRPDLHIEVWFTRHLIDPHRQLGNALI 201 - 240
     n
RMPDLPVMLYSNSADLNLINNPEIFTHAKENYVIPDVKTT 241 - 280

SDFSVTILSMDATTEGTYIWRVVNTKTKNVISEHSITVTT 281 - 320

YYRPNITVVGDPVLTGQTYAAYCNVSKYYPPHSVRVRWTS 321 - 360
    n                  n
RFGNIGKNFITDAIQEYANGLFSYVSAVRIPQQKQMDYPP 361 - 400

PAIQCNVLWIRDGVSNMKYSAVVTPDVYPFPNVSIGIIDG 401 - 440
                               n
HIVCTAKCVPRGVVHFVWWVNDSPINHENSEITGVCDQNK 441 - 480

RFVNMQSSCPTSELDGPITYSCHLDGYPKKFPPFSAVYTY 481 - 520

DASTYATTFSVVAVIIGVISILGTLGLIAVIATLCIRCCS 521 - 560
          ========================
```

Fig. 1 Annotated amino acid sequence of VZV gC. Putative signal (*dashed line*) and transmembrane (*double dashed line*) domains were predicted with SignalP and TMPred software, respectively. Potential N- and O-linked glycosylation sites are marked by "n" or "o." These were predicted by NetNGlyc and NetOGlyc software, respectively

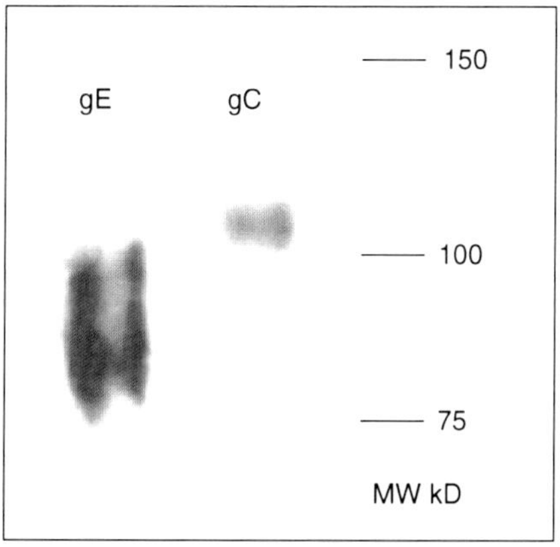

Fig. 2 Immunoblotting of VZV gE and gC proteins. A VZV-infected lysate was separated by gel electrophoresis using native nonreducing conditions. After transfer of the proteins, they were immunoblotted with mouse monoclonal antibodies to gE and gC. The gC protein migrates more slowly (105 kD) than the prominent and broad gE complex (88–98 kD)

3 Delayed Expression of VZV gC Protein in Cultured Cells

As an alphaherpesvirus, VZV is considered to have a 12–18 h replication cycle (Grose and Ng 1992). In most VZV experiments, the inoculum consists of trypsin-dispersed infected cells from a monolayer with advanced CPE (Grose et al. 1979b). Because the titer of input virus is very low, several replication cycles are required before cytopathology is visible in the newly infected monolayers. While examining glycoprotein expression at increasing intervals after infection, we observed that all syncytia at early time points (24 and 48 hpi) were clearly labeled with a monoclonal antibody against gE, but only an occasional VZV induced syncytium was labeled with anti-gC antibody (Storlie et al. 2006). Since the inoculum consisted of sonicated infected cells, input syncytia were destroyed; therefore, syncytia seen in the newly infected monolayer were the result of new virus replication. The center of each newly formed syncytium includes an aggregation of Golgi; thus, the newly synthesized gC protein in a positive syncytium was located mainly in Golgi and not on the cell surface. Even though conditions of cell associated VZV infection are asynchronous, by 48 hpi all input viral populations would have undergone 3–4 replication cycles. VZV gE and gH were easily detectable; yet, VZV gC was invariably difficult to detect by confocal microscopy at 48 hpi (Fig. 3). In all our prior studies of VZV glycoproteins, there was no precedence for absence of any VZV structural protein so late after infection (Grose 1990b).

In order to demonstrate that gC expression and localization were not simply outside of the plane of our original confocal images, we obtained multiple serial images of our infected cells at various time points. We also included antibody probes for other VZV proteins such as gE, gH, and major capsid protein (MCP; ORF40). This series of optical sections collected at different levels perpendicular to the optical axis (Z-stack) was further processed into a three-dimensional representation of the infected cell using three-dimensional projection options for ImageJ and LSM Image Browser. By examining infected monolayers between 4 and 48 hpi, we were able to document that small amounts of gC could be found on some inoculum cells but were rarely detectable in the newly infected monolayer in the first 48 hpi.

To further investigate the expression of gC as compared with gE by another method, we infected replicate monolayers and harvested them over a 96 h time period. When examined by immunoblotting, this experiment confirmed that gC protein expression was markedly delayed when compared with gE, namely, abundant amounts of gC protein were not observed until 72 hpi (Storlie et al. 2008a). This time point is equivalent to 5–6 replication cycles. Between 72 and 96 hpi, increasing amounts of gC expression were finally detected in the infected monolayers. In contrast, western blotting showed that gE was present in abundance at 24 hpi. Finally, in order to eliminate the possibility that gC was actually being expressed at earlier time points but not in a form recognizable by the anti-gC MAb, we generated a rabbit polyclonal monospecific anti-gC antibody; the results were comparable. The latter result indicated that weak affinity of the anti-gC mouse monoclonal antibody was not an explanation for the immunoblotting data.

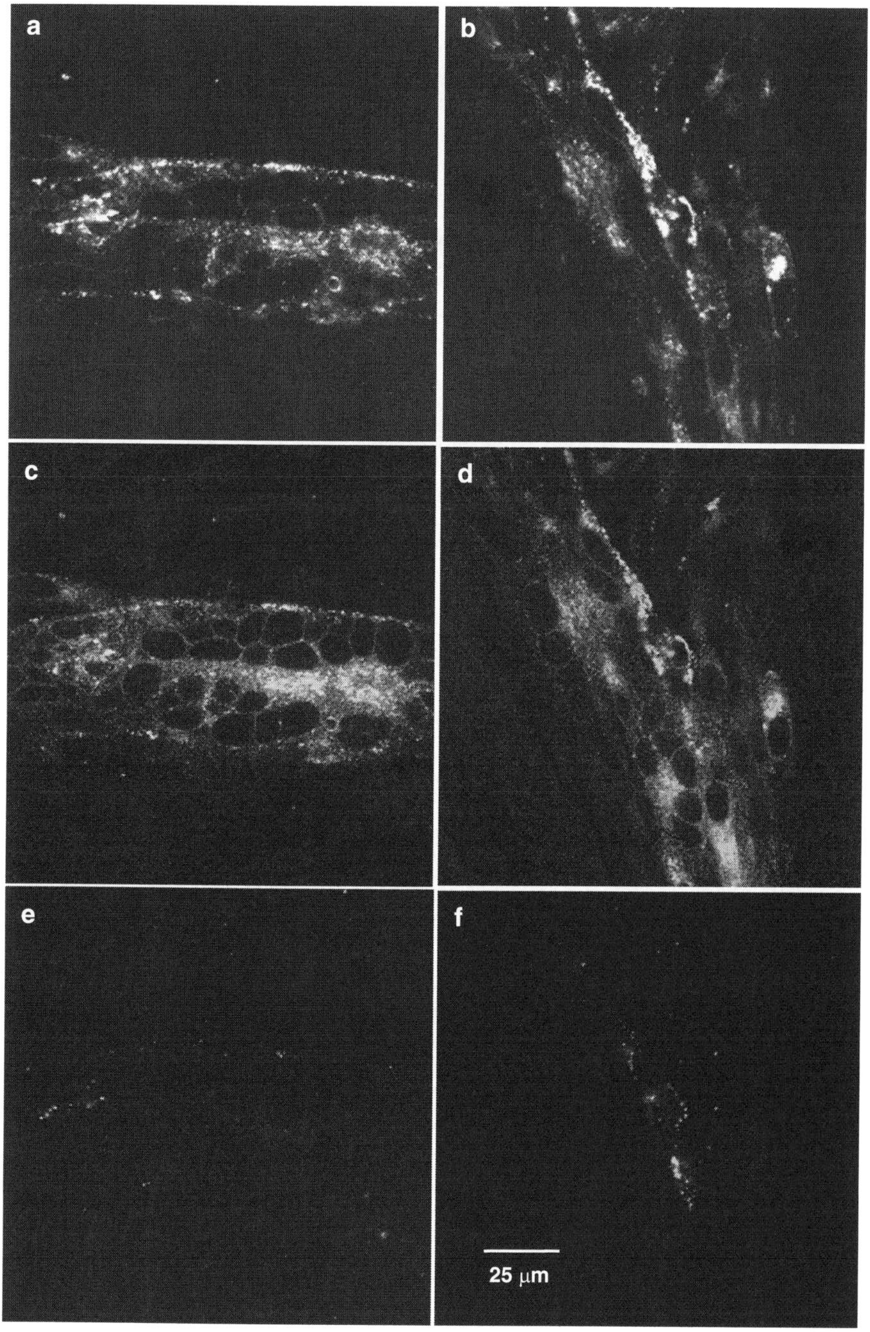

Fig. 3 Delayed expression of VZV gC when compared with gH and gE expression in infected cells. VZV-infected monolayers were fixed after 72 h and immunolabeled with antibodies to gE, gH, and gC. Panels (**a–b**) show gE protein; (**c–d**) show gH protein and (**e–f**) show gC protein. All images were taken on a Zeiss LSM 510 confocal microscope at a magnification of 630×. The panels document that gC expression is greatly delayed compared with either gE or gH

Furthermore, two other explanations for the above gC effect were excluded. First, we investigated whether the effect was specific to one VZV strain. Our initial experiments were performed with the VZV-32 strain used in this laboratory for many years (Grose et al. 1981). This strain has now been completely sequenced and contains an intact genome with no major genetic changes when compared with VZV Dumas (Peters et al. 2006). We also examined gC expression after infection of cells with two additional strains, including VZV-MSP and VZV-Ellen. VZV-MSP has also been completely sequenced and found to have an intact gC gene. The results with the latter two strains demonstrated a delayed gC biosynthesis similar to VZV-32 strain. Secondly, we investigated whether the effect was specific to one cell line: Our initial experiments were performed in human melanoma cells, a substrate highly susceptible to VZV infection (Harson and Grose 1995). To this end, we infected simian derived Vero cells and found that gC production was similarly delayed. In short, the marked delay in gC biosynthesis in VZV-infected cells was not restricted to one VZV strain or one cell line.

4 Delayed VZV gC Transcription in Cultured Cells

Earlier researchers who had reported an apparent absence of gC production from certain VZV strains hypothesized a transcriptional mechanism for this absence (Kinchington et al. 1990a). Since our protein studies also suggested an effect on gC transcription, we tested to see if there was delay in the gC transcription that could account for the delay in protein expression. To investigate further, we used real-time Real Time-PCR to determine if this delay in VZV gC expression resulted from a parallel delay in gC transcription. In order to reduce the effects of heterogeneity on infections resulting from the presence of inoculum cells, we selected cell-free inoculum to infect the monolayers that we analyzed (Fig. 4). Real Time-PCR revealed a delay in the appearance of gC transcripts that mirrored a similar delay in the appearance of gC protein (Storlie et al. 2008a). Glycoprotein C transcript levels rose in untreated infected cells by 72 hpi, and reached a peak at 96 hpi. We also discovered that gC transcription was accelerated by the addition of HMBA to the culture medium (Storlie et al. 2006, 2008a). HMBA had been previously noted to enhance the replication of Marek's disease herpesvirus (Denesvre et al. 2007; Parcells et al. 2001). In contrast to the gC mRNA results, gE transcript levels were easily detectable in the first 24 hpi (Fig. 4).

Thus, there are marked differences between the kinetics of VZV and HSV gC transcription (Levine et al. 1990; Zhang and Wagner 1987). Easily detectable HSV gC transcription certainly occurs within the first replication cycle in cultured cells, with the subsequent production of HSV gC protein. That process is markedly delayed in VZV-infected cells. Early investigations found that the nucleotide binding domains in the 5′ and 3′ UTR regions of VZV ORF14 were different from those in the nucleotide sequence of HSV gC, suggesting a fundamental difference in transcriptional control between the two alpha herpesviruses

Fig. 4 Transcription differences between VZV gC and gE in cultured cells. Monolayers were inoculated with cell-free virus. Real Time-PCR was performed in quadruplicate to detect mRNA in monolayers at increasing intervals after infection. Data were analyzed with Applied Biosystems RQ Manager 1.2 software. *Error bars* indicate 95% confidence intervals. Note minimal gC transcription until 72 hpi. *x*-axis = hpi; *y*-axis = glycoprotein-specific mRNA levels

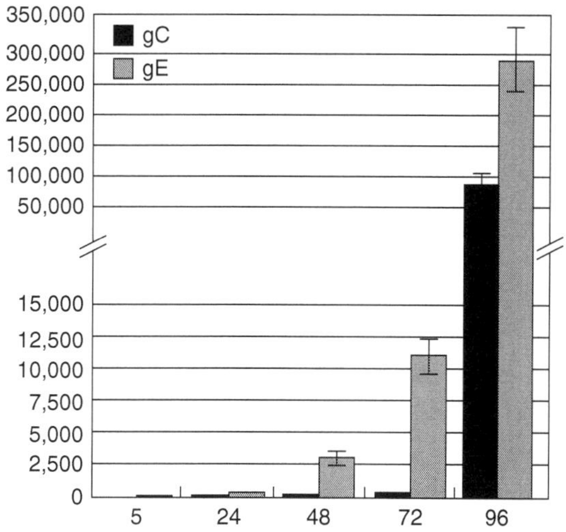

(Ling et al. 1991). Interestingly, the discovery of HOXA9 binding sites in the 5′ UTR of gC suggested that cellular factors may activate gC transcription (Storlie et al. 2006).

5 Role of gC in Low VZV Infectivity

The role of VZV gC is confounded by the fact that earlier studies were performed before we knew that very little gC protein was produced in infected cells. For example, using a gC-null recombinant virus, VZV gC was found to be dispensable for VZV growth in cultured cells (Cohen and Seidel 1994). VZV gC also may not be required for establishment or maintenance of VZV latency in neurons (Chen et al. 2003; Grinfeld et al. 2004). But in all of the above studies, the wild type control virus was likely producing very small amounts of gC; thus, the conclusions must be carefully considered.

HSV gC appears to have at least two functions as a member of the viral envelope: one, attaching to cells via heparan sulfate and two, binding to the complement component C3b and preventing the generation of C5 convertase and thus stopping the complement cascade from destroying the infected cell (Spear 2004). Spear has suggested that the latter function is the more important function of HSV gC. In fact, researchers have identified a shortened version of HSV gC that is secreted from the infected cell, apparently to engage the complement response extracellularly (Sedlackova et al. 2008).

Recently, we have discovered that VZV grown in cultured cells has an extremely high particle:PFU ratio of 40, 000:1 (Carpenter et al. 2009). In other words, many viral particles are assembled in cell culture, but very few are infectious.

In fact, most viral particles resemble light (L) particles, in that they have an outer envelope but lack an internal capsid (Carpenter et al. 2008). Since very little gC is produced during the first 48–72 hpi, we suggest that many envelopes (although assembled) also lack gC. Further studies are required to assess whether lack of gC during VZV assembly is a contributing factor toward the invariably low titer and high number of aberrant particles.

6 VZV gC and gH Proteins in Human Zoster Vesicles

The skin vesicle is the final site of virion assembly during chickenpox and zoster. Since complete viral particles are easily detectable in the human vesicle (Fig. 5), we investigated whether gC was expressed in these vesicles. We immunolabeled a

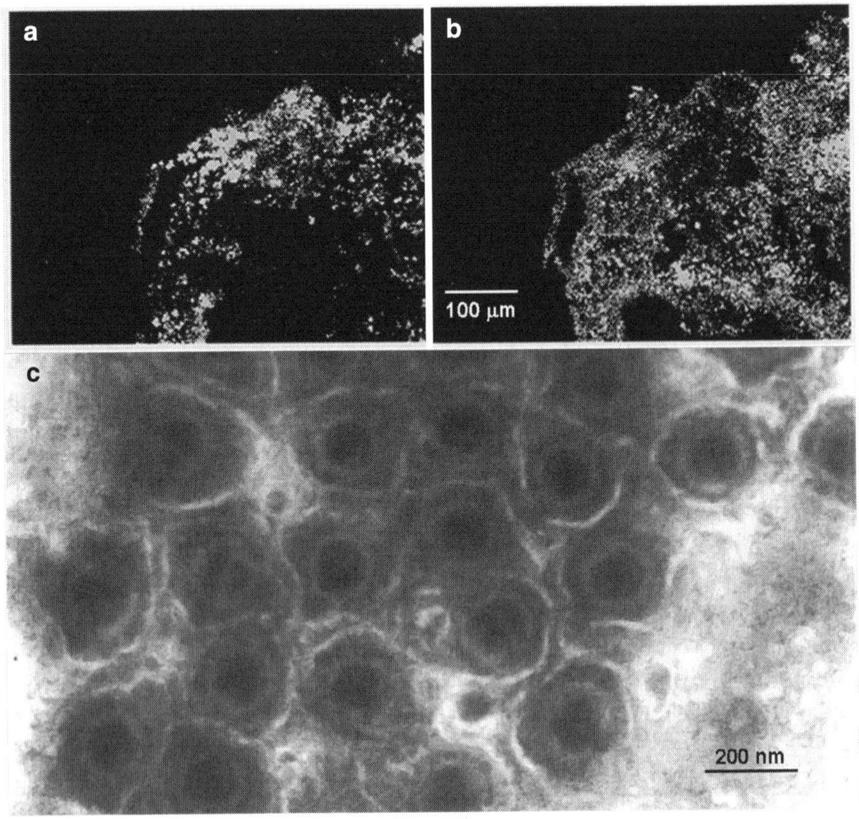

Fig. 5 Detection of abundant VZV gC as well as viral particles within a biopsy of a human zoster vesicle. Two sections of a vesicle biopsy (**a–b**) were immunolabeled with mouse monoclonal antibody for gC and linked with the Alexa 488 fluorescent antibody. They were viewed by confocal microscopy. Another section of a zoster vesicle was examined by transmission electron microscopy (**c**), in order to illustrate the presence of numerous enveloped VZV particles

biopsy specimen removed from a human zoster vesicle with antibodies to various viral proteins; then we attached fluorescently labeled secondary antibodies to those primary antibodies in order to identify location and approximate concentration of glycoproteins. We had previously documented that gE and gH were present in abundance in human skin vesicles (Weigle and Grose 1983). In the present study (Storlie et al. 2008a), we also observed similar abundance of intense spherical gC immunofluorescent labeling in zoster vesicles (Fig. 5). This pattern was of particular interest to us, since this type of labeling is commensurate with the type of signal produced by dozens of fluorescently labeled glycoproteins packed together on the envelope of a viral particle. This vesicle study also implies that the delay in gC expression seen in cultured cells is not present in the infected human. Of interest, VZV gC protein has also been detected in mature human dendritic cells, following cocultivation of the dendritic cells with VZV-infected human fibroblast cells in a culture dish (Abendroth et al. 2001; Morrow et al. 2003).

7 VZV gH and gL Structure

VZV glycoproteins gH (118 kD) and gL (18 kD) are encoded by open reading frames 37 and 60, respectively (Forghani et al. 1994; Grose 1990a). The gH sequence is highly conserved among herpesviruses, whereas function but not sequence is conserved among the herpesvirus gL proteins, many of which were initially identified by genome position rather than sequence homology (Forghani et al. 1994; Kaye et al. 1992; Spaete et al. 1993; Yaswen et al. 1993). Similar to gE and in contrast to gC, abundant amounts of VZV gH are produced in infected cells (Fig. 3). However, early attempts to study herpesviral gH function independent of infection in cell culture were largely unsuccessful until it was discovered that gH associated with a second glycoprotein called gL (Duus et al. 1995; Hutchinson et al. 1992). Subsequent coexpression, mutation, and trafficking studies revealed that VZV gH also requires a chaperone activity provided by the gL glycoprotein for efficient cell membrane expression of fusion-competent gH molecules (Duus et al. 1995; Duus and Grose 1996; Maresova et al. 2000; Pasieka et al. 2003). More recent studies demonstrated that the fusogenic activity of gH and ultimately the incorporation of this important glycoprotein into the virion envelope is further regulated by endocytosis of the mature protein once it has reached the cell membrane (Maresova et al. 2005; Pasieka et al. 2003, 2004).

8 Interaction and Trafficking of the VZV gH/gL Complex

Interactions between gH and gL have been extensively investigated. Immunoprecipitation of the gL protein from cells coexpressing gH with polyclonal anti-gL serum also pulled down pre-gH molecules, indicating that a pre-gH:gL complex was formed (Duus et al. 1995; Maresova et al. 2000). Only pre-gH:mature gL complexes were immunoprecipitated from cell lysates expressing both proteins; no complexes with mature gH molecules or immature gL (pre-gL) molecules were detected (Duus et al. 1995; Duus and Grose 1996). Disrupting the intramolecular cysteine disulfide bonds of the gL molecule by mutating different cysteine residues to alanines resulted in the loss of the pre-gH:gL complex formation, and loss of mature, functional gH molecule expression in immunoprecipitation, immunofluorescence, and fusion assays (Duus et al. 1995), providing an additional line of evidence for a gL chaperone function. Monensin treatment of gL-expressing cells did not alter the glycosylation of gL, indicating that the processing of gL is completed in the cis/medial Golgi, and does not require trafficking to the trans-Golgi network (TGN; Duus and Grose 1996).

A mechanism for the retrograde trafficking of gL from the Golgi apparatus back to the ER was suggested by Maresova and others (Maresova et al. 2000), who proposed that the hydrophobic ER-targeting motif in gL is masked by association with gH, and unmasked when gH is mature and the gH:gL complex dissociates in the Golgi. Collectively, these studies lead to a model of gH:gL processing and trafficking illustrated in Fig. 6. In this schema, the immature gL traffics as far as the cis/medial Golgi, where it matures and then returns to the ER, associates with pre-gH, and escorts it to the cis/medial and trans Golgi. Somewhere between the cis/medial and trans Golgi, the mature gH and gL dissociate and gL either returns to the ER or is targeted to a lysosome, where it is degraded. Meanwhile, mature gH moves on to the plasma membrane. In the presence of other viral glycoproteins (gE or gI), pre-gH is still able to traffic to the cell membrane, where it forms cap-like structures but is not able to mediate cell-to-cell fusion.

9 Fusion Mediated by VZV gH and gL Coexpression

VZV gH transfected alone in cell lines does not result in detectable levels of the protein on outer cell membranes, and only a 97 kD immature, endoH-sensitive precursor gH (pre-gH) is immunoprecipitated with monoclonal antibodies against gH (Duus et al. 1995; Maresova et al. 2000). However, in the presence of coexpressed gL, high levels of mature, endo-H resistant 118 kD gH molecules are detected at the cell membrane. Furthermore, the cells in the gH/gL-expressing monolayer form polykaryons containing 6–24 nuclei in confocal microscopy fusion assays (Duus et al. 1995; Maresova et al. 2000; Pasieka et al. 2004). This gH-mediated fusion was inhibited by the same monoclonal antibody shown to inhibit syncytia formation in VZV-infected melanoma monolayers (see Fig. 7 below), providing the first direct evidence that the mechanism of cell fusion in VZV infection was mediated by coexpression of the gH and gL glycoproteins.

VZV gL is also glycosylated, presumably on its one predicted N-linked glycosylation site. Although processing of gH was shown to be dependent upon gL expression, gL processing does not appear to require coexpression of gH, as gL

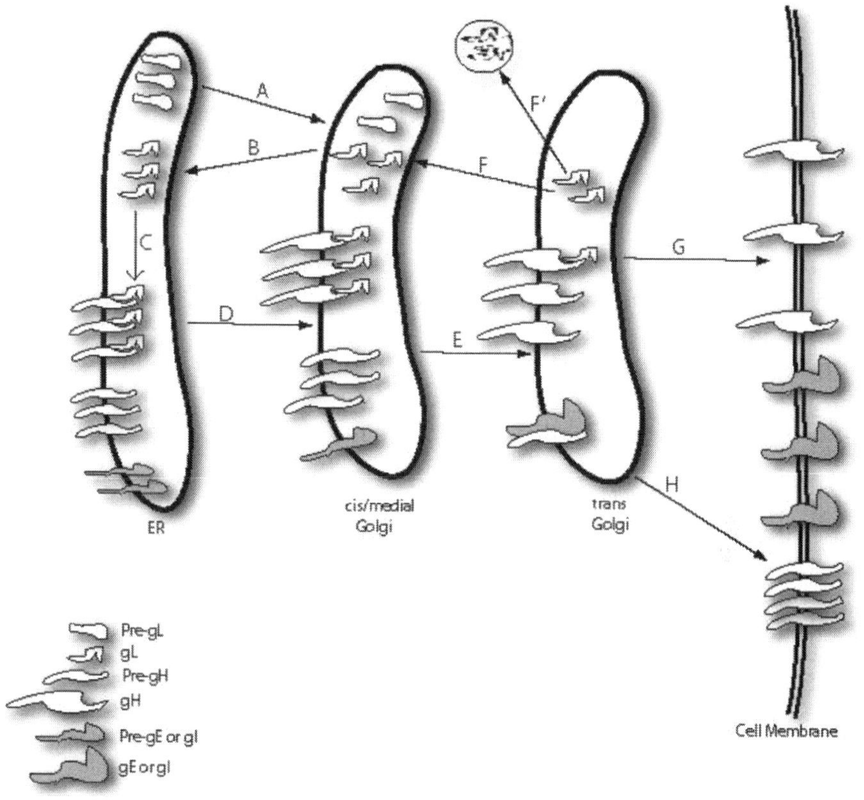

Fig. 6 Model of VZV gH trafficking with maturation regulation by gL. (A) Immature pre-gL is glycosylated in the cis/medial Golgi, and (B) gL traffics back to the ER. (C) In the ER, mature gL associates with pre-gH, and (D–E) gL chaperones pre-gH through the Golgi, whereupon gL either traffics (F) back to the ER, or (F′) is degraded. (G) Mature gH traffics to the cell membrane, where it functions as a major VZV fusogen. In the presence of VZV glycoproteins gE or gI, pre-gH is able to associate with these proteins in the Golgi and (H) traffic to the cell membrane, where gH forms cap-like structures

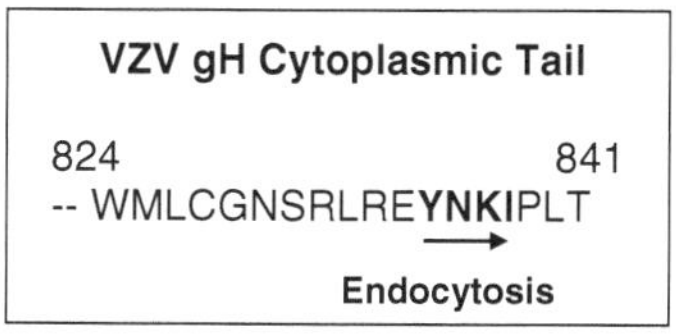

Fig. 7 Endocytosis motif of VZV gH. The entire gH sequence consists of 841 residues, with a functional internalization motif in the endodomain. The other component gL of the gH/gL complex consists of 159 amino acids but without an endocytosis motif

expressed alone was shown to be endo-H resistant and endo-F sensitive (Duus et al. 1995). The endo-H sensitive pre-gL is 14 kD, while the endo-F sensitive mature gL is 18 kD (Maresova et al. 2000).

VZV gL is unique among the VZV glycoproteins in its ability to facilitate gH maturation and cell membrane expression. Coexpression of either of two other VZV glycoproteins, gE or gI, with gH did not support the complete maturation of gH, although the presence of either of these glycoproteins did allow aggregates of pre-gH molecules to reach cell membrane, where they were detected by immunofluorescence as a cap-like structure (Duus et al. 1995). Collectively this set of studies suggested that gL provides a chaperone function to facilitate gH-mediated fusion. Consistent with this idea, the gL protein is known to contain a peptide sequence closely matching a 16-residue hydrophobic ER-targeting motif encoded by many cellular ER resident proteins known to be involved in glycoprotein processing (Nigam et al. 1994).

10 Endocytosis-Dependent Regulation of VZV gH-Mediated Fusion

Endocytosis motifs are located in the cytoplasmic tails of VZV gE, gI, and gB (Grose et al. 2006). Of note, the short gH cytoplasmic tail also contains a functional YNKI motif, which mediates clathrin-dependent and antibody-independent gH endocytosis in the presence of gL (Fig. 8). This clathrin-dependent endocytosis event occurs in both gH/gL-transfected and VZV-infected cells, where the majority of the endocytosis occurred within 24 h post-infection (Pasieka et al. 2003). A subsequent analysis of gH endocytosis on virus-mediated cell–cell fusion revealed that regulation of fusion occurs by this mechanism (Pasieka et al. 2004). In addition, Pasieka et al. (2004) demonstrated that coexpression of gE further modulates gH-mediated fusion activity by coendocytosis of a gH:gE complex from the cell membrane. Thus, endocytosis is certainly an important component of gH trafficking within an infected cell.

11 VZV gH Neutralization Epitope

All monoclonal antibodies produced in this laboratory have been tested in a complement-independent neutralization assay (Grose and Litwin 1988). The titer in a complement-independent neutralization assay correlates with a titer, is a fluorescent antibody to membrane antigen assay, which in turn correlates with a protective antibody response (Grose et al. 1979a). The glycoprotein-specific monoclonal antibody that has the highest neutralization titer is that against gH.

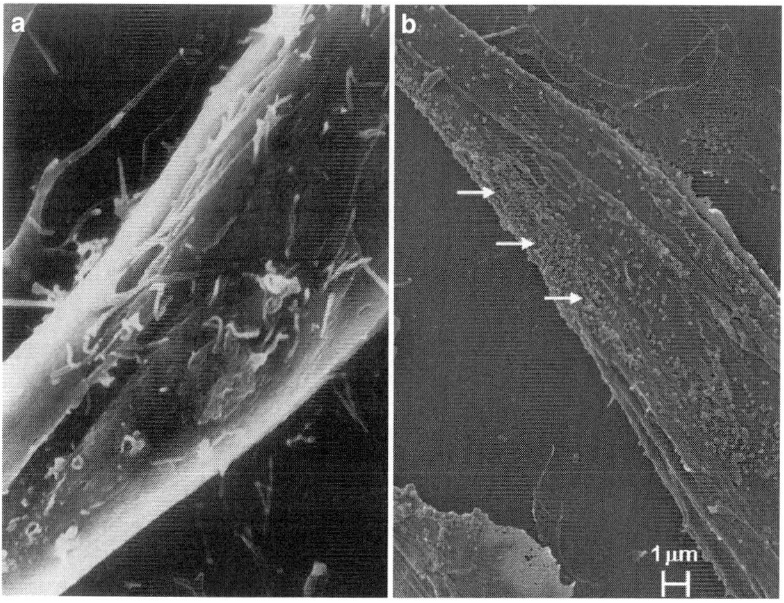

Fig. 8 Inhibition of viral egress by neutralization antibody to VZV gH. Monolayers were inoculated with VZV cell-free virus and then incubated in the presence of high titered anti-gH monoclonal antibody in the culture medium. After 48 h, no viral egress was evident (**a**). However, when the antibody was removed from the culture medium, viral egress (*arrows*) was evident after another 48 h of incubation (**b**). The cells in monolayer in panel (**b**) also show signs of extensive virus-induced fusion

In a subsequent experiment, the effect of anti-gH antibody on the spread of infection was evaluated by high resolution scanning electron microscopy. Two antibodies were initially selected, the first being an anti-gH antibody clone 206 with a 1:40,000 titer of complement-independent neutralization activity in a plaque reduction assay (Rodriguez et al. 1993). The second control antibody called 5C6 precipitates the VZV ORF 62 regulatory protein, the homolog of the HSV ICP4 protein, and has no neutralization activity.

Both antibodies were tested to determine their effect when placed in medium overlying an infected culture. For this experiment, the monolayers were inoculated with cell-free virus. Thereafter, the medium in each test culture was supplemented with either antibody 206 or antibody 5C6 (1:50 dilution of monoclonal ascites). The cultures were incubated for 48 h at 32°C. No cytopathology developed in the cultures treated with anti-gH antibody, while those incubated with antibody 5C6 developed syncytia with viral egress as seen in untreated infected monolayers. When viewed by scanning electron microscopy, cultures treated with anti-gH antibody lacked visual evidence of syncytia and had almost no viral particles on the cell surface (Fig. 7a). When the anti-gH antibody was removed from the culture medium, viral egress was seen in another 48 h (Fig. 7b). Recently, the gH conformational neutralization epitope has been extensively mapped with a panel of ten antibodies (Akahori et al. 2009).

Acknowledgments This research was supported in part by NIH grant AI22795. We acknowledge the important contributions of former graduate students and postdoctoral fellows, whose research is cited in this chapter.

References

Abendroth A, Morrow G, Cunningham AL, Slobedman B (2001) Varicella-zoster virus infection of human dendritic cells and transmission to T cells: implications for virus dissemination in the host. J Virol 75:6183–6192

Akahori Y, Suzuki K, Daikoku T, Iwai M, Yoshida Y, Asano Y, Kurosawa Y, Shiraki K (2009) Characterization of neutralizing epitopes of varicella-zoster virus glycoprotein H. J Virol 83:2020–2024

Carpenter JE, Hutchinson JA, Jackson W, Grose C (2008) Egress of light particles among filopodia on the surface of varicella-zoster virus-infected cells. J Virol 82:2821–2835

Carpenter JE, Henderson EP, Grose C (2009) Enumeration of an extremely high particle to PFU ratio for Varicella zoster virus. J Virol 83(13):6917–6921

Chen JJ, Gershon AA, Li ZS, Lungu O, Gershon MD (2003) Latent and lytic infection of isolated guinea pig enteric ganglia by varicella zoster virus. J Med Virol 70(Suppl 1):S71–S78

Cohen JI, Seidel KE (1994) Absence of varicella-zoster virus (VZV) glycoprotein V does not alter growth of VZV in vitro or sensitivity to heparin. J Gen Virol 75(Pt 11):3087–3093

Davison AJ, Scott JE (1986) The complete DNA sequence of varicella-zoster virus. J Gen Virol 67:1759–1816

Denesvre C, Blondeau C, Lemesle M, Le Vern Y, Vautherot D, Roingeard P, Vautherot JF (2007) Morphogenesis of a highly replicative EGFPVP22 recombinant Marek's disease virus in cell culture. J Virol 81:12348–12359

Duus KM, Grose C (1996) Multiple regulatory effects of varicella-zoster virus (VZV) gL on trafficking patterns and fusogenic properties of VZV gH. J Virol 70:8961–8971

Duus KM, Hatfield C, Grose C (1995) Cell surface expression and fusion by the varicella-zoster virus gH:gL glycoprotein complex: analysis by laser scanning confocal microscopy. Virology 210:429–440

Forghani B, Ni L, Grose C (1994) Neutralization epitope of the varicella-zoster virus gH:gL glycoprotein complex. Virology 199:458–462

Grinfeld E, Sadzot-Delvaux C, Kennedy PG (2004) Varicella-zoster virus proteins encoded by open reading frames 14 and 67 are both dispensable for the establishment of latency in a rat model. Virology 323:85–90

Grose C (1990a) Glycoproteins of varicella-zoster virus and their herpes simplex homologs. Rev Infect Dis 13:S960–S963

Grose C (1990b) Glycoproteins encoded by varicella-zoster virus: biosynthesis, phosphorylation, and intracellular trafficking. Annu Rev Microbiol 44:59–80

Grose C, Litwin V (1988) Immunology of the varicella-zoster virus glycoproteins. J Infect Dis 157:877–881

Grose C, Ng TI (1992) Intracellular synthesis of varicella-zoster virus. J Infect Dis 166(Suppl 1): S7–S12

Grose C, Edmond BJ, Brunell PA (1979a) Complement-enhanced neutralizing antibody response to varicella-zoster virus. J Infect Dis 139:432–437

Grose C, Perrotta DM, Brunell PA, Smith GC (1979b) Cell-free varicella-zoster virus in cultured human melanoma cells. J Gen Virol 43:15–27

Grose C, Edmond BJ, Friedrichs WE (1981) Immunogenic glycoproteins of laboratory and vaccine strains of varicella-zoster virus. Infect Immun 31:1044–1053

Grose C, Maresova L, Medigeshi G, Scott G, Thomas G (2006) Endocytosis of varicella-zoster virus glycoproteins: virion envelopment and egress. In: Sandri-Goldin R (ed) Advances in alpha herpesviruses. Academic, London

Harson R, Grose C (1995) Egress of varicella-zoster virus from the melanoma cell: a tropism for the melanocyte. J Virol 69:4994–5010

Hutchinson L, Browne H, Wargent V, Davis-Poynter N, Primorac S, Goldsmith K, Minson AC, Johnson DC (1992) A novel herpes simplex virus glycoprotein, gL, forms a complex with glycoprotein H (gH) and affects normal folding and surface expression of gH. J Virol 66: 2240–2250

Kaye JF, Gompels UA, Minson AC (1992) Glycoprotein H of human cytomegalovirus (HCMV) forms a stable complex with the HCMV UL115 gene product. J Gen Virol 73(Pt 10): 2693–2698

Kinchington PR, Remenick J, Ostrove JM, Straus SE, Ruyechan WT, Hay J (1986) Putative glycoprotein gene of varicella-zoster virus with variable copy numbers of a 42-base-pair repeat sequence has homology to herpes simplex virus glycoprotein C. J Virol 59:660–668

Kinchington PR, Ling P, Pensiero M, Gershon A, Hay J, Ruyechan WT (1990a) A possible role for glycoprotein gpV in the pathogenesis of varicella-zoster virus. Adv Exp Med Biol 278:83–91

Kinchington PR, Ling P, Pensiero M, Moss B, Ruyechan WT, Hay J (1990b) The glycoprotein products of varicella-zoster virus gene 14 and their defective accumulation in a vaccine strain (Oka). J Virol 64:4540–4548

Levine M, Krikos A, Glorioso JC, Homa FL (1990) Regulation of expression of the glycoprotein genes of herpes simplex virus type 1 (HSV-1). Adv Exp Med Biol 278:151–164

Ling P, Kinchington PR, Ruyechan WT, Hay J (1991) A detailed analysis of transcripts mapping to varicella zoster virus gene 14 (glycoprotein V). Virology 184:625–635

Maresova L, Kutinova L, Ludvikova V, Zak R, Mares M, Nemeckova S (2000) Characterization of interaction of gH and gL glycoproteins of varicella-zoster virus: their processing and trafficking. J Gen Virol 81:1545–1552

Maresova L, Pasieka TJ, Homan E, Gerday E, Grose C (2005) Incorporation of three endocytosed varicella-zoster virus glycoproteins, gE, gH, and gB, into the virion envelope. J Virol 79: 997–1007

Morrow G, Slobedman B, Cunningham AL, Abendroth A (2003) Varicella-zoster virus productively infects mature dendritic cells and alters their immune function. J Virol 77:4950–4959

Nigam SK, Goldberg AL, Ho S, Rohde MF, Bush KT, Sherman M (1994) A set of endoplasmic reticulum proteins possessing properties of molecular chaperones includes Ca(2+)-binding proteins and members of the thioredoxin superfamily. J Biol Chem 269:1744–1749

Parcells MS, Lin SF, Dienglewicz RL, Majerciak V, Robinson DR, Chen HC, Wu Z, Dubyak GR, Brunovskis P, Hunt HD, Lee LF, Kung HJ (2001) Marek's disease virus (MDV) encodes an interleukin-8 homolog (vIL-8): characterization of the vIL-8 protein and a vIL-8 deletion mutant MDV. J Virol 75:5159–5173

Pasieka TJ, Woolson RF, Grose C (2003) Viral induced fusion and syncytium formation: measurement by the Kolmogorov-Smirnov statistical test. J Virol Methods 111:157–161

Pasieka TJ, Maresova L, Shiraki K, Grose C (2004) Regulation of varicella-zoster virus-induced cell-to-cell fusion by the endocytosis-competent glycoproteins gH and gE. J Virol 78: 2884–2896

Peters G, Tyler S, Grose C, Severini A, Gray M, Upton C, Tipples G (2006) A full genome phylogenetic analysis of varicella-zoster virus reveals a novel origin of replication-based genotyping scheme and evidence of recombination between major circulating clades. J Virol 80:9850–9860

Rodriguez JE, Moninger T, Grose C (1993) Entry and egress of varicella virus blocked by same anti-gH monoclonal antibody. Virology 196:840–844

Sedlackova L, Perkins KD, Lengyel J, Strain AK, van Santen VL, Rice SA (2008) Herpes simplex virus type 1 ICP27 regulates expression of a variant, secreted form of glycoprotein C by an intron retention mechanism. J Virol 82:7443–7455

Spaete RR, Perot K, Scott PI, Nelson JA, Stinski MF, Pachl C (1993) Coexpression of truncated human cytomegalovirus gH with the UL115 gene product or the truncated human fibroblast growth factor receptor results in transport of gH to the cell surface. Virology 193:853–861

Spear PG (2004) Herpes simplex virus: receptors and ligands for cell entry. Cell Microbiol 6: 401–410

Storlie J, Jackson W, Hutchinson J, Grose C (2006) Delayed biosynthesis of varicella-zoster virus glycoprotein C: upregulation by hexamethylene bisacetamide and retinoic acid treatment of infected cells. J Virol 80:9544–9556

Storlie J, Carpenter JE, Jackson W, Grose C (2008a) Discordant varicella-zoster virus glycoprotein C expression and localization between cultured cells and human skin vesicles. Virology 382:171–181

Storlie J, Maresova L, Jackson W, Grose C (2008b) Comparative analyses of the 9 glycoprotein genes found in wild-type and vaccine strains of varicella-zoster virus. J Infect Dis 197(Suppl 2): S49–S53

Weigle KA, Grose C (1983) Common expression of varicella-zoster viral glycoprotein antigens in vitro and in chickenpox and zoster vesicles. J Infect Dis 148:630–638

Yaswen LR, Stephens EB, Davenport LC, Hutt-Fletcher LM (1993) Epstein-Barr virus glycoprotein gp85 associates with the BKRF2 gene product and is incompletely processed as a recombinant protein. Virology 195:387–396

Zhang YF, Wagner EK (1987) The kinetics of expression of individual herpes simplex virus type 1 transcripts. Virus Genes 1:49–60

Analysis of the Functions of Glycoproteins E and I and Their Promoters During VZV Replication In Vitro and in Skin and T-Cell Xenografts in the SCID Mouse Model of VZV Pathogenesis

Ann M. Arvin, Stefan Oliver, Mike Reichelt, Jennifer F. Moffat, Marvin Sommer, Leigh Zerboni, and Barbara Berarducci

Contents

Abstract The two VZV glycoproteins, gE and gI, are encoded by genes that are designated open reading frames, ORF67 and ORF68, located in the short unique region of the VZV genome. These proteins have homologs in the other alphaherpesviruses. Like their homologues, VZV gE and gI exhibit prominent co-localization in infected cells and form heterodimers. However, VZV gE is much larger than its homologues because it has a unique N-terminal domain, consisting of 188 amino

A.M. Arvin (✉)
Stanford University School of Medicine, G311, Stanford, CA 94305, USA
e-mail: aarvin@stanford.edu

S. Oliver, M. Reichelt, and M. Sommer
Stanford University School of Medicine, S356, Stanford, CA 94305, USA
e-mail: sloliver@stanford.edu; reichelt@stanford.edu; msommer@stanford.edu

L. Zerboni
Stanford University School of Medicine, S366, Stanford, CA 94305, USA
e-mail: zerboni@stanford.edu

J.F. Moffat
SUNY Upstate Medical University, Rm. 2215, 750 East Adams Street, Syracuse, NY, 13210, USA
e-mail: moffatj@upstate.edu

B. Berarducci
Départment de Virologie, Institut Pasteur, 25 Rue du Docteur Roux, 75015, Paris, France
e-mail: barbara.berarducci@pasteur.fr

A.M. Arvin et al. (eds.), *Varicella-zoster Virus*,
Current Topics in Microbiology and Immunology 342, DOI 10.1007/82_2009_1

acids that are not present in these other gene products. VZV gE also differs from the related gE proteins, in that it is essential for viral replication. Targeted mutations of gE that are compatible with VZV replication in cultured cells have varying phenotypes in skin and T-cell xenografts in the SCID mouse model of VZV pathogenesis *in vivo*. While gI is dispensable for growth in cultured cells *in vitro*, this glycoprotein is essential for VZV infection of differentiated human skin and T cells *in vivo*. The promoter regions of gE and gI are regulated by the cellular transactivator, specificity protein factor 1 (Sp1) in combination with the major VZV transactivator in reporter construct experiments and some Sp1 promoter elements are important for VZV virulence *in vivo*. Further analysis of VZV gE and gI functions and their interactions with other viral and host cell proteins are important areas for studies of VZV replication and pathogenesis.

1 Introduction

VZV gE and gI are encoded by the open reading frames ORF68 and ORF67, which are located in the short unique region of the VZV genome (Fig. 1). VZV gE and gI are both type 1 membrane proteins, like their homologs that are found in all of the other alphaherpesviruses (Cole and Grose 2003). In contrast to the other alphaherpesviruses, the VZV genome has only these two glycoprotein genes, ORF67 and ORF68, in the short unique region. Notably VZV lacks a gene for gD, which serves as a key glycoprotein in virion entry in the other alphaherpesviruses, including the closely related human herpes simplex viruses, HSV-1 and HSV-2.

VZV gE is the most abundant glycoprotein in infected cells and is a major component of the virion envelope. Glycoproteins, both gE and gI undergo post-translational modifications; gE is extensively phosphorylated and gI also has residues that are phosphorylated, as shown in a series of reports from the Grose laboratory, (Cole and Grose 2003; Olson et al. 1997; Olson and Grose 1997, 1998;

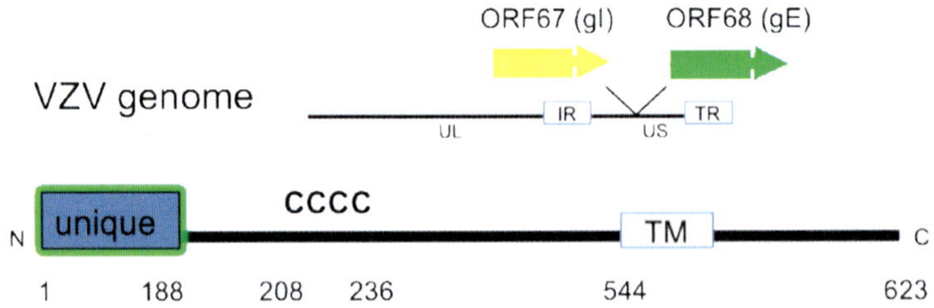

Fig. 1 Schematic of the VZV genome and gE and gI. The upper section of the diagram shows location of gE (ORF68) and gI (ORF67) in the short unique segment of the VZV genome. The gI promoter is upstream of ORF68 and the gE promoter is in the intragenic region between ORF67 and ORF68. The lower section shows gE (623 amino acids) with the unique region (1–188) and a cysteine-rich region (206–236) motifs identified in the N terminus. The C terminus has an endocytosis motif, YAGL (582–585), a TGN-targeting motif, AYRV (568–571), and an "acid cluster" phosphorylation motif, SSTT (588–601)

Pasieka et al. 2004). gE and gI are typically present as heterodimers in infected cells although gE dimer formation is also observed. Both gE and gI traffic to the plasma membranes of infected cells – where gE is expressed as the predominant VZV glycoprotein – and are also sorted to the trans-Golgi network (TGN). gE has an endocytosis motif (Olson and Grose 1997, 1998; Pasieka et al. 2004) and a TGN-signaling motif that functions to retrieve gE from the cell membrane and gI has residues direct trafficking to the TGN (Cole and Grose 2003). In addition, gI has chaperone functions that contribute to correct gE trafficking and gI/gE interactions are involved in gE maturation. When expressed in an inducible cell line in the absence of gI, gE colocalized with a TGN marker in perinuclear sites but did not reach plasma membranes (Mo et al. 2002). In polarized cells, gE expression accelerated the establishment of functional tight junctions, facilitating epithelial cell–cell contacts, which might enhance mucosal infection (Mo et al. 2000). Recently, gE was reported to bind to the cellular protein, insulin degrading enzyme (IDE), which was proposed to function as a cell surface receptor for VZV entry (Ali et al. 2009; Li et al. 2006, 2007). Our analysis of the kinetics of viral protein expression, genome synthesis, and virion assembly at the single-cell level showed that whereas gE has been considered to be a late gene product, gE was detectable in a Golgi-compartment-like cytoplasmic distribution by 4 h; gE was expressed extensively on plasma membranes by 9 h (Reichelt et al. 2009) (Fig. 2).

Our laboratory has focused on the analysis of gE and gI functions using molecular genetics methods to introduce deletions and targeted mutations into the VZV genome. The objective of this chapter is to summarize what we have learned about gE and gI functions by mutagenesis of ORF68 and ORF67 and the promoter regions that regulate their transcription in the context of the viral genome. We have also investigated the roles that these glycoproteins play during VZV replication in cultured cells *in vitro* and during VZV pathogenesis *in vivo* using our SCID mouse model (Arvin 2006). In this model, human T cell and skin tissue xenografts are established in SCID mice. These xenografts are infected with intact VZV or selected VZV mutants to define the functions of viral proteins, their subdomains, or their promoter motifs in the infection of differentiated human cells within their intact tissue microenvironment *in vivo*. SCID mice with human xenografts make it possible to define genetic determinants of VZV virulence in differentiated cell types that are infected in the course of primary VZV infection and VZV recurrences in the human host (Ku et al. 2005). In the SCID model, VZV infects CD4 and CD8 T cells in thymus–liver xenografts and dermal and epidermal cells in skin xenografts (Arvin 2006). VZV skin tropism requires polykaryocyte formation, and our experiments in the SCIDhu skin model demonstrate that VZV mutants retain some infectivity, as long as cell fusion is preserved, even when virion formation and secondary envelopment are severely compromised. Infection of T-cell xenografts appears to require assembly and release of infectious virus particles because T-cell fusion is not observed. Infection of both skin and T-cell xenografts occurs in the absence of an adaptive immune response, which allows an assessment of the intrinsic cellular response to the virus in infected and neighboring cells within the tissue microenvironment. In evaluating VZV mutants in this model, viruses

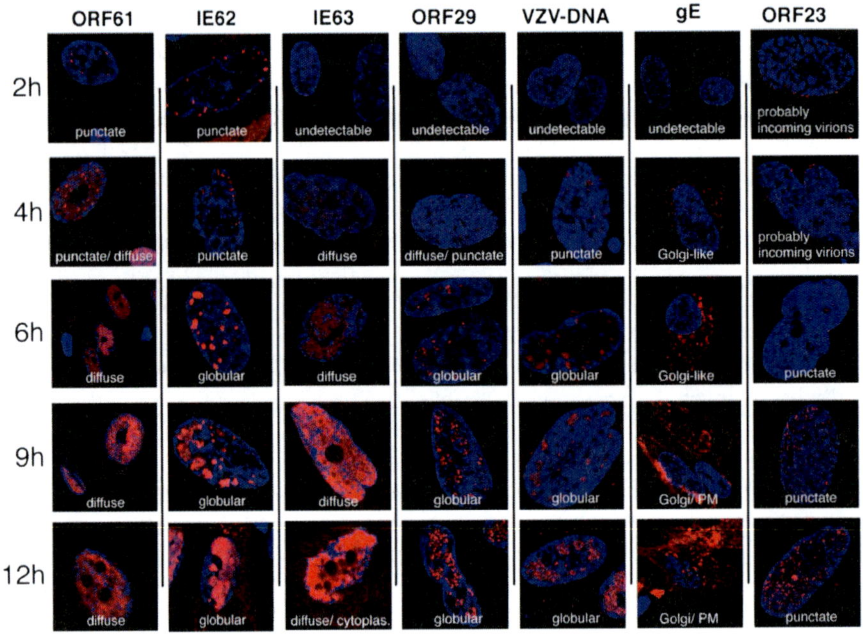

Fig. 2 The patterns of spatiotemporal expression of selected VZV immediate-early, early, and late proteins and viral genomic DNA. Human fibroblasts were fixed at 0 h (not shown), 2, 4, 6, 9, and 12 h after inoculation with input cells labeled with green CellTracker. The cells were stained with specific antibodies for VZV proteins (*red*, from *left* to *right*) ORF61, IE62, IE63, ORF29, gE, and ORF23 followed by Texas Red-conjugated secondary antibodies. VZV DNA was detected by DNA in situ hybridization (*red*). Cell nuclei were counterstained with Hoechst 22358. Ten fields of 30–50 output cells were scanned to determine the representative staining patterns in newly infected cells at each time point. Single representative cells or nuclei are shown. *PM* plasma membrane, *cytoplas.* cytoplasm (Fig. 2 from Reichelt et al. (2009); reproduced with permission)

recovered from the tissue xenografts are sequenced to document the persistence of the expected mutations and to exclude a potential reversion to virulence *in vivo*. Experiments using our gE and gI mutants and the promoter mutants to evaluate their contributions to VZV neurotropism in the SCID mouse model of VZV neuropathogenesis are described in chapter 16.

2 gE Functions

Of interest, before any targeted mutagenesis of gE had been undertaken, a naturally occurring VZV strain, VZV-MSP, was discovered by Grose and colleagues (Santos et al. 2000). VZV-MSP had an alteration in the gE ectodomain and showed a different pattern of cellular egress as well as enhanced cell–cell spread *in vitro* The variant gE has a substitution in codon 150 that replaces a guanine with an adenine, changing the amino acid from aspartic acid to asparagine. Notably, experiments in the SCID model showed that VZV-MSP also exhibited accelerated spread in skin xenografts, which documented that a naturally occurring gE mutation could alter VZV pathogenesis *in vivo*.

In undertaking targeted mutagenesis of gE to identify its important functional domains, we first investigated whether deletion of ORF68 alone from the VZV genome prevented the recovery of infectious virus from cosmid transfections (Mo et al. 2002; Table 1). Removing ORF68 was lethal but infectious virus was generated when ORF68 was re-introduced at the non-native AvrII site. Thus, in contrast to the other alphaherpesviruses, gE is an essential protein in VZV. Of interest, expression of gE as a fusion protein with luciferase in the context of the VZV genome was associated with rapid excision of the foreign gene sequence from the virus after short-term passage of the gE-luciferase recombinant virus in cultured cells (Oliver et al. 2008). While gE in the other alphaherpesviruses is encoded within the unique short region of the genome, analysis of ORF68 reveals that whereas most of the sequence is also in the unique short region, a segment of 113 nucleotides at its $3'$ end extends into the terminal repeat region and is duplicated in the IRS. It seems likely that this repeat sequence allowed the efficient repair of the gE-luciferase recombinant. We speculate that this capacity of VZV to maintain the correct ORF68 sequence in the genome is important because gE is essential in VZV.

Based on the finding that gE is required for VZV replication, our initial ORF68 mutagenesis strategies were designed to examine the functions of the subdomains of the gE C terminus for replication *in vitro* and VZV pathogenesis in skin and T cells *in vivo* (Moffat et al. 2004). gE has a short C-terminal region of 62 amino acids that contains three functional motifs as shown by their analysis in expression constructs (Cole and Grose 2003; Olson and Grose 1997). These included an endocytosis motif, YAGL, at amino acids 582–585, a TGN-targeting motif, AYRV, at amino acids 568–571, and an "acid cluster" phosphorylation motif, SSTT at amino acids 588–601. Cosmid mutagenesis was used to make a complete deletion of the gE C terminus after the transmembrane region and to introduce specific amino acid substitutions in these motifs; the mutations were Y582G in YAGL, Y569A in AYRV, and S593A, S595A, T596A, and T598A in SSTT. The C-terminal deletion was lethal and experiments with the cosmid that encoded the Y582G mutation showed that this result was a consequence of disrupting the YAGL endocytosis motif. These observations indicated that gE endocytosis from the plasma membrane is critical for VZV replication, probably to achieve correct gE localization at the site of secondary envelopment of virions. In contrast, the AYRV and SSTT mutations did not impair VZV replication *in vitro*. When these mutants, rOka-gE-AYRV and rOka-gE-SSTT, were assessed in skin and T-cell xenografts, the gE TGN-targeting motif in the C terminus proved to be important for VZV virulence in skin and to a lesser extent in T cells (Fig. 3). However, altering the SSTT residues in the "acid patch" domain did not affect VZV pathogenesis in skin or T-cell xenografts in the SCID mouse model. Only 4 of 12 skin xenografts (33%) infected with rOka-gE-AYRV produced infectious virus compared to 8 of 12 infected with rOka and 7 of 12 infected with rOka-gE-SSTT; when rOka-gE-AYRV was recovered, the titers were also significantly lower than both those of rOka

Table 1 Phenotypes of VZV recombinants with gE promoter and gE mutations *in vitro* and *in vivo*

Virus	Replication kinetics *in vitro*	Plaque size	gE/gI binding	gE/IDE binding	T-cell entry *in vitro*	Infection of xenografts *in vivo*	
						Skin	T cells
gE promoter							
gE proΔI	NL	NL			NL	NL	
gE proΔII	NL	NL			NL	NL	
gE Sp1A	Lethal						
gE Sp1B	Delayed	Decrease					
gE							
ΔgE	Lethal						
gE C terminus							
gE-SSTT	Increase	NL				NL	
gE-AYRV	Increase	NL				Decrease	
gE N terminus							
gE S31A	Slight decrease	NL	+	+		Decrease	NL
gE S49A	NL	NL				NL	
gE ∇P27	Delayed	Slight decrease	+	+		Slight decrease	
gE ∇Y51	NL	NL	+	+		NL	NL
gE ∇G90	NL	NL					
gE ∇I146	NL	NL					
gE ∇P187	NL	NL					
gE ΔP27-Y51	NL	Decrease	+	+		Decrease	
gE ΔY51-G90	NL	Decrease	+	+	+	NL	
gE ΔP27-G90	Slight decrease	Decrease	+	–	+	Decrease	NL
gE ΔP27-P187	Lethal						
gE ΔY51-P187	Decrease	Decrease	+	+/–	+	Lethal	Lethal
gE Δcys208-236	NL	Decrease	–	NL	NL	Severe decrease	NL

NL equivalent to parent virus, *blank* not tested, ∇ linker insertion, Δ deletion

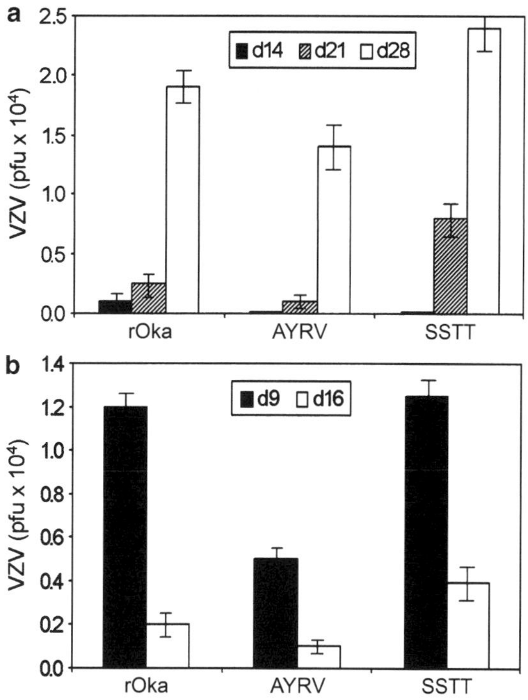

Fig. 3 Effects of mutations in the gE C terminus on VZV replication in skin and T-cell xenografts in SCIDhu mice. (**a**) Skin xenografts were inoculated with rOka or rOka-gE-AYRV or rOka-gE-SSTT mutant. Each bar represents the mean titer of infectious virus recovered from skin implants harvested at day 14, 21, or 28 after inoculation; error bars indicate the standard error. (**b**) Mean titer of infectious virus found when T-cell xenografts were inoculated with rOka, rOka-gE-AYRV, or rOka-gE-SSTT and harvested at day 9 or 16 after infection. Inoculum titers were 2.5×10^4 to 7.0×10^4 PFU/ml. Significance was determined by Student's *t* test ($P < 0.05$) (Fig. 6 from Moffat et al. (2002); reproduced with permission)

and rOka-gE-SSTT at 28 days ($P < 0.05$; Fig. 3a). Peak titers of rOka-gE-AYRV were also diminished in T-cell xenografts compared to rOka and rOka-gE-SSTT ($P < 0.05$; Fig. 3b) but most T-cell xenografts (89%) yielded some virus. Thus, gE endocytosis is required for VZV replication, and proper TGN targeting is important within the skin and T-cell tissue microenvironments where VZV must replicate *in vivo*.

Examination of the gE N terminus using bioinformatics methods and comparisons with the gE N-terminal regions of the homologs in other alphaherpesviruses revealed that gE has a substantial component, consisting of its first 188 residues, that is not present in related proteins (Berarducci et al. 2006). Therefore, this non-conserved, unique N-termimal region was of particular interest for its potential contributions to VZV replication. Linker insertions, serine-to-alanine mutations, and deletions were introduced in the gE N-terminal region by cosmid mutagenesis as illustrated in Fig. 4. The effects of these mutations on virus replication and cell–cell spread, gE trafficking, and localization and virion formation were assessed in cultured cells (Fig. 5). These experiments showed no recovery of virus from cosmid transfections when the complete unique gE N terminus (except for the putative signal peptide from 1 to 16) was deleted. When expressed from a plasmid, the mutant gE was present in TGN in transfected cells, but no trafficking to plasma membranes occurred even when it was co-expressed with gI. When smaller

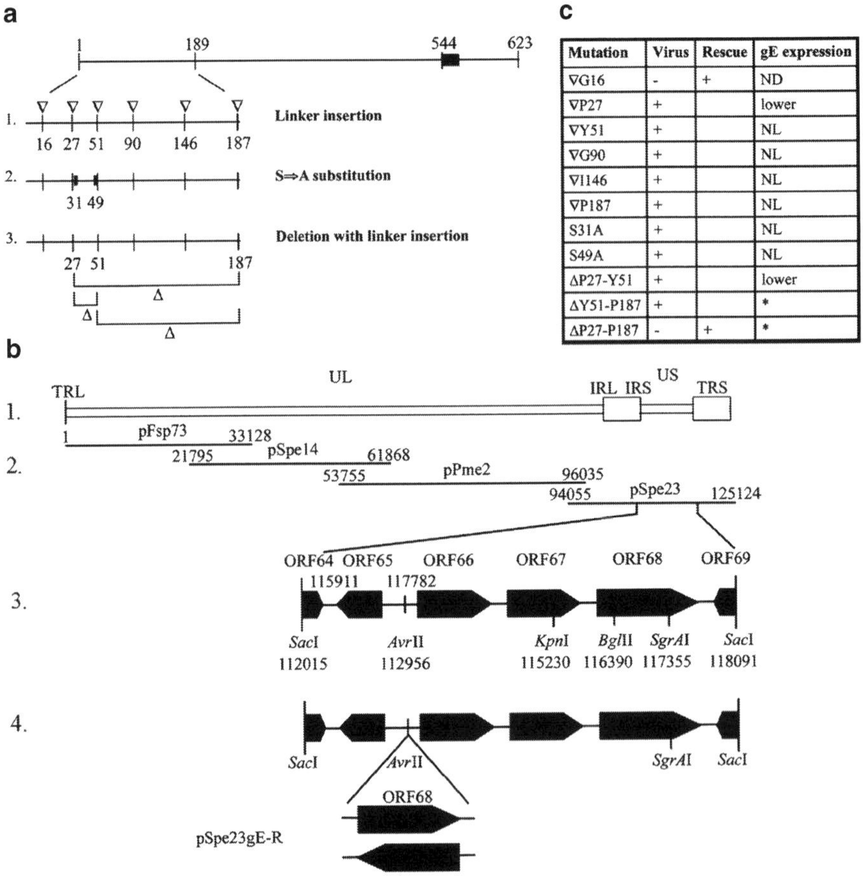

Mutation	Virus	Rescue	gE expression
∇G16	-	+	ND
∇P27	+		lower
∇Y51	+		NL
∇G90	+		NL
∇I146	+		NL
∇P187	+		NL
S31A	+		NL
S49A	+		NL
ΔP27-Y51	+		lower
ΔY51-P187	+		*
ΔP27-P187	-	+	*

Fig. 4 Mutagenesis of the unique gE N terminus. (**a**) Schematic representation of the gE mutations. The gE glycoprotein and the positions of the mutations are represented. The position of the linker insertions (*inverted triangle*, line 1), serine-to-alanine alterations (*thick bar*, line 2), and deletions (D, line 3) are indicated. Amino acids are numbered from the N terminus to the C terminus of the gE protein. The *black box* represents the transmembrane domain. (**b**) Schema of the cosmid system. Line 1, schematic representation of the VZV genome; line 2, diagram of the overlapping cosmids containing the VZV genome (parental Oka); line 3, the SacI fragment from the pSpe23 cosmid containing the ORF68 gene; line 4, insertion of the gE rescue cassette in the AvrII site. pSpe23gE-R indicates the rescued gE mutants. (**c**) Summary of the results from gE mutagenesis. Recovery of recombinant virus, rescue of the mutation with lethal phenotype, and level of gE expression are indicated for each mutation. *ND* not done, *NL* normal. An *asterisk* indicates that the expression of mutant ΔY51-P187 and ΔP27-P187 gE was analyzed by immunofluorescence only *TRL* terminal repeat long, *UL* unique long region, *IRL* internal repeat long, *IRS* inverted repeat short, *US* unique short region, *TRS* terminal repeat short (Fig. 2 from Berarducci et al. (2006); reproduced with permission)

deletions were made, virus was recovered but rOka-ΔΠ27-Y51 and rOka-ΔY51-P187 mutants showed less cell–cell spread and rOka-ΔY51-P187 titers were lower compared to rOka, in association with less gE present in infected cells. Deleting

Y51-P187 resulted in major disruption of secondary envelopment of virions and accumulation of cytoplasmic aggregates *in vitro*. VZV was recovered when the mutations of the unique gE N terminus consisted of linker insertions, which tends to suggest that functions of this region are not highly conformation-dependent.

In vivo experiments in skin xenografts showed that the Y51-P187 deletion was lethal for VZV replication (Fig. 6). No virus was recovered from any xenografts that were infected with rOka-ΔY51-P187 and evaluated after 10 and 22 days. Other

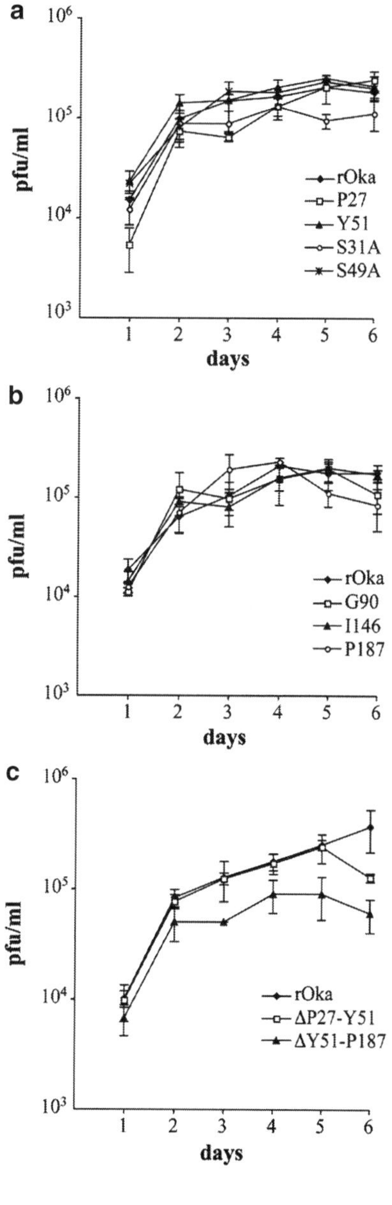

Fig. 5 Replication of the gE N-terminal mutants in vitro. Melanoma cells were inoculated on day 0 with 1×10^3 PFU of rOka, rOka-P27, rOka-Y51, rOka-S31A, and rOka-S49A (**a**); rOka, rOka-G90, rOka-I146, and rOka-P187 (**b**); or rOka-Δ P27-Y51 and ΔY51-P187 (**c**). Aliquots were harvested from day 1 to day 6. Each point represents the mean of three wells. The error bars represent the standard deviations. Significance between the rOka and the rOka mutants was determined by Student's *t* tests (Fig. 5 from Berarducci et al. (2006); reproduced with permission)

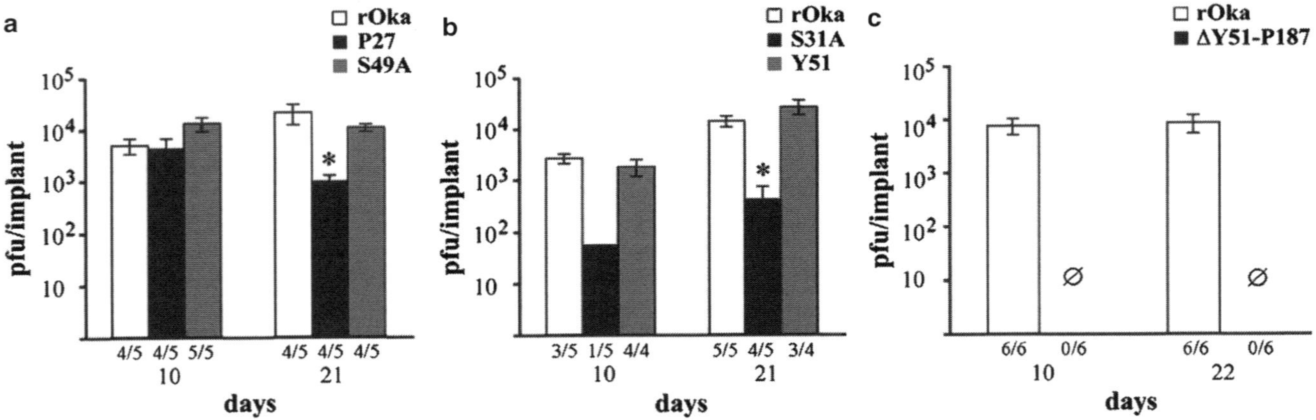

Fig. 6 Effect of the gE N-terminal mutations on VZV replication in skin xenografts. Skin xenografts were inoculated with MRC-5 cells infected with rOka, rOka-P27, and rOka-S49A (**a**); rOka, rOka-Y51, and rOka-S31A (**b**); and rOka and rOka-ΔY51-P187 (**c**). The infected xenografts were collected at days 10 and 21 or days 10 and 22 for the rOka-ΔY51-P187 mutant. The number of samples from which infectious virus was recovered per total number of xenografts inoculated is indicated on the horizontal axis. Each bar represents the mean titer, and the error bar indicates the standard error. The *asterisk* indicates significance ($P < 0.05$) (Fig. 8 from Berarducci et al. (2006); reproduced with permission)

mutants that were evaluated in skin included four mutants that had somewhat diminished replication *in vitro*, three of which, rOka-P27, rOka-S31A, and rOka-ΔY51-P187, showed no defect in skin pathogenesis. Two other mutants, rOka-Y51 and rOka-S49A, that were equivalent to rOka *in vitro* were also not impaired for skin infection when evaluated at day 10 and day 21 after inoculation (Fig. 6a, b). In contrast, both the P27 mutant and the S31 mutant showed significant reductions in growth in skin. Titers in rOka-P27-infected xenografts were similar to rOka at day 10 but less at day 21 ($P = 0.017$; Fig. 6a). Virus was recovered from only one of five xenografts infected with rOka-S31A and tested at day 10 and titers remained very low after 21 days (Fig. 6b) indicating that the serine at position 31 is an important determinant of virulence *in vivo*. As we have observed with other mutants, this growth defect in differentiated human cells within the intact skin microenvironment *in vivo* occurred even though this mutant had only minimal differences from intact VZV when tested for its capacity to replicate in cultured cells *in vitro*. rOka-S31A was only slightly impaired in its peak titer compared to rOka and had only slightly less gE production. We speculate that this S31 residue may be important for interaction with other viral and cellular proteins during VZV infection of epidermal cells *in vivo*. Thus, experiments deleting the gE region between residues 51 and 187 showed that this region was critical for virion assembly *in vitro* and was required for VZV tropism for human epidermal and dermal cells *in vivo*, indicating that the unique VZV gE N terminus is an important part of the gE ectodomain. Various subdomains of the unique gE N-terminal region were identified as having specific roles in viral replication, cell–cell spread, and secondary envelopment *in vitro* and several specific residues and regions of the protein contribute to or are essential for VZV infection of skin xenografts *in vivo* (Table 1).

Further analysis of the residues in the gE ectodomain showed that a cysteine-rich region was located from amino acids 208 to 236 (Fig. 1). This motif is within the conserved component of gE and cysteine-rich elements are also present in gE encoded by other alphaherpesviruses. Since all of these viruses share the characteristic gE/gI heterodimer formation, we postulated that this motif might be important for gE binding to gI in VZV-infected cells (Berarducci et al. 2009). When the coding sequence for this region was deleted from ORF68 in the cosmid, infectious virus was recovered, showing that it was not essential for VZV replication *in vitro*. However, cell–cell spread of the resulting mutant, rOka-gE-ΔCys, was reduced significantly and as predicted, eliminating the cysteine-rich region disrupted the binding of the mutant gE protein to gI. gE/gI heterodimers were not detected in infected cells. In association with this mutation, the gE expression at the plasma membrane of infected cells was punctate, resembling the pattern of gE distribution observed in cells infected with the gI deletion mutant. Of interest, blocking gE/gI binding altered the maturation of gI and reduced gI incorporation into viral particles (Fig. 7). The gE cysteine-rich region was not necessary for gE binding to IDE, indicating that this interaction is independent of its interaction with gE when the proteins are expressed in the context of VZV replication, as had been shown using expression constructs (Li et al. 2006). The rOka-ΔCys mutant retained the capacity to infect human tonsil T cells *in vitro* and was not necessary for infection of melanoma

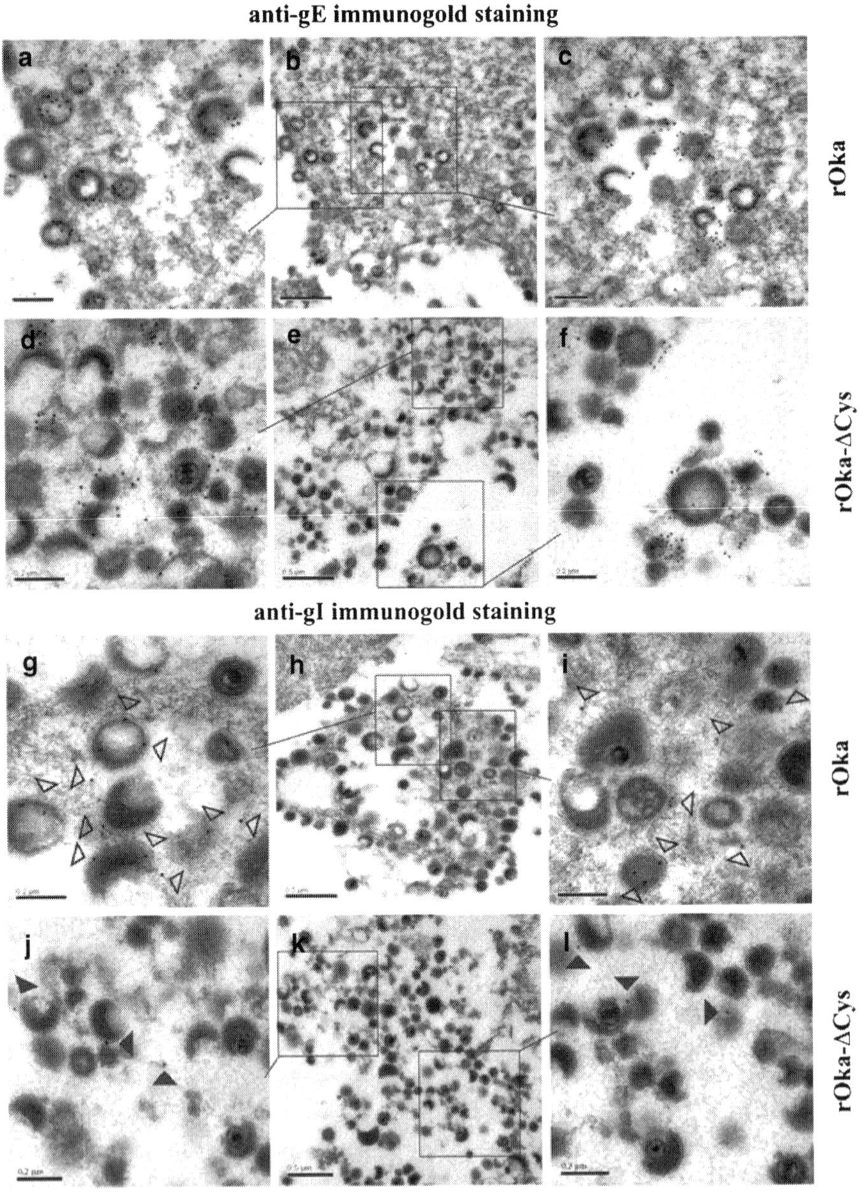

Fig. 7 Incorporation of gE and gI in the envelopes of rOka-ΔCys mutant virions. Immunogold labeling and TEM of rOka-infected (**a–c** and **g–i**) and rOka-ΔCys-infected (**d–f** and **j–l**) HELF. Labeling was performed using the anti-gE MAb from Chemicon to detect gE (A–F) and the rabbit polyclonal anti-gI to detect gI (**g–l**). The *open arrowheads* in panels (**g**) and (**i**) and the *black arrowheads* in panels (**j**) and (**l**) indicate gold-decorated particles. Scale bars, 0.2 μm (**a**, **c**, **d**, **f**, **g**, **i**, **j**, and **l**) and 0.5 μm (**b**, **e**, **h**, and **k**). The *boxed areas* in the *middle column* are enlarged in the *right* and *left columns* as indicated (Fig. 5 from (Berarducci et al. (2009); reproduced with permission)

cells with a cell-free inoculum, despite the failure to detect gI incorporation into the viral envelope in the absence of gE/gI binding. Nevertheless, gE/gI heterodimer formation proved to be a major determinant of VZV virulence in skin xenografts in the SCID mouse model (Berarducci et al. 2009). When the ability of the rOka-ΔCys mutant to replicate in skin xenografts *in vivo* was assessed, the mutant virus was recovered from only 2 of 21 xenografts. In two separate experiments, 1 of 10 skin xenografts was infected at day 10, and 1 of 11 xenografts inoculated with rOka-ΔCys produced infectious virus at day 21. In both cases, the titers were lower than those in skin tissue infected with rOka. Both of the viruses isolated from rOka-ΔCys-infected skin maintained the mutation and exhibited the small plaque size observed before inoculation of the xenografts. This severe impairment of virulence in skin that was produced by mutating a cysteine-rich region in the gE ectodomain can be explained by the markedly decreased capacity for cell–cell spread associated with disrupting gE/gI binding. These observations suggest that the cell–cell fusion and polykaryon formation in VZV skin infection depends on the capacity of gE and gI to form heterodimers, an interaction which is also prominent in HSV-infected cells.

Since gE binding to IDE is another prominent interaction that was mapped to the gE N terminus using mutagenesis of expression constructs, but was preserved in the viruses that we had made in the course of our first gE targeted mutagenesis experiments, we made additional mutations using the cosmid system in order to map this interaction in the context of VZV replication (Berarducci et al. 2010). These experiments showed that residues from P27-G90 were necessary for gE binding to IDE in infected cells, which was consistent with the reports from the Cohen laboratory that deleting amino acids 32–71 blocked the gE/IDE interaction (Ali et al. 2009; Li et al. 2006, 2007). While compatible with viral replication, the rOka-ΔP27-G90 mutant exhibited reduced cell–cell spread *in vitro*. Furthermore, its virulence in skin xenografts was somewhat diminished, although much less so than was observed with the rOka-gE-ΔCys mutant that prevented gE/gI binding. We also noted that the gE point mutation, linker insertions, and partial deletions in the aa27–90 region, as well as the deletion of a large portion of the unique N-terminal region, aa52–187, which did not block gE/IDE binding had similar or more severe effects on VZV replication *in vitro* and *in vivo*. Notably, the rOka-ΔP27-G90 mutant showed no impairment of virus entry into T cells *in vitro* and its capacity to infect T-cell xenografts in the SCID mouse model was also not affected by disrupting gE/IDE binding. These observations are of interest because, as noted, VZV-infected T cells do not undergo cell–cell fusion in T-cell xenografts; in contrast to skin infection, virus spread depends on the release of infectious virus particles and entry into other T cells in the xenograft. The fact that the gE residues that bind IDE are not required for T-cell tropism suggests that this interaction has a limited role in T-cell entry. However, the larger deletion of the unique N terminus created in the rOka-gE-ΔY51-P187 mutant was lethal for VZV infection of skin as well as T-cell xenografts. Further study of the gE functions that are encoded in this region will be of interest for understanding determinants of VZV virulence present in the unique gE N terminus.

3 gE Promoter Functions

The analysis of the gE promoter using bioinformatics methods revealed that like gI, this promoter also contained binding sites predicted to bind to the cellular transactivator, specificity protein 1 (Sp1) (Berarducci et al. 2007). To explore the hypothesis that cellular transactivators can function together with VZV specific viral transactivators and act as cell type specific determinants of VZV virulence in differentiated human cells *in vivo*, the functional analysis of these two motifs was undertaken using gE promoter-luciferase constructs. These reporter construct experiments showed that deletion or point mutations of the two Sp1 sites blocked Sp1 binding as well as IE62-mediated transactivation of the gE promoter. Having established their relevance for gE transcription *in vitro*, these mutations were introduced into the gE promoter sequence in the viral genome using VZV cosmids (Table 1). When both of the Sp1 sites, designated Sp1A and Sp1B, were disrupted, either by deletion or point mutation, VZV replication was blocked. Further mutagenesis using cosmids showed that eliminating the Sp1 site that was atypical when compared to the usual Sp1 motif in human gene promoters, was lethal. In contrast, while mutation of the second Sp1 reduced VZV replication, it was compatible with recovery of infectious virus; however, replication was delayed, plaques were smaller, and less gE was produced. Thus, there are functional differences in the roles that the two Sp1 binding sites in the gE promoter play in VZV replication. Bioinformatics inspection also indicated the presence of possible binding sites for other cellular transcriptional factors in the gE promoter. However, mutations of these motifs in the context of the VZV genome did not affect replication in cultured cells or in T cells *in vitro* and also had no effect on VZV pathogenesis in skin xenografts in the SCID mouse model. These experiments indicate IE62 alone is not sufficient for activation of the gE promoter but requires the cellular transactivator, Sp1 as a co-factor for upregulation of the promoter of this critical VZV gene during VZV pathogenesis.

4 gI Functions

Initial mutagenesis experiments using the VZV cosmid method showed that deleting both ORF67 and ORF68 was lethal but a single ORF67 deletion, eliminating gI expression was compatible with VZV replication in cell culture (Mallory et al. 1997), as was also shown by Cohen and Nguyen (Cohen and Nguyen 1997). However, removing ORF67 was associated with reduced viral titers and strikingly aberrant syncytial formation, which is a hallmark of VZV replication (Table 2). Notably, in the absence of gI, infected cells showed an unusual punctate distribution of gE on plasma membranes, and diminished synthesis of the mature 94 kDa form of gE. Similar phenotypes were conferred by partial deletions of either the N-terminal and C-terminal regions of gI. When cells infected with VZV mutants that had complete or partial gI deletions were examined by electron microscopy, adjacent cisternae were shown to be adherent, with marked distortions of the TGN; VZ

Table 2 Phenotypes of VZV recombinants with gI promoter and gI mutations *in vitro* and *in vivo*

Virus	Replication kinetics *in vitro*	Plaque size	gE/gI binding	Infection of xenografts *in vivo*	
				Skin	T cells
gI promoter					
gI-Sp1	NL	NL		Decrease	Decrease
gI-Ap1	NL	NL		NL	NL
gI-USF	NL	NL		Slight decrease	NL
gI-Sp1/USF	NL	Decrease		Lethal	Decrease
gI-RE-1	Decrease days 1–3	NL		Slight decrease	NL
gI-RE-2	Decrease days 1–3	NL		NL	NL
gI-RE-3	Decrease days 1–3	NL		Decrease	Decrease
gI					
ΔgI	Decrease	Decrease		Lethal	Lethal
ΔgI-C	Decrease	Decrease			
ΔgI-N	Decrease	Decrease			

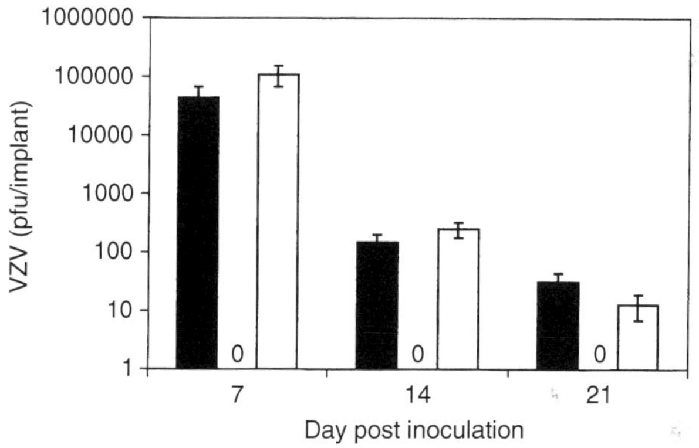

Fig. 8 Replication of VZV rOka, rOkaΔgI, and rOka-gI@Avr in human T cells. Thymus/liver implants infected with rOka (*black*), rOkaΔgI (*zeros*), and rOka-gI@Avr (*white*) were harvested at 7, 14, and 21 days after infection. The number of PFU/implant was determined using an infectious focus assay. Each *bar* represents the mean and standard deviation from three to five implants from two separate experiments (Fig. 3 from Moffat et al. (2002); reproduced with permission)

virions were not identified in post-Golgi compartments in cells infected with gI mutants as shown by the Gershon laboratory (Wang et al. 2001). Although gI was dispensable in cultured cells, evaluation of rOka-ΔgI, the complete gI deletion mutant, in the SCID mouse model demonstrated that gI was essential for VZV replication in skin as well as T-cell xenografts (Moffat et al. 2002). No virus was recovered at any time point after inoculation of the xenografts, in contrast to the expected patterns of infection observed with intact VZV derived from cosmids and a repaired virus, in which ORF67 was inserted into a non-native AvrII cloning site between ORF65 and ORF66 (Fig. 8). Since gE expression was preserved in the rOka-ΔgI mutant, these experiments indicated that gI has a direct role in VZV virulence *in vivo*.

5 gI Promoter Functions

The effects of targeted mutations in nucleotides of the gI promoter were examined by generating VZV recombinants with these changes that were designed based on information about functional motifs as identified in reporter construct assays by the Ruyechan laboratory (He et al. 2001). All VZV promoters require activation by the major immediate-early viral transactivator, IE62. In addition, gI transcription was modulated by an activating upstream sequence containing motifs for binding the cellular transcription factors, Sp1, upstream stimulatory factor (USF), and activator protein 1 (Ap-1). The gI promoter also has an element that responds to the ORF29 DNA binding protein, called 29RE, which mediates enhancement of IE62-induced transcription by ORF29 protein (ref). When 2-bp substitutions were made using VZV cosmids to create viruses designated rOKAgI-Sp1, rOKAgI-Ap1, and rOKAgI-USF, no differences in replication were identified compared to rOKA *in vitro* (Ito et al. 2003). The mutation of the Ap1 site had no effect on skin or T-cell tropism in the SCID mouse model. However, the mutant with disruption of the Sp1 motif in the gI promoter showed a significant reduction of growth in skin and T-cell xenografts *in vivo*. Synergy between these motifs was evident in that rOKAgI-Sp1/USF, which had substitutions in both Sp1 and USF motifs, did not replicate at all in skin, although some growth occurred in T-cell xenografts. The role of these binding sites for cellular transactivators was confirmed by the normal infectivity of a repaired virus, rOKAgI:rep-Sp1/USF. These experiments provided first evidence of the importance of cell regulatory proteins in modulating VZV virulence *in vivo*, even though such effects were not apparent *in vitro*, and showed that the contribution of or in some cases, the requirement for cellular transactivators was different, depending upon whether the host target cell type was skin or T cells. Mutations in the 29RE of the gI promoter were made by substituting each of four 10-bp blocks in this region with a 10-bp sequence, GATAACTACA that was predicted to interfere with the enhancer effects of the ORF29 protein. Of these mutants, rOKAgI-29RE-3 was associated with diminished replication in skin and T cells, indicating that the enhancement of IE62 activation of the gI promoter by ORF29 protein modulates VZV virulence. Thus, gI promoter mutants that had variably diminished gI expression showed phenotypes consistent with the requirement of gI expression for VZV pathogenesis in skin and T cells, as observed with the gI-null mutant. More generally, the experiments offered a proof of concept that mutagenesis of motifs in VZV gene promoters may confer altered virulence and could be a novel approach to engineering attenuated vaccine viruses.

6 Summary

VZV gE and its ubiquitous binding partner, gI, are critically important glycoproteins that are essential either for any replication, in the case of gE, or for infection of human skin and T cells *in vivo* in the case of gI. VZV gE is notable for its large unique N-terminal region not present in the homologous proteins of other alpha-herpesviruses. The capacity of gE to bind gI, in contrast to its interactions with IDE, is a determinant of its virulence during VZV pathogenesis in the SCID mouse model of VZV skin and T-cell tropism. Further analysis of the functional domains in the gE residues in the unique region as well as its interactions with viral proteins other than gI and with host cell proteins, are important areas for further investigation.

Acknowledgments This work on the functions of gE and gI in the SCID mouse model of VZV pathogenesis was supported by an NIH grant, AI20459.

References

Ali MA, Li Q, Fischer ER, Cohen JI (2009) The insulin degrading enzyme binding domain of varicella-zoster virus (VZV) glycoprotein E is important for cell-to-cell spread and VZV infectivity, while a glycoprotein I binding domain is essential for infection. Virology 386: 270–279

Arvin AM (2006) Investigations of the pathogenesis of Varicella zoster virus infection in the SCIDhu mouse model. Herpes 13:75–80, Review

Berarducci B, Ikoma M, Stamatis S, Sommer M, Grose C, Arvin AM (2006) Essential functions of the unique N-terminal region of the varicella-zoster virus glycoprotein E ectodomain in viral replication and in the pathogenesis of skin infection. J Virol 80:9481–9496

Berarducci B, Rajamani J, Zerboni L, Che X, Sommer M, Arvin AM (2010) Functions of the unique N-terminal region of glycoprotein E in the pathogenesis of varicella-zoster virus infection. Proc Natl Acad Sci USA 107:282–287

Berarducci B, Sommer M, Zerboni L, Rajamani J, Arvin AM (2007) Cellular and viral factors regulate the varicella-zoster virus gE promoter during viral replication. J Virol 81: 10258–10267

Berarducci B, Rajamani J, Reichelt M, Sommer M, Zerboni L, Arvin AM (2009) Deletion of the first cysteine-rich region of the varicella-zoster virus glycoprotein E ectodomain abolishes the gE and gI interaction and differentially affects cell-cell spread and viral entry. J Virol 83:228–240

Cohen JI, Nguyen H (1997) Varicella-zoster virus glycoprotein I is essential for growth of virus in Vero cells. J Virol 71:6913–6920

Cole NL, Grose C (2003) Membrane fusion mediated by herpesvirus glycoproteins: the paradigm of varicella-zoster virus. Rev Med Virol 13:207–222

He H, Boucaud D, Hay J, Ruyechan WT (2001) Cis and trans elements regulating expression of the varicella zoster virus gI gene. Arch Virol Suppl 17:57–70

Ito H, Sommer MH, Zerboni LH, He H, Boucaud D, Hay J, Ruyechan W, Arvin AM (2003) Promoter sequences of varicella-zoster virus glycoprotein I targeted by cellular transcription factors, Sp1 and USF, determine virulence for human skin and T cells in the SCIDhu mouse in vivo. J Virol 77:489–498

Ku C-C, Besser J, Abendroth A, Grose C, Arvin AM (2005) Varicella-zoster virus pathogenesis and immunobiology: new concepts emerging from investigations in the SCIDhu mouse model. J Virol 79:5, 2651–2658

Li Q, Ali MA, Cohen JI (2006) Insulin degrading enzyme is a cellular receptor mediating varicella-zoster virus infection and cell-to-cell spread. Cell 127:305–316

Li Q, Krogmann T, Ali MA, Tang WJ, Cohen JI (2007) The amino terminus of varicella-zoster virus (VZV) glycoprotein E is required for binding to insulin-degrading enzyme, a VZV receptor. J Virol 81:8525–8532

Mallory S, Sommer M, Arvin AM (1997) Mutational analysis of the role of glycoprotein I in varicella-zoster virus replication and its effects on glycoprotein E conformation and trafficking. J Virol 71:8279–8288

Maresova L, Pasieka TJ, Homan E, Gerday E, Grose C (2005) Incorporation of three endocytosed varicella-zoster virus glycoproteins, gE, gH, and gB, into the virion envelope. J Virol 79: 997–1007

Mo C, Schneeberger EE, Arvin AM (2000) Glycoprotein E of varicella-zoster virus enhances cell-cell contact in polarized epithelial cells. J Virol 74:11377–11387

Mo C, Lee J, Sommer M, Grose C, Arvin AM (2002) The requirement of varicella zoster virus glycoprotein E (gE) for viral replication and effects of glycoprotein I on gE in melanoma cells. Virology 304:176–186

Moffat J, Ito H, Sommer M, Taylor S, Arvin AM (2002) Glycoprotein I of varicella-zoster virus is required for viral replication in skin and T cells. J Virol 76:8468–8471

Moffat J, Mo C, Cheng JJ, Sommer M, Zerboni L, Stamatis S, Arvin AM (2004) Functions of the C-terminal domain of varicella-zoster virus glycoprotein E in viral replication in vitro and skin and T-cell tropism in vivo. J Virol 78:12406–12415

Oliver SL, Zerboni L, Sommer M, Rajamani J, Arvin AM (2008) Development of recombinant varicella-zoster viruses expressing luciferase fusion proteins for live in vivo imaging in human skin and dorsal root ganglia xenografts. J Virol Methods 154:182–193

Olson JK, Grose C (1997) Endocytosis and recycling of varicella-zoster virus Fc receptor glycoprotein gE: internalization mediated by a YXXL motif in the cytoplasmic tail. J Virol 71:4042–4054

Olson JK, Grose C (1998) Complex formation facilitates endocytosis of the varicella-zoster virus gE:gI Fc receptor. J Virol 72:1542–1551

Olson K, Bishop GA, Grose C (1997) Varicella-zoster virus Fc receptor gE glycoprotein: serine/threonine and tyrosine phosphorylation of monomeric and dimeric forms. J Virol 71:110–119

Pasieka TJ, Maresova L, Shiraki K, Grose C (2004) Regulation of varicella-zoster virus-induced cell-to-cell fusion by the endocytosis-competent glycoproteins gH and gE. J Virol 78: 2884–2896

Reichelt M, Brady J, Arvin AM (2009) The replication cycle of varicella-zoster virus: analysis of the kinetics of viral protein expression, genome synthesis, and virion assembly at the single-cell level. J Virol 83:3904–3918

Santos RA, Hatfield CC, Cole NL, Padilla JA, Moffat JF, Arvin AM, Ruyechan WT, Hay J, Grose C (2000) Varicella-zoster virus gE escape mutant VZV-MSP exhibits an accelerated cell-to-cell spread phenotype in both infected cell cultures and SCID-hu mice. Virology 275: 306–317

Wang Z-H, Gershon MD, Lungu O, Zhenglun Z, Mallory S, Arvin AM, Gershon AA (2001) Essential role played by the C-terminal domain of gI in the envelopment of varicella zoster virus in the trans-Golgi network: interactions of glycoproteins with tegument. J Virol 75:323–340

Varicella-Zoster Virus Glycoprotein M

Yasuko Mori and Tomohiko Sadaoka

Contents

Abstract Glycoprotein M (gM) is conserved among herpesviruses. Important features are its 6–8 transmembrane domains without a large extracellular domain, localization to the virion envelope, complex formation with another envelope glycoprotein, glycoprotein N (gN), and role in virion assembly and egress. In varicella-zoster virus (VZV), the gM homolog is encoded by ORF50. VZV gM is predicted to be an eight-transmembrane envelope glycoprotein with a complex N-linked oligosaccharide. It mainly localizes to the *trans*-Golgi network, where final virion envelopment occurs. Studies in which VZV gM or its partner gN were disrupted suggest that the gM/gN complex plays an important role in cell-to-cell spread. Here, we summarize the biological features of VZV gM, including our recent findings on its characterization and function.

Y. Mori (✉) and T. Sadaoka
Division of Clinical Virology, Kobe University Graduate School of Medicine, 7-5-1, Kusunoki-cho, Chuo-ku, Kobe 650-0017, Japan
e-mail: ymori@med.kobe-u.ac.jp

Laboratoy of Virology and Vaccinology, Division of Biomedical Research, National Institute of Biomedical Innovation, 7-6-8, Saito-Asagi, Ibaraki, Osaka 567-0085, Japan

A.M. Arvin et al. (eds.), *Varicella-zoster Virus*,
Current Topics in Microbiology and Immunology 342, DOI 10.1007/82_2010_30

1 Introduction

The herpesvirus genome encodes various envelope glycoproteins, which are involved in viral entry into the host cell (Spear and Longnecker 2003) and in the envelopment and maturation of progeny virions (Mettenleiter 2004). The VZV envelope contains conserved glycoproteins common to all herpesviruses, including glycoproteins B, H, L, M, and N (gB, gH, gL, gM, and gN) (Cohen et al. 2006).

Glycoprotein M (gM) is one of the most conserved of all the herpesvirus glycoproteins and functions mainly in virion maturation (Baines et al. 2007; Dijkstra et al. 1996; Jons et al. 1998; Mach et al. 2005; Osterrieder et al. 1996). In general, the gM proteins encoded by herpesviruses are highly hydrophobic and predicted to contain six to eight transmembrane domains. They also have a potential N-glycosylation site within the first extracellular domain and a conserved cysteine residue within the same loop, which is predicted to form a disulfide bond with the relevant glycoprotein N (gN) (Dijkstra et al. 1996; Jons et al. 1998).

The gM cytoplasmic tail encodes several predicted trafficking motifs, whose conservation in alpha-, beta-, and gamma-herpesviruses suggests that they have important functions. The gM in HCMV (Hobom et al. 2000) and MDV-1 (Tischer et al. 2002) is essential for viral replication. The gM/gN complex is dispensable for alphaherpesvirus replication in cell culture, although the disruption of gM or gN reduces viral growth (Baines and Roizman 1991; Dijkstra et al. 1996, 1998; Osterrieder et al. 1996). Thus, the functional roles of gM in viral replication may be different among herpesviruses. In HCMV, the cytoplasmic tail of gM is required for viral trafficking during viral particle assembly (Krzyzaniak et al. 2007). Antibodies against the HCMV gM/gN complex were shown to have a neutralizing function in HCMV infection (Shimamura et al. 2006), suggesting that the gM/gN complex is also involved in viral entry into host cells.

In contrast to the information available for other herpesviruses, there have been only a few reports on the VZV gM homolog (Yamagishi et al. 2008). The gN homolog encoded by VZV has been reported only to be nonessential for viral replication in cell culture, although its disruption impairs viral growth and syncytia formation (Ross et al. 1997)

The gM of VZV is encoded by the ORF50 gene. The predicted nucleotide length of the ORF is 1,308 bp, and the number of amino acids is 435. VZV gM is also predicted to be an integral membrane protein with eight transmembrane domains, a putative N-glycosylation site, and a cysteine residue within the first ectodomain that could form a disulfide bond, like the gM of other herpesviruses (Fig. 1).

The characterization and function of gM during VZV infection has recently been reported by our group (Yamagishi et al. 2008). Here we review the reported data regarding VZV gM and our recent findings.

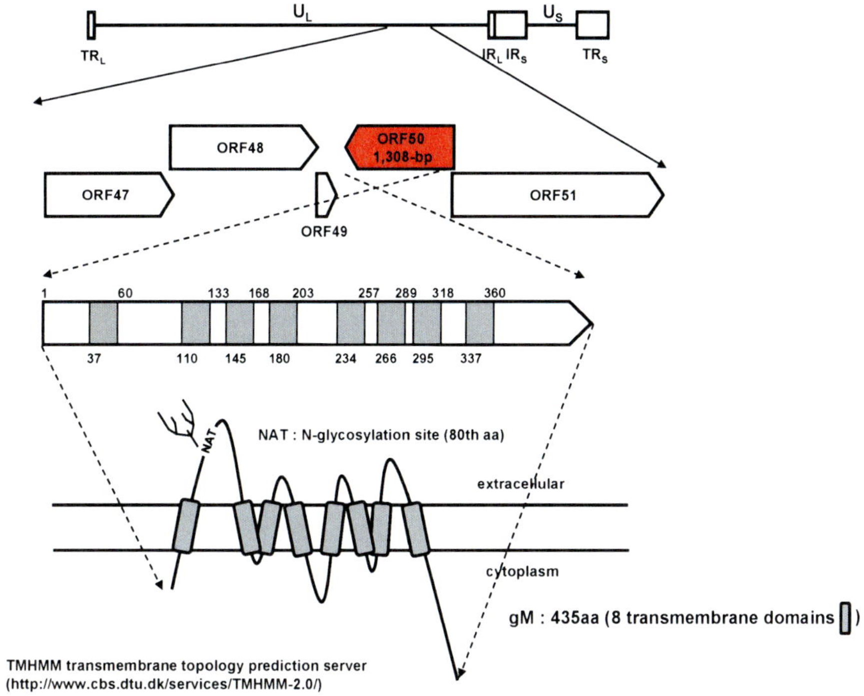

Fig. 1 Schematic representation of the gM-encoding VZV ORF50 gene and the predicted topology of gM. The predicted VZV gM topology was performed using the TMHMM transmembrane topology prediction server (http://www.cbs.dtu.dk/services/TMHMM-2.0/). A schematic representation of the predicted gM protein is shown. The location of an N-linked glycosylation site (NAT, branched structure: N-linked sugar chain) is indicated

2 VZV gM Is an Envelope Glycoprotein Modified with a Complex N-Linked Oligosaccharide

VZV gM is detected in both infected cells and virus particles as a broad band with a molecular mass of 42–48 kDa. In infected cells but not virus particles, it has an additional band with a molecular mass of 37 kDa (Fig. 2). The 42–48 kDa band is shifted to approximately 35 kDa by PNGase F but not endo H digestion (Fig. 2) (Yamagishi et al. 2008). However, the 37 kDa band detected only in infected cells is digested by both endo H and PNGase F (Fig. 2). These results indicate that the VZV gM, like other gMs, is a membrane glycoprotein modified with a complex N-linked oligosaccharide and exists in virus particles as an envelope glycoprotein, and that the 37 kDa form is a precursor form of gM. Since VZV gM is also found incorporated into virus particles, it is thought to function in viral entry into host cells and virion maturation.

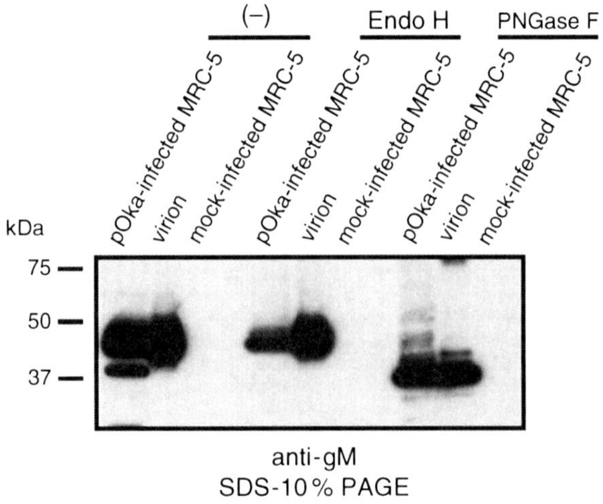

Fig. 2 VZV gM is an envelope glycoprotein. Infected cells and purified virions were subjected to Western blotting with an anti-gM antibody. Lysates were digested with Endo H or PNGase F, and resolved by SDS–PAGE under reducing conditions. Molecular masses are indicated at *left*

3 The Main Location of the gM in VZV-Infected Cells Is the *trans*-Golgi Network

At the late phase of VZV infection, the gM colocalizes with p230, a marker for the *trans*-Golgi network (TGN), indicating that VZV gM mainly localizes to the Golgi apparatus and TGN in VZV-infected MeWo cells (Fig. 3). In addition, the VZV gM partially colocalizes with lamp1, a marker of late endosomes and lysosomes, indicating that gM also localizes to late endosomes (Yamagishi et al. 2008). Since VZV has been shown to bud into TGN-derived vesicles (Gershon et al. 1994), gM may be incorporated into virions at the membrane derived from the TGN.

4 VZV gM Plays a Role in the Cell-to-Cell Spread of Virus

A deletion mutant of the gM gene was constructed by mutagenesis of a BAC clone harboring the VZV parental Oka strain genome (pOka-BAC) (Nagaike et al. 2004; Yamagishi et al. 2008). Amino acid positions 136–353 of the gM were replaced in the pOka-BAC clone, and the reconstituted virus with a defective gM gene was produced in MRC-5 or MeWo cells (rpOkaΔgM).

The growth of the gM-defective VZV (rpOkaΔgM) was examined by plaque formation assay. Compared to the wild-type VZV-infected cells, the plaque formation was impaired in the rpOkaΔgM-infected cells (Fig. 4), indicating

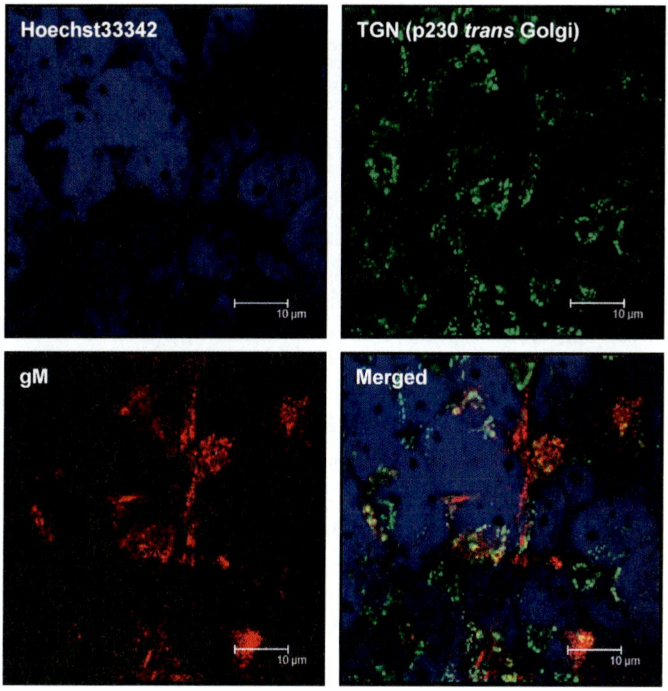

Fig. 3 VZV gM localizes to the TGN. VZV-infected MeWo cells were stained with a rabbit anti-gM antibody and mouse anti-p230 monoclonal antibody. Nuclei were stained with Hoechst 33342. The *red* and *green* colors indicate gM and p230, a marker for TGN, respectively. *Yellow* indicates colocalization in the merged image

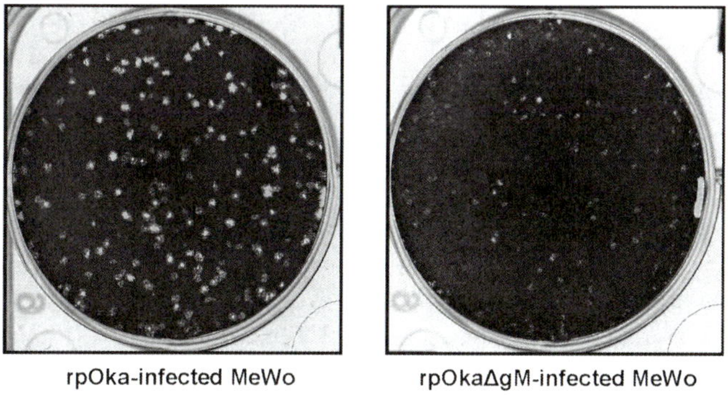

Fig. 4 The plaque sizes are reduced in gM-defective VZV-infected cells. MeWo cells were infected with rpOka or rpOkaΔgM virus and cultured for 7 days. The cells were fixed and stained with 3% crystal violet/70% ethanol

that the disruption of gM causes a reduction in plaque formation in virus-infected cells. Moreover, the virus spread was also impaired in the gM-defective VZV, as shown by an infectious center assay. By electron microscopic examination, numerous aberrant vacuoles containing electron-dense material were found in the cytoplasm of gM-defective VZV-infected cells; however, light (L) particles were observed on the host-cell surface, and capsids were observed in the nucleus (Yamagishi et al. 2008). Taken together, these data indicate that VZV gM functions in viral growth and cell-to-cell spread, although the final envelopment and capsid formation in the nucleus occur even without the expression of gM.

As described earlier, gM forms a complex with gN, which is also conserved in all herpesviruses (Jons et al. 1998; Koyano et al. 2003; Liang et al. 1996; Mach et al. 2005; Wu et al. 1998), via a disulfide bond, and this interaction is required for the glycosylation of gM to its mature form and for the efficient transportation of the gM/gN complex to the Golgi apparatus for viral assembly and egress (Koyano et al. 2003; Lake et al. 1998; Mach et al. 2000, 2005, 2007). In HCMV infection, the gM/gN complex formation is mediated not only by the conserved disulfide bond involving the conserved cysteine residue at the second extracellular loop but also by another, noncovalent linkage (Mach et al. 2005).

In VZV, the gM also forms a complex with gN, and this interaction is required for the efficient glycosylation of gM and the transportation of the gM/gN complex. The VZV gM/gN complex formation is not mediated via a disulfide bond despite the presence of the conserved cysteine residues in gM and gN (presented by T. Sadaoka, K. Yamanishi, and Y. Mori at the International Herpesvirus Workshop 2008, Estoril, Portugal; manuscript in preparation). This noncovalent linkage requires two amino acid residues of gM, valine at position 42 and glycine at 301, and the glycine at 301 is conserved in all human herpesviruses in the putative seventh transmembrane region (presented by T. Sadaoka, K. Yamanishi, and Y. Mori at the International Herpesvirus Workshop 2009, Ithaca, NY; manuscript in preparation). This finding suggests that the binding mechanism of the gM/gN complex is a noncovalent linkage containing the conserved glycine residue.

A VZV gM mutant virus that does not undergo maturation showed that the maturation of gM is not essential for viral replication or virion morphogenesis, although this mutant virus does not cause efficient cell-to-cell spread by inducing syncytia formation (presented by T. Sadaoka, K. Yamanishi, and Y. Mori at the International Herpesvirus Workshop 2009, Ithaca, NY; manuscript in preparation). The gMs of alphaherpesviruses are reported to suppress the transfection-based membrane fusion mediated by several viral glycoproteins besides gM (Crump et al. 2004; Klupp et al. 2000; Koyano et al. 2003). However, since the disruption of VZV gN also reduces syncytia formation (Ross et al. 1997), the gM/gN complex in VZV is thought to enhance syncytia formation. The VZV gM/gN complex expressed on the viral envelope may therefore function to enhance cell–cell fusions, which are induced by gB, gH/gL, and gE (Fig. 5).

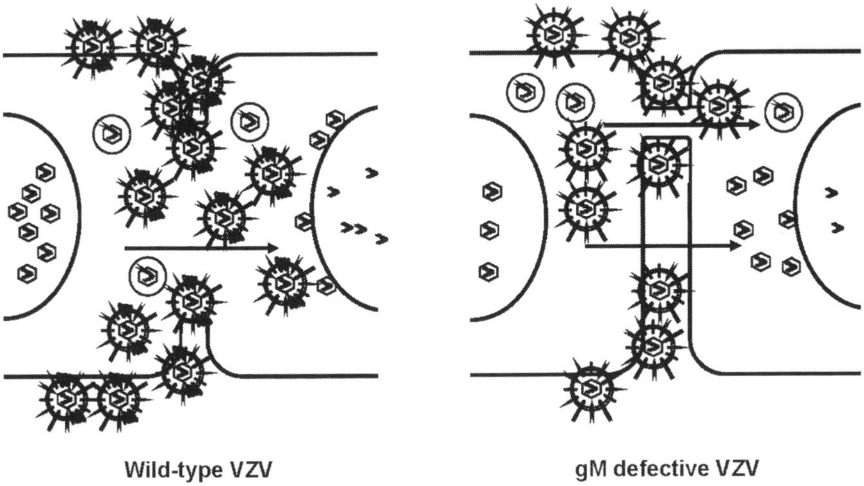

Fig. 5 Model of the function of the VZV gM/gN complex expressed on the viral envelope. The gM/gN complex expressed on the viral envelope of VZV may function to enhance cell–cell fusion, which is induced by gB, gH/gL, and gE, and by which the produced progeny virus particles may be transferred to uninfected neighboring cells

5 Alternative Splicing of the VZV ORF50 Gene

As described earlier, VZV gM is encoded by the ORF50 gene of VZV. We found that the ORF50 gene expression is regulated by alternative splicing. Three transcripts, excluding the nonspliced one that encodes gM, arise from the ORF50 gene by using two GT splicing donors and two AG splicing acceptors. All of the spliced mRNAs are expressed during lytic infection, but are rarely translated into proteins *in vitro* (presented by T. Sadaoka, K. Yamanishi, and Y. Mori at the International Herpesvirus Workshop 2009, Ithaca, NY; manuscript in preparation). However, the functions of these spliced mRNAs have been unidentified.

Recent investigations have revealed several interesting properties of VZV gM and its possible roles in VZV infection, some of which differ from the gMs of other herpesviruses. Further studies should elucidate the detailed functions and mechanism of VZV gM in the viral life cycle.

References

Baines JD, Roizman B (1991) J Virol 65:938–944

Baines JD, Wills E, Jacob RJ, Pennington J, Roizman B (2007) J Virol 81:800–812

Cohen JI, Straus SE, Arvin AM (2006) Varicella-Zoster virus, replication, pathogenesis, and management. In: Knipe DM, Howley PM, Griffin DE, Lamb RA, Martin MA, Roizman B, Straus SE (eds) Fields virology. Lippincott Williams & Wilkins, Philadelphia, pp 2774–2818

Crump CM, Bruun B, Bell S, Pomeranz LE, Minson T, Browne HM (2004) J Gen Virol 85:3517–3527
Dijkstra JM, Visser N, Mettenleiter TC, Klupp BG (1996) J Virol 70:5684–5688
Dijkstra JM, Brack A, Jons A, Klupp BG, Mettenleiter TC (1998) J Gen Virol 79(Pt 4):851–854
Gershon AA, Sherman DL, Zhu Z, Gabel CA, Ambron RT, Gershon MD (1994) J Virol 68:6372–6390
Hobom U, Brune W, Messerle M, Hahn G, Koszinowski UH (2000) J Virol 74:7720–7729
Jons A, Dijkstra JM, Mettenleiter TC (1998) J Virol 72:550–557
Klupp BG, Nixdorf R, Mettenleiter TC (2000) J Virol 74:6760–6768
Koyano S, Mar EC, Stamey FR, Inoue N (2003) J Gen Virol 84:1485–1491
Krzyzaniak M, Mach M, Britt WJ (2007) J Virol 81:10316–10328
Lake CM, Molesworth SJ, Hutt-Fletcher LM (1998) J Virol 72:5559–5564
Liang X, Chow B, Raggo C, Babiuk LA (1996) J Virol 70:1448–1454
Mach M, Kropff B, Dal Monte P, Britt W (2000) J Virol 74:11881–11892
Mach M, Kropff B, Kryzaniak M, Britt W (2005) J Virol 79:2160–2170
Mach M, Osinski K, Kropff B, Schloetzer-Schrehardt U, Krzyzaniak M, Britt W (2007) J Virol 81:5212–5224
Mettenleiter TC (2004) Virus Res 106:167–180
Nagaike K, Mori Y, Gomi Y, Yoshii H, Takahashi M, Wagner M, Koszinowski U, Yamanishi K (2004) Vaccine 22:4069–4074
Osterrieder N, Neubauer A, Brandmuller C, Braun B, Kaaden OR, Baines JD (1996) J Virol 70:4110–4115
Ross J, Williams M, Cohen JI (1997) Virology 234:186–195
Shimamura M, Mach M, Britt WJ (2006) J Virol 80:4591–4600
Spear PG, Longnecker R (2003) J Virol 77:10179–10185
Tischer BK, Schumacher D, Messerle M, Wagner M, Osterrieder N (2002) J Gen Virol 83:997–1003
Wu SX, Zhu XP, Letchworth GJ (1998) J Virol 72:3029–3036
Yamagishi Y, Sadaoka T, Yoshii H, Somboonthum P, Imazawa T, Nagaike K, Ozono K, Yamanishi K, Mori Y (2008) J Virol 82:795–804

Varicella Zoster Virus Immune Evasion Strategies

Allison Abendroth, Paul R. Kinchington, and Barry Slobedman

Contents

Abstract The capacity of varicella zoster virus (VZV) to cause varicella (chickenpox) relies upon multiple steps, beginning with inoculation of the host at mucosal sites with infectious virus in respiratory droplets. Despite the presence of a powerful immune defense system, this virus is able to disseminate from the site of initial infection to multiple sites, resulting in the emergence of distinctive cutaneous vesiculopustular lesions. Most recently, it has been proposed that the steps leading

A. Abendroth (✉)
Department of Infectious Diseases and Immunology, University of Sydney, Blackburn Building, Room 601, Camperdown, NSW 2006, Australia
e-mail: allison.abendroth@sydney.edu.au
Centre for Virus Research, Westmead Millennium Institute, Westmead, NSW 2145, Australia

B. Slobedman
Centre for Virus Research, Westmead Millennium Institute, Westmead, NSW 2145, Australia
e-mail: barry.slobedman@sydney.edu.au

P.R. Kinchington
Department of Ophthalmology, School of Medicine, University of Pittsburgh, Pittsburgh, USA
Department of Molecular Microbiology and Genetics, School of Medicine, University of Pittsburgh, Pittsburgh, USA
e-mail: kinchingtonp@upmc.edu

A.M. Arvin et al. (eds.), *Varicella-zoster Virus*,
Current Topics in Microbiology and Immunology 342, DOI 10.1007/82_2010_41

to cutaneous infection include VZV infecting human tonsillar $CD4^+$ T cells that express skin homing markers that allow them to transport VZV directly from the lymph node to the skin during the primary viremia. It has also been proposed that dendritic cells (DC) of the respiratory mucosa may be among the first cells to encounter VZV and these cells may transport virus to the draining lymph node. These various virus-host cell interactions would all need to occur in the face of an intact host immune response for the virus to successfully cause disease. Significantly, following primary exposure to VZV, there is a prolonged incubation period before emergence of skin lesions, during which time the adaptive immune response is delayed. For these reasons, it has been proposed that VZV must encode functions which benefit the virus by evading the immune response. This chapter will review the diverse array of immunomodulatory mechanisms identified to date that VZV has evolved to at least transiently limit immune recognition.

1 Introduction

Primary varicella zoster virus (VZV) infection leading to varicella (chickenpox) is initiated by inoculation of mucosal sites, most frequently the upper respiratory tract, with infectious virus in respiratory droplets (Grose 1981). From the time of virus inoculation of the host to development of the cutaneous rash, there are multiple steps and many virus–cell interactions. For example, VZV may enter directly into a primary viremia, following infection of human tonsillar $CD4^+$ T cells that express skin homing markers that allow them to transport VZV directly from the lymph node to the skin during the primary viremia (Ku et al. 2002, 2004). Furthermore, it has been proposed that dendritic cells (DCs) of the respiratory mucosa may be among the first cells to encounter VZV during primary infection and may serve as a means for viral transport to the draining lymph node (Abendroth et al. 2001b; Morrow et al. 2003), where a primary cell-associated viremia initiates, during which time virus is transported to the reticuloendothelial organs where it undergoes another period of replication that results in a secondary cell-associated viremia and virus transport via T cells to the skin (Arvin et al. 1996; Grose 1981). At the skin, deep cutaneous infections are established which are maintained and contained during the incubation period, followed by emergence at the surface into distinctive vesiculopustular lesions. VZV also gains access to nerve axonal termini, and establishes a persistent latent state in which some viral antigens appear to be expressed. These different virus–host cell interactions would all need to occur in the face of an intact host immune response for the virus to successfully cause disease. It is, therefore, reasonable to predict that VZV encoded immune evasion mechanisms manifested during the first stages of primary infection as well as during latency and following reactivation so as to benefit virus by limiting and/or delaying immune recognition. Indeed, after primary exposure to VZV there is a prolonged incubation period of 10–21 days before appearance of skin lesions, during which time the adaptive response is delayed or virus remains initially undetected by the developing adaptive immune response. This suggests that VZV must encode immunomodulatory strategies to delay immune detection (Arvin 2001). This chapter will review the diverse array of mechanisms identified to date that VZV has evolved to at least transiently evade the immune response.

2 VZV Interference with Interferons

Immune control of productive VZV infection involves both innate and adaptive host responses (Abendroth and Arvin 1999). The innate host immune response, involving natural killer (NK) cells, NK-T cells, and type 1 (α and β) and type II (γ) interferon (IFN), is an important early host response designed to prevent or limit virus spread within the host (Abendroth and Arvin 2001; Arvin et al. 1986). In response to virus infection, a variety of signal transduction pathways are activated by the innate responses, including the expression of type I IFNs (i.e., IFN α/β), that induce a large number of IFN-stimulated genes, leading to an antiviral state. These genes include those encoding proteins such as protein kinase R (PKR), 2–5 oligoadenylate synthetase (2–5 OAS), and the Mx proteins that induce an antiviral response by interfering with viral transcription, translation, and likely other viral processes such as viral DNA replication and assembly (Muller et al. 1994; Sadler and Williams 2008). Many viruses have evolved mechanisms that impair the synthesis of IFNs or interfere with the downstream antiviral effects of IFNs (Garcia-Sastre and Biron 2006; Haller et al. 2006; Katze et al. 2002).

Both type I and type II IFNs can inhibit VZV replication in vitro (Balachandra et al. 1994; Desloges et al. 2005). The importance of type I IFNs in vivo is highlighted by the observation that treatment of immunocompromised individuals with IFNα can reduce the severity of varicella (Arvin et al. 1982). Thus, VZV induced control of expression of IFNs and/or the downstream IFN signaling events would be likely to provide a survival advantage to the virus during replication. Ku et al. assessed IFNα expression during VZV infection of human skin xenografts in SCID mice (Ku et al. 2004) by immunohistochemical detection of IFNα and reported its expression in epidermal skin cells in uninfected skin (Ku et al. 2004). In contrast, IFNα expression was downregulated in VZV infected cells within VZV infected skin xenografts, and upregulated in bystander uninfected epidermal cells, suggesting that local proximity to the infection were in an antiviral state, but that VZV modulated the expression of IFNα. To verify block in IFNα signaling in VZV infected skin cells, immunohistochemical staining for Stat-1 phosphorylation revealed that Stat1 was phosphorylated and translocated to the nuclei in the neighboring uninfected cells, but not in the VZV-infected cells. In combination, these observations suggest that VZV infection of human skin cells impairs Stat1 activation and IFNα production in VZV infected skin cells, but may not be able to prevent signaling to bystander cells, which in turn may contain the spreading VZV lesion. This is supported by the observation that inhibition of Type I IFN activity by administration of an IFN α/β receptor neutralizing antibody (Colamonici and

Domanski 1993) to SCID-hu skin mice resulted in more extensive viral replication and lesion formation compared with mice that received no antibody (Ku et al. 2004). These findings demonstrate that IFNα can modulate cutaneous VZV replication in vivo but that VZV can inhibit IFNα and in doing so enhance the capacity to replicate in the skin.

In the context of immune function in the presence of IFNα, Ambagala and Cohen (2007) reported that VZV IE63 is required to inhibit IFNα induced antiviral responses in vitro (Ambagala and Cohen 2007). This study utilized a viable ORF63 deletion virus to infect human melanoma cells and showed that this virus was hypersensitive to the antiviral effects of human IFNα (i.e., replication was severely inhibited in the presence of IFNα), compared to parent virus or other viral gene mutants. The ORF63 deletion mutant was hypersensitive to IFNα but not IFNγ, with IFNα inhibiting viral gene expression at a posttranscriptional level in ORF63 deletion virus-infected cells.

An important component of the innate response which is enhanced by the activity of IFNs is signaling by the double stranded RNA sensor PKR. Unless blocked, activated PKR phopshorylates the α-subunit of eukaryotic initiation factor 2 (eIF-2α) and effectively inhibits initiation of translation. In the closely related herpesvirus HSV-1, PKR is blocked by several viral genes, predominantly by the γ34.5 gene, which redirects the protein phosphatase 2 (PP2A) to dephosphorylate EIF2α. An increased level of phosphorylated eIF-2α was reported in VZV ORF63 deletion virus-infected cells compared to those infected with parent virus (Ambagala and Cohen 2007). In the same study, cells transiently transfected with a plasmid expressing ORF63 showed a decrease in basal levels of phosphorylated eIF-2α, demonstrating that IE63 is sufficient to inhibit this phosphorylation. Taken together, these results indicated that IE63 play a role in modulating innate immune response to VZV by interfering with IFNα induced signaling and the activity of the PKR sensor. Interestingly, Desloges et al. reported that productive VZV infection does not significantly alter levels of PKR in human melanoma cells (Desloges et al. 2005). Thus, there is uncertainty as to how VZV induces phosphorylation of eIF-2α in the absence of IE63, but it has been postulated that VZV ORF63 mediated disruption of eIF-2α phosphorylation may occur in a PKR-independent manner (Ambagala and Cohen 2007).

Thus, it appears that VZV can both impair expression of IFNα as well as inhibit antiviral signaling induced by IFNα. Additional studies will be required to identify the viral gene(s) that suppress IFNα expression. Similarly, further work examining PKR expression and function during infection with ORF63 deleted virus will be required to reveal the precise mechanism by which this viral gene product disrupts eIF-2α phosphorylation.

VZV ORF66 protein has been shown to block the induction of the IFN signaling pathway in T cells following IFNγ exposure (Schaap et al. 2005). Schaap et al. (2005) reported that following IFNγ treatment Stat-1 phosphorylation was significantly reduced in T cells infected with parental virus as compared to cells infected without a functional ORF66 (Schaap et al. 2005). The mechanism by which ORF66 modulates IFNγ signaling has yet to be reported, although it may be possible that

ORF66 functions in a manner similar to its HSV-1 related gene product Us3 which phosphorylates the IFNg receptor (Liang and Roizman 2008).

3 Interference with Antigen Presentation by VZV

During primary infection, both VZV specific $CD8^+$ and $CD4^+$ T cells develop and function in the resolution of varicella (Abendroth and Arvin 1999). Individuals with impaired cell-mediated immunity have an increased risk of more severe varicella (Gershon et al. 1997; Jura et al. 1989). A decline of cell-mediated immunity has also been associated with the increased risk of herpes zoster in the elderly, highlighting the significance of VZV-specific T cells in reactivation from latency (Levin and Hayward 1996). Evaluation of the kinetics of the VZV specific $CD4^+$ T cell response during varicella revealed that VZV specific T cells were rarely detected until varicella rash onset (Arvin et al. 1986). These observations are consistent with the hypothesis that VZV evades host recognition by T cells during the prolonged incubation period following initial infection (Abendroth and Arvin 2001), enhancing virus access to skin sites of replication, thus enabling transmission to others. A delay in the acquisition of VZV-specific T cells for >72 h was associated with persistent viremia, more lesions, and in extreme cases, potentially fatal virus dissemination (Abendroth and Arvin 2001). These observations have led to the analysis of major histocompatibility complex (MHC) class I and class II expression in VZV infected cells.

3.1 Downregulation of MHC Class I Molecules by VZV

Surface MHC class I, consisting of heterotrimers of a membrane bound heavy chain (αC), a light chain $\beta_2$microglobulin (β_2m), and antigenic peptides, are required for target cell recognition by $CD8^+$ T cells (Hansen and Bouvier 2009). Based upon the hypothesis that VZV may evade immune detection by T cells, Cohen (1998) reported that both wild-type VZV (Emily) and the vaccine virus (Oka) could downregulate cell-surface levels of MHC class I heavy chain on human fibroblasts (HFs) (Cohen 1998). Radioactive labeling experiments and western blotting for MHC class heavy chain revealed that the amount of newly synthesized and total cellular MHC class I protein was comparable in VZV-infected and uninfected cells, suggesting that VZV may interfere with the MHC class I biosynthesis pathway at a posttranslational level (Cohen 1998).

Similar findings were reported by Abendroth et al. following VZV infection of HFs and T cells (Abendroth et al. 2001a). Flow cytometric analysis revealed that a clinical VZV strain (Schenke) and the vaccine virus (Oka) selectively downregulated cell-surface expression of MHC class I on cultured HFs and also on VZV-infected T cells derived from infected SCID-hu thymus/liver mice. To identify potential mechanisms, biochemical analyses and immunofluorescent staining and confocal microscopy of VZV infected HFs were used. These approaches revealed that VZV interferes with MHC class I transport from the Golgi to the cell-surface (Abendroth et al. 2001a), suggesting the pathway by which VZV downmodulates cell-surface MHC class I expression is different from that of other α-herpesviruses such as herpes simplex virus (HSV) and bovine herpes virus (BHV). The latter express proteins (HSV ICP47 and BHV UL49.5) that interfere with the transporter associated with antigen presentation (TAP), thus blocking the transport of antigenic peptides from the cytoplasm into the endoplasmic reticulum (ER) lumen (Ahn et al. 1996; Koppers-Lalic et al. 2005, 2008; Tomazin et al. 1996; Verweij et al. 2008).

To elucidate viral gene classes responsible for MHC class I downmodulation in VZV infected cells, phosphonoacetic acid (PAA), an inhibitor of viral DNA replication and VZV late gene expression, was added to VZV infected cells and cell-surface MHC class I expression was measured by flow cytometry. In the presence of PAA, MHC class I expression at cell surfaces was not reduced compared to untreated cells, suggesting that VZV immediate early or early gene product(s), or a virion component(s) maybe involved in downregulation of cell-surface MHC class I on infected cells. Furthermore, a transient transfection approach using a variety of VZV expression constructs in HFs was then utilized to better define the viral gene responsible for this phenotype. In cells transiently transfected with a plasmid encoding the VZV ORF66 protein kinase, there was a significant decrease in cell-surface MHC class I expression, suggesting that ORF66 was sufficient to downregulate cell-surface MHC class I expression (Abendroth et al. 2001a).

Eisfeld et al. (2007) went on to examine the effects of the ORF66 protein kinase on cell surface MHC class I expression alone and in the context of VZV infection using a panel of recombinant replication defective adenoviruses and VZV expressing functional or altered ORF66 protein kinase genes tagged with green fluorescent protein (Eisfeld et al. 2007). In VZV infected MRC-5 cells, downregulation of cell-surface MHC class I required the expression of a functional ORF66 protein kinase domain. This represents a novel role for VZV ORF66 protein kinase in immune evasion. The expression of a functional ORF66 kinase impaired MHC class I maturation in the absence of influencing MHC class I synthesis, degradation or association with β2m, and suggested the kinase induced a delay in the processing of MHC class I trimeric complexes through the Golgi to an endoglycosidase H resistant form. Using a combination of immunoprecipitation experiments and immunofluorescent staining and confocal microscopy there was little evidence suggesting a close or direct association of ORF66 with folded MHC class I molecules or MHC class I heavy chains (Eisfeld et al. 2007). However, VZV lacking the kinase activity still downmodulated MHC class I expression to a lesser extent, suggesting additional proteins may be involved. In the same study, the role of VZV ORF9a protein, which is analogous to the BHV UL49.5 gene which modulates MHC class I, was also assessed. HEK293T cells transiently transfected with an ORF9a expression construct showed no decrease in cell-surface MHC class I, indicating that the MHC class I downmodulatory function of ORF9a is not conserved between BHV and VZV. Additional detail on the function of the ORF66 protein kinase is provided in (Erazo et al. 2009).

VZV specific cell-mediated immunity would be well established in the skin during the development of varicella skin lesions and ORF66 encoded MHC class I downregulation may play a role in allowing skin cells to transiently evade $CD8^+$ T cell surveillance, facilitating local virus replication and transmission during the first few days of cutaneous lesion formation. The assessment of MHC class I expression on human skin cells following inoculation of SCID-hu skin mice or of human skin explants with an ORF66 mutant virus would assist in elucidating the in vivo immunomodulatory roles of ORF66. It also remains to be determined whether ORF66 causes the downregulation of MHC class I observed on VZV infected T cells (Abendroth et al. 2001a), but it is interesting to note that ORF66 mutant viruses are impaired for growth in T cells and this has been attributed to an increased susceptibility of infected T cells to apoptosis (Schaap et al. 2005). Thus, ORF66 may employ multiple mechanisms during VZV infection to enable immune evasion and promote T cell survival to enable virus spread and transfer to the skin.

VZV and other viruses that modulate cell-surface MHC class I expression may evade $CD8^+$ T cell recognition; however, the overall reduction of cell-surface MHC class I may make these infected cells more sensitive to NK cell mediated killing (Farrell and Davis-Poynter 1998; Tortorella et al. 2000). In this respect, VZV infected HFs are susceptible to NK cell mediated lysis (Bowden et al. 1985; Ihara et al. 1984; Ito et al. 1996), although any contribution of downregulated MHC class I to this killing remains to be examined. Prior studies on human cytomegalovirus (HCMV) and murine cytomegalovirus (MCMV) have revealed that these herpesviruses have evolved a variety of mechanisms to combat NK cell recognitions as well as $CD8^+$ T cell recognition and killing (Miller-Kittrell and Sparer 2009; Wilkinson et al. 2008). In contrast, to date VZV has not been reported to encode an MHC class I homolog, although it remains possible that VZV may subvert NK cell-mediated killing either via expression of an as yet unidentified MHC class I homolog or downmodulation of selective MHC class I alleles. Human immunodeficiency virus (HIV) has been previously shown to selectively downregulate the cell-surface expression of specific MHC class I alleles and not others, preventing NK mediated killing (Cohen et al. 1999). To date, there have been no reports examining whether VZV may also cause an allele specific downmodulation of MHC class I molecules which may enable VZV to evade NK cell mediated killing.

3.2 VZV Interference with MHC Class II Expression

Unlike MHC class I molecules, constitutive expression of MHC class II is restricted to B cells, monocytes, DCs, and thymic epithelium. However, IFNγ treatment can stimulate MHC class II expression by many cell types, including HFs (Collins et al. 1984; Pober et al. 1983). The importance of $CD4^+$ T cells for resolution of varicella lead to the postulation that VZV may encode an immunomodulatory function that allows the virus to inhibit the induction of MHC class II expression by IFNγ (Abendroth et al. 2000). VZV strain Schenke infected HFs were treated with IFNγ to stimulate MHC class II expression and then analyzed by flow cytometry. Upregulation of cell-surface MHC class II expression was impaired in VZV infected cells compared with mock infected counterparts. In contrast, cells that were treated with IFNγ prior to VZV infection expressed comparable levels of MHC class II to mock infected cells treated with IFNγ. Taken together, these results demonstrated that VZV inhibited IFNγ-mediated upregulation of MHC class II but could not downregulate MHC class II already induced by IFNγ. Northern blot and in situ hybridization for MHC class II α-chain transcripts in infected cells treated with IFNγ revealed that VZV suppressed upregulation of MHC class II at the level of mRNA transcription (Abendroth et al. 2000). VZV infection inhibited the expression of Stat1α and Jak2 proteins, but had no effect on Jak1. Furthermore, VZV infection inhibited transcription of the interferon regulatory factor (IRF-1) and class II transactivator (CIITA). Collectively these data demonstrated that VZV encodes an immunomodulatory function which directly interferes with the IFNγ signal transduction via the Jak/Stat pathway and enabled the virus to inhibit IFNγ induction of cell-surface MHC class II expression (Abendroth et al. 2000). The significance of these in vitro based studies was further confirmed by examination of varicella and herpes zoster skin biopsies for MHC class II and VZV RNA synthesis by in situ hybridization. These experiments demonstrated that during natural cutaneous infection, dermal and epidermal cells infected with VZV do not express MHC class II transcripts in vivo, whereas MHC class II transcripts were readily detected in the uninfected bystander uninfected cells (Abendroth et al. 2000).

More recently, analyses of VZV encoded MHC class II modulation was extended to VZV infected human keratinocytes. Black et al. (2009) demonstrated that immortalized human keratinocytes infected with cell-free VZV virus (Oka) and subsequently treated with IFNγ failed to upregulate MHC class II molecules to the same level as the IFNγ treated mock infected keratinocytes (Black et al. 2009). This supports the notion that despite attenuation, the vaccine virus still retains its capacity to interfere with IFNγ induced MHC class II upregulation, as is the case for downregulation of MHC class I (Abendroth et al. 2001a; Cohen 1998). Significantly, VZV infected keratinocytes treated with IFNγ were impaired in their ability to stimulate antigen specific $CD4^+$ and $CD8^+$ T cells in vitro compared with IFN treated uninfected keratinocytes (Black et al. 2009). The mechanisms involved in modulating the IFNγ upregulation of MHC class II in keratinocytes may be similar to that reported in VZV infected HFs (Abendroth et al. 2000), although this remains to be determined. Furthermore, the viral gene product(s) responsible for inhibiting IFNγ induced MHC class II in keratinocytes or HFs has yet to be elucidated. It has been shown that ORF66 blocks the induction of IFN signaling in human T cells following IFN treatment (Schaap et al. 2005). Given that HFs and keratinocytes are important cell types for viral replication in the skin (Nikkels et al. 1995), the capacity of VZV to inhibit IFNγ induced upregulation of MHC class II in these cells is likely to provide the virus with an important strategy for evasion of $CD4^+$ T cell recognition during both varicella and herpes zoster.

4 VZV Interference with the NFκB Pathway and Intercellular Adhesion Molecule 1 Expression

NFκB is a potent transcription factor that is normally present within cells in an inactivate state in the cytoplasm due to its association with the inhibitory protein IκBα. Degradation of IκBα results in translocation of NFκB proteins to the nucleus where they stimulate expression of a wide range of genes, including those involved in the immune response (Ghosh and Hayden 2008; Hayden and Ghosh 2008). In a microarray based study of VZV infected HFs, Jones and Arvin (2005) reported that many NFκB responsive genes were downregulated following infection (Jones and Arvin 2005). In a subsequent report, the same authors performed an analysis of the NFκB activation pathway in VZV infected HFs and revealed that after a transient nuclear localization VZV interferes with this pathway by sequestering NFκB proteins (p50 and p65) in the cytoplasm of VZV infected cells (Jones and Arvin 2006). The cytoplasmic sequestration of p50 and p65 required VZV protein expression, as UV-inactivated virus did not inhibit the nuclear translocation of these NFκB proteins. In addition, while IκBα is normally degraded to enable translocation of NFκB proteins to the nucleus, VZV infection of HFs inhibited this degradation. The inhibition of nuclear import of NFκB proteins was confirmed in vivo, where in epidermal cells in skin xenografts of SCID-hu mice infected with VZV, p50 and p65 remained in the cytoplasm of VZV infected cells, yet neighboring uninfected epidermal cells displayed normal nuclear accumulation of these NFκB proteins (Jones and Arvin 2006). This finding is consistent with other work from the Arvin group demonstrating that VZV infected skin cells in vivo lacked IFNα expression and Stat-1 remained localized to the cytoplasm, whereas surrounding uninfected bystander cells expressed IFNα and Stat-1 was phosphorylated and translocated into the nucleus (Ku et al. 2004). Given that NFκB is a major inducer of IFNα transcription, viral modulation of the NFκB signaling pathway would likely limit IFNα production within VZV infected cells. The VZV gene product(s) responsible for the modulation of the NFκB pathway in HFs or any other cell-type is yet to be elucidated. Given the pivotal role the NFκB signaling pathway plays in both the innate and adaptive arms of the immune response, it is likely that regulating the actions of this transcription factor may be central in other cell-types infected with VZV such as DCs.

In the presence of proinflammatory cytokines such as IFNγ and tumor necrosis factor (TNF), keratinocytes can be induced to express not only MHC class II molecules but also surface intercellular adhesion molecule (ICAM-1) which is the ligand for leukocyte function antigen (LFA-1) expressing T cells (Rothlein et al. 1986). Nikkels et al. (2004) performed an immunohistochemical analysis for a variety of different immune cell markers and cytokines on frozen sections from herpes zoster skin biopsies. Despite increased expression of IFNγ, TNFα, and IL-6 in VZV infected skin there was a decrease in expression of both ICAM-1 and MHC class II in VZV infected keratinocytes within the center of the herpes zoster lesions (Nikkels et al. 2004). This was the first demonstration that VZV could modulate ICAM-1 expression. Black et al. (2009) also examined ICAM-1 expression on uninfected keratinocytes in comparison to VZV infected keratinocytes in vitro. Similar to their assessment of cell-surface MHC class II expression, human keratinocytes infected with VZV inhibited IFNγ mediated upregulation of ICAM-1. These reports demonstrate that VZV encodes an immunoevasive strategy targeting the expression of ICAM-1 in both keratinocytes in vitro and in vivo and identifies an additional mechanism by which VZV may evade T cell clearance.

The molecular basis of VZV mediated ICAM-1 inhibition was assessed in infected melanoma cells (MeWo) and human MRC5 cells (El Mjiyad et al. 2007). Consistent with the study by Jones and Arvin (2006), nuclear translocation of p50 was strongly decreased by VZV infection, although inhibition of p65 translocation was less marked as nuclear translocation of this subunit was still observed. Interestingly, using a coimmunoprecipitation approach, VZV infection induced the nuclear accumulation of the NFκB inhibitor p100. Significantly, in addition to the demonstration of inhibition of ICAM-1 mRNA synthesis, analysis of TNFα treated VZV infected cells using an electrophoretic mobility shift assay (EMSA) revealed that NFκB subunits present in VZV infected cells were unable to bind to ICAM-1 or IL-8 promoters, thus providing a mechanistic basis for the inhibition of ICAM-1 expression (and possibly other genes) by VZV mediated interference with NFκB activation.

5 Impact of VZV on Human dendritic cells

DCs are bone-marrow derived potent antigen presenting cells (APCs) that are located at many sites, including the skin, blood, lymph, and mucosal surfaces (Banchereau et al. 2000; Banchereau and Steinman 1998; Klagge and Schneider-Schaulies 1999). DCs uptake and process antigen in the periphery and transport viral antigens to T-cell rich areas of the lymph nodes, where they display MHC–peptide complexes together with costimulatory molecules. This results in the activation of naïve and resting antigen-specific T cells and effector T cell differentiation (Banchereau et al. 2000). The hypothesis that DCs of the respiratory mucosa may be the first cell type to encounter VZV during primary infection led to studies to investigate VZV–DC interactions. Human immature monocyte derived DCs were shown to be fully permissive to a productive VZV infection as immediate early (IE), early (E), and late (L) viral gene products are made in $CD1a^+$ DCs and infectious virus can be recovered (Fig. 2). VZV infected immature DCs showed no significant decrease in cell viability or evidence of apoptosis and did not exhibit altered cell surface levels of the immune molecules MHC class I, MHC class II, CD86, CD40, or CD1a. Significantly, when autologous T cells were incubated with VZV infected DCs, VZV antigens were readily detected in $CD3^+$ T cells and infectious virus was recovered from these cells (Abendroth et al. 2001b). This work provided the first evidence that immature DCs were permissive to VZV and that DC infection could lead to virus transmission to T cells, supporting the hypothesis that DC may mediate virus dissemination in the initial viremia following infection (Fig. 1).

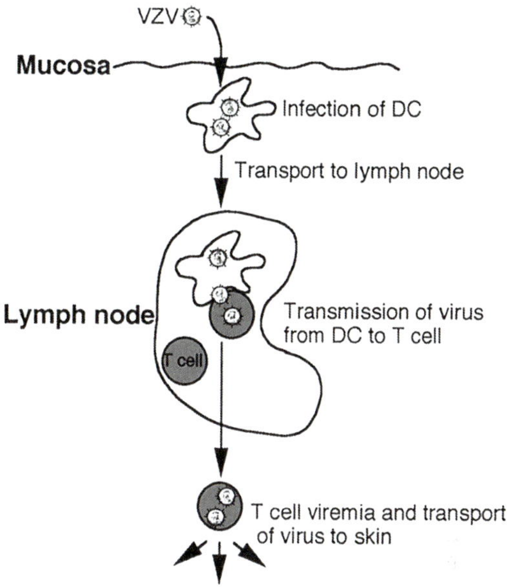

Fig. 1 Proposed model of VZV transport from mucosal sites to the T cells in lymph nodes during primary VZV infection. Copyright © American Society for Microbiology, J Virol, July 2001, pp. 6183–6192, vol. 75, no. 13. DOI: 10.1128/JVI.75.13.6183-6192.2001

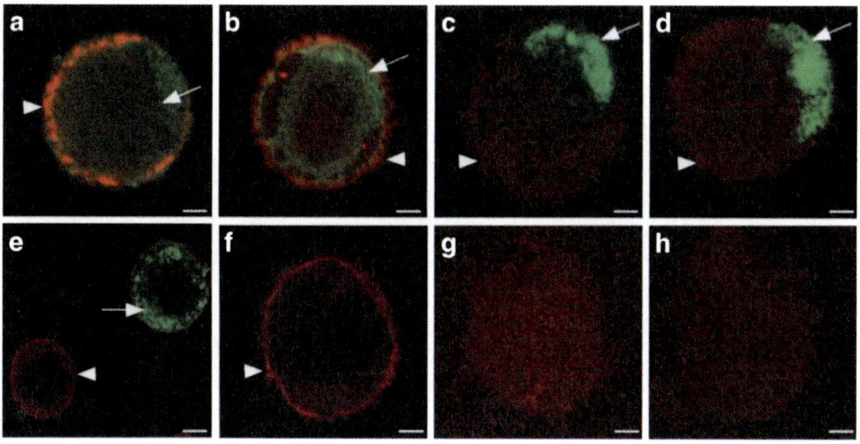

Fig. 2 Varicella zoster virus infection of human immature DCs in vitro. Immunofluorescent staining of immature DCs at 48 h postinfection (**a–e**, **g**) and mock infected (**f** and **h**) for CD1a (*red staining*) and VZV antigens ORF62 (**a** and **f**), ORF4 (**b**), ORF29 (**c**), ORF61 (**d**), and glycoprotein C (**e**). Copyright © American Society for Microbiology, J Virol, July 2001, pp. 6183–6192, vol. 75, no. 13. DOI: 10.1128/JVI.75.13.6183-6192.2001

As immature DC become mature their ability to stimulate T cells increases as a result of upregulation of MHC I, MHC II, CD86, CD83, and CD80 (Banchereau and Steinman 1998). Maturation of DC leading to migration and increased T cell stimulation is essential for initiation of the immune response (Banchereau et al. 2000; Banchereau and Steinman 1998), and so may represent an area targeted for disruption by VZV. While the impact of VZV infection on the maturation of DCs has not yet been published, this is currently an area of active investigation. As mature DCs are potent APCs essential for initiating successful antiviral immune responses (Banchereau and Steinman 1998), they would also serve as an ideal target for viruses such as VZV seeking to evade or delay the immune response by disrupting their immune function. In this respect, Morrow et al. showed that VZV productively infects mature monocyte derived DC, and in doing so, impairs their ability to function properly by the selective downregulation of functionally important cell-surface immune molecules MHC class I, CD80, CD83 and CD86 (Morrow et al. 2003). Importantly, the same study demonstrated that VZV infection of mature DCs significantly reduced their capacity to stimulate the proliferation of allogeneic T cells. Although the precise molecular mechanisms responsible for VZV mediated downregulation of cell-surface MHC class I, CD80, CD83, and CD86 remain to be elucidated, current evidence suggests that direct virus infection and not soluble factors are required for this phenotype (Morrow et al. 2003). The identity of the VZV gene(s) responsible for the disruption of immune molecules during productive infection of mature DCs has yet to be determined.

Hu and Cohen (2005) utilized a VZV ORF47 deletion virus to show that ORF47 was critical for virus replication in immature DC, but not in mature DC. VZV ORF47 was however, required for the full replicative cycle and for transmission of virus from these cells to other cells. This study also demonstrated that both immature and mature DCs infected with VZV had reduced surface Fas expression, which may protect infected cells from apoptosis, a finding consistent with the previous report of a lack of apoptosis in VZV infected immature DCs (Abendroth et al. 2001b).

The skin is a critical site of VZV infection during both varicella and herpes zoster, and skin DCs play a pivotal role in the induction of antiviral immunity, so there is good reason to study infection and modulation of DCs in human skin during natural cutaneous VZV infection. The two subsets of DC that are normally present in human skin and which may therefore be involved in the pathogenesis of VZV infection are the Langerhans cells (LC) of the epidermis and dermal DC (DDC) (Valladeau and Saeland 2005). These $CD1a^+$ DC are found in an immature state in uninfected skin and following antigen capture have the capacity to migrate from the periphery to the lymph nodes, where they interact with T cells to initiate an immune response (Valladeau and Saeland 2005). There are also other types of DC, such as the blood derived myeloid DC (MDC) and plasmacytoid DC (PDC) that may also play a role in the pathogenesis of VZV infection. Of particular interest is the importance of PDC in innate antiviral immune responses due to their ability to recruit to sites of inflammation and secrete high levels of IFNα (Liu 2005; Siegal et al. 1999). PDC also participate in adaptive immune responses through their secretion of cytokines and chemokines that promote activation of effector cells, including NK, NKT, B, and T cells, and also antigen presentation to T cells (Colonna et al. 2004; Salio et al. 2004; Zhang and Wang 2005).

There have been several studies examining the impact of VZV infection on different DC subsets. It has been reported that the frequency of $CD1a^+$ DCs is reduced in VZV infected skin epidermis compared to uninfected skin, demonstrating an alteration in the distribution of DC in response to VZV infection of the skin (Nikkels et al. 2004). It has also been reported that PDCs infiltrate into the dermis of varicella lesions, suggesting that these cells contribute to the immune control of VZV infection (Gerlini et al. 2006).

Recent work from the Abendroth group has examined in more detail DC subsets in skin biopsies from varicella and herpes zoster cases (Huch et al. 2010). Immunostaining and microscopy analysis of VZV infected lesions of both varicella and herpes zoster showed that in comparison to normal uninfected skin, the proportion of cells expressing DC-SIGN (DDC marker) or DC-LAMP and CD83 (mature DC markers) were not significantly altered. In contrast, the frequency of LCs was significantly decreased in VZV infected skin, concomitant with a striking influx of PDC into VZV infected skin. The authors suggested that this loss of LC from the skin was most likely to be a consequence of migration of these cells to distal sites, such as lymph nodes. Within infected skin the LCs and PDC were closely associated with VZV antigen positive cells, and a small proportion of both LC and PDC showed evidence of VZV infection. Despite only sporadic detection of VZV infected PDC and LC during natural cutaneous infection this finding is thought to be important when considered in the context of the frequency of infection of other cell types which have been shown to play crucial roles in the course of natural VZV infection. In this respect, the proportion of VZV infected lymphocytes in peripheral blood during natural VZV infection is very small, with estimates in the range of 1 in 100,000 PBMCs from healthy varicella patients becoming infected, yet the role of peripheral blood T cells in transporting virus to distal sites is regarded as a significant event in VZV pathogenesis (Koropchak et al. 1989, 1991).

In an extension of these in vivo observations, PDC isolated from human blood and LC derived from the MUTZ-3 cell line were shown to be permissive to VZV infection (Huch et al. 2010). Interestingly, significant induction of IFNα by PDC did not occur following VZV infection and infected PDC cultures remained refractory to IFNα induction even when stimulated with a TLR9 agonist which stimulates IFNα production by PDC. Additional work will be required to define the mechanism of VZV encoded modulation of IFNα production by PDC and identification of any viral gene(s) that encode this function. Furthermore, it remains to be determined whether infected PDC or LC are impaired in other functions such as antigen presentation.

In summary, definition of changes that occur to the distribution of multiple DC subsets in the skin of individuals suffering from primary and recurrent VZV disease, with the identification of LC and PDC as subsets most affected during infection, implicates these DCs as playing important roles in VZV pathogenesis. Furthermore, the capacity of VZV to infect and impair function of different DC subsets highlights VZV mediated immune control of these cells.

6 Concluding Remarks and Future Perspectives

Modulation of immune function has emerged as a powerful strategy by which the virus is likely to evade or delay host defenses during critical stages of infection. This chapter has highlighted the plethora of VZV encoded immune evasion strategies that are particularly relevant to those who suffer from either varicella or herpes zoster, and ongoing investigations will better define the relationship between VZV and the host immune system. A significant outcome of elucidating mechanism and identifying viral genes that modulate host immune surveillance and infection will be for development of a better "second generation" vaccine against VZV disease. This vaccine should consist of specifically targeted modifications, such that it replicates at the inoculation site without causing a lesion and lacks viral genes that permit evasion of host defense mechanisms or infectivity for DCs, yet induces immunity as effectively as natural infection.

The study of VZV encoded immune modulation poses several significant challenges. Firstly, the high species specificity and the lack animal models to study VZV infection in the context of a fully in intact immune response limits the capacity to study VZV control of immune function. Secondly, the study of naturally infected individuals is complicated by the difficulty in obtaining tissues from different anatomical sites and/or low levels of infection, e.g., T cells in the blood. For these reasons, experimental models of infection using primary cultured human cell types or tissues implanted into SCID mice will continue to play a critical role in the analysis of VZV mediated immunomodulation and in driving analysis of naturally infected cell and tissue samples.

Acknowledgments AA and BS were supported by NHMRC grant 457356. PRK acknowledges support for this work by Public Health Service NIH grants NS064022 and EY08098, and funds from the Research to Prevent Blindness Inc. and the Eye and Ear Institute of Pittsburgh.

References

Abendroth A, Arvin A (1999) Varicella-zoster virus immune evasion. Immunol Rev 168:143–156

Abendroth A, Arvin AM (2001) Immune evasion as a pathogenic mechanism of varicella zoster virus. Semin Immunol 13:27–39

Abendroth A, Slobedman B, Lee E, Mellins E, Wallace M, Arvin AM (2000) Modulation of major histocompatibility class II protein expression by varicella-zoster virus. J Virol 74:1900–1907

Abendroth A, Lin I, Slobedman B, Ploegh H, Arvin AM (2001a) Varicella-zoster virus retains major histocompatibility complex class I proteins in the Golgi compartment of infected cells. J Virol 75:4878–4888

Abendroth A, Morrow G, Cunningham AL, Slobedman B (2001b) Varicella-zoster virus infection of human dendritic cells and transmission to T cells: implications for virus dissemination in the host. J Virol 75:6183–6192

Ahn K, Meyer TH, Uebel S, Sempe P, Djaballah H, Yang Y, Peterson PA, Fruh K, Tampe R (1996) Molecular mechanism and species specificity of TAP inhibition by herpes simplex virus ICP47. EMBO J 15:3247–3255

Ambagala AP, Cohen JI (2007) Varicella-zoster virus IE63, a major viral latency protein, is required to inhibit the alpha interferon-induced antiviral response. J Virol 81:7844–7851
Arvin A (2001) Varicella zoster virus. In: Knipe DaHP (ed) Fields virology, vol 2, 4th edn. Lippincott Williams and Wilkins, Philadelphia, pp 2731–2767
Arvin AM, Schmidt NJ, Cantell K, Merigan TC (1982) Alpha interferon administration to infants with congenital rubella. Antimicrob Agents Chemother 21:259–261
Arvin AM, Koropchak CM, Williams BR, Grumet FC, Foung SK (1986) Early immune response in healthy and immunocompromised subjects with primary varicella-zoster virus infection. J Infect Dis 154:422–429
Arvin AM, Moffat JF, Redman R (1996) Varicella-zoster virus: aspects of pathogenesis and host response to natural infection and varicella vaccine. Adv Virus Res 46:263–309
Balachandra K, Thawaranantha D, Ayuthaya PI, Bhumisawasdi J, Shiraki K, Yamanishi K (1994) Effects of human alpha, beta and gamma interferons on varicella zoster virus in vitro. Southeast Asian J Trop Med Public Health 25:252–257
Banchereau J, Steinman RM (1998) Dendritic cells and the control of immunity. Nature 392: 245–252
Banchereau J, Briere F, Caux C, Davoust J, Lebecque S, Liu YJ, Pulendran B, Palucka K (2000) Immunobiology of dendritic cells. Annu Rev Immunol 18:767–811
Black AP, Jones L, Malavige GN, Ogg GS (2009) Immune evasion during varicella zoster virus infection of keratinocytes. Clin Exp Dermatol 34(8):941–944
Bowden RA, Levin MJ, Giller RH, Tubergen DG, Hayward AR (1985) Lysis of varicella zoster virus infected cells by lymphocytes from normal humans and immunosuppressed pediatric leukaemic patients. Clin Exp Immunol 60:387–395
Cohen JI (1998) Infection of cells with varicella-zoster virus down-regulates surface expression of class I major histocompatibility complex antigens. J Infect Dis 177:1390–1393
Cohen GB, Gandhi RT, Davis DM, Mandelboim O, Chen BK, Strominger JL, Baltimore D (1999) The selective downregulation of class I major histocompatibility complex proteins by HIV-1 protects HIV-infected cells from NK cells. Immunity 10:661–671
Colamonici OR, Domanski P (1993) Identification of a novel subunit of the type I interferon receptor localized to human chromosome 21. J Biol Chem 268:10895–10899
Collins T, Korman AJ, Wake CT, Boss JM, Kappes DJ, Fiers W, Ault KA, Gimbrone MA Jr, Strominger JL, Pober JS (1984) Immune interferon activates multiple class II major histocompatibility complex genes and the associated invariant chain gene in human endothelial cells and dermal fibroblasts. Proc Natl Acad Sci USA 81:4917–4921
Colonna M, Trinchieri G, Liu YJ (2004) Plasmacytoid dendritic cells in immunity. Nat Immunol 5:1219–1226
Desloges N, Rahaus M, Wolff MH (2005) Role of the protein kinase PKR in the inhibition of varicella-zoster virus replication by beta interferon and gamma interferon. J Gen Virol 86:1–6
Eisfeld AJ, Yee MB, Erazo A, Abendroth A, Kinchington PR (2007) Downregulation of class I major histocompatibility complex surface expression by varicella-zoster virus involves open reading frame 66 protein kinase-dependent and -independent mechanisms. J Virol 81:9034–9049
El Mjiyad N, Bontems S, Gloire G, Horion J, Vandevenne P, Dejardin E, Piette J, Sadzot-Delvaux C (2007) Varicella-zoster virus modulates NF-kappaB recruitment on selected cellular promoters. J Virol 81:13092–13104
Erazo A, Kinchington PR (2009) Varicella-zoster virus open reading frame 66 protein kinase and its relationship to alphaherpesvirus US3 kinases. Curr Top Microbiol Immunol, DOI 10.1007/82_2009_7
Farrell HE, Davis-Poynter NJ (1998) From sabotage to camouflage: viral evasion of cytotoxic T lymphocyte and natural killer cell-mediated immunity. Semin Cell Dev Biol 9:369–378
Garcia-Sastre A, Biron CA (2006) Type 1 interferons and the virus–host relationship: a lesson in detente. Science 312:879–882
Gerlini G, Mariotti G, Bianchi B, Pimpinelli N (2006) Massive recruitment of type I interferon producing plasmacytoid dendritic cells in varicella skin lesions. J Invest Dermatol 126:507–509

Gershon AA, Mervish N, LaRussa P, Steinberg S, Lo SH, Hodes D, Fikrig S, Bonagura V, Bakshi S (1997) Varicella-zoster virus infection in children with underlying human immunodeficiency virus infection. J Infect Dis 176:1496–1500

Ghosh S, Hayden MS (2008) New regulators of NF-kappaB in inflammation. Nat Rev Immunol 8:837–848

Grose C (1981) Variation on a theme by Fenner: the pathogenesis of chickenpox. Pediatrics 68:735–737

Haller O, Kochs G, Weber F (2006) The interferon response circuit: induction and suppression by pathogenic viruses. Virology 344:119–130

Hansen TH, Bouvier M (2009) MHC class I antigen presentation: learning from viral evasion strategies. Nat Rev Immunol 9:503–513

Hayden MS, Ghosh S (2008) Shared principles in NF-kappaB signaling. Cell 132:344–362

Hu H, Cohen JI (2005) Varicella-zoster virus open reading frame 47 (ORF47) protein is critical for virus replication in dendritic cells and for spread to other cells. Virology 337:304–311

Huch J, Cunningham A, Arvin A, Nasr N, Santegoets S, Slobedman E, Slobedman B, Abendroth A (2010) Impact of varicella zoster virus on dendritic cell subsets in human skin during natural infection. J Virol 84:4060–4072

Ihara T, Starr SE, Ito M, Douglas SD, Arbeter AM (1984) Human polymorphonuclear leukocyte-mediated cytotoxicity against varicella-zoster virus-infected fibroblasts. J Virol 51: 110–116

Ito M, Watanabe M, Kamiya H, Sakurai M (1996) Inhibition of natural killer (NK) cell activity against varicella-zoster virus (VZV)-infected fibroblasts and lymphocyte activation in response to VZV antigen by nitric oxide-releasing agents. Clin Exp Immunol 106:40–44

Jones JO, Arvin AM (2005) Viral and cellular gene transcription in fibroblasts infected with small plaque mutants of varicella-zoster virus. Antiviral Res 68:56–65

Jones JO, Arvin AM (2006) Inhibition of the NF-kappaB pathway by varicella-zoster virus in vitro and in human epidermal cells in vivo. J Virol 80:5113–5124

Jura E, Chadwick EG, Josephs SH, Steinberg SP, Yogev R, Gershon AA, Krasinski KM, Borkowsky W (1989) Varicella-zoster virus infections in children infected with human immunodeficiency virus. Pediatr Infect Dis J 8:586–590

Katze MG, He Y, Gale M Jr (2002) Viruses and interferon: a fight for supremacy. Nat Rev Immunol 2:675–687

Klagge IM, Schneider-Schaulies S (1999) Virus interactions with dendritic cells. J Gen Virol 80:823–833

Koppers-Lalic D, Reits EA, Ressing ME, Lipinska AD, Abele R, Koch J, Marcondes Rezende M, Admiraal P, van Leeuwen D, Bienkowska-Szewczyk K, Mettenleiter TC, Rijsewijk FA, Tampe R, Neefjes J, Wiertz EJ (2005) Varicelloviruses avoid T cell recognition by UL49.5-mediated inactivation of the transporter associated with antigen processing. Proc Natl Acad Sci USA 102:5144–5149

Koppers-Lalic D, Verweij MC, Lipinska AD, Wang Y, Quinten E, Reits EA, Koch J, Loch S, Marcondes Rezende M, Daus F, Bienkowska-Szewczyk K, Osterrieder N, Mettenleiter TC, Heemskerk MH, Tampe R, Neefjes JJ, Chowdhury SI, Ressing ME, Rijsewijk FA, Wiertz EJ (2008) Varicellovirus UL 49.5 proteins differentially affect the function of the transporter associated with antigen processing, TAP. PLoS Pathog 4:e1000080

Koropchak CM, Solem SM, Diaz PS, Arvin AM (1989) Investigation of varicella-zoster virus infection of lymphocytes by in situ hybridization. J Virol 63:2392–2395

Koropchak CM, Graham G, Palmer J, Winsberg M, Ting SF, Wallace M, Prober CG, Arvin AM (1991) Investigation of varicella-zoster virus infection by polymerase chain reaction in the immunocompetent host with acute varicella. J Infect Dis 163:1016–1022

Ku CC, Padilla JA, Grose C, Butcher EC, Arvin AM (2002) Tropism of varicella-zoster virus for human tonsillar CD4(+) T lymphocytes that express activation, memory, and skin homing markers. J Virol 76:11425–11433

Ku CC, Zerboni L, Ito H, Graham BS, Wallace M, Arvin AM (2004) Varicella-zoster virus transfer to skin by T Cells and modulation of viral replication by epidermal cell interferon-alpha. J Exp Med 200:917–925, Epub 2004 Sep 27

Levin MJ, Hayward AR (1996) The varicella vaccine. Prevention of herpes zoster. Infect Dis Clin North Am 10:657–675

Liang L, Roizman B (2008) Expression of gamma interferon-dependent genes is blocked independently by virion host shutoff RNase and by US3 protein kinase. J Virol 82:4688–4696

Liu YJ (2005) IPC: professional type 1 interferon-producing cells and plasmacytoid dendritic cell precursors. Annu Rev Immunol 23:275–306

Miller-Kittrell M, Sparer TE (2009) Feeling manipulated: cytomegalovirus immune manipulation. Virol J 6:4

Morrow G, Slobedman B, Cunningham AL, Abendroth A (2003) Varicella-zoster virus productively infects mature dendritic cells and alters their immune function. J Virol 77:4950–4959

Muller U, Steinhoff U, Reis LF, Hemmi S, Pavlovic J, Zinkernagel RM, Aguet M (1994) Functional role of type I and type II interferons in antiviral defense. Science 264:1918–1921

Nikkels AF, Debrus S, Sadzot-Delvaux C, Piette J, Rentier B, Pierard GE (1995) Localization of varicella-zoster virus nucleic acids and proteins in human skin. Neurology 45:S47–S49

Nikkels AF, Sadzot-Delvaux C, Pierard GE (2004) Absence of intercellular adhesion molecule 1 expression in varicella zoster virus-infected keratinocytes during herpes zoster: another immune evasion strategy? Am J Dermatopathol 26:27–32

Pober JS, Gimbrone MA Jr, Cotran RS, Reiss CS, Burakoff SJ, Fiers W, Ault KA (1983) Ia expression by vascular endothelium is inducible by activated T cells and by human gamma interferon. J Exp Med 157:1339–1353

Rothlein R, Dustin ML, Marlin SD, Springer TA (1986) A human intercellular adhesion molecule (ICAM-1) distinct from LFA-1. J Immunol 137:1270–1274

Sadler AJ, Williams BR (2008) Interferon-inducible antiviral effectors. Nat Rev Immunol 8:559–568

Salio M, Palmowski MJ, Atzberger A, Hermans IF, Cerundolo V (2004) CpG-matured murine plasmacytoid dendritic cells are capable of in vivo priming of functional CD8 T cell responses to endogenous but not exogenous antigens. J Exp Med 199:567–579

Schaap A, Fortin JF, Sommer M, Zerboni L, Stamatis S, Ku CC, Nolan GP, Arvin AM (2005) T-cell tropism and the role of ORF66 protein in pathogenesis of varicella-zoster virus infection. J Virol 79:12921–12933

Siegal FP, Kadowaki N, Shodell M, Fitzgerald-Bocarsly PA, Shah K, Ho S, Antonenko S, Liu YJ (1999) The nature of the principal type 1 interferon-producing cells in human blood. Science 284:1835–1837

Tomazin R, Hill AB, Jugovic P, York I, van Endert P, Ploegh HL, Andrews DW, Johnson DC (1996) Stable binding of the herpes simplex virus ICP47 protein to the peptide binding site of TAP. EMBO J 15:3256–3266

Tortorella D, Gewurz BE, Furman MH, Schust DJ, Ploegh HL (2000) Viral subversion of the immune system. Annu Rev Immunol 18:861–926

Valladeau J, Saeland S (2005) Cutaneous dendritic cells. Semin Immunol 17:273–283

Verweij MC, Koppers-Lalic D, Loch S, Klauschies F, de la Salle H, Quinten E, Lehner PJ, Mulder A, Knittler MR, Tampe R, Koch J, Ressing ME, Wiertz EJ (2008) The varicellovirus UL49.5 protein blocks the transporter associated with antigen processing (TAP) by inhibiting essential conformational transitions in the 6+6 transmembrane TAP core complex. J Immunol 181:4894–4907

Wilkinson GW, Tomasec P, Stanton RJ, Armstrong M, Prod'homme V, Aicheler R, McSharry BP, Rickards CR, Cochrane D, Llewellyn-Lacey S, Wang EC, Griffin CA, Davison AJ (2008) Modulation of natural killer cells by human cytomegalovirus. J Clin Virol 41:206–212

Zhang Z, Wang FS (2005) Plasmacytoid dendritic cells act as the most competent cell type in linking antiviral innate and adaptive immune responses. Cell Mol Immunol 2:411–417

VZV Infection of Keratinocytes: Production of Cell-Free Infectious Virions In Vivo

Michael D. Gershon and Anne A. Gershon

Contents

Abstract Varicella-zoster virus (VZV) is the cause of varicella (chickenpox) and zoster (shingles). Varicella is a primary infection that spreads rapidly in epidemics while zoster is a secondary infection that occurs sporadically as a result of the reactivation of previously acquired VZV. Reactivation is made possible by the establishment of latency during the initial episode of varicella. The signature lesions of both varicella and zoster are cutaneous vesicles, which are filled with a clear fluid that is rich in infectious viral particles. It has been postulated that the skin is the critical organ in which both host-to-host transmission of VZV and the infection of neurons to establish latency occur. This hypothesis is built on evidence that the large cation-independent mannose 6-phosphate receptor (MPR^{ci}) interacts with VZV in virtually all infected cells, except those of the suprabasal epidermis, in a way that prevents the release of infectious viral particles. Specifically, the virus is diverted in an MPR^{ci}-dependent manner from the secretory pathway to late endosomes where VZV is degraded. Because nonepidermal cells are thus prevented from releasing infectious VZV, a slow process, possibly involving fusion of infected cells with their neighbors, becomes the means by which VZV is disseminated. In the epidermis, however, the maturation of keratinocytes to give rise to corneocytes in the suprabasal epidermis is associated uniquely with a downregulation of the MPR^{ci}. As a result, the diversion of VZV to late endosomes

M.D. Gershon (✉) and A.A. Gershon
Department of Pathology and Cell Biology, College of Physicians and Surgeons, Columbia University, 630 West 168th Street, New York, NY 10032, USA
Department of Pediatrics, College of Physicians and Surgeons, Columbia University, 630 West 168th Street, New York, NY 10032, USA
email: mdg4@columbia.edu; aag1@columbia.edu

A.M. Arvin et al. (eds.), *Varicella-zoster Virus*,
Current Topics in Microbiology and Immunology 342, DOI 10.1007/82_2010_13

does not occur in the suprabasal epidermis where vesicular lesions occur. The formation of the waterproof, chemically resistant barrier of the epidermis, however, requires that constitutive secretion outlast the downregulation of the endosomal pathway. Infectious VZV is therefore secreted by default, accounting for the presence of infectious virions in vesicular fluid. Sloughing of corneocytes, aided by scratching, then aerosolizes the virus, which can float with dust to be inhaled by susceptible hosts. Infectious virions also bathe the terminals of those sensory neurons that innervate the epidermis. These terminals become infected with VZV and provide a route, retrograde transport, which can conduct VZV to cranial nerve (CNG), dorsal root ganglia (DRG), and enteric ganglia (EG) to establish latency. Reactivation returns VZV to the skin, now via anterograde transport in axons, to cause the lesions of zoster. Evidence in support of these hypotheses includes observations of the VZV-infected human epidermis and studies of guinea pig neurons in an in vitro model system.

1 Introduction

Varicella-zoster virus (VZV) is one of the eight herpesviruses that infect humans and is, arguably, the most infectious (Gershon et al. 2008a; Ross et al. 1962). Primary VZV infection is manifested as varicella (chickenpox). During this illness, VZV becomes latent in dorsal root (DRG), cranial nerve (CNG), and enteric ganglia (EG; see below) where VZV remains for life. In about 30% of individuals with latent VZV, the virus reactivates to cause a secondary infection, zoster (shingles) (Gershon et al. 2008a). Varicella can thus be acquired from patients with zoster, but zoster, which arises only after reactivation, cannot be acquired from patients with varicella or zoster. Latency and reactivation provide VZV with an evolutionary advantage. Because varicella is so infectious, it sweeps rapidly through populations in epidemics that exhaust the supply of susceptibles. Latency, however, allows VZV to persist in a host for years despite the host's immunity. Reactivation at a later date, after a new supply of susceptibles has been regenerated, allows VZV to emerge and transmit infection to a new generation. In this way VZV, which does not have an animal reservoir, can perpetuate itself.

VZV is a highly successful parasite. It is in its evolutionary interest, as it is in the interest of any parasite, to do as little damage as possible to the host that provides it with a residence. Despite its highly infectious nature, VZV spreads very slowly within infected individuals, allowing time for adaptive immunity to develop before the infection crosses the clinical threshold. Immunity of the hosts, within which VZV is latent, helps to ensure that providing VZV with a safe haven does not compromise the host's survival. The highly infectious nature of VZV makes its slow in vivo spread and long incubation period seem counterintuitive, even paradoxical. In fact, however, the behavior of VZV when grown in vitro bears no resemblance to its ferocious host-to-host spread. The same virus that sweeps with amazing speed and vehemence through a susceptible population of children

(Ross et al. 1962) spreads almost reluctantly and slowly through a population of cultured cells. Most of the cells that are infected with VZV in vitro, moreover, do not secrete infectious virions; in vitro spread requires the very slow process of cell-to-cell contact (Gershon et al. 2007), which may or may not involve fusion of infected cells with their neighbors (Reichelt et al. 2009). Medium in which infected cells are grown is not infectious (Gershon et al. 2007). A similar form of slow cell-to-cell transmission could account for the slow intrahost dissemination of VZV in vivo.

What accounts for the profound difference between the rapid host-to-host and the slow cell-to-cell spread of VZV? It has been proposed that free virions are responsible for host-to-host transmission of VZV, while cell-to-cell contact is the means by which VZV grows and disseminates within an infected individual (Chen et al. 2004). This hypothesis implies that infectious viral particles emerge only at the site where VZV is transmitted to new hosts, which is the skin. Infected keratinocytes are postulated to be veritable VZV factories, which uniquely churn out infectious virions. The survival advantage to VZV is that slow spread within a newly infected host prevents the virus from overwhelming that host with a flood of infectious particles. The slow intrahost dissemination delays the emergence of dangerous free virions until the adaptive immune response has been marshaled. By the time infectious virions are being mass-produced in keratinocytes, the defenses of the host are sufficiently ready to assure that the host, together with its reservoir of latent virions, survives. On the other hand, sloughing of infected keratinocytes allows a cornucopia of infectious virions to waft away and drift in the wind to where they can be inhaled by unwitting hosts.

The hypothesis that cell-to-cell spread is responsible for dissemination of VZV within infected hosts began with observations that sought to understand the profound cell association of VZV in vitro. Morphological and other studies demonstrated that nucleocapsids assemble within the nuclei of infected cells. Virions were found to acquire an envelope by budding through the inner nuclear membrane (Gershon et al. 1994). This process delivers the just-enveloped virion to the perinuclear cisterna, which is continuous with the rough endoplasmic reticulum (RER). Curiously, however, electron microscopic radioautographic studies revealed that the virions in this location were not labeled after any time of chase following exposure to a pulse of ^{3}H-mannose (Zhu et al. 1995). This observation suggests that the glycoproteins (gps) with N-linked oligosaccharides of the final viral envelope are not present in virions that emerge from the nuclei of infected cells. In contrast, the membranes of the RER were labeled by ^{3}H-mannose almost immediately, suggesting that N-glycosylation of viral gps, like that of cellular gps, occurs cotranslationally in the RER. At later times of chase, no movement of radioactivity back to the nucleus was ever detected; rather, later times of chase revealed a movement of labeled gps to the Golgi apparatus and *trans*-Golgi network, where virions became labeled. The virions that bud out of the nucleus to reach the cisternal space were observed to fuse almost immediately with the RER. This fusion delivers the nucleocapsids to the cytosol, while the original nuclear membrane-provided viral envelope is incorporated into the RER. This process requires that VZV undergo a secondary envelopment during which tegument proteins, which lack signal sequences and are synthesized by free polyribosome in the cytosol, can be incorporated into the completed virions.

VZV receives tegument and its final envelope in the TGN. The TGN is thus a meeting place of the envelope proteins that have been transported from the RER with tegument proteins and nucleocapsids. The secondary envelopment of VZV that occurs in the TGN critically involves VZV glycoprotein I (gI) (Wang et al. 2001). Cellular and viral proteins are sorted within the cisternae of the TGN, which are caused to form thin semicircular sacs. Viral proteins segregate to the concave surfaces of these sacs and cellular proteins, including large cation-independent mannose 6-phosphate receptors (MPR^{ci}s) (Goda and Pfeffer 1988), segregate to the convex surface (Gershon et al. 1994; Zhu et al. 1995). These TGN sacs wrap around nucleocapsids and fuse to envelop them. The original concave surface of the TGN wrapping cisternae becomes the viral envelope, while the convex surface becomes an MPR^{ci}-rich transport vesicle. The newly enveloped virions are then diverted to late endosomes. The environment that the virus encounters within late endosomes is acidic, which is deadly to VZV (Gershon et al. 1994). The morphology of VZV is degraded within the late endosomes, and subsequent exocytosis leads to the release of degraded virions, which can be visualized extracellularly by electron microscopy (EM) (Carpenter et al. 2009; Gershon et al. 1994). The post-Golgi degradation of the virions thus appears to render them noninfectious when they are ultimately released.

The critical step in the process of viral exit is the diversion of newly enveloped VZV from the secretory pathway to late endosomes, which may be due to the presence of mannose 6-phosphpate (Man 6-P) residues in at least four gps of the viral envelope (Gabel et al. 1989). Man 6-P interacts with the MPR^{ci}. MPR^{ci}s are present in the transport vesicles that arise with the virions in specialized cisternae of the TGN where VZV is enveloped (Chen et al. 2004; Gershon et al. 2008b). By interacting with MPR^{ci}s, newly enveloped virions can follow the MPR^{ci}'s itinerary during their post-Golgi transport through the cytoplasm of infected cells. MPR^{ci}s traffic from the TGN, not only to late endosomes, but also to the plasma membrane (Brown et al. 1986; Gabel and Foster 1986; Kornfeld 1987), where they act as receptors that mediate endocytosis of ligands, such as lysosomal enzymes, that contain Man 6-P residues (Gabel and Foster 1986; Yadavalli and Nadimpalli 2009).

The idea that MPR^{ci}s play a critical role in the intracellular trafficking of VZV has been supported by experiments that utilized expression of antisense cDNA or siRNA to generate five stable human cell lines that were deficient in MPRs (Chen et al. 2004). All five of these lines secreted lysosomal enzymes, indicating that their ability to divert Man 6-P-bearing proteins from the secretory pathway was defective. All five lines also released infectious virions when infected with cell-associated VZV. The addition of the simple sugar Man 6-P, moreover, was found to inhibit the infection of cells by cell-free VZV, which is consistent with the idea that Man 6-P can compete with the gps of the viral envelope for binding to cell surface MPR^{ci}s (Gabel et al. 1989). Man 6-P, however, cannot interfere with the infection of cells by cell-associated VZV, which probably requires cell fusion, and has been linked to the interaction of VZV gE with insulin degrading enzyme (Li et al. 2006). All five of the MPR-deficient lines generated with antisense cDNA or siRNA resisted infection by cell-free but not cell-associated VZV (Chen et al. 2004). That observation supports the idea that plasmalemmal MPR^{ci}s are necessary for infection of cells by cell-free VZV. MPR^{ci}s are known to be responsible for the receptor-mediated endocytosis of Man 6-P-bearing ligands, such as lysosomal enzymes (Canfield et al. 1991; Duncan and Kornfeld 1988; Johnson et al. 1990; Kornfeld 1987). Cell surface MPR^{ci}s may also mediate the endocytosis of cell-free VZV, which is critically involved in viral entry (Hambleton et al. 2007). It is therefore probable that intracellular MPR^{ci}s divert newly enveloped VZV to late endosomes, while MPR^{ci}s of the plasma membrane participate in infection of target cells by cell-free VZV.

The function of MPR^{ci}s relates to the transmission of VZV because MPR^{ci}s are naturally downregulated in the suprabasal layers of the epidermis in human skin (Chen et al. 2004) (Fig. 1). This downregulation is a property of normal keratinocyte maturation, which is a complex process that VZV has evolved to exploit. Keratinocytes are generated through the proliferation and differentiation of stem cells in the basal layer (stratum germinativum) of the epidermis (Koster 2009). The stem cells give rise to transit amplifying cells, also in the basal layer, which undergo multiple rounds of mitoses before they initiate the process of terminal differentiation. That process occurs suprabasally and begins when the transit amplifying cells permanently withdraw from the cell cycle. The postmitotic keratinocytes continue to mature as they move suprabasally in columns called epidermal proliferation units. Maturing keratinocytes ultimately differentiate into corneocytes or squames, which form the cellular component of the waterproofing barrier of the skin (Proksch et al. 2008). Mature corneocytes are the nonliving remains of cells that have jettisoned both their nucleus and their cytoplasmic

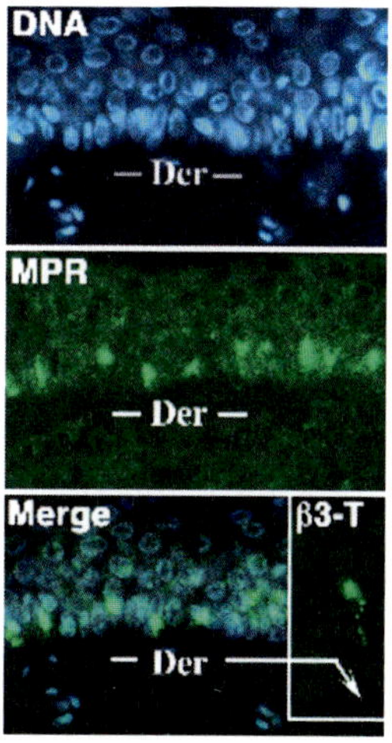

Fig. 1 The MPR^{ci} is downregulated during natural maturation of keratinocytes in the suprabasal epidermis. A section of skin is stained to demonstrate DNA (*blue fluorescence*) with bisbenzimide and MPR^{ci} immunoreactivity (*green fluorescence*). Note the absence of MPR^{ci} immunoreactivity in the superficial layers of the suprabasal epidermis and its strong presence in the basal layer. *Inset*: Intraepidermal nerve fiber demonstrated with antibodies to the neuronal marker β3-tubulin

organelles. They are essentially envelopes that are filled with arrays of intermediate filaments (composed of cytokeratins), which are organized into tight bundles by the essential protein, fillagrin. Corneocytes are stably connected to one another by corneodesmosomes. The cornified cell envelope has both protein and lipid components. Proteins include involucrin, loricrin, and trichohyalin, which are tightly crosslinked both by disulfide bonds and by N-ε-(γ-glutamyl)lysine isopeptide bonds. The formation of these bonds is catalyzed by transglutaminases. The lysosomal enzyme, cathepsin D, which is an aspartase protease, is required to cleave transglutaminase 1 to liberate its active 35-kDa form.

Enveloped corneocytes adhere strongly, not only to one another, but also to the intercellular matrix that their predecessors have secreted and to which the cornified envelope becomes covalently linked. These predecessors, in the upper stratum spinosum and the stratum granulosum, contain cells with so-called lamellar bodies, which contain lipid components of the epidermal barrier. Lamellar bodies are Golgi-derived vesicles that contain polar lipids, glycophospholipids, free sterols, phospholipids, catabolic enzymes, and β-defensin 2. The contents of lamellar bodies are secreted by exocytosis within the stratum granulosum despite the continuing progression of the cells of this layer toward their impending death. The lipids secreted from lamellar bodies become modified and arranged into lamellae that are parallel to the surfaces of cells in the matrix. Extracellular enzymes convert secreted polar lipids into nonpolar products, secreted glycosphingolipids into ceramides [amide-linked fatty acids containing a long-chain amino alcohol (sphingoid base)], and secreted phospholipids into free fatty acids. The stratum corneum ultimately acquires at least nine ceramides, which become covalently bound to the cornified envelope, primarily to involucrin. The cornified envelope and the extracellular matrix thus forms a tightly linked scaffold that waterproofs the epidermis and confers upon it an extraordinary degree of mechanical stability and chemical resistance. The process that constructs this essential barrier between the body and the outside world is thus a remarkable process in which secretion continues despite the inactivation of housekeeping functions in cells marked for death. The default secretory pathway thus outlasts MPR^{ci} receptors. This is the phenomenon that evidently provides VZV with the escape hatch that it needs to evade diversion to endosomes and to emerge intact. Despite their anchoring, moreover, corneocytes are dead cells that must be replaced. Desquamation is thus an on-going process. A person moving on leaves an invisible cloud of shed skin behind. If the epidermis is infected with VZV, this cloud is laden with infectious virions.

Biopsies from VZV-infected human skin, both in varicella and zoster, have revealed that although VZV is concentrated and degraded in the endosomes of basal cells, within which MPR^{ci}-mediated diversion to late endosomes is intact, VZV is not degraded in the infected cells of the suprabasal epidermis (Fig. 2), within which MPR^{ci} have been downregulated (Chen et al. 2004). Instead, intact virions are packaged singly in small vesicles (Fig. 2) within the MPR^{ci}-deficient suprabasal cells and are released intact into the intercellular space (Fig. 3). These spaces expand to form the vesicular lesions of varicella and zoster, within which

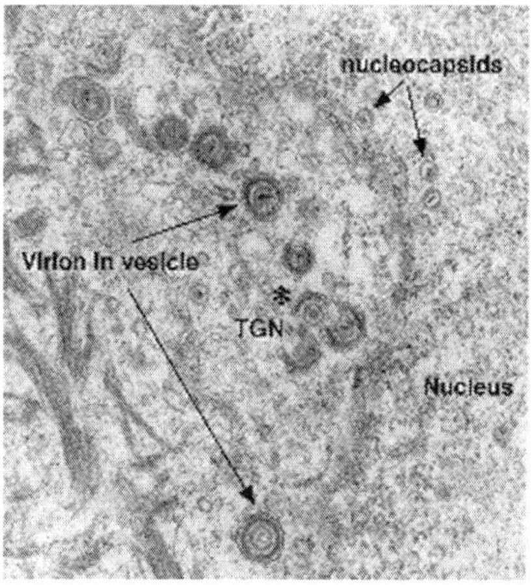

Fig. 2 Within superficial keratinocytes of the VZV-infected suprabasal epidermis, VZV is enveloped in the TGN and packaged singly in transport vesicles. An electron micrograph illustrates the suprabasal epidermis of the biopsied skin of a patient with varicella. All stages of viral envelopment can be seen in the infected keratinocytes. Unenveloped nucleocapsids can be seen in the nucleus and nucleocapsids are being enveloped (*asterisk*) by specialized cisternae of the TGN. Transport vesicles (*arrows*) each contain a single enveloped virion. Accumulating cytokeratin filaments are visible at the *lower left* in the cytoplasm of the keratinocytes

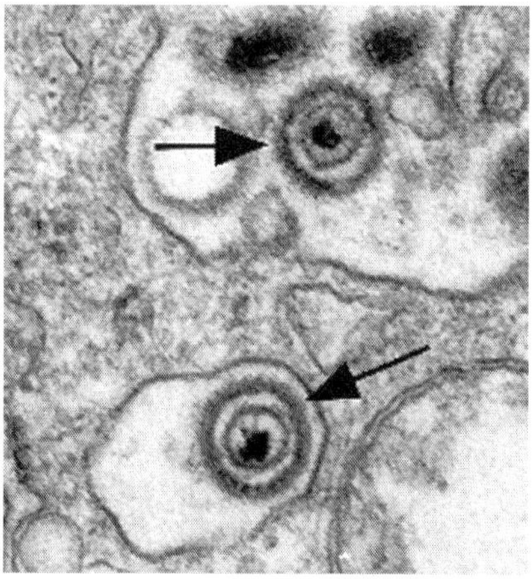

Fig. 3 Keratinocytes of the suprabasal layers of the VZV-infected epidermis secrete intact virions to the extracellular space. Two extracellular virions are illustrated (*arrows*)

perfectly formed intact virions that are highly infectious are concentrated. Consistent with the release of VZV from maturing suprabasal keratinocytes, these intraepidermal lesions, particularly early ones, are typically found within the suprabasal layers of the epidermis (Fig. 4). Until intraepidermal vesicles break down, the waterproofing provided by the corneocytyes of the stratum corneum provides a roof over intravesicular pockets of virion-rich intercellular fluid. Lesions may expand or be enlarged by superinfection, but most epidermal vesicles do not include the basal layer of the epidermis, the underlying basal lamina, or the dermis. These

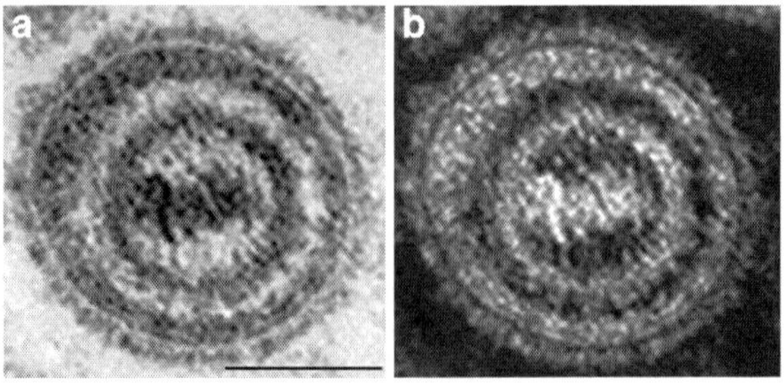

Fig. 4 Intact virions are found in the fluid extracted from within vesicular epidermal lesions of patients with varicella. Vesicular fluid was drawn from an epidermal vesicle of a patient with varicella and prepared for electron microscopic examination. A virion is illustrated at high magnification and imaged in brightfield (**a**) and darkfield (**b**) to emphasize the near-perfect viral morphology. Note the spikes on the viral envelope, the complexity of the underlying tegument, and the apparent connections between the nucleocapsid and the tegument. The marker = 100 nm

observations support the hypothesis outlined earlier that the natural loss of MPR^{ci} expression in the maturing cells of the suprabasal epidermis causes these cells to fail to divert VZV to late endosomes. Because constitutive secretion is the default pathway of post-TGN transport, the downregulation of the MPR^{ci} causes maturing keratinocytes to secrete infectious VZV constitutively instead of diverting newly enveloped virions to late endosomes. Desquamation of the epidermis, natural or induced by scratching, would break the last remaining roof of corneocytes over intraepidermal vesicles and allow masses of accumulated infectious virions to aerosolize and be inhaled by a susceptible host. This process, if correct, would account for the highly effective and rapid host-to-host spread of VZV. The contrasting slow cell-to-cell spread of VZV in all sites except the superficial epidermis restrains what would otherwise be a lethal VZV infection and provides the host with the time needed to mount an adaptive immune response. Varicella is thus usually a relatively benign disease, although it is certainly not benign in adults (Gregorakos et al. 2002; Wallace et al. 1993), individuals who are immunodeficient and unable to mount an immune response (Wood et al. 2008), or in pregnant mothers who transmit VZV to fetuses in whom adaptive immunity is immature (Auriti et al. 2009; Gershon et al. 2008a; Mustonen et al. 2001; Sauerbrei and Wutzler 2000).

Because neurons from DRG and CNG send axonal projections into the epidermis (McArthur et al. 1998), the intraepidermal axons of DRG and CNG neurons are in a strategically favored place from the point of view of VZV. The infectious virions that suprabasal keratinocytes release, of necessity, bathe the nerve endings that happen to be included within the suprabasal vesicular lesions that form in varicella. By infecting these nerve fibers, VZV can gain access via retrograde transport to the cell bodies of DRG and the CNG, such as the trigeminal, which project to the epidermis. Once in the DRG and CNG, VZV can establish latency. Intraepidermal neuronal projections, therefore, can provide conduits between the skin and ganglia that VZV can utilize for travel secure from the predations of immune effectors. As long as VZV remains latent in neurons, moreover, the virus is shielded from the host's immune response. This protection requires that the latently infected neurons survive and not express viral antigens on their surface. Until reactivation occurs, the latent virus and the host can coexist in harmony. After reactivation, however, VZV can return to the epidermis by way of its axonal conduit, this time by anterograde transport. The epidermis again becomes infected, now with zoster. Viral antigens are expressed on the surfaces of neurons (Lungu et al. 1995), which probably die as a result of reactivation (Chen et al. 2003; Gowrishankar et al. 2007; Nagashima et al. 1975; Zerboni et al. 2005). Certainly, in model systems, reactivation of VZV is rapidly lethal to the neurons in which reactivation takes place (Chen et al. 2003; Gershon et al. 2008b). The infection of the epidermis that results from reactivation is usually limited by the host's immune response, but it is painful and can be followed by the neuropathic pain of post-herpetic neuralgia (PHN) (Mueller et al. 2008; Nagel and Gilden 2007). The reactivated virus, moreover, is once again able to exploit the downregulation of MPR in the epidermis. It can again undergo constitutive secretion, reach vesicular fluid, and spread via desquamation from the skin to find new hosts. The disorder leaves the original host highly discomforted but alive.

In adults and children, varicella is spread predominately by the airborne route from the epidermal lesions of individuals with varicella or zoster (Breuer and Whitley 2007). Spread from the respiratory tract cannot be ruled out, but unlike respiratory infections that are transmitted by droplets (Morawska 2006; Yu et al. 2004), varicella is not accompanied by coughing and sneezing, which propel infectious droplets into the air (Gershon et al. 2008a). VZV is difficult to isolate from respiratory secretions although it is easily isolated from fresh skin vesicles (Gold 1966). Virions in skin vesicles are of an appropriate size (200 nm) to travel in air. Although VZV DNA has been identified in air samples from patient rooms by PCR for many days (Sawyer et al. 1993; Yoshikawa et al. 2001), there is no evidence that this DNA is infectious. Finally, only vaccinated persons with skin lesions are capable of transmitting the Oka vaccine strain to other individuals (Sharrar et al. 2000; Tsolia et al. 1990). Infectious virions, produced, as described earlier, in the superficial epidermis, shed from the desquamating skin and travel through the air with dust particles to reach respiratory surfaces of susceptible hosts to cause infection. Infection is initiated in tonsilar T cells (Ku et al. 2002) and may spread to tonsilar keratinocytes for additional multiplication.

During an incubation period of 10–21 days, during which the location of VZV is not completely clear (although the virus is traditionally said to proliferate in the lungs and the liver), a viremia occurs; VZV circulates within infected T lymphocytes (Koropchak et al. 1989). The T cells home to the skin and infect the stem cells and transit amplifying cells of the stratum germinativum, from which infection can disseminate by cell-to-cell spread to reach the suprabasal layers of the epidermis. Modulation of viral replication by innate immunity in the skin has also been postulated to prevent VZV infection from overwhelming its host (Ku et al. 2004). The intraepidermal proliferation of VZV, with associated inflammation and death of infected cells, causes the vesicular rash of varicella.

It has been proposed that VZV establishes latency because the free virions are uniquely released from the infected suprabasal keratinocytes that have downregulated MPR^{ci} expression but still secrete, bathe, and infect intraepidermal sensory nerve endings. Virions (or infectious components derived from them) are transported in the retrograde direction to nerve cell bodies in DRG and CNG where they establish latency (Gershon et al. 2008b). Zoster occurs when VZV reactivates from latency and proliferates in neuronal perikaya. Virions, or, as in the case of herpes simplex, viral components that can be assembled into virions within axon terminals (Diefenbach et al. 2008), are transported, now in the anterograde direction, back down the axons to the epidermis, where they again infect keratinocytes and give rise to a vesicular rash. Viremic dissemination, however, is unusual in zoster because reactivation usually occurs in at least partially immune host.

Ganglion cells within which VZV is latent, can survive for years. Only a limited set of viral genes is transcribed during latency; these encode immediate early and early proteins (ORF4, 21, 29, 62, 63, 40, 66) but gps are not expressed (Cohrs et al. 1995, 1996, 2003; Cohrs and Gilden 2003; Croen et al. 1988; Gershon et al. 2008b; Grinfeld and Kennedy 2004; Grinfeld et al. 2004; Kennedy et al. 2000, 2001; Lungu et al. 1998). Neurons in which VZV reactivates display lytic infection. They thus transcribe a full set of viral genes, including ORF61 and gps; then they die (Chen et al. 2003; Head and Campbell 1900). The death of neurons in which VZV reactivates causes sensory deficits (Head and Campbell 1900), although these are usually mild and compensated by expansion of the receptive fields of remaining neurons (Feller et al. 2005; Gershon et al. 2008b). Neuronal death, however, is followed, about 15% of the time (depending on the age of the patient), by the debilitating and often long-lasting neuropathic pain of PHN (Dworkin et al. 2007; Feller et al. 2005; Fields et al. 1998).

Research on the consequences of VZV infection in neurons has been hampered by the lack of a suitable animal model. Investigators have had limited success in infecting guinea pig DRG (Arvin et al. 1987; Lowry et al. 1992, 1993; Matsunga et al. 1982; Myers et al. 1980, 1985, 1991; Sato et al. 2003) and, although latent infection of ganglia has been achieved in cotton rats (Cohen et al. 2007) and rats (Rentier et al. 1996), VZV has not been reactivated in either of these animals. We have furthered our own ability to analyze the life cycle of VZV by developing an in vitro model that utilizes enteric neurons isolated from the guinea pig small intestine (Chen et al. 2003; Gershon et al. 2008b). The advantage of the ENS for this purpose is that its structure resembles that of the brain (Gershon 2005). Fibroblasts and connective tissue are not present within EG. Instead, enteric neurons are supported by glia, which resemble CNS astrocytes (Gershon and Rothman 1991). As a result of their glia-dependent integrity, ganglia remain intact when the bowel is dissociated with collagenase (Chen et al. 2003; Gershon et al. 2008b). EG can then be selected from a collagenase-dissociated suspension of enteric cells with a micropipette and isolated as relatively pure preparations of neurons and glia. When isolated EG are exposed to cell-free VZV, latent infection of neurons results. The same VZV transcripts and proteins that are expressed during latency in DRG and CNG are also expressed in latently infected enteric neurons, which survive for as long as cultures can be maintained. In contrast, when connective tissue cells are present at the time of inoculation (either because infection is passed with cell-associated VZV or because enteric fibroblasts have been included in the original cultures), lytic infection of the neurons occurs. Such neurons now express the full set of VZV proteins and die within 48–72 h.

After latent VZV infection has been established in cultured enteric neurons, reactivation cannot be initiated by adding fibroblasts to the cultures (Chen et al. 2003; Gershon et al. 2008b). The ability of fibroblasts to cause VZV infection of enteric neurons to be lytic rather than latent requires that the fibroblasts be present at the time neurons are first exposed to VZV. It has been postulated that fibroblasts that are present at the time of inoculation become infected with VZV and then fuse with neighboring neurons. Such a fusion would allow a nonstructural protein to gain entrance to the infected neurons at the moment of their infection. This idea was tested and supported by expressing the nonstructural VZV protein, ORF61p (Kinchington et al. 1995), or its HSV orthologue, ICP0 (Moriuchi et al. 1993), in latently infected enteric neurons. Both ORF61p and ICP0 proved to be able to reactivate VZV (Chen et al. 2003; Gershon et al. 2008b); the late VZV gene products, including the gps, were now produced and neurons, which had harbored latent VZV for weeks, died within 48–72 h of reactivation. Before their death, neurons were found by EM to contain intact virions (identified as VZV by showing gE immunoreactivity in their envelopes); moreover, following reactivation, but not during latency, enteric neurons were able to transmit VZV infection to cocultures of human melanoma cells (MeWo) in culture. It has been suggested that ORF61p acts as a switch that triggers the lytic cascade (Gershon et al. 2008b). The observation that ORF61p contributes to the virulence of VZV at cutaneous sites of VZV replication is consistent with this suggestion (Wang et al. 2009).

It might seem odd that VZV should recapitulate its life cycle in cultures of enteric neurons, which do not project to the epidermis, rather than in those of DRG or CNG, which do so. At first, it was thought that EG could be infected with VZV because the ENS, like DRG and CNG, contains primary afferent neurons (Wang et al. 2009), for which VZV was considered to be trophic. This idea turned out to be incorrect. VZV infects all enteric neurons, not only those that display sensory markers (Chen et al. 2003; Gershon et al. 2008b). The key is that enteric neurons can be isolated free of contaminating connective tissue cells, whereas neurons from DRG and CNG cannot. Further studies, moreover, established unexpectedly that the ENS actually is a natural target of VZV and becomes infected in human hosts (Chen et al. 2009; Gershon et al. 2008b). Transcripts encoding VZV proteins that are expressed during latency (ORFps 4, 21, 62, or 63) were detectable in 88% of 31 specimens of adult human gut obtained during surgery for unrelated clinical indications. This observation has recently been confirmed by demonstrating VZV DNA, as well as transcripts encoding latency-associated proteins, in specimens of gut obtained at surgery from children with a history of varicella and from those with a history of varicella vaccination (Gershon et al. 2009). In contrast, no VZV DNA or transcripts encoding VZV were detected in specimens of newborn bowel. These observations suggest that VZV can and does become latent in human enteric neurons after varicella or after administration of the varicella vaccine. The observation that varicella vaccination, which is typically delivered to the skin of the arm, can lead to latency in EG strongly suggests that, in contrast to expectations, vaccination is associated with a viremia. This suggestion is further supported by observations made in autopsy specimens obtained from vaccinated children dying suddenly of unrelated causes, that viral DNA and transcripts are found not only in ganglia innervating the vaccination site, but also bilaterally in distant ganglia (Gershon et al. 2009). Still to be determined, however, is whether the VZV that establishes latent infection can only be delivered to ganglia by axons that acquire VZV released as infectious particles in the skin, or whether they can also become infected by the T lymphocytes that carry VZV during a viremia (Abendroth et al. 2001; Asanuma et al. 2000; Koropchak et al. 1989; Ku et al. 2002; Moffat et al. 1995; Schaap et al. 2005). There is evidence that VZV-infected lymphocytes release infectious cell-free viral particles (Moffat et al. 1995), although this evidence is not universally accepted and contrary data have also been reported (Soong et al. 2000). If VZV-infected T lymphocytes do release infectious virions, then they should be as able as keratinocytes to infect neurons in a manner that can establish latency. If not, then either cell-free VZV is not required for the establishment of latency in DRG and CNG, or a lymphocyte-borne viremia might cause subclinical infections of keratinocytes in many regions of the skin. Should that occur, then sufficient virions might be secreted by VZV-infected keratinocytes to infect intraepidermal nerves, even in the absence of rash.

Acknowledgments Supported by NIH grants NS12969, AI27187, and AI24021.

References

Abendroth A, Lin I, Slobedman B, Ploegh H, Arvin AM (2001) Varicella-zoster virus retains major histocompatibility complex class I proteins in the Golgi compartment of infected cells. J Virol 75:4878–4888

Arvin AM, Solem S, Koropchak C, Kinney-Thomas E, Paryani SG (1987) Humoral and cellular immunity to varicella-zoster virus glycoprotein, gpI, and to a non-glycosulated protein p170, in strain 2 guinea pigs. J Gen Virol 68:2449–2454

Asanuma H, Sharp M, Maecker HT, Maino VC, Arvin AM (2000) Frequencies of memory T cells specific for varicella-zoster virus, herpes simplex virus, and cytomegalovirus by intracellular detection of cytokine expression. J Infect Dis 181:859–866

Auriti C, Piersigilli F, De Gasperis MR, Seganti G (2009) Congenital varicella syndrome: still a problem? Fetal Diagn Ther 25:224–229

Breuer J, Whitley R (2007) Varicella zoster virus: natural history and current therapies of varicella and herpes zoster. Herpes 14(Suppl 2):25–29

Brown WJ, Goodhouse J, Farquhar MG (1986) Mannose-6-phosphate receptors for lysosomal enzymes cycle between the Golgi complex and endosomes. J Cell Biol 103:1235–1247

Canfield WM, Johnson KF, Ye RD, Gregory W, Kornfeld S (1991) Localization of the signal for rapid internalization of the bovine cation-independent mannose 6-phosphate/insulin-like growth factor-II receptor to amino acids 24–29 of the cytoplasmic tail. J Biol Chem 266:5682–5688

Carpenter JE, Henderson EP, Grose C (2009) Enumeration of an extremely high particle-to-PFU ratio for varicella-zoster virus. J Virol 83:6917–6921

Chen J, Gershon A, Silverstein SJ, Li ZS, Lungu O, Gershon MD (2003) Latent and lytic infection of isolated guinea pig enteric and dorsal root ganglia by varicella zoster virus. J Med Virol 70: S71–S78

Chen JJ, Zhu Z, Gershon AA, Gershon MD (2004) Mannose 6-phosphate receptor dependence of varicella zoster virus infection in vitro and in the epidermis during varicella and zoster. Cell 119:915–926

Chen JJ, Gershon MD, Wan S, Cowles RA, Ruiz-Elizalde A, Bischoff S, Gershon AA (2009) Latent infection of the human enteric nervous system by varicella zoster virus (VZV). Gastroenterology 136:W1689

Cohen JI, Krogmann T, Pesnicak L, Ali MA (2007) Absence or overexpression of the varicella-zoster virus (VZV) ORF29 latency-associated protein impairs late gene expression and reduces VZV latency in a rodent model. J Virol 81:1586–1591

Cohrs RJ, Gilden DH (2003) Varicella zoster virus transcription in latently-infected human ganglia. Anticancer Res 23:2063–2069

Cohrs RJ, Barbour MB, Mahalingam R, Wellish M, Gilden DH (1995) Varicella-zoster virus (VZV) transcription during latency in human ganglia: prevalence of VZV gene 21 transcripts in latently infected human ganglia. J Virol 69:2674–2678

Cohrs RJ, Barbour M, Gilden DH (1996) Varicella-zoster virus (VZV) transcription during latency in human ganglia: detection of transcripts to genes 21, 29, 62, and 63 in a cDNA library enriched for VZV RNA. J Virol 70:2789–2796

Cohrs RJ, Gilden DH, Kinchington PR, Grinfeld E, Kennedy PG (2003) Varicella-zoster virus gene 66 transcription and translation in latently infected human ganglia. J Virol 77:6660–6665

Croen KD, Ostrove JM, Dragovic LY, Straus SE (1988) Patterns of gene expression and sites of latency in human ganglia are different for varicella-zoster and herpes simplex viruses. Proc Natl Acad Sci USA 85:9773–9777

Diefenbach RJ, Miranda-Saksena M, Douglas MW, Cunningham AL (2008) Transport and egress of herpes simplex virus in neurons. Rev Med Virol 18:35–51

Duncan JR, Kornfeld S (1988) Intracellular movement of two mannose 6-phosphate receptors: return to the Golgi apparatus. J Cell Biol 106:617–628

Dworkin RH, Johnson RW, Breuer J, Gnann JW, Levin MJ, Backonja M, Betts RF, Gershon AA, Haanpaa ML, McKendrick MW, Nurmikko TJ, Oaklander AL, Oxman MN, Pavan-Langston D, Petersen KL, Rowbotham MC, Schmader KE, Stacey BR, Tyring SK, van Wijck AJ, Wallace MS, Wassilew SW, Whitley RJ (2007) Recommendations for the management of herpes zoster. Clin Infect Dis 44(Suppl 1):S1–S26

Feller L, Jadwat Y, Bouckaert M (2005) Herpes zoster post-herpetic neuralgia. S Afr Dent J 60 (432):436–437

Fields HL, Rowbotham M, Baron R (1998) Postherpetic neuralgia: irritable nociceptors and deafferentation. Neurobiol Dis 5:209–227

Gabel CA, Foster SA (1986) Mannose 6-phosphate receptor-mediated endocytosis of acid hydrolases: internalization of beta-glucuronidase is accompanied by a limited dephosphorylation. J Cell Biol 103:1817–1827

Gabel C, Dubey L, Steinberg S, Gershon M, Gershon A (1989) Varicella-zoster virus glycoproteins are phosphorylated during posttranslational maturation. J Virol 63:4264–4276

Gershon MD (2005) Nerves, reflexes, and the enteric nervous system: pathogenesis of the irritable bowel syndrome. J Clin Gastroenterol 39:S184–S193

Gershon MD, Rothman TP (1991) Enteric glia. Glia 4:195–204

Gershon AA, Sherman DL, Zhu Z, Gabel CA, Ambron RT, Gershon MD (1994) Intracellular transport of newly synthesized varicella-zoster virus: final envelopment in the trans-Golgi network. J Virol 68:6372–6390

Gershon A, Chen J, LaRussa P, Steinberg S (2007) Varicella-zoster virus. In: Murray PR, Baron E, Jorgensen J, Landry M, Pfaller M (eds) Manual of clinical microbiology, 9th edn. ASM Press, Washington, D.C, pp 1537–1548
Gershon A, Takahashi M, Seward J (2008a) Live attenuated varicella vaccine. In: Plotkin S, Orenstein W, Offit P (eds) Vaccines, 5th edn. WB Saunders, Philadelphia, pp 915–958
Gershon AA, Chen J, Gershon MD (2008b) A model of lytic, latent, and reactivating varicella-zoster virus infections in isolated enteric neurons. J Infect Dis 197(Suppl 2):S61–S65
Gershon AA, Chen J, Davis L, Krinsky C, Cowles R, Gershon MD (2009) In: Presented at the 34th international herpesvirus workshop, Ithaca, NY
Goda Y, Pfeffer SR (1988) Selective recycling of the mannose 6-phosphate/IGF-II receptor to the trans Golgi network in vitro. Cell 55:309–320
Gold E (1966) Serologic and virus-isolation studies of patients with varicella or herpes zoster infection. N Engl J Med 274:181–185
Gowrishankar K, Slobedman B, Cunningham AL, Miranda-Saksena M, Boadle RA, Abendroth A (2007) Productive varicella-zoster virus infection of cultured intact human ganglia. J Virol 81:6752–6756
Gregorakos L, Myrianthefs P, Markou N, Chroni D, Sakagianni E (2002) Severity of illness and outcome in adult patients with primary varicella pneumonia. Respiration 69:330–334
Grinfeld E, Kennedy PG (2004) Translation of varicella-zoster virus genes during human ganglionic latency. Virus Genes 29:317–319
Grinfeld E, Sadzot-Delvaux C, Kennedy PG (2004) Varicella-zoster virus proteins encoded by open reading frames 14 and 67 are both dispensable for the establishment of latency in a rat model. Virology 323:85–90
Hambleton S, Steinberg SP, Gershon MD, Gershon AA (2007) Cholesterol dependence of varicella-zoster virion entry into target cells. J Virol 81:7548–7558
Head H, Campbell AW (1900) The pathology of herpes zoster and its bearing on sensory localization. Brain 23:353–523
Johnson CF, Chan W, Kornfeld S (1990) Cation-dependent mannose 6-phosphate receptor contains two internalization signals in its cytoplasmic domain. Proc Natl Acad Sci USA 87:10010–10014
Kennedy PGE, Grinfeld E, Bell JE (2000) Varicella-zoster virus gene expression in latently infected and explanted human ganglia. J Virol 74:11893–11898
Kennedy PG, Grinfeld E, Bontems S, Sadzot-Delvaux C (2001) Varicella-Zoster virus gene expression in latently infected rat dorsal root ganglia. Virology 289:218–223
Kinchington PR, Bookey D, Turse SE (1995) The transcriptional regulatory proteins encoded by varicella-zoster virus are open reading frames (ORFs) 4 and 63, but not ORF 61, are associated with purified virus particles. J Virol 69:4274–4282
Kornfeld S (1987) Trafficking of lysosomal enzymes. FASEB J 1:462–468
Koropchak C, Solem S, Diaz P, Arvin A (1989) Investigation of varicella-zoster virus infection of lymphocytes by in situ hybridization. J Virol 63:2392–2395
Koster MI (2009) Making an epidermis. Ann N Y Acad Sci 1170:7–10
Ku CC, Padilla JA, Grose C, Butcher EC, Arvin AM (2002) Tropism of varicella-zoster virus for human tonsillar CD4(+) T lymphocytes that express activation, memory, and skin homing markers. J Virol 76:11425–11433
Ku CC, Zerboni L, Ito H, Graham BS, Wallace M, Arvin AM (2004) Varicella-zoster virus transfer to skin by T cells and modulation of viral replication by epidermal cell interferon-{alpha}. J Exp Med 200:917–925
Li Q, Ali MA, Cohen JI (2006) Insulin degrading enzyme is a cellular receptor mediating varicella-zoster virus infection and cell-to-cell spread. Cell 127:305–316
Lowry PW, Solem S, Watson BN, Koropchak C, Thackeray H, Kinchington P, Ruyechan W, Ling P, Hay J, Arvin A (1992) Immunity in strain 2 guinea pigs inoculated with vaccinia virus recombinants expressing varicella-zoster virus glycoproteins I, IV, V, or the protein product of the immediate early gene 62. J Gen Virol 73:811–819

Lowry PW, Sabella C, Koropchek C, Watson BN, Thackray HM, Abruzzi GM, Arvin AM (1993) Investigation of the pathogenesis of varicella-zoster virus infection in guinea pigs by using polymerase chain reaction. J Infect Dis 167:78–83
Lungu O, Annunziato P, Gershon A, Stegatis S, Josefson D, LaRussa P, Silverstein S (1995) Reactivated and latent varicella-zoster virus in human dorsal root ganglia. Proc Natl Acad Sci USA 92:10980–10984
Lungu O, Panagiotidis C, Annunziato P, Gershon A, Silverstein S (1998) Aberrant intracellular localization of varicella-zoster virus regulatory proteins during latency. Proc Natl Acad Sci USA 95:7080–7085
Matsunga Y, Yamanishi K, Takahashi M (1982) Experimental infection and immune responses of guinea pigs with varicella zoster virus. Infect Immun 37:407
McArthur JC, Stocks EA, Hauer P, Cornblath DR, Griffin JW (1998) Epidermal nerve fiber density: normative reference range and diagnostic efficiency. Arch Neurol 55:1513–1520
Moffat JF, Stein MD, Kaneshima H, Arvin AM (1995) Tropism of varicella-zoster virus for human CD4+ and CD8+ T lymphocytes and epidermal cells in SCID-hu mice. J Virol 69:5236–5242
Morawska L (2006) Droplet fate in indoor environments, or can we prevent the spread of infection? Indoor Air 16:335–347
Moriuchi H, Moriuchi M, Straus S, Cohen J (1993) Varicella-zoster virus (VZV) open reading frame 61 protein transactivates VZV gene promoters and enhances the infectivity of VZV DNA. J Virol 67:4290–4295
Mueller NH, Gilden DH, Cohrs RJ, Mahalingam R, Nagel MA (2008) Varicella zoster virus infection: clinical features, molecular pathogenesis of disease, and latency. Neurol Clin 26:675–697
Mustonen K, Mustakangas P, Valanne L, Professor MH, Koskiniemi M (2001) Congenital varicella-zoster virus infection after maternal subclinical infection: clinical and neuropathological findings. J Perinatol 21:141–146
Myers M, Duer HL, Haulser CK (1980) Experimental infection of guinea pigs with varicella-zoster virus. J Infect Dis 142:414–420
Myers M, Stanberry L, Edmond B (1985) Varicella-zoster virus infection of strain 2 guinea pigs. J Infect Dis 151:106–113
Myers MG, Connelly B, Stanberry LR (1991) Varicella in hairless guinea pigs. J Infect Dis 163:746–751
Nagashima K, Nakazawa M, Endo H (1975) Pathology of the human spinal ganglia in varicella-zoster virus infection. Acta Neuropathol 33:105–117
Nagel MA, Gilden DH (2007) The protean neurologic manifestations of varicella-zoster virus infection. Cleve Clin J Med 74:489–494, 496, 498–499 passim
Proksch E, Brandner JM, Jensen JM (2008) The skin: an indispensable barrier. Exp Dermatol 17:1063–1072
Reichelt M, Brady J, Arvin AM (2009) The replication cycle of varicella-zoster virus: analysis of the kinetics of viral protein expression, genome synthesis, and virion assembly at the single-cell level. J Virol 83:3904–3918
Rentier B, Debrus S, Sadzot-Delvaux C, Nikkels A, Piette J, Mahalingam R, Wellish M, Cohrs R, Gilden DH, Lungu O, LaRussa P, Silverstein S, Annunziato P, Gershon A (1996) In: Presented at the keystone symposium on virus entry, replication, and pathogenesis, Santa Fe, NM, pp 10–16
Ross AH, Lencher E, Reitman G (1962) Modification of chickenpox in family contacts by administration of gamma globulin. N Engl J Med 267:369–376
Sato H, Pesnicak L, Cohen JI (2003) Use of a rodent model to show that varicella-zoster virus ORF61 is dispensable for establishment of latency. J Med Virol 70(Suppl 1):S79–S81
Sauerbrei A, Wutzler P (2000) The congenital varicella syndrome. J Perinatol 20:548–554
Sawyer M, Chamberlin C, Wu Y, Aintablian N, Wallace M (1993) Detection of varicella-zoster virus DNA in air samples from hospital rooms. J Infect Dis 169:91–94

Schaap A, Fortin JF, Sommer M, Zerboni L, Stamatis S, Ku CC, Nolan GP, Arvin AM (2005) T-cell tropism and the role of ORF66 protein in pathogenesis of varicella-zoster virus infection. J Virol 79:12921–12933

Sharrar RG, LaRussa P, Galea S, Steinberg S, Sweet A, Keatley M, Wells M, Stephenson W, Gershon A (2000) The postmarketing safety profile of varicella vaccine. Vaccine 19:916–923

Soong W, Schultz JC, Patera AC, Sommer MH, Cohen JI (2000) Infection of human T lymphocytes with varicella-zoster virus: an analysis with viral mutants and clinical isolates. J Virol 74:1864–1870

Tsolia M, Gershon A, Steinberg S, Gelb L (1990) Live attenuated varicella vaccine: evidence that the virus is attenuated and the importance of skin lesions in transmission of varicella-zoster virus. J Pediatr 116:184–189

Wallace MR, Bowler WA, Oldfield EC 3rd (1993) Treatment of varicella in the immunocompetent adult. J Med Virol (Suppl 1):90–92

Wang L, Sommer M, Rajamani J, Arvin AM (2009) Regulation of the ORF61 promoter and ORF61 functions in varicella-zoster virus replication and pathogenesis. J Virol 83:7560–7572

Wang Z-H, Gershon MD, Lungu O, Zhu Z, Mallory S, Arvin A, Gershon A (2001) Essential role played by the C-terminal domain of glycoprotein I in envelopment of varicella-zoster virus in the trans-Golgi network: interactions of glycoproteins with tegument. J Virol 75:323–340

Wood SM, Shah SS, Steenhoff AP, Rutstein RM (2008) Primary varicella and herpes zoster among HIV-infected children from 1989 to 2006. Pediatrics 121:e150–e156

Yadavalli S, Nadimpalli SK (2009) Role of cation independent mannose 6-phosphate receptor protein in sorting and intracellular trafficking of lysosomal enzymes in chicken embryonic fibroblast (CEF) cells. Glycoconj J

Yoshikawa T, Ihira M, Suzuki K, Suga S, Tomitaka A, Ueda H, Asano Y (2001) Rapid contamination of the environment with varicella-zoster virus DNA from a patient with herpes zoster. J Med Virol 63:64–66

Yu IT, Li Y, Wong TW, Tam W, Chan AT, Lee JH, Leung DY, Ho T (2004) Evidence of airborne transmission of the severe acute respiratory syndrome virus. N Engl J Med 350:1731–1739

Zerboni L, Ku CC, Jones CD, Zehnder JL, Arvin AM (2005) Varicella-zoster virus infection of human dorsal root ganglia in vivo. Proc Natl Acad Sci USA 102:6490–6495

Zhu Z, Gershon MD, Gabel C, Sherman D, Ambron R, Gershon AA (1995) Entry and egress of VZV: role of mannose 6-phosphate, heparan sulfate proteoglycan, and signal sequences in targeting virions and viral glycoproteins. Neurology 45:S15–S17

Varicella-Zoster Virus T Cell Tropism and the Pathogenesis of Skin Infection

Ann M. Arvin, Jennifer F. Moffat, Marvin Sommer, Stefan Oliver, Xibing Che, Susan Vleck, Leigh Zerboni, and Chia-Chi Ku

Contents

Abstract Varicella-zoster virus (VZV) is a medically important human alphaherpesvirus that causes varicella and zoster. VZV initiates primary infection by inoculation of the respiratory mucosa. In the course of primary infection, VZV establishes a life-long persistence in sensory ganglia; VZV reactivation from latency may result in zoster in healthy and immunocompromised patients. The VZV genome has at least 70 known or predicted open reading frames (ORFs), but understanding how these gene products function in virulence is difficult because VZV is a highly human-specific pathogen. We have addressed this obstacle by investigating VZV infection of human tissue xenografts in the severe combined immunodeficiency mouse model. In studies relevant to the pathogenesis of primary VZV infection, we have examined VZV infection of human T cell (thymus/liver) and skin xenografts. This work supports a new paradigm for VZV pathogenesis in which VZV T cell tropism provides a mechanism for delivering the virus to skin.

A.M. Arvin (✉), M. Sommer, S. Oliver, X. Che, S. Vleck and L. Zerboni
Departments of Pediatrics and Microbiology and Immunology, Stanford University School of Medicine, Stanford, CA, USA
e-mail: aarvin@stanford.edu

J.F. Moffat
Department of Microbiology, University of New York, Syracuse, NY, USA

C.-C. Ku
The Graduate Institute of Immunology, College of Medicine, National Taiwan University, Taipei, Taiwan

A.M. Arvin et al. (eds.), *Varicella-zoster Virus*,
Current Topics in Microbiology and Immunology 342, DOI 10.1007/82_2010_29

We have also shown that VZV-infected T cells transfer VZV to neurons in sensory ganglia. The construction of infectious VZV recombinants that have deletions or targeted mutations of viral genes or their promoters and the evaluation of VZV mutants in T cell and skin xenografts has revealed determinants of VZV virulence that are important for T cell and skin tropism *in vivo*.

1 Introduction

Varicella-zoster virus (VZV) is the causative agent of varicella, which is recognized by its characteristic vesicular exanthem. In the course of primary infection, VZV reaches cranial nerve and dorsal root sensory ganglia (DRG) where it establishes latency; VZV reactivation from latency may result in the clinical syndrome of herpes zoster in healthy and immunocompromised patients (Cohen et al. 2007; Zerboni and Arvin 2008). Since VZV is highly host-restricted, events in VZV pathogenesis have been deduced from clinical observations about primary and recurrent infections in its native human host (Gilden et al. 2000; Arvin 2001a). Epidemiologic studies indicate that primary infection is initiated by mucosal inoculation, which is followed by a viremic phase, allowing viral transport to skin sites of replication. Clinical evidence that VZV viremia is cell-associated consists of recovery of the virus from peripheral blood mononuclear cells (PBMC) obtained just before and after the appearance of the varicella rash (Arvin 2001a). In our early studies, VZV DNA was detected by in situ hybridization in one in 30,000–100,000 PBMC from healthy individuals with acute varicella, and infected cells appeared to be lymphocytes (Koropchak et al. 1989). The capacity of VZV to cause viremia can result in viral dissemination to the lungs, liver, and other organs and in life-threatening varicella in immunocompromised children, unless antiviral therapy is given.

To overcome the challenge of the host restriction of VZV, we have developed methods to study VZV pathogenesis using human tissue xenografts, including thymus/liver (T cell), skin, and DRG, in mice with severe combined immunodeficiency (SCID) (reviewed in Ku et al. 2005; Arvin et al. 2006; Zerboni et al. 2005a, b; Zerboni and Arvin 2008). VZV cannot infect mouse tissues and the foreign tissue grafts are maintained without rejection in these animals. This system also allows the investigation of VZV infection of differentiated human cells in their tissue microenvironment *in vivo* in the absence of any adaptive immune response. In the first report using this model, Moffat et al. showed that VZV inoculation of skin xenografts produced infection of epidermal and dermal cells similar to those observed in clinical biopsies of VZV lesions; these experiments also demonstrated that VZV was highly infectious for human T cells in thymus/liver xenografts (Moffat et al. 1995). Importantly, this T cell tropism differentiates VZV from the other human alphaherpesviruses, herpes simplex virus (HSV) 1 and 2, which cause mucocutaneous lesions without evidence of systemic spread through a cell-associated viremia.

Studies of VZV pathogenesis in the SCID mouse model have been enhanced by the parallel development of methods to introduce targeted mutations into the VZV genome using cosmids consisting of overlapping fragments of the VZV genome (Cohen and Seidel 1993; Mallory et al. 1997; Kemble et al. 2000; Niizuma et al. 2003); VZV mutagenesis is also now accomplished using bacterial artificial chromosome (BAC) techniques (Zhang et al. 2007; Tischer et al. 2007). When deletions or nucleotide substitutions in the region of interest are not lethal for VZV replication, VZV recombinants can be exploited to identify determinants of VZV virulence in human tissue xenografts in the SCID mouse model and to assess differential requirements for particular gene products and functional domains within protein and promoter elements in T cell, skin, and DRG xenografts.

VZV T cell tropism in T cell xenografts in SCID mouse model. Our initial experiments in human T cell xenografts in SCID mice *in vivo* demonstrated the synthesis of VZV DNA and viral proteins in both CD4 and CD8 T cell subpopulations (Moffat et al. 1995). VZV infection of T cells *in vivo* also results in robust virion formation and the appearance of complete virus particles on T cell surfaces (Schaap et al. 2005; Schaap-Nutt et al. 2006). Important characteristics of T cell tropism that emerged from these studies are that VZV-infected T cells do not fuse with adjacent uninfected T cells and the progression of VZV infection in T cell xenografts was associated with the formation of complete VZ virions and release of infectious virus (Fig. 1). This pattern differs from the polykaryocyte or syncytia formation, which is the hallmark of VZV replication in skin and cultured cells *in vitro* (Fig. 2). Thus, VZV infection of T cells, in contrast to skin, appears to require efficient virion formation and egress for transfer to uninfected cells.

VZV tropism for tonsil T cells in vitro. While T cells can become infected with VZV when PBMC cultures are inoculated with the virus, the percentage of infected cells is low (Koropchak et al. 1989; Soong et al. 2000). Based on the evidence that VZV pathogenesis begins with inoculation of mucosal epithelial cells of the respiratory tract, we speculated that VZV could gain access to T cells in the lymphoid tissue that comprises the tonsils and other components of Waldeyer's ring (Fig. 3) and that these T cells would be more susceptible to VZV infection than circulating T cells. More than 20% of mononuclear cells in the tonsils are CD3 T cells. In addition, tonsils have a surface layer of respiratory epithelial cells, these cells penetrate into the tonsillar crypts, and T cells as well as B cells migrate across the epithelial cell layer into and out of the tonsils. As predicted, tonsil T cells were highly susceptible to VZV infection *in vitro* (Ku et al. 2002). Thus, VZV targets tonsil T cells in a manner that is analogous to the tropism of Epstein-Barr virus for tonsil B cells. Of interest, the tonsil T cell populations that were most likely to be infected were activated CD4 T cells expressing CD69 and other activation markers and were predominantly memory T cells; 20–25% of CD4 T cells were infected compared to 10–15% of CD8 T cells. During natural infection, this pattern of tropism for CD4 T cells in tonsils would be expected to result in higher absolute numbers of infected CD4 T cells because two-thirds of tonsil T cells are in this subpopulation. Activated memory CD4 T cells are also common in tonsil T cell populations, presumably because of continuous exposure to various antigens and

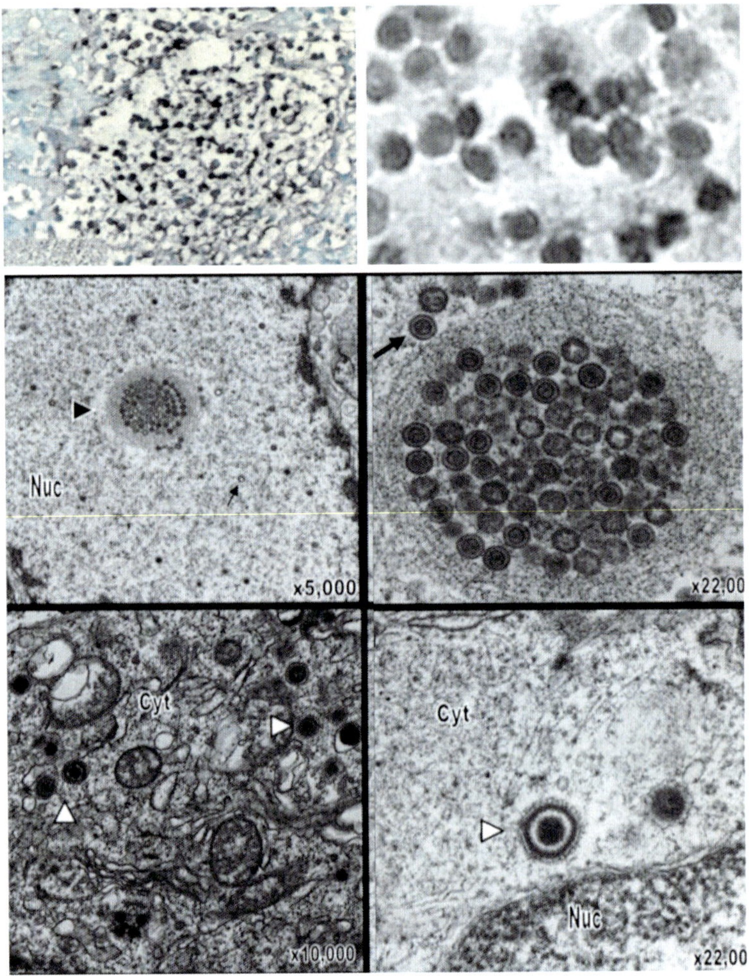

Fig. 1 VZV infection of T cells in thymus/liver (T cell) xenografts in the SCID mouse model. On day 7 after infection, infected T cell xenografts were tested for VZV DNA by in situ hybridization; darkly stained cells indicate VZV DNA in T cells visualized at lower (*left*) and higher magnification (*right*; ×786). By electron microscopy on day 14 after infection, VZV nucleocapsids were present in the nuclei with most containing a dark VZV DNA core (*large arrows*); some empty capsids were also seen (*small arrows*). Virions were abundant in T cell nuclei, both individually and in clusters (*arrowhead*), also shown at higher magnification. Complete virions were found in the cytoplasm. Magnification and locations of cytoplasm (cyt) and nucleus (nuc) are as indicated. From Moffat et al. 1995 and Schaap et al. 2005; reproduced with permission

the cytokine-rich milieu. When infected T cells were treated with phorbol ester, the frequency of VZV-positive T cells increased twofold, indicating that VZV DNA was present and that expression of VZV genes of the putative α, β, and γ kinetic classes was inducible by T cell activation. VZV also preferentially infected memory T cell subpopulations that expressed the skin homing markers, cutaneous

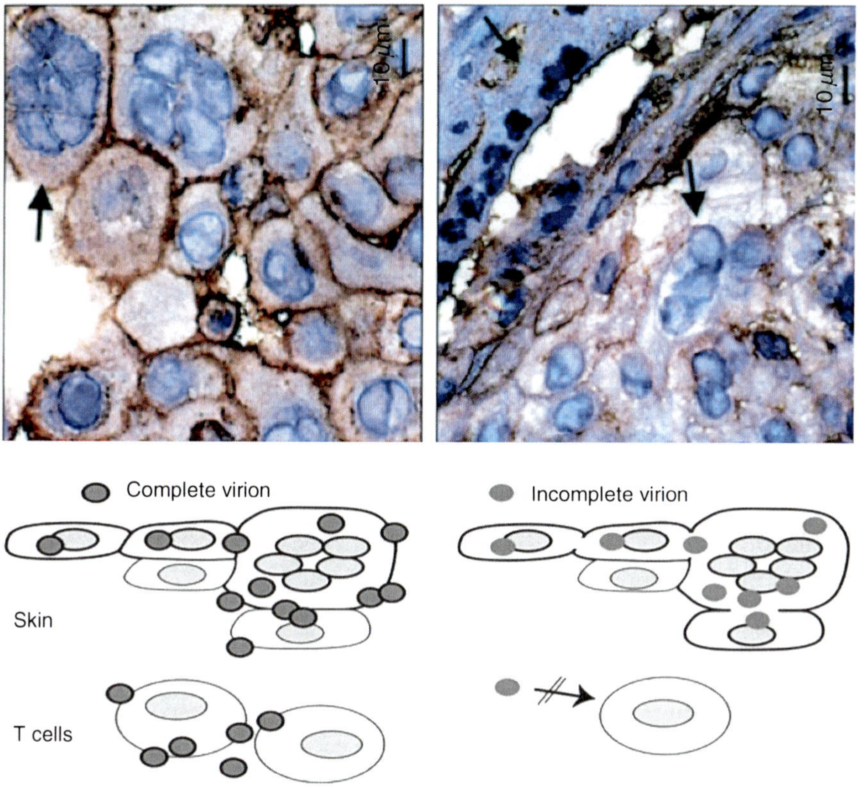

Fig. 2 Role of cell fusion and polykaryocyte formation in VZV infection of skin xenografts. The *upper panels* show virion formation in skin xenografts infected with VZV (*left panel*) and kinase-defective ORF47 mutant (*right panel*) and examined at day 21 by staining with antibody to VZV gE and hematoxylin. Transmission microscopy demonstrated many intact viral particles in rOka-infected skin cells. Viral particles were incomplete and fewer in rOka47D-N infected skin cells. *Arrows* indicate polykaryocytes with dark staining of gE in plasma membranes of rOka-infected cells and persistent formation of polykaryocytes in skin infected with the ORF47 kinase mutant despite minimal gE localization to plasma membranes. As illustrated, virus-induced cell fusion, with the characteristic formation of polykaryocytes, supports VZV infection in skin but VZV infection does induce cell fusion. ORF47 kinase mutants do not infect T cells, suggesting that VZV T cell tropism depends upon efficient assembly and release of complete virions. (Adapted from Besser et al. 2004; figures reproduced with permission)

leukocyte antigen (CLA), and chemokine receptor 4 (CCR4). Targeting memory T cells that are programmed for immune surveillance, and particularly those that have skin homing potential, would be expected to enhance T cell-mediated transport of VZV from the initial site of inoculation in the naïve host to the skin.

Transport of VZV by infected T cells to skin and the formation of skin lesions. Subsequent experiments combining our SCID mouse skin xenograft model with the method to infect tonsil T cells *in vitro* confirmed that VZV-infected tonsil T cells that are introduced into the mouse circulation by tail vein inoculation can migrate

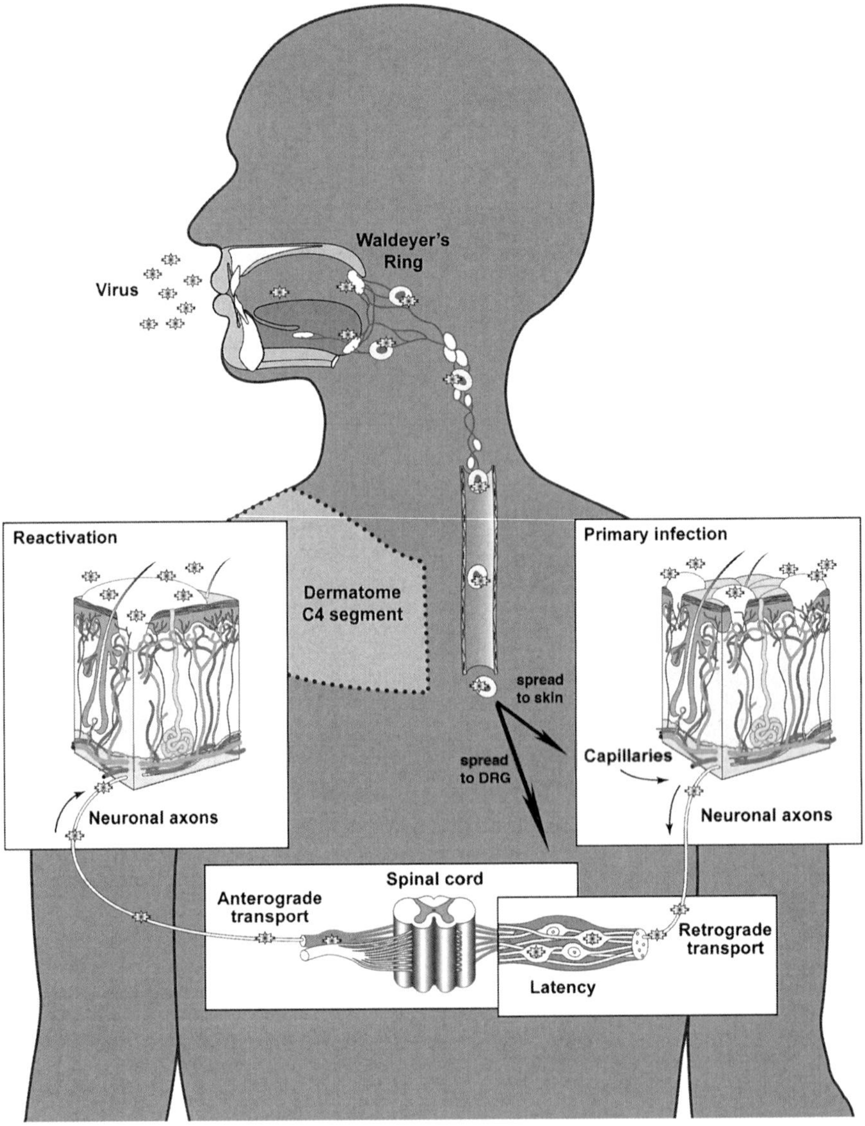

Fig. 3 A model of the pathogenesis of primary and recurrent VZV infection. VZV infection is acquired by inoculation of mucosal epithelial cells via the respiratory route, transfer across epithelial layers allows infection of T cells in tonsils and other lymphoid tissue that comprise Waldeyer's ring and allows transport to the skin via a T cell-associated viremia. Infection of skin produces the vesicular rash characteristic of varicella. VZV may reach sensory ganglia by T cell viremia or during skin infection, VZ virions may gain access to the sensory nerve cell body by retrograde axonal transport; lifelong latent infection is established in sensory ganglia in the course of primary VZV infection. Clinical reactivation of latent VZV results in herpes zoster, during which VZ particles gain access to skin via anterograde axonal transport. Adapted from Zerboni and Arvin 2008

into skin xenografts and deliver infectious virus to epidermal cells in which replication and lesion formation ensues (Ku et al. 2004). T cells were detected in skin xenografts, particularly near hair follicles, within 24 h after being introduced intravenously. Memory CD4 T cells predominated in skin xenografts after infusion of uninfected or infected cells, further supporting the concept that VZV exploits immune surveillance by migratory T cells for viral transport to skin. Thus, the cell-associated viremia necessary for the pathogenesis of primary VZV infection and the occurrence of the characteristic varicella skin lesions appears to depend upon VZV T cell tropism and is enhanced by preferential infection of T cell subpopulations that are most likely to cross the capillary endothelium into skin by the diapedesis mechanisms that these cells use for immune surveillance functions (Fig. 4).

When the progression of VZV infection in skin xenografts after transfer of virus by T cells was examined, we observed that the appearance of skin lesions at the surface of the skin xenografts required an interval of 10–21 days. Infected epidermal cells were detected in small foci that enlarged gradually over this prolonged period, which was in marked contrast to the typical lysis of VZV-infected cells *in vitro* within 48–72 h. Since SCID mice lack the capacity to mount an adaptive

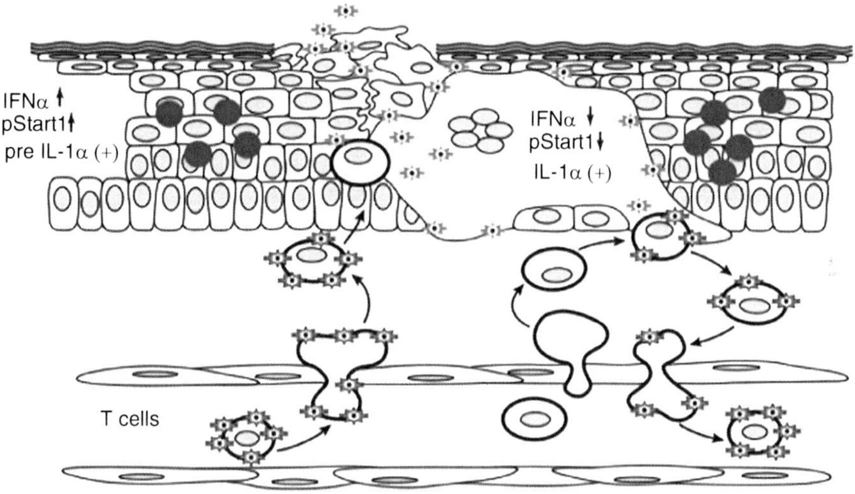

Fig. 4 Events in the pathogenesis of VZV skin infection. According to the model of primary VZV pathogenesis, T cells within the lymphoid tissues of Waldeyer's ring become infected by VZV transfer following the initial inoculation of respiratory epithelial cells with the virus. Infected T cells enter the circulation and rapidly transport VZV to the skin, exiting from capillaries across endothelial cells by diapedesis, as occurs during the usual trafficking of T cells through tissues. VZV is then released by the infected T cells at skin sites of replication. The 10–21 day incubation period observed after exposure of naïve individuals to VZV is the interval required for VZV to overcome the innate IFN-α response mounted by epidermal cells and create a lesion that reaches the skin surface, IFN-α production by epidermal cells that surround those infected with VZV prevents a rapid, uncontrolled cell–cell spread that would otherwise incapacitate the host. Viremia may be amplified by trafficking of uninfected T cells through areas where skin lesions are forming. (Adapted from Ku et al. 2005; figure reproduced with permission)

immune response, we predicted that this phenomenon might be the result of innate host antiviral responses. Induction of the interferon (IFN) pathway is well known to be highly restrictive of herpes viruses. Evaluation of VZV-infected skin xenografts revealed that IFN-α was not expressed in the clusters of epidermal cells that were actively infected, as shown by expression of VZV protein markers (Ku et al. 2004) (Fig. 5). In contrast, IFN-α production was prominent in epidermal cells within the xenograft that did not exhibit cyopathic changes or viral protein synthesis; up-regulation of the IFN pathway in these cells was also documented by the phosphorylation and nuclear translocation of Stat1. The relevance of this response for the pathogenesis of VZV skin infection was documented by inoculating skin xenografts with VZV and treating the mice with antibody to block the human IFN-α/b receptor or a control antibody. Interference with the up-regulation of the IFN response was associated with a tenfold increase in infectious virus recovery and with accelerated and extensive skin lesion formation in treated animals. In other experiments, we showed that VZV also inhibits the NF-kB-mediated innate

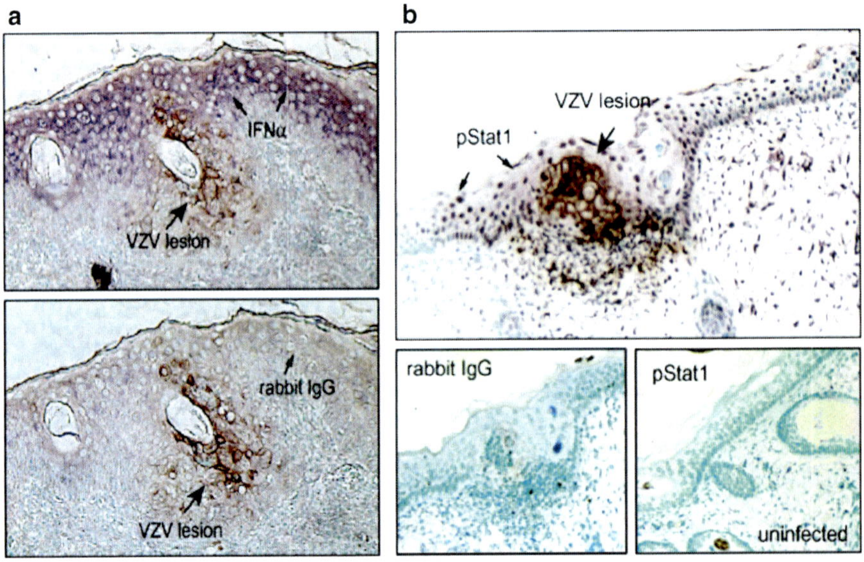

Fig. 5 IFN-α expression and Stat1 phosphorylation in VZV-infected and uninfected epidermal cells. Formalin-fixed, paraffin-embedded sections of VZV-infected skin xenografts were pretreated and double stained with anti-VZV IgG, detected with DAB (*brown*), and anti-IFNα or anti-pStat1 antibody detected with Vector VIP (*purple*). Sections are shown at magnification ×200. (**a**) Infected cells were identified by VZV protein (*brown*) expression. In double labeled skin sections, IFN-α (*purple*) expression was prominent in adjacent uninfected cells but not in VZV-infected cells (*top*) compared with sections stained with rabbit IgG as a control for IFN-α (*bottom*). (**b**) In double labeled sections, phosphorylated Stat1 (*purple*) was up-regulated in adjacent uninfected cells but absent in VZV-infected cells (*top*); pStat1 was not detected in uninfected skin (*bottom right*) compared with rabbit IgG control stain (*bottom left*). From Ku et al. 2004; reproduced with permission

response in VZV-infected skin cells while this pathway is up-regulated in adjacent uninfected cells (Jones and Arvin 2003, 2006).

The analysis of VZV-infected skin xenografts made it possible to evaluate whether VZV replication up-regulated the expression of endothelial cell adhesion molecules known to enhance the entry of effector cells that mediate the adaptive immune response. Comparing levels of E-selectin, ICAM-1, VCAM-1, and chemoattractant receptors in cells within VZV-infected skin xenografts with expression in cells in lesion biopsies taken from healthy individuals with varicella or zoster showed that these proteins were detected abundantly in VZV lesion biopsy specimens but not in skin xenografts. Thus, while virus-infected T cells deliver VZV to skin, initial VZV replication in skin does not trigger expression of molecules that would immediately attract immune cells with the capacity to lyse infected cells and interrupt cell–cell spread. Nevertheless, the usual migration of immune monitoring T cells through skin could amplify the numbers of VZV-infected T cells that then traffic to new skin sites before adaptive immunity is elicited (Fig. 4). Amplification of cell-associated viremia may also occur in other reticuloendothelial tissues, for example, liver and spleen, during this interval.

The role of T cell tropism in the pathogenesis of VZV infection. These observations suggest that VZV infects T cells in tonsils and other regional lymphoid tissues of the upper respiratory tract, and that these infected T cells can transport VZV to skin within a short time after entering the circulation. These experiments make it possible to attribute the relatively prolonged interval between exposure of the naïve host to VZV and the first appearance of the varicella rash to the potent innate responses of epidermal cells that inhibit the cell-to-cell spread of VZV in skin; after VZV is delivered to skin, it must overcome these barriers before lesions are evident at the skin surface (Figs. 3 and 4). By this model, the detection of cell-associated viremia in the last few days of the incubation period does not result in a very rapid formation of skin lesions; instead, it is likely to represent the infection of T cells as they enter and exit from skin sites of active VZV replication during the period that VZV is spreading from cell to cell towards the skin surface. This highly modulated process can be expected to facilitate and enhance the persistence of VZV in the population, since an infection that incapacitates the host would be predicted to reduce its opportunities for transmission to other susceptibles.

2 Defining Determinants of VZV Tropism for T cells and Skin in the SCID Model *In Vivo*

Combining the SCID mouse model to study VZV pathogenesis in human T cell and skin xenografts with the capacity to make targeted mutations in the VZV genome has allowed investigations of the genetic requirements for VZV virulence in these target cells within their tissue microenvironment *in vivo*. As summarized later, we have evaluated how mutations that block expression alter the levels of expression or disrupt functional domains of the VZV glycoprotein, gB, the tegument/regulatory proteins, IE62 and IE63, the viral kinases encoded by ORF47 and ORF66, and genes of the conserved ORF9-12 cluster, as well as mutations of VZV gene promoter elements affect VZV tropism for T cells and skin (Table 1). These mutants were also characterized for their effects for VZV replication in cultured cells, which are described in detail in each report cited; consequences for growth kinetics and plaque size are summarized in Table 1. We have also used the *in vitro* T cell assay to identify gene products required for T cell entry in experiments with these mutants, as indicated in Table 1. The contributions of VZV gH was examined using anti-gH antibody-mediated interference with its functions in skin pathogenesis. Mutational analyses of the functions of gE, gI, and their promoters and their consequences for VZV replication *in vitro* and virulence in skin and T cells *in vivo* are described in the chapter on VZV gE and gI.

Table 1 Effects of selected mutations in VZV genes and promoters on the tropism of varicella-zoster virus for T cells and skin in the SCID mouse model

	Cell culture *in vitro*				SCIDhu mice *in vivo*	
	Virus	Growth kinetics	Plaque size	T cell entry	Skin xenografts	T cell xenografts
Recombinant Oka	+	+++	Normal		++	++
Glycoproteins						
ΔgB 491-RSSR-494	+	+++	Normal		+	
gB 428-GSGG-431	+	+++	Normal		+	
Viral kinases						
ORF47- ΔC (C-terminal deletion)	+	++	Normal		+	–
ORF47P-S1 (kinase motif mutant)	+	++	Normal		++	
ORF47P-S2 (kinase motif mutant)	+	++	Normal		++	
ORF47D-N (kinase motif mutant)	+	++	Normal		+	–
ORF66S (stop codon mutant)	+	+				+
ORF66 S48A (kinase domain)	+					
ORF66 S331A (kinase domain)	+					
ORF66 G102A (kinase domain)	+	++			+	+/–
ORF66 S250P (kinase domain)	+	++			++	++
Tegument/regulatory proteins						
ΔORF62 (complete deletion)	+	+++	Normal		++	
	+	+++	Normal		++	

(*continued*)

Table 1 (continued)

	Cell culture *in vitro*				SCIDhu mice *in vivo*	
	Virus	Growth kinetics	Plaque size	T cell entry	Skin xenografts	T cell xenografts
ΔORF71 (complete deletion)						
ΔORF62/71 (double deletion)	–					
Δ63/70 (double deletion)	–					
Δ63 (single copy deletion)	+	++	Normal		++	++
ORF63 T171 (phosphorylation)	+	+	Small		+	++
ORF63 S181 (phosphorylation)	+	+	Small		+	++
ORF63 S185 (phosphorylation)	+	+	Small		+	++
ORF9-10 cluster						
ORF9	–					
ΔORF10 (complete deletion)	+	+++	Normal		+	++
ORF10 P28A (acidic domain)	+	+++	Normal		++	
ORF10 P28S (acidic domain)	+	+++	Normal		++	
ΔORF11	+	+++	Normal	+	+/–	
ΔORF12	+	+++	Normal	+	++	
ΔORF10/11	+	+++	Normal		+/–	
ΔORF10/11/12	+	+++	Normal	+	+/–	
ΔORF11/12	+	+++	Normal		+/–	
Promoter mutants						
ORF10-proΔ+39/+63	+	+++	Normal		++	
ORF10-proΔUSF	+	+++	Normal		+	

3 The Role of VZV Glycoproteins in T cell and Skin Infection

VZV gB. Information about VZV gB is quite limited. However, we found that modeling predicts a structure similar to HSV gB (Oliver et al. 2009). Nevertheless, in contrast to HSV gB, VZV gB has a furin recognition motif, which is predicted to result in gB cleavage. Substitutions in the primary fusion loop, W180G and Y185G, the deletion Δ^{491}RSRR^{494}, and point mutations ^{491}GSGG^{494} in the furin recognition motif did not affect gB expression or cellular localization. VZV was recovered from pOka BACs that had either the Δ^{491}RSRR^{494} or ^{491}GSGG^{494} mutations (Table 1) but not when the point mutations W180G and Y185G were introduced. Thus, mutagenesis demonstrated that residues in the primary fusion loop of gB were essential but gB cleavage was not. Virion morphology, protein localization, and replication were unaffected for the gBΔ^{491}RSRR^{494} and gB^{491}GSGG^{494} viruses *in vitro*. Nevertheless, deleting the furin recognition motif and blocking gB cleavage caused attenuation in skin xenografts *in vivo*. Viral titers were lower at both 10 and 21 days after inoculation of skin xenografts with pOka-gBΔ^{491}RSRR^{494} compared to wild-type pOka; the titer reductions associated with mutagenesis of the cleavage site were 1.5 $\log_{10}$ pfu (31-fold; day 10) and 1.0 $\log_{10}$ pfu (tenfold; day 21). These experiments provided the first evidence that cleavage of a herpes virus fusion protein contributes to pathogenesis *in vivo*, as has been observed for fusion proteins in other virus families.

VZV gH. The glycoprotein gH is a highly conserved herpes virus protein that has been shown to have functions in virus entry and cell–cell spread when evaluated in other herpes viruses. Little is known about the functions of VZV gH. In experiments to study gH functions in VZV pathogenesis, we took advantage of the availability of the murine monoclonal antibody against gH, MAb 206, which is known to neutralize VZV and inhibit cell fusion *in vitro*. When given to SCID mice that had skin xenografts infected with VZV before treatment, we found that anti-gH antibody administration starting 6 h after inoculation and continuing for 12 days reduced the number of skin xenografts that became infected by 60%; in xenografts that became infected, virus titers, genome copies, and lesion extent were reduced (Vleck et al. 2010) (Fig. 6). In contrast, initiating anti-gH antibody 4 days after inoculation suppressed but did not block VZV replication. *In vitro*, anti-gH antibody bound to plasma membranes and to surface virions. Notably, anti-gH antibody was also internalized into vacuoles within infected cells, associated with intracellular virions and colocalized with markers for early endosomes and the multivesicular body pathway but not the trans-Golgi network. Thus, antibody-mediated interference with gH has the capacity to block virion transfer into uninfected cells and cell–cell fusion and anti-gH antibody can enter cells where it may bind to gH or virions, which may be targeted for degradation. As a consequence, anti-gH antibody can prevent (if given very early) or modulate VZV skin infection *in vivo*.

The role of VZV regulatory/tegument proteins in VZV pathogenesis. Like other herpes viruses, the VZV tegument contains important regulatory proteins that are available to initiate replication when the infectious particle is uncoated after entering the target cell. Of these, we have examined the contributions of IE62 and IE63 to T cell and skin tropism.

VZV IE62. IE62 is a major immediate early transactivating protein that induces many VZV genes that have been evaluated; it is encoded by duplicated genes, ORFs 62 and 71 (Cohen et al. 2007). When the effects of mutations in these genes on VZV replication *in vitro* were evaluated, transfections using pOka cosmids from which ORF62, ORF71, or the ORF62/71 gene pair were deleted showed that at least one copy of ORF62 was required (Sato et al. 2003). Inserting ORF62 from pOka or vOka into a non-native site in U_S allowed VZV growth in cell culture *in vitro*, although plaque size and virus titers were decreased (Table 1). Targeted mutations in binding sites reported to affect interactions with IE4 or a putative ORF9 protein binding site were compatible with replication. In single deletions of ORF62 or ORF71, recombination events repaired the defective repeat region in some progeny viruses *in vitro* and in some skin xenografts. Although insertion of ORF62 into the non-native *Avr*II

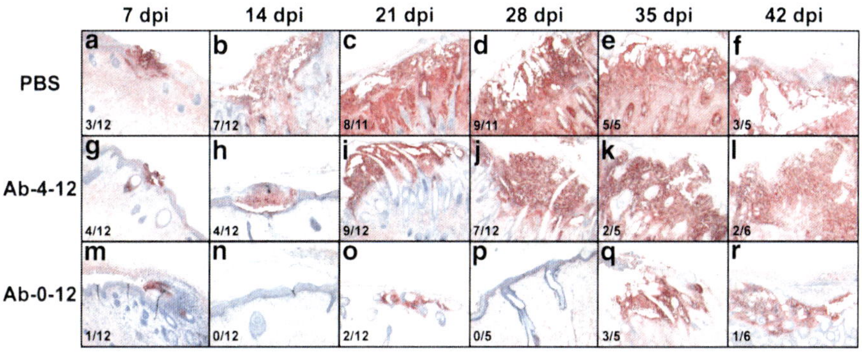

Fig. 6 The formation of lesions in VZV-infected human skin xenografts treated with the anti-gH mAb 206 for 0–12 days or 4–12 days post inoculation. Lesions in skin xenografts inoculated with pOka-infected HELF were identified by VZV gE expression, and were counterstained with hematoxylin. Representative lesions are shown at each timepoint: 7 dpi, A, G, M. 14 dpi, B, H, N. 21 dpi, C, I, O. 28 dpi, D, J, P. 35 dpi, E, K, Q. 42 dpi, F, L, R. 42 The total number of xenografts with lesions is shown in the lower 1 left of each panel. Lesions were identified in xenografts from the PBS (A–F) and Ab-4-12 (G–L) groups at all time points. Lesions were only observed in xenografts from the Ab-0-12 group (M–R) at 7, 21, 35 and 42 dpi. Representative uninfected xenografts are shown for the Ab-0-12 group at 14 and 28 dpi. Syncytia were seen in all lesions. Magnification: ×50. From Vleck et al. 2010, reproduced with permission

site permitted growth in cell culture, IE62 expression from its native gene locus was necessary for cell–cell spread in skin xenografts *in vivo*.

VZV IE63. In addition to IE62, we have used cosmid mutagenesis to analyze the functional domains of IE63, also an immediate early regulatory/tegument protein, which is encoded by duplicate genes, ORFs 63, and 70, (Baiker et al. 2004). In this work, 22 ORF63 mutations were also evaluated for effects on IE63 binding to IE62 and on IE63 phosphorylation and nuclear localization in transient transfections. By sequence, IE63 was shown to resemble the HSV-1 $U_S1.5$ protein, which is expressed colinearly with ICP22 (U_S1). The IE62 binding site was mapped to IE63 aa55-67, with R59/L60 being critical residues. Alanine substitutions within the IE63 center region showed that S165, S173, and S185 were phosphorylated by cellular kinases. Four mutations that changed two putative NLS sequences altered IE63 distribution to a cytoplasmic/nuclear pattern. Only three of 22 mutations in ORF63 were compatible with recovery of infectious VZV, although IE63 is not absolutely essential when IE62 is provided as an ORF62 plasmid (Cohen et al. 2005). The three viable IE63 mutants had alanine substitutions altering T171, S181, or S185 (Table 1). These mutants produced less infectious virus and had decreased plaque sizes *in vitro*. ORF47 kinase and gE synthesis was reduced, showing that IE63 contributes to optimal expression of early and late gene products. The virulence of the three IE63 mutants was markedly attenuated in skin xenografts but not in T cell xenografts, indicating that the contribution of the IE63 tegument/regulatory protein depends upon the differentiated human cell type that is being targeted within the intact tissue microenvironment during VZV pathogenesis (Fig. 7).

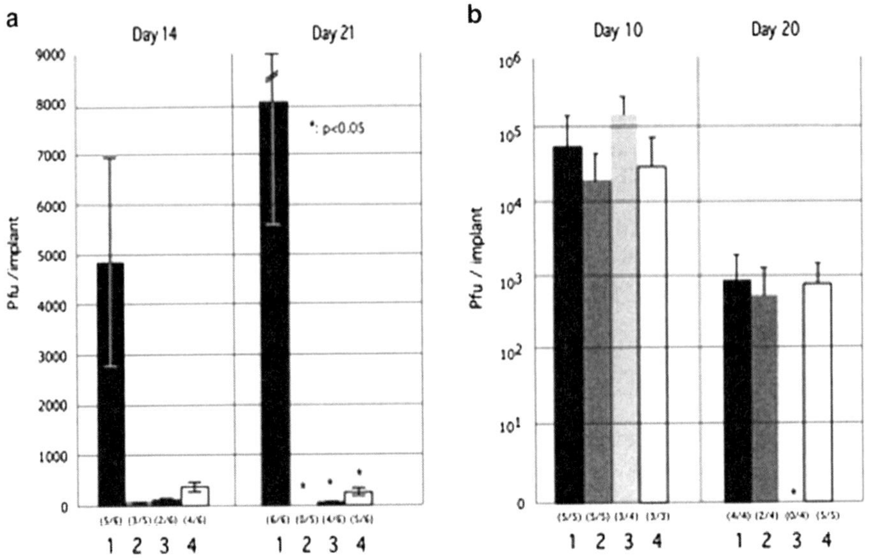

Fig. 7 Replication of IE63 mutant viruses in skin and T cell xenografts in SCID-hu mice. Skin xenografts in SCID mice were injected with (*left to right*) rOKA, rOKA/ORF63rev[T171], rOKA/ORF63rev[S181], or rOKA/ORF63rev[S185] having equivalent inoculum titers. Virus titers in skin xenografts were assessed after harvest at day 14 (A, *left panel*) and day 21 (A, *right panel*) after inoculation and were graphed as mean titers for xenografts that yielded infectious virus, with lines indicating standard errors. The number of xenografts from which infectious virus was recovered per number that were inoculated is given in parentheses below the horizontal axis. The *P* values were <0.05 when titers of rOka and each of the IE63 mutant viruses were compared at day 21. Replication of VZV recombinants in T cells was assessed at days 10 and 20 (B). *Lines* indicate the standard errors. From Baiker et al. 2004; reproduced with permission

The role of the VZV kinases in T cell and skin tropism. VZV encodes two viral kinases, ORF47 and ORF66. ORF47 is a conserved herpes virus gene and ORF66 is present only in the alphaherpesviruses. Deletion of both of these viral kinases is compatible with VZV replication in cultured cells (Cohen et al. 2007); however, both have required functions for VZV pathogenesis in T cells and skin. Initial experiments with ORF47 and ORF66 stop codon mutants demonstrated that inoculation of T cell and skin xenografts showed that ORF47 protein was required for viral growth in human T cells and skin; virulence was restored when ORF47 was expressed from a non-native site in the VZV genome (Moffat et al. 1998a, b). Blocking ORF66 expression impaired VZV infection of T cells but not skin compared with intact VZV.

ORF47 protein. We investigated the role of the ORF47 protein kinase further by constructing VZV recombinants with targeted mutations in conserved motifs of ORF47 and expressing a truncated ORF47 (Table 1) (Besser et al. 2003). Both the C-terminal truncation mutant, rOka47ΔC, and the mutation disrupting the DYS kinase motif in rOKa47Δ-N eliminated ORF47 kinase activity and caused extensive nuclear retention of ORF47 and IE62 proteins *in vitro*. Disrupting ORF47 kinase function also resulted in a marked decrease in VZV replication and cutaneous lesion formation in skin xenografts *in vivo*. However, infection *in vivo* was not blocked completely as long as ORF47 protein binding to IE62 protein was preserved; this function was mapped to the ORF47 N-terminus. These experiments indicated that ORF47 kinase activity is critically important for VZV infection and cell–cell spread in human skin *in vivo*, but suggested that complex formation with IE62, rather than kinase activity, is the essential contribution of ORF47 protein to VZV replication *in vivo*.

In subsequent experiments, we found a differential requirement for cell fusion and virion formation in the pathogenesis of VZV infection in skin and T cells (Besser et al. 2004). The rOka47ΔC and rOka47D-N mutants did not infect human T cell xenografts but remained infectious in skin. Epidermal cell fusion persisted and some VZV polykaryocytes were generated in skin infected with rOka47ΔC and rOka47D-N (Fig. 2). Virion assembly was impaired *in vitro*, but cell fusion continued, causing characteristic VZV syncytia in cultured cells infected with rOka47ΔC or rOka47D-N. Intracellular trafficking of envelope gE and the ORF47 and IE62 proteins was aberrant without ORF47 kinase activity. In skin, ORF47 mutants exhibited cell–cell spread even though EM studies showed markedly defective virion formation. In contrast, VZV-infected T cells do not undergo cell fusion, and the impaired virion assembly due to ORF47 mutations effectively eliminated T cell infection. Thus, we concluded that skin infection can proceed if some cell fusion occurs, whereas VZV T cell tropism is much more dependent on full assembly of infectious virus particles.

ORF66 protein. As noted earlier, VZV infection of T cells is associated with robust virion production and modulation of apoptosis and IFN pathways (Schaap et al. 2005). The serine/threonine protein kinase encoded by ORF66 is needed for efficient vOka replication in T cells (Table 1). Preventing ORF66 expression by a stop codon impaired pOka growth in T cell xenografts *in vivo,* reduced VZ virion formation in T cells, increased the susceptibility of infected T cells to apoptosis, and reduced the capacity of the virus to interfere with induction of the IFN signaling pathway following exposure to IFNγ (Fig. 8). However, preventing ORF66 protein expression reduced growth in cultured cells only slightly and did not diminish virion formation *in vitro*. The ORF66 stop codon mutant also showed only a slight growth defect in skin compared with pOka. These observations suggested that ORF66 kinase has a unique role during infection of T cells and supports VZV T cell tropism by enhancing survival of infected T cells and contributing to immune evasion.

Subsequent experiments showed that an ORF66 mutant with a G102A substitution had defective kinase function, which was demonstrated as a block in autophosphorylation (Schaap-Nutt et al. 2006). Blocking ORF66 kinase function also prevented late IE62 localization to the cytoplasm *in vitro*, whereas an S250P substitution had no effect on either of these ORF66 functions. Both kinase domain mutants replicated to titers equivalent to pOka *in vitro* (Table 1). pOka66G102A had slightly reduced growth in skin, which was comparable to the reduction observed with the stop codon mutant, pOka66S. In contrast, infection of T cell xenografts with pOka66G102A was associated with a significant decrease in

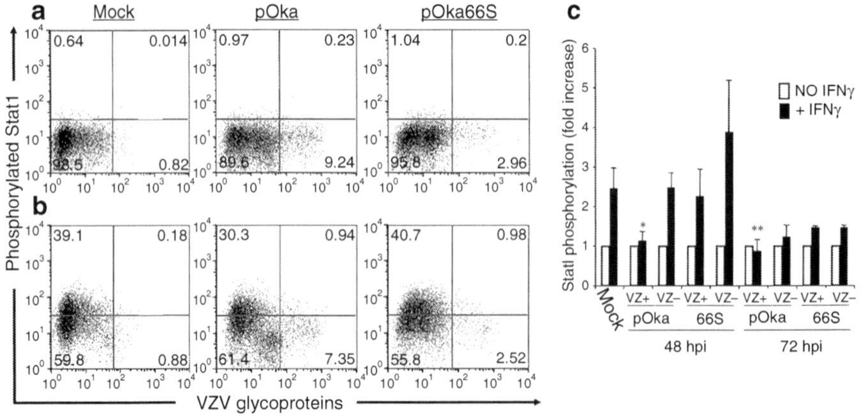

Fig. 8 Flow cytometric analysis of Stat1 phosphorylation in human tonsil T cells stimulated with IFN-γ. Column-purified human tonsil T cells were cocultured with VZV-infected HEL monolayers. After 48 h, cells were removed from the monolayer and either left unstimulated (**a**) or stimulated with recombinant human IFN-γ for 10 min at 37°C (**b**). Cells were fixed in paraformaldehyde, stained with antibodies and fluorescent conjugates to VZV proteins, permeabilized in methanol, and then stained with antibodies to phospho-Stat1 and CD3. FACS plots show anti-phospho-Stat1 versus anti-VZV staining of uninfected (*left plots*), pOka-infected (*middle plots*), and pOka66S-infected (*right plots*) tonsil T cells (gated on $CD3^+$ cells) without stimulation (**a**) or following stimulation with IFN-γ (**b**). (**c**) Data from T cells cultured for 48 or 72 h with infected HEL monolayers are shown as the average fold increase in phospho-Stat1 fluorescence intensity following IFN-γ treatment in VZV-infected and uninfected T cells for four independent experiments done after 48 h and two experiments combined for 72 h. *One asterisk* indicates a fold increase that is significantly different from that of uninfected cells from the same culture with $P = 0.01$, while *two asterisks* indicate a difference with $P = 0.02$. From Schaap et al. 2005; reproduced with permission

infectious virus production equivalent to the impaired T cell tropism found with pOka66S. Of interest, disrupting kinase activity with the G102A mutation did not alter IE62 cytoplasmic localization in VZV-infected T cells, suggesting that decreased T cell tropism is due to other ORF66 protein functions. The G102A mutation reduced the antiapoptotic effects of VZV infection of T cells. Thus, VZV T cell tropism depends upon ORF66 kinase activity.

Contributions of the ORF9-12 gene cluster to VZV pathogenesis. VZV has a conserved gene cluster found in other herpes viruses that includes ORF9, ORF10, ORF11, and ORF12. ORF9 is an essential gene, but deleting the other genes in the cluster, either alone or in combination, is compatible with VZV replication *in vitro*.

ORF10 protein. VZV ORF10 encodes a tegument protein that enhances transactivation of VZV genes and has some homology to HSV-1 VP16 (α-TIF). While VP16 is essential for HSV replication, ORF10 is dispensable *in vitro*. To analyze ORF10 functions, we made pOkaΔ10 and mutated the acidic activation domain and the putative motif for binding human cellular factor-1 (HCF-1) (Che et al. 2006). No effects on VZV replication, IE gene transcription, or virion assembly occurred *in vitro*, but deleting ORF10 reduced infection in skin xenografts *in vivo* (Table 1).

Epidermal cells infected with pOkaΔ10 had fewer DNA-containing nucleocapsids and complete virions; extensive aggregates of intracytoplasmic viral particles were also observed. In contrast, deleting ORF10 did not impair VZV T cell tropism *in vivo*. Altering the activation or putative HCF-1 domains of ORF10 protein had no consequences *in vivo*. Thus, ORF10 protein is required for efficient VZ virion assembly and is a determinant of VZV virulence in skin *in vivo*. Mutagenesis of the ORF10 promoter showed that an USF binding site was important for skin tropism, although no effect on VZV replication was observed in cultured cells (Che et al. 2007).

Further analysis of the ORF9-12 gene cluster showed that recombinants lacking ORF10/11, ORF11/12, or ORF10/11/12 had normal growth, virion assembly, and infectivity in tonsil T cells (Table 1) (Che et al. 2008). Deleting ORF12 had no effect in skin but ORF11, 10/11, and 11/12 mutants were attenuated and virus was not recovered after deleting ORF10/11/12, indicating that the intact ORF10-ORF12 cluster is important in skin *in vivo* (Fig. 9). ORF9 was found to be incorporated into virion tegument and coprecipitated with gE. Work in progress indicates that ORF11 is an even more critical gene product for the pathogenesis of skin infection than ORF10, whereas ORF12 appears to be completely dispensable.

Comparison of the T cell and skin tropims of parent Oka and vaccine Oka viruses. VZV is the only human herpes virus for which licensed vaccines are available. These vaccines are made from the parent Oka (pOka) virus, a Japanese clinical isolate that was attenuated empirically by passage in human and guinea pig embryo fibroblasts, yielding vaccine Oka (vOka) (Gershon 2001). This virus stock is used to make varicella vaccines and the high potency live attenuated vaccine for zoster prevention. We used the SCID mouse model to compare the virulence of vOka in skin xenografts with pOka, another low-passage clinical isolate, and the Ellen strain, which is a highly passaged laboratory virus (Moffat et al. 1998a, b). Although replication of these viruses was not different in cultured cells, pOka and the other clinical isolate showed the most extensive lesion formation and produced the highest titers of infectious virus in skin xenografts. These experiments provided the first evidence of a genetic basis for vOka attenuation and suggested that it results from the accumulation of mutations resulting from tissue culture passage (Arvin 2001b). Of interest, the highly passaged Ellen strain was avirulent in skin xenografts, indicating that prolonged adaptation *in vitro* can completely abrogate the pathogenic potential of VZV.

The genetic basis of vOka attenuation in skin xenografts was analyzed further by making chimeras from five fragment pOka and vOka cosmids (Zerboni et al. 2005a, b). The virulence of these pOka/vOka chimeras was variable in skin, suggesting that vOka attenuation is due to mutations in various VZV genes that are located in different regions of the genome. Extensive sequencing of the vaccine Oka stock has demonstrated that it represents a mixture of genomes with different mutations, as described in the chapters on VZV and vaccine genomics. Importantly, the comparative evaluation of pOka and vOka replication in T cell xenografts showed no attenuation of T cell tropism *in vivo*, which is consistent with the capacity of the vaccine virus to cause viremia and a varicella-like illness in immunocompromised

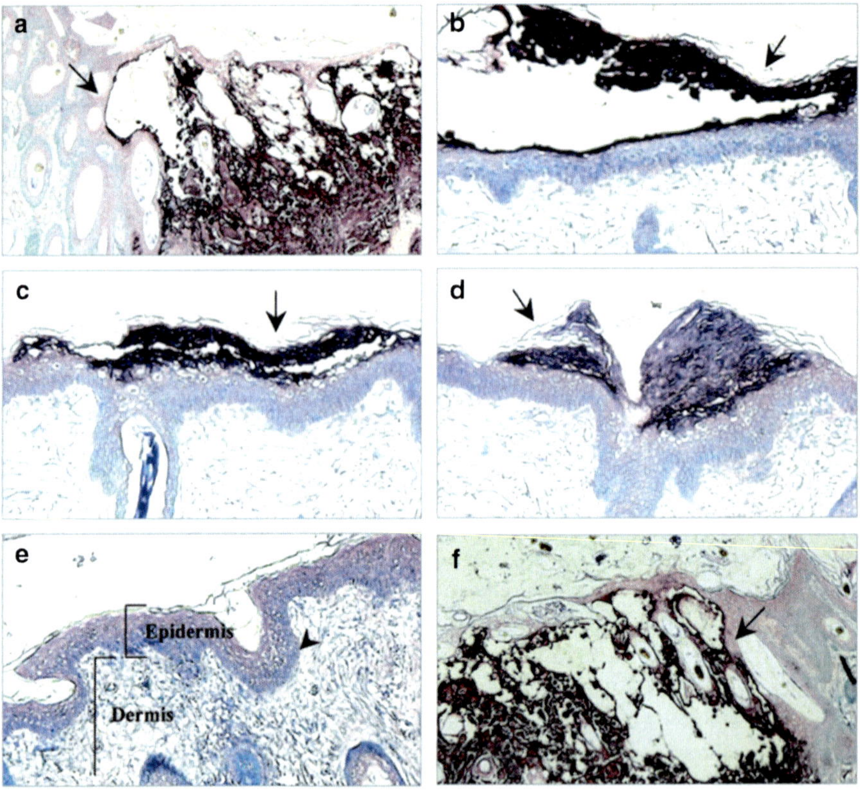

Fig. 9 Lesion formation in skin xenografts infected with ORF10-to-ORF12 cluster gene mutants. Deparaffinized skin sections harvested at days 21 after infection were incubated with human polyclonal anti-VZV IgG and stained with biotinylated secondary antibody (magnification, ×10). Sections representative of skin xenografts infected with each virus are shown. Large lesions were observed for skin infected with POKA (**a**) and POKAΔ12 (**f**). Small lesions (*arrows*) restricted to the epidermal layer were observed for skin infected with POKAΔ11 (**b**), POKAΔ10/11 (**c**), and POKAΔ11/12 (**d**). POKAΔ10/11/12 (**e**) did not produce detectable skin lesions. The normal structures of the epidermis, the dermis, and the basement membrane that separates the epidermis from the underlying dermis are indicated in this section. From Che et al. 2008; reproduced with permission

children (Gershon 2001). These experiments suggest that functions required for T cell infection are not affected by attenuation achieved through passage in fibroblasts.

4 Conclusion

VZV remains a medically important pathogen despite the availability of live attenuated vaccines and antiviral drugs. Information about the determinants of VZV tropism for T cells and skin has the potential to yield new approaches to a "genetically engineered" vaccine and new targets for antiviral drugs. In particular, a vaccine based on a VZV recombinant virus that has impaired infectivity for T cells should be less likely to disseminate in high risk patients or to establish latency in individuals.

Acknowledgments This summary describes the accomplishments of the graduate students, post-doctoral fellows, and research staff whose investigations of VZV molecular virology and pathogenesis are reported in detail in their published work. These studies were supported by NIH grants AI053846, AI20459, and CA49605.

References

Arvin AM (2001a) Varicella-zoster virus. In: Knipe DM, Howley P (eds) Fields' virology, 4th edn. Lippincott-Williams & Wilkins, Philadelphia, pp 2731–2768

Arvin AM (2001b) Varicella vaccine: genesis, attenuation and efficacy. Virology 284:153–158

Arvin AM, Schaap AC, Ku C-C, Jones JO, Sommer M, Zerboni Z (2006) Investigations of the molecular mechanisms of varicella-zoster virus pathogenesis. In: Sandri-Golden R (ed) The alphaherpesviruses. Horizon Press Inc, UK

Baiker A, Bagowski C, Ito H, Sommer M, Zerboni L, Fabell K, Hay J, Ruyechan W, Arvin AM (2004) The immediate early 63 protein of varicella-zoster virus: analysis of functional domains required for replication *in vitro* and for T cell and skin tropism in the SCID hu model *in vivo*. J Virol 78:1181–1194

Besser J, Sommer MH, Zerboni L, Bagowski C, Ito H, Moffat J, Ku C-C, Arvin AM (2003) Differentiation of varicella-zoster virus ORF47 protein kinase and IE62 protein binding domains and their contributions to replication in human skin xenografts in the SCID-hu mouse. J Virol 77:5964–5974

Besser J, Ikoma M, Fabel K, Sommer MH, Zerboni L, Grose C, Arvin AM (2004) Differential requirement for cell fusion and virion formation in varicella-zoster virus infection of skin and T cells. J Virol 78:13293–13305

Che X, Zerboni L, Sommer MH, Arvin AM (2006) Varicella-zoster virus open reading frame 10 is a virulence determinant in skin cells but not in T-cells *in vivo*. J Virol 7:3238–3248

Che X, Berarducci B, Sommer M, Ruyechan WT, Arvin AM (2007) The ubiquitous cellular transcriptional factor USF targets the varicella-zoster virus open reading frame 10 promoter and determines virulence in human skin xenografts in SCIDhu mice *in vivo*. J Virol 81:3229–3239

Che X, Reichelt M, Sommer MH, Rajamani J, Zerboni L, Arvin AM (2008) Functions of the ORF9-to-ORF12 gene cluster in varicella-zoster virus replication and in the pathogenesis of skin infection. J Virol 82:5825–5834

Cohen JI, Seidel KE (1993) Generation of VZV and viral mutants from cosmid DNAs: VZV thymidine kinase is not essential for replication *in vitro*. Proc Natl Acad Sci USA 90:7376–7380

Cohen JI, Krogmann T, Bontems S, Sadzot-Delvaux C, Pesnicak L (2005) Regions of the varicella-zoster virus open reading frame 63 latency-associated protein important for replication *in vitro* are also critical for efficient establishment of latency. J Virol 79: 5069–5077

Cohen J, Straus S, Arvin A (2007) Varicella-zoster virus. In: Knipe DM, Howley P (eds) Fields' virology, 5th edn. Lippincott-Williams & Wilkins, Philadelphia, pp 2547–2586

Gershon AA (2001) Live attenuated varicella vaccine. Infect Dis Clin N Am 15:65–81

Gilden DH, Kleinschmidt-DeMasters BK, LaGuardia JJ, Mahalingam R, Cohrs RJ (2000) Neurologic complications of the reactivation of varicella-zoster virus. N Engl J Med 342:635–645

Jones JO, Arvin AM (2003) Microarray analysis of host cell gene transcription in response to varicella-zoster virus infection of human T cells and fibroblasts *in vitro* and SCID hu skin xenografts *in vivo*. J Virol 77:1268–1280

Jones JO, Arvin AM (2006) Inhibition of the NF-κB pathway by varicella-zoster virus *in vitro* and in human epidermal cells *in vivo*. J Virol 80:5113–5124

Kemble GW, Annunziato P, Lungu O, Winter RE, Cha TA, Silverstein SJ, Spaete RR (2000) Open reading frame S/L of varicella- zoster virus encodes a cytoplasmic protein expressed in infected cells. J Virol 74:11311–11321

Koropchak CM, Solem S, Diaz PS, Arvin AM (1989) Investigation of varicella-zoster virus infection of lymphocytes by *in situ* hybridization. J Virol 63:2392–2395

Ku C-C, Padilla J, Grose C, Butcher EC, Arvin AM (2002) Tropism of varicella-zoster virus for human tonsillar CD4+ T lymphocytes that express activation, memory and skin homing markers. J Virol 76:11425–11433

Ku C-C, Zerboni L, Ito H, Wallace M, Graham B, Arvin AM (2004) Transport of varicella-zoster virus to skin by infected CD4 T cells and modulation of viral replication by epidermal cell interferon-a. J Exp Med 200:917–925

Ku C-C, Besser J, Abendroth A, Grose C, Arvin AM (2005) Varicella-zoster virus pathogenesis and immunobiology: New concepts emerging from investigations in the SCIDhu mouse model. J Virol 79(5):2651–2658

Mallory S, Sommer M, Arvin AM (1997) Mutational analysis of the role of glycoprotein I in varicella-zoster virus replication and its effects on glycoprotein conformation and trafficking. J Virol 71:8279–8288

Moffat JF, Stein MD, Kaneshima H, Arvin AM (1995) Tropism of varicella-zoster virus for human $CD4^+$ and $CD8^+$ T-lymphocytes and epidermal cells in SCID-hu mice. J Virol 69:5236–5242

Moffat J, Zerboni L, Stein M, Grose C, Kaneshima H, Arvin A (1998a) The attenuation of the vaccine Oka strain of varicella-zoster virus and the role of glycoprotein C in alphaherpesvirus virulence demonstrated in the SCID-hu mouse. J Virol 72:965–974

Moffat JF, Zerboni L, Sommer MH, Heineman TC, Cohen JI, Kaneshima H, Arvin AM (1998b) The ORF47 and ORF66 putative protein kinases of varicella-zoster virus determine tropism for human T cells and skin in the SCID-hu mouse. Proc Natl Acad Sci USA 95:11969–11974

Niizuma T, Sommer MH, Ito H, Hinchliffe S, Zerboni L, Arvin AM (2003) Mutational analysis of varicella-zoster virus ORF65 protein and its role in infection of human skin and T cell xenografts in the SCIDhu mouse model. J Virol 77:6062–6065

Oliver SL, Sommer M, Zerboni L, Rajamani J, Grose C, Arvin AM (2009) Mutagenesis of varicella zoster virus glycoprotein B: putative fusion loop residues are essential for viral replication, and the furin cleavage motif contributes to pathogenesis in skin tissue *in vivo*. J Virol 83:7495–7506

Sato B, Ito H, Hinchliffe S, Sommer MH, Zerboni L, Arvin AM (2003) Effects of mutations in open reading frames 62 and 71, encoding the immediate early transactivating protein, IE62, of varicella-zoster virus on replication *in vitro* and in skin xenografts in the SCID-hu mouse. J Virol 77:5607–5620

Schaap A, Fortin J-F, Sommer M, Zerboni L, Stamatis S, Ku C-C, Arvin AM (2005) T cell tropism and the role of ORF66 protein in the pathogenesis of varicella-zoster virus infection. J Virol 79:12921–12933

Schaap-Nutt A, Sommer M, Che X, Zerboni L, Arvin AM (2006) ORF66 protein kinase function is required for T cell tropism of varicella-zoster virus *in vivo*. J Virol 80:11806–11816

Soong W, Schultz JC, Patera AC, Sommer MH, Cohen JI (2000) Infection of human T lymphocytes with varicella-zoster virus: an analysis with viral mutants and clinical isolates. J Virol 74:1864–1870

Tischer BK, Kaufer BB, Sommer M, Wussow F, Arvin AM, Osterrieder N (2007) A self-excisable infectious bacterial artificial chromosome clone of varicella-zoster virus allows analysis of the essential tegument protein encoded by ORF9. J Virol 81:13200–13208

Vleck SE, Oliver SL, Reichelt M, Rajamani J, Zerboni L, Jones C, Zehnder J, Grose C, Arvin AM (2010) Anti-glycoprotein H antibody impairs the pathogenicity of varicella-zoster virus in skin xenografts in the SCID Mouse Model. J Virol 84:141–152

Zerboni L, Arvin AM (2008) The pathogenesis of varicella zoster virus neurotropism and infection. In: Carol R (ed) Neurotropic viral infections. Cambridge Press, New York, pp 225–250

Zerboni L, Hinchliffe S, Sommer MH, Ito H, Besser J, Stamatis S, Cheng J, DiStefano D, Kraiouchkine N, Shaw A, Arvin AM (2005a) Analysis of varicella-zoster virus attenuation by evaluation of chimeric parent Oka/vaccine Oka recombinant viruses in skin xenografts in the SCIDhu mouse model. Virology 332:337–346

Zerboni L, Ku C-C, Jones C, Zehnder J, Arvin AM (2005b) Varicella-zoster virus infection of human dorsal root ganglia *in vivo*. Proc Natl Acad Sci USA 102:6490–6495

Zhang X, Rowe J, Wang W, Sommer M, Arvin A, Moffat J, Zhu H (2007) Genetic analysis of varicella zoster virus ORF0 to 4 using a novel luciferase bacterial artificial chromosome system. J Virol 81:9024–9033

Experimental Models to Study Varicella-Zoster Virus Infection of Neurons

Megan Steain, Barry Slobedman, and Allison Abendroth

Contents

Abstract Varicella zoster virus (VZV) infection results in the establishment of latency in human sensory neurons. Reactivation of VZV leads to herpes zoster which can be followed by persistent neuropathic pain, termed post-herpetic neuralgia (PHN). Humans are the only natural host for VZV, and the strict species specificity of the virus has restricted the development of an animal model of infection which mimics all phases of disease. In order to elucidate the mechanisms which control the establishment of latency and reactivation as well as the effect of VZV replication on neuronal function, *in vitro* models of neuronal infection have been developed. Currently these models involve culturing and infecting dissociated

M. Steain and A. Adendroth (✉)
Department of Infectious Diseases and Immunology, University of Sydney, Blackburn Building D06, Camperdown, NSW, 2006, Australia
e-mail: allison.abendroth@sydney.edu.au

B. Slobedman
Centre for Virus Research, Westmead Millennium Institute, Westmead, NSW 2145, Australia

A.M. Arvin et al. (eds.), *Varicella-zoster Virus*,
Current Topics in Microbiology and Immunology 342, DOI 10.1007/82_2010_15

human fetal neurons, with or without their supporting cells, an intact explant fetal dorsal root ganglia (DRG) model, neuroblastoma cell lines and rodent neuronal cell models. Each of these models has distinct advantages as well as disadvantages, and all have contributed towards our understanding of VZV neuronal infection. However, as yet none have been able to recapitulate the full virus lifecycle from primary infection to latency through to reactivation. The development of such a model will be a crucial step towards advancing our understanding of the mechanisms involved in VZV replication in neuronal cells, and the design of new therapies to combat VZV-related disease.

1 Introduction

During primary varicella-zoster virus (VZV) infection (varicella), the virus establishes a latent infection in sensory ganglia, mainly dorsal root ganglia (DRG) and trigeminal ganglia (TG), from where it can reactivate years later, resulting in re-initiation of productive replication with a consequence of the development of herpes zoster (shingles). Neurons within sensory ganglia are the primary site of latent infection, although there have also been sporadic reports of latency in ganglionic satellite cells (Hyman et al. 1983; Gilden et al. 1987; Croen et al. 1988; Schmidbauer et al. 1992; Lungu et al. 1995, 1998; Kennedy et al. 1998; LaGuardia et al. 1999; Levin et al. 2003). Sensory ganglia are part of the peripheral nervous system and sensory neurons can respond to mechanical, thermal, or chemical stimuli (Lawson 2005). Despite the critical importance of sensory ganglia in the lifecycle of VZV infection, not much is understood about the interactions between VZV, neurons, and their supporting cells. Specifically, factors controlling the establishment, maintenance, and reactivation from latency, as well as the impact of VZV replication on normal neuronal function remain poorly defined. This is largely due to the high species specificity of VZV, meaning the virus does not progress to a full cycle of lytic infection in most non-human cells (Weller and Stoddard 1952). This chapter will overview current experimental models that have been developed to study the interaction of VZV with ganglionic cells, with a particular focus on cell culture-based models of VZV infection as animal models of infection are detailed elsewhere in this volume.

2 Experimental VZV Infection of Primary Human Neurons from Dissociated Neural Tissue

The first *in vitro* study to examine VZV replication in human cells derived from neural tissue used cultures of human brain and ganglia cells. These cultures contained a mixed population of uncharacterised cells of mesenchymal and neuroglial origin, although due to multiple passaging neurons were not present (Gilden et al. 1978). VZV was successfully able to infect these cells and the infection occurred in a similar fashion to human lung fibroblasts, with the exception that cells of neural origin exhibited the formation of large intracytoplasmic vacuoles (Gilden et al. 1978).

Because of the inherent difficulty in obtaining fresh adult human sensory ganglia to derive primary neurons, several studies have utilised fetal sensory ganglia-derived neurons to study VZV neuropathogenesis. Although the capacity to obtain fresh fetal sensory ganglia remains a limiting factor, this approach is advantageous as humans are the only natural host for VZV and the virus generally does not replicate efficiently in cells of non-human origin (Weller and Stoddard 1952). Aborted fetal samples between 8 and 20 weeks of gestation are generally obtained and DRG harvested. Dissociation of the neurons within DRG is achieved usually after physically removing some of the surrounding epineurium before treatment with trypsin or collagenase to disrupt cell–cell junctions. To further dissociate the cells, they have been triturated through fire-polished pipettes or a fine gauge needle to obtain a uniform single cell suspension. Mitotic inhibitors have then often been employed to eliminate dividing non-neuronal cells such as fibroblasts, resulting in a relatively pure population of neurons, although some satellite cells often remain. After plating onto an extracellular matrix such as collagen, these cells can then be cultured in the presence of nerve growth factor (NGF) to ensure their survival (Wigdahl et al. 1986; Assouline et al. 1990; Somekh et al. 1992; Somekh and Levin 1993; Hood et al. 2003, 2006).

The first studies to utilise human fetal neurons were performed by Wigdahl et al. (1986) and Assouline et al. (1990). Using cell-associated and cell-free methods of infections, respectively, both studies showed that fetal neurons could be infected *in vitro* as demonstrated by the detection of viral proteins using immunofluorescence and electron microscopy to show the presence of viral particles within infected neurons (Wigdahl et al. 1986; Assouline et al. 1990). A cytopathic effect characterised by cellular degeneration and detachment of the cells was also observed within these infected neuronal cultures, consistent with productive replication. However, both studies also noted that infection of neurons progressed more slowly than infection of fibroblasts, a cell type highly permissive to productive VZV infection. Assouline et al. (1990) went on to characterise the temporal cascade of VZV protein expression and showed that the early protein encoded by open reading frame (ORF) 36, deoxypyrimidine kinase, accumulated to detectable levels prior to the immediate early gene IE62 and remained predominant in the cytoplasm, which was in contrast to the infection of fibroblasts where ORF36 was detected predominantly in the nucleus (Assouline et al. 1990). These studies showed that there are likely to be important differences between the nature of VZV replication in neurons compared to fibroblasts and the authors suggested that neurons may be able to control some aspects of viral replication to limit virus-induced damage (Wigdahl et al. 1986). Somekh and Levin (1993) also utilised dissociated fetal neurons to demonstrate infection with the attenuated vaccine strain vOKA; however, infection with the wild-type strain resulted in a much greater percentage of infected neurons, implying that the vaccine strain has attenuated neurotropism (Somekh and Levin 1993).

2.1 Modulation of Neuronal Apoptosis by VZV

Our group used primary human fetal DRG-derived neurons to examine the interaction of VZV with this cell type (Hood et al. 2003) (Fig. 1). It was shown that during productive infection with VZV, apoptosis in fibroblasts increased proportionally to the number of infected cells. In contrast, VZV infected neurons appeared to be protected from apoptosis, highlighting a functional difference between VZV replication in neurons and fibroblasts (Hood et al. 2003). In a follow-up study we demonstrated a role for the VZV protein IE63 in the protection of neurons from apoptosis during productive infection (Hood et al. 2006). IE63 is encoded by ORF63 and is duplicated in the VZV genome, with this additional copy known as ORF70. In comparison to wild-type parental virus infection, infection of primary human neurons with either an ORF63 or an ORF70 deletion virus resulted in an increase in the percentage of neurons undergoing apoptosis (Hood et al. 2006). This protection from apoptosis by IE63 was further supported by experiments showing that the induction of apoptosis by withdrawal of NGF from cultured rat neurons was suppressed when these cells were first transfected with a plasmid encoding IE63 (Hood et al. 2006). At present the mechanism by which IE63 inhibits neuronal apoptosis has not been elucidated. In addition to IE63, other VZV genes may also play a function to prevent apoptosis. For example, the ORF66 protein product, which is expressed during productive infection as well as during latent infection of neurons (Cohrs et al. 2003), has been shown to play a role in preventing apoptosis in VZV-infected T cells (Schaap et al. 2005), although additional studies will be required to determine whether this gene product functions in an anti-apoptotic manner in neurons.

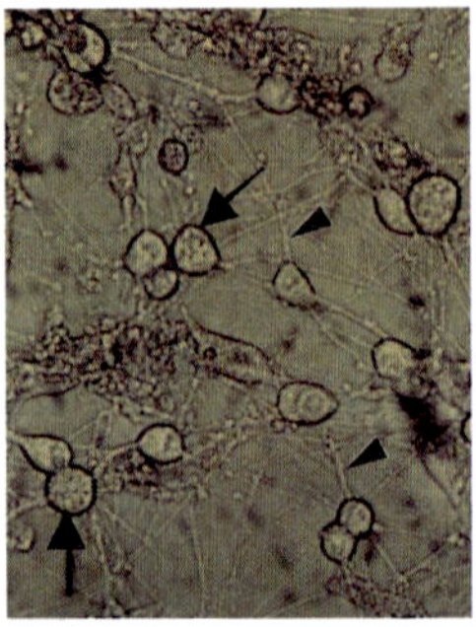

Fig. 1 Light microscopy of dissociated human fetal dorsal root ganglia (DRG) neuronal cell culture. After culture of dissociated ganglionic cells in media containing nerve growth factor for 3 days, neurons (*arrows*) develop extensive axonal networks (*arrowheads*). Copyright © American Society for Microbiology, Hood et al. (2003), DOI: 10.1128/JVI.77.23.12852-12864.2003

Protecting neurons from apoptosis during the critical first stages of virus reactivation would likely allow for greater production of new virions for axonal transport to the skin and herpes zoster lesion formation (Hood et al. 2003). It may also be of benefit to the virus to actively resist the induction of apoptosis during the latent phase of infection. It can also be argued that resistance of neuronal apoptosis may also benefit the host, given that neurons are post-mitotic and therefore are not replaced following cell death. Interestingly, however, observations in human ganglia obtained post-mortem from patients suffering from herpes zoster and post-herpetic neuralgia (PHN) have revealed that these ganglia display regions of altered morphology, where damage to neuronal tissue and cell loss has occurred as a result of VZV reactivation (Head and Campbell 1900; Denny-Brown et al. 1944; Smith 1978; Watson et al. 1988, 1991; Steain et al. 2009). Thus, mechanisms responsible for this damage *in vivo* may be a consequence of the inflammatory process rather than VZV-induced apoptosis (Hood et al. 2003). Understanding the mechanisms that underlie VZV-induced cellular damage and PHN will ultimately require extension of *in vitro*-based findings to additional analysis of naturally infected human ganglia. In this respect, our group has gained access to adult human ganglia removed post-mortem from people suffering from herpes zoster at or near the time of death. Experiments to define the nature of the ganglionic cellular infiltrate and neuronal damage that accompanies VZV reactivation are currently underway, and it is hoped that these analyses will provide a better understanding of the factors that influence reactivation and the development of PHN.

3 Models of Latent Infection of Human Neurons

Another important aspect of the life cycle of VZV is the establishment of latency. In an attempt to create an *in vitro* model of latency, Somekh et al. (1992) treated neuronal cultures as well as satellite cell and mixed (neurons and satellite cells) cultures with the anti-herpes arabinosyl nucleoside analogue bromovinyl arabinosyl uracil (BVaraU) before infecting with cell-free VZV. They found that BVaraU was able to prevent a full productive infection in these cultures, and that no viral replication occurred even after BVaraU was removed 7 days post-infection. However, infectious VZV was able to be recovered from some of the mixed (neuronal and satellite cell) cultures a week after infection and BVaraU withdrawal, when the cells were trypsinized and co-cultured with human embryonic lung fibroblasts (HELFs) (Somekh et al. 1992). It was concluded from these experiments that satellite cells may play an important role in the establishment and/or maintenance of latency. However, important differences exist between this model and latency in vivo due to the requirement of BVaraU to establish latency *in vitro* and the absence of IE62, which has been detected in neurons during latency *in vivo* (Lungu et al. 1998). Since this initial report, there has been no new published work utilising cell-culture based models of latent VZV infection using cultures of dissociated human fetal DRG. Rather, the published studies of VZV latency in human neurons have relied predominantly on examination of naturally infected adult ganglia obtained post mortem (Gilden et al. 1983, 2001; Croen et al. 1988; Vafai et al. 1988; Mahalingam et al. 1992, 1993, 1996; Schmidbauer et al. 1992; Lungu et al. 1998; LaGuardia et al. 1999; Theil et al. 2003; Cohrs et al. 2003, 2005; Hufner et al. 2006; Cohrs and Gilden 2007; Verjans et al. 2007).

4 Experimental VZV Infection of Intact Ganglia

Dissociation of ganglia results in separation of neurons from satellite cells, which may affect their function and slow VZV spread *in vitro*, given that *in vivo* neurons are tightly enclosed by multiple satellite cells (Hanani 2005). Hanani et al. noted that "Isolated intact sensory ganglia should be the first choice for studying SGC (satellite glial cell)–neuron interactions as there is minimal tissue disruption" (Hanani 2005). Thus, to study fetal neurons in the context of a more intact anatomical architecture, two different approaches have been developed. The Arvin group reported the development of an *in vivo* model whereby severe combined immunodeficiency (SCID) mice were used following implantation of intact human fetal DRG under the kidney capsule. This model has provided a very useful means to study different routes of infection of DRG, and has been used to provide evidence for a role of VZV-infected T cells in the transfer of virus to the DRG (Zerboni et al. 2005). VZV also establishes a persistent infection within these DRG, which can last for up to 8 weeks post-infection in the absence of detectable infectious virus production (Zerboni et al. 2005). Additional studies using this model are detailed in (Zerboni et al. 2010). Disadvantages of the SCID-hu mouse model, however, are the extended period of time required to allow for the establishment of the DRG xenograph and costs associated with housing animals.

Our laboratory has established an intact explant fetal DRG culture model (Gowrishankar et al. 2007). By extracting fetal DRG from aborted fetuses between the ages of 14–20 weeks gestation and removing the surrounding epineurium, DRG can be cultured directly on glass coverslips in the presence of NGF. After 2–3 days post-explant extensive axonal growth can be seen protruding from the entire DRG (Gowrishankar et al. 2007). Characterization of the ganglia post-explanting has shown that the architecture of the ganglia is preserved, with neurons being surrounded by supporting satellite cells. Immunohistochemical staining has also demonstrated the presence of ganglionic cell markers, with neurons staining positive for the neural cell adhesion molecule (NCAM) and satellite cells positive for S100B (Gowrishankar et al. 2007), similar to normal human adult ganglia. This model allows VZV infection of neurons and satellite cells to be studied in the context of an intact ganglion *in vitro*. Following 48 h of cell-associated infection with VZV strain Schenke, discrete VZV glycoprotein-positive neurons were detected throughout cultured ganglia, suggesting productive infection and axonal transport of the virus from the periphery (Fig. 2). The number of infected neurons

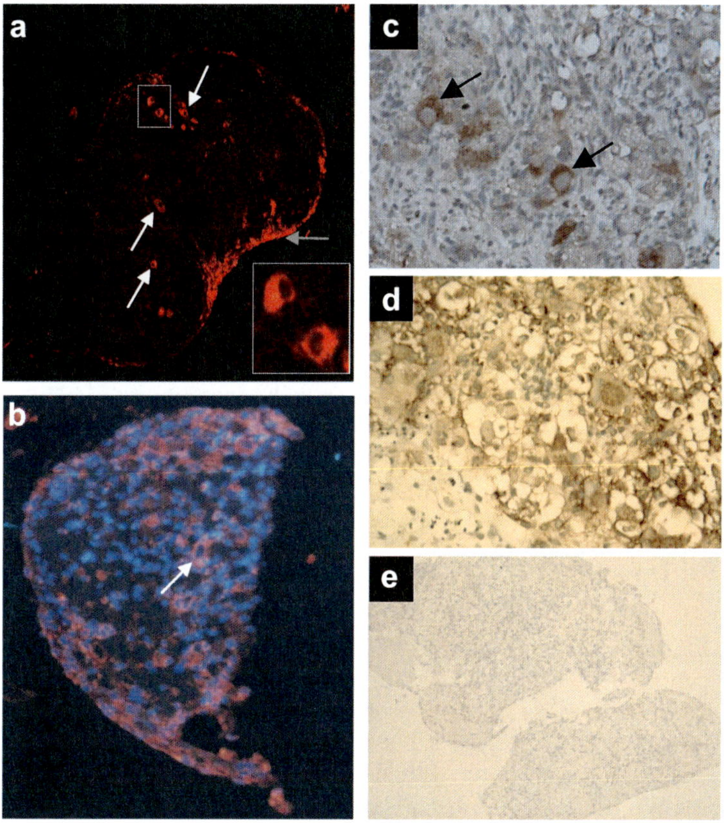

Fig. 2 VZV antigen expression in infected intact explant human dorsal root ganglia (DRG). (**a**) Immunofluorescent staining of a DRG 48 h post infection with human VZV hyper immune serum and secondary antibody consisting of fluorescently conjugated anti-human AlexaFluor 594 (*red staining*). VZV antigen expression on the surfaces of distinct, scattered neurons (*white arrows*) and around the DRG body (*green arrows*). *Boxed inset* shows magnified image of VZV antigen positive neurons. (**b**) Immunofluorescent staining of a DRG at 48 h post infection with mouse anti-VZV gE monoclonal antibody and secondary antibody consisting of fluorescently conjugated anti-mouse AlexaFluor 594 (*red staining*). Nuclear blue DAPI staining is indicated. (**c–e**) Immunohistochemical detection of VZV infected DRG stained with mouse anti-VZV gB monoclonal antibody at 48 h post infection (**c**) and 72 h post infection (**d**) (*brown staining*) or mock infected DRG (**e**). *Black arrows* indicate infected neurons. Sections were lightly counterstained with hematoxylin (*blue staining*). Copyright © American Society for Microbiology, Gowrishankar et al. (2007), DOI: 10.1128/JVI.02793-06

and supporting cells increased over time, indicating that cell-to-cell spread of the virus likely occurred within the infected DRG (Gowrishankar et al. 2007). The presence of virus particles in neurons, as well as in the extracelluar space, was also demonstrated by transmission electron microscopy (TEM) (Gowrishankar et al. 2007). Importantly, cell-free virus was shown to be released from these cultures, which is a unique feature of this model, as VZV typically remains highly cell-associated in culture (Weller 1953). The explant DRG model provides a number of opportunities to study various aspects of VZV replication within DRG, including innate immune responses to infection and effects of viral replication on neuronal function. This model could also be used to screen potential new vaccine candidates for neurotropism, and determine the basis of VZV neuropathogenesis using viral gene deletion viruses. This model may also have other uses such as studying neuronal signaling during VZV infection and anterograde and retrograde transport of VZV proteins in axons, which could be done in combination with explanted skin sections. This approach has previously been used in pioneering work by the Cunningham group to study axonal transport of herpes simplex virus (HSV) from human DRG (Penfold et al. 1994, 1996; Holland et al. 1998, 1999; Mikloska et al. 1999; Mikloska and Cunningham 2001), and a similar approach could be used to study VZV virion transport and assembly. Such studies may lead to the identification of new drug targets to limit axonal transport of the virus, which may prevent the establishment of latency or aid in the treatment of herpes zoster.

5 VZV Infection of Neuronal Cell Lines

An alternative approach for studying VZV infection of primary neurons has been to examine infection of neuroblastoma cell lines. Such a surrogate model of neuronal infection affords distinct advantages over working with primary neurons in terms of cost, availability, cell number, ease of manipulation, a lack of donor variation, and a reduced risk of contamination with other infectious agents. Further, by using neuroblastomas, ethical issues that arise when using fetal tissues are avoided. Cell lines in general are also easier to genetically manipulate, and thus neuroblastomas have the potential to be used to evaluate the role of cellular genes in VZV infection. Although a caveat of using neuroblastoma cell lines is whether they adequately mimic primary neurons, it remains rather surprising that there have been so few reports exploring the nature and properties of VZV infection in neuroblastomas.

VZV infection of the human derived IMR-32 and the murine neuro-2A neuroblastoma cell lines have been studied (Bourdon-Wouters et al. 1990). Infection of IMR-32 cells, in both a cell-free and a cell-associated manner using infected MRC-5 cells, resulted in a cytopathic effect, cell death, and the release of cell-free virus. In contrast, infection of neuro-2A cells using a similar approach was non-productive, with no viral antigens detected within cells, despite persistence of VZV DNA, as detected by in situ hybridization (ISH) (Bourdon-Wouters et al. 1990). These differences may be due to the non-human origin of the neuro-2A cells (Bourdon-Wouters et al. 1990).

The SK-N-SH neuroblastoma cell line, which is also of human origin, as well as one of its derivatives SH-SY5Y (Biedler et al. 1973; Ross et al. 1983), has been used for infection with VZV (Cohen and Nguyen 1998; Cohen et al. 2001; Sato et al. 2002); however, the infection of these neuroblastomas has not yet been fully characterised. SH-SY5Y cells can be readily differentiated using retinoic acid and brain-derived neurotrophic factor (BDNF) to produce growth-arrested cells which have extended neuritic processes and express neuronal markers, rendering them similar to primary neurons (Encinas et al. 2000). SH-SY5Y cells have been shown to be more permissive to HSV-1 infection in their differentiated form due to an upregulation of Nectin-1 and -2 and HVEM (Gimenez-Cassina et al. 2006). Unpublished data from our group indicates that infection of both undifferentiated and differentiated SH-SY5Y cells with VZV results in a full productive infection, which shares many characteristics of infection with dissociated primary human fetal neurons. Mechanisms involved in the development of neuropathic pain during post-herpetic neuralgia are poorly understood. SY-SY5Y cells can be stably transfected to express transient receptor potential (TRP) vanilliod receptors (Lam et al. 2007) and the tetrodotoxin-resistant voltage-gated sodium channel alpha-subunit Nav1.8 (Dekker et al. 2005), giving properties of nociceptor (pain sensing) neurons. Therefore, these cells could be used to study the effects of VZV infection on neurotransmitter release and neuronal function, which may have implications for PHN. Thus the use of neuroblastomas may serve as a more practical alternative to using fetal neurons, at least for initial studies of various aspects of VZV infection, which could then be confirmed and extended using primary neuronal cultures or naturally infected tissue samples.

6 VZV Infection of Rodent Neurons

Neurons of rodent origin have also been used to study VZV infection. However, as VZV is highly species specific and humans are the only natural host for the virus, infection of cells of non-human origin may not accurately reflect the virus lifecycle *in vivo*. Ganglionic neurons of rodent origin, however, are much easier to obtain than human fetal ganglia, and this has been the predominant driving force in pursuing this approach.

The first study of VZV infection using animal neurons *in vitro* was performed by Merville-Louis et al. (1989). A mixed population of dissociated neurons and supporting cells was established, with neurons consisting of approximately 10% of the culture. Cells were infected in either a cell-free or cell-associated manner, and in both cases no cytopathic effect was observed up to 10 days post-infection. In addition, no cell-free virus was released and virus was not able to be transferred to permissive MRC-5 cells. Viral antigens were detected in a small percentage of neurons, using human serum which was known to contain anti-VZV antibodies; however, antigen was detected only up until day 5 post-infection. Despite the lack of de novo virus production, VZV nucleic acids were detected within neurons by ISH, and this increased from 20% of neurons being positive at day 1 to 50% at day 6 post-infection. No ISH signal was detected in any of the non-neuronal cells. ISH specific for RNA transcripts of immediate early, early, and late genes showed that genes from all the temporal classes were expressed in neurons, although the IE63 transcript was found to give a stronger hybridization signal. From this study the authors concluded that the infection of rat neurons resulted in a non-productive, persistent infection, and that the IE63 gene may play a role in repressing a productive infection (Merville-Louis et al. 1989).

Subsequent studies using rat DRG neurons *in vitro* were able to achieve a productive infection using a "microculture" system (Kress and Fickenscher 2001; Schmidt et al. 2003). Neurons and infected fibroblasts were concentrated into very small volumes of media and incubated together for 4 h before additional culture media was added. Presumably this increased cell-to-cell contact between infected fibroblasts and neurons facilitated the spread of VZV. IE62, IE63, and glycoprotein E could be detected within neurons by immunofluorescence. Within 3–4 days post-infection, the majority of productively infected rat neurons died. This was characterised by a loss of viable cells within the culture, which was in contrast to control uninfected cultures, which appeared healthy. These studies went on to further examine the effects of VZV infection on rat neurons with respect to neuronal function and showed that VZV infection results in a gain-of-function conferring sensitivity to adrenergic agonists, including norepinephrine which is associated with pain (Kress and Fickenscher 2001; Schmidt et al. 2003). It was also shown that this effect was greatly reduced when rat neurons were infected with the vaccine OKA strain, despite the vaccine strain being able to infect an equal percentage of neurons as the wild-type strain (Schmidt et al. 2003). This contrasted with a prior study using human fetal neuron cultures, which showed that the vaccine strain did not infect neurons to the same degree as did the wild-type VZV strains (Somekh and Levin 1993). It was proposed that the induction of sensitivity to adrenergic stimulation following VZV infection could be responsible for heat hyperalgesia, which is commonly linked with PHN (Schmidt et al. 2003). In a separate study using rat neurons, a VZV IE63 expression construct was transfected into neurons derived from rat embryos to show that this caused an increase in calcitonin gene-related peptide (CGRP) release, which *in vivo* can increase the sensation of pain (Hamza et al. 2007). Thus expression of IE63 in neurons may play a role in the development of pain that is experienced during herpes zoster and PHN.

Isolated guinea pig and mouse enteric neurons have also been used to establish a model of lytic, latent, and reactivating VZV infection *in vitro* (Chen et al. 2003; Gershon et al. 2008). Enteric ganglia differ from dorsal root ganglia in that they not only contain afferent (sensory) neurons, but also efferent (motor) neurons and interneurons. To validate the use of enteric neurons for VZV infection, it has been shown *in vivo* that enteric neurons can harbor latent VZV (Gershon et al. 2009). Using this guinea pig-derived model it was demonstrated that infection of guinea pig enteric neuronal cultures with cell-free VZV resulted in an apparent latent infection. No cytopathic effect was observed, and transcripts for the latency-associated ORFs 4, 21, 29, 40, 62 and 63 could be detected (Chen et al. 2003). In addition, several proteins arising from these transcripts were detected in neurons by immunofluorescence where they were found in the cytoplasm of cells, which is indicative of latency (Lungu et al. 1998; Chen et al. 2003). Immunofluorescence staining also revealed that sensory neurons present in the culture, as well as other types of neurons, were able to be infected in a nonproductive manner, and these infected neurons could survive in culture for weeks, similar to uninfected controls

(Chen et al. 2003). Interestingly, when these neurons were infected in a cell-associated manner, or when other non-neuronal cells from the bowel wall were present, a productive infection resulted (Chen et al. 2003). In this scenario, glycoproteins, which have not been reported to be produced during latency, were detected and the neurons died rapidly (Chen et al. 2003). It was proposed that the absence of the non-structural ORF61 protein may enable the virus to establish a latent infection when cell-free virus was used as the inoculum. It was subsequently shown that superinfection of latently infected enteric neurons with a viral vector expressing the ORF61 protein resulted in an apparent reactivation of the virus, with concomitant expression of viral glycoprotein and neuronal death (Gershon et al. 2008). This model has also been used to demonstrate that the cellular localization of IE63 and ORF29p is cell-type dependant, with both proteins accumulating in the cytoplasm of guinea pig enteric neurons and the nucleus of epithelial cells, and that the expression of ORF61p in neurons can drive their nuclear import in a protease-dependent fashion (Stallings et al. 2006; Walters et al. 2008). Infected guinea pig enteric neurons were also used to show a link between IE63 and the human antisilencing function 1 protein, which may influence transcription of genes during VZV infection (Ambagala et al. 2009).

In addition to the use of dissociated rodent neurons to study VZV infection, a number of studies have sought to utilise whole-animal models of infection. In this respect, *in vivo* VZV infection of weanling guinea pigs via various routes of inoculation, including intranasally, subcutaneously, intramuscularly, and via corneal infection has been reported to result in viremia and nasopharyngeal shedding of virus, which can lead to transmission of VZV to other animals (Myers et al. 1980, 1985; Matsunaga et al. 1982). In addition, the presence of VZV DNA in DRG of infected guinea pigs was also demonstrated by PCR, indicating that virus can access sensory ganglia in these animals (Lowry et al. 1993). Likewise, there has also been a report of VZV infection of mice via a corneal inoculation, resulting in viremia and detection of VZV DNA in various tissues including the trigeminal ganglia, the brain stem, and the spleen, up to 33 days post-infection, with the mice not suffering from any apparent VZV-related illness (Wroblewska et al. 1993). There has also been interest in creating an *in vivo* model of VZV infection using rats, and these have been used to study VZV latency and as a behavioral model of VZV-induced pain (Sadzot-Delvaux et al. 1990, 1995, 1998; Annunziato et al. 1998; Fleetwood-Walker et al. 1999; Kennedy et al. 2001; Sato et al. 2003; Grinfeld et al. 2004; Garry et al. 2005). The development and uses of rat models of latency is discussed further in (Cohen 2010).

7 VZV Infection of Ganglionic Satellite Cells

Neurons within the DRG and TG are surrounded by satellite cells, and in addition to providing physical support to neurons, satellite cells appear to be capable of affecting the microenvironment of the ganglia and may even play a role in neuronal signaling (review in Hanani 2005). Recently, satellite cells of the TG have also been shown to share some properties of antigen presenting cells, and therefore may also influence the immune response within ganglia (van Velzen et al. 2009). It is not surprising that satellite cells are required to perform immune surveillance, and antigen presentation within the ganglia, given neurons within sensory ganglia, like neurons of the central nervous system, fail to express major histocompatibility complex (MHC) molecules (Turnley et al. 2002). This may allow VZV to persist latently within neurons without recognition by the immune system. This would be beneficial for both the virus and the host as neurons are post-mitotic cells and therefore will not be replaced if killed as a result of infection (Joly et al. 1991). VZV DNA and RNA have been detected within satellite cells during latency (Croen et al. 1988; Schmidbauer et al. 1992; Kennedy et al. 1998; Lungu et al. 1998; LaGuardia et al. 1999). VZV antigens within satellite cells have also been reported in studies of explanted VZV-infected fetal ganglia (Gowrishankar et al. 2007) and in infected fetal ganglia xenographed into SCID mice (Zerboni et al. 2007; Reichelt et al. 2008). However, further studies examining the impact of VZV infection on satellite cell function are needed due to the important role these cells play in supporting neurons within the ganglia.

8 Conclusions and Perspectives

Productive, latent, and reactivated VZV infection of sensory neurons and their supporting satellite cells may result in alterations to normal cellular processes, which in combination with an immune-mediated response to viral replication may contribute towards the pain experienced during herpes zoster and subsequently during PHN. Thus, a greater understanding of how VZV interacts with cells of the sensory ganglia is required to develop new therapeutic strategies to prevent or manage these conditions.

The lack of an adaptive immune response is the greatest disadvantage to *in vitro* models of VZV neuronal infection, as well as the *in vivo* SCID-hu mouse model, and without an animal model that can mimic all phases of disease progression in the context of an immune response, studies of post-mortem ganglia affected by varicella, herpes zoster, and PHN will remain critical.

From the *in vitro* studies that have been conducted thus far, it is becoming clear that distinct differences exist between VZV replication in neuronal cells compared to other diploid cells. These differences could play a key role in enabling VZV to successfully replicate in neurons within sensory ganglia, for example, during the initial stages of virus reactivation, leading to herpes zoster. The precise viral and cellular mechanisms that underpin virus replication within this cell type, as well as those that facilitate latency, remain to be established.

Differences in VZV replication *in vitro* are also evident between studies and models, and which model is the most accurate representation of VZV infection *in vivo* remains unknown. Like during natural VZV replication in human skin, in both the explant fetal DRG model and in IMR32 neuroblastoma cells, cell-free virus release has been demonstrated (Bourdon-Wouters et al. 1990; Gowrishankar et al. 2007). This is in stark contrast to infection of other cell types *in vitro* (Weller 1953), including other neuroblastoma-based culture systems, where virus remains cell-associated (Abendroth, unpublished data). This may reflect differences in the virus lifecycle in these cells and perhaps a difference in the expression of mannose-6-phosphate receptor, which has been shown to affect cell-free virus release (Chen et al. 2004). Studies comparing neuronal infection *in vitro* to naturally infected human ganglia derived post-mortem or surgically are therefore needed to establish the robustness of these models, as well as to establish the role of the adaptive immune response.

Another difference observed between models of VZV neuronal infection *in vitro* is that VZV-infected rat and guinea pig enteric neurons do not survive infection (Merville-Louis et al. 1989; Chen et al. 2003; Gershon et al. 2008), which is in contrast to studies of primary human neuronal cells, in which VZV infection has been shown not to induce apoptosis (Hood et al. 2003). The IE63 protein has been shown to protect human neurons from apoptosis; however, the precise mechanism behind this protection is unknown (Hood et al. 2006). Further studies of the different models, animal vs. human neurons, may help to elucidate how IE63 modulates the apoptotic response of neurons.

Pain as a result of herpes zoster and especially PHN can be disabling and have a major negative impact on patient's quality of life (Dworkin et al. 2001). In coming years the number of individuals suffering from herpes zoster and PHN is expected to rise, concomitant with the increasing number of patients who are elderly or immunosuppressed due to infection or therapies for cancer or transplantations. Therefore, VZV and associated diseases are likely to place a large burden on health care systems, and adequate treatment and prevention strategies are needed. To achieve this, further studies into the mechanisms involved in VZV replication in neuronal cells are necessary. In addition, given that VZV is able to readily infect neurons, like HSV, it may be a suitable vector for gene delivery into neuronal cells for the treatment of various nervous system disorders. This could be achieved only by further studies into the interaction between VZV and neuronal cells, and suitable attenuation of the virus to limit replication in other cell types.

References

Ambagala AP, Bosma T, Ali MA, Poustovoitov M, Chen JJ, Gershon MD, Adams PD, Cohen JI (2009) Varicella-zoster virus immediate-early 63 protein interacts with human antisilencing function 1 protein and alters its ability to bind histones h3.1 and h3.3. J Virol 83(1):200–209

Annunziato P, LaRussa P, Lee P, Steinberg S, Lungu O, Gershon AA, Silverstein S (1998) Evidence of latent varicella-zoster virus in rat dorsal root ganglia. J Infect Dis 178(Suppl 1): S48–S51

Assouline JG, Levin MJ, Major EO, Forghani B, Straus SE, Ostrove JM (1990) Varicella-zoster virus infection of human astrocytes, Schwann cells, and neurons. Virology 179(2):834–844

Biedler JL, Helson L, Spengler BA (1973) Morphology and growth, tumorigenicity, and cytogenetics of human neuroblastoma cells in continuous culture. Cancer Res 33(11):2643–2652

Bourdon-Wouters C, Merville-Louis MP, Sadzot-Delvaux C, Marc P, Piette J, Delree P, Moonen G, Rentier B (1990) Acute and persistent varicella-zoster virus infection of human and murine neuroblastoma cell lines. J Neurosci Res 26(1):90–97

Chen JJ, Gershon AA, Li ZS, Lungu O, Gershon MD (2003) Latent and lytic infection of isolated guinea pig enteric ganglia by varicella zoster virus. J Med Virol 70(Suppl 1):S71–S78

Chen JJ, Zhu Z, Gershon AA, Gershon MD (2004) Mannose 6-phosphate receptor dependence of varicella zoster virus infection *in vitro* and in the epidermis during varicella and zoster. Cell 119(7):915–926

Cohen JI (2010) Rodent models of varicella-zoster virus neurotropism. Curr Top Microbiol Immunol doi 10.1007/82_2010_11

Cohen JI, Nguyen H (1998) Varicella-zoster virus ORF61 deletion mutants replicate in cell culture, but a mutant with stop codons in ORF61 reverts to wild-type virus. Virology 246 (2):306–316

Cohen JI, Sato H, Srinivas S, Lekstrom K (2001) Varicella-zoster virus (VZV) ORF65 virion protein is dispensable for replication in cell culture and is phosphorylated by casein kinase II, but not by the VZV protein kinases. Virology 280(1):62–71

Cohrs RJ, Gilden DH (2007) Prevalence and abundance of latently transcribed varicella-zoster virus genes in human ganglia. J Virol 81(6):2950–2956

Cohrs RJ, Gilden DH, Kinchington PR, Grinfeld E, Kennedy PG (2003) Varicella-zoster virus gene 66 transcription and translation in latently infected human Ganglia. J Virol 77(12):6660–6665

Cohrs RJ, Laguardia JJ, Gilden D (2005) Distribution of latent herpes simplex virus type-1 and varicella zoster virus DNA in human trigeminal Ganglia. Virus Genes 31(2):223–227

Croen KD, Ostrove JM, Dragovic LJ, Straus SE (1988) Patterns of gene expression and sites of latency in human nerve ganglia are different for varicella-zoster and herpes simplex viruses. Proc Natl Acad Sci USA 85(24):9773–9777

Dekker LV, Daniels Z, Hick C, Elsegood K, Bowden S, Szestak T, Burley JR, Southan A, Cronk D, James IF (2005) Analysis of human Nav1.8 expressed in SH-SY5Y neuroblastoma cells. Eur J Pharmacol 528(1–3):52–58

Denny-Brown D, Adam R, Fitzgerald P (1944) Pathological features of herpes zoster. Am Med Assoc Arch Dermatol 75:193–196

Dworkin RH, Nagasako EM, Johnson RW, Griffin DR (2001) Acute pain in herpes zoster: the famciclovir database project. Pain 94(1):113–119

Encinas M, Iglesias M, Liu Y, Wang H, Muhaisen A, Cena V, Gallego C, Comella JX (2000) Sequential treatment of SH-SY5Y cells with retinoic acid and brain-derived neurotrophic factor gives rise to fully differentiated, neurotrophic factor-dependent, human neuron-like cells. J Neurochem 75(3):991–1003

Fleetwood-Walker SM, Quinn JP, Wallace C, Blackburn-Munro G, Kelly BG, Fiskerstrand CE, Nash AA, Dalziel RG (1999) Behavioural changes in the rat following infection with varicella-zoster virus. J Gen Virol 80(Pt 9):2433–2436

Garry EM, Delaney A, Anderson HA, Sirinathsinghji EC, Clapp RH, Martin WJ, Kinchington PR, Krah DL, Abbadie C, Fleetwood-Walker SM (2005) Varicella zoster virus induces neuropathic changes in rat dorsal root ganglia and behavioral reflex sensitisation that is attenuated by gabapentin or sodium channel blocking drugs. Pain 118(1–2):97–111

Gershon AA, Chen J, Davis L, Krinsky C, Cowles R, Gershon MD (2009) Distribution of latent varicella zoster virus (VZV) in sensory ganglia and gut after vaccination and wild-type infection: evidence for viremic spread. The 34th Annual International Herpesvirus Workshop. Ithaca, New York, USA

Gershon AA, Chen J, Gershon MD (2008) A model of lytic, latent, and reactivating varicella-zoster virus infections in isolated enteric neurons. J Infect Dis 197(Suppl 2):S61–S65

Gilden DH, Gesser R, Smith J, Wellish M, Laguardia JJ, Cohrs RJ, Mahalingam R (2001) Presence of VZV and HSV-1 DNA in human nodose and celiac ganglia. Virus Genes 23(2):145–147

Gilden DH, Rozenman Y, Murray R, Devlin M, Vafai A (1987) Detection of varicella-zoster virus nucleic acid in neurons of normal human thoracic ganglia. Ann Neurol 22(3):377–380

Gilden DH, Vafai A, Shtram Y, Becker Y, Devlin M, Wellish M (1983) Varicella-zoster virus DNA in human sensory ganglia. Nature 306(5942):478–480

Gilden DH, Wroblewska Z, Kindt V, Warren KG, Wolinsky JS (1978) Varicella-zoster virus infection of human brain cells and ganglion cells in tissue culture. Arch Virol 56(1–2):105–117

Gimenez-Cassina A, Lim F, Diaz-Nido J (2006) Differentiation of a human neuroblastoma into neuron-like cells increases their susceptibility to transduction by herpesviral vectors. J Neurosci Res 84(4):755–767

Gowrishankar K, Slobedman B, Cunningham AL, Miranda-Saksena M, Boadle RA, Abendroth A (2007) Productive varicella-zoster virus infection of cultured intact human ganglia. J Virol 81(12):6752–6756

Grinfeld E, Sadzot-Delvaux C, Kennedy PG (2004) Varicella-Zoster virus proteins encoded by open reading frames 14 and 67 are both dispensable for the establishment of latency in a rat model. Virology 323(1):85–90

Hamza MA, Higgins DM, Ruyechan WT (2007) Two alphaherpesvirus latency-associated gene products influence calcitonin gene-related peptide levels in rat trigeminal neurons. Neurobiol Dis 25(3):553–560

Hanani M (2005) Satellite glial cells in sensory ganglia: from form to function. Brain Res Brain Res Rev 48(3):457–476

Head H, Campbell W (1900) The pathology of herpes zoster and its bearing on sensory localization. Brain 23:353–523

Holland DJ, Cunningham AL, Boadle RA (1998) The axonal transmission of herpes simplex virus to epidermal cells: a novel use of the freeze substitution technique applied to explant cultures retained on cover slips. J Microsc 192(Pt 1):69–72

Holland DJ, Miranda-Saksena M, Boadle RA, Armati P, Cunningham AL (1999) Anterograde transport of herpes simplex virus proteins in axons of peripheral human fetal neurons: an immunoelectron microscopy study. J Virol 73(10):8503–8511

Hood C, Cunningham AL, Slobedman B, Arvin AM, Sommer MH, Kinchington PR, Abendroth A (2006) Varicella-zoster virus ORF63 inhibits apoptosis of primary human neurons. J Virol 80(2):1025–1031

Hood C, Cunningham AL, Slobedman B, Boadle RA, Abendroth A (2003) Varicella-zoster virus-infected human sensory neurons are resistant to apoptosis, yet human foreskin fibroblasts are susceptible: evidence for a cell-type-specific apoptotic response. J Virol 77(23):12852–12864

Hufner K, Derfuss T, Herberger S, Sunami K, Russell S, Sinicina I, Arbusow V, Strupp M, Brandt T, Theil D (2006) Latency of alpha-herpes viruses is accompanied by a chronic inflammation in human trigeminal ganglia but not in dorsal root ganglia. J Neuropathol Exp Neurol 65(10): 1022–1030

Hyman RW, Ecker JR, Tenser RB (1983) Varicella-zoster virus RNA in human trigeminal ganglia. Lancet 2(8354):814–816

Joly E, Mucke L, Oldstone MB (1991) Viral persistence in neurons explained by lack of major histocompatibility class I expression. Science 253(5025):1283–1285

Kennedy PG, Grinfeld E, Bontems S, Sadzot-Delvaux C (2001) Varicella-Zoster virus gene expression in latently infected rat dorsal root ganglia. Virology 289(2):218–223

Kennedy PG, Grinfeld E, Gow JW (1998) Latent varicella-zoster virus is located predominantly in neurons in human trigeminal ganglia. Proc Natl Acad Sci USA 95(8):4658–4662

Kress M, Fickenscher H (2001) Infection by human varicella-zoster virus confers norepinephrine sensitivity to sensory neurons from rat dorsal root ganglia. FASEB J 15(6):1037–1043

LaGuardia JJ, Cohrs RJ, Gilden DH (1999) Prevalence of varicella-zoster virus DNA in dissociated human trigeminal ganglion neurons and nonneuronal cells. J Virol 73(10):8571–8577

Lam PM, Hainsworth AH, Smith GD, Owen DE, Davies J, Lambert DG (2007) Activation of recombinant human TRPV1 receptors expressed in SH-SY5Y human neuroblastoma cells

increases [Ca(2+)](i), initiates neurotransmitter release and promotes delayed cell death. J Neurochem 102(3):801–811
Lawson SN (2005) The peripheral sensory nervous system: dorsal root ganglion neurons. In: Dyck PJ, Thomas PK (eds) Peripheral Neuropathy. Elsevier Saunders, Philadelphia, PA, pp 163–202
Levin MJ, Cai GY, Manchak MD, Pizer LI (2003) Varicella-zoster virus DNA in cells isolated from human trigeminal ganglia. J Virol 77(12):6979–6987
Lowry PW, Sabella C, Koropchak CM, Watson BN, Thackray HM, Abbruzzi GM, Arvin AM (1993) Investigation of the pathogenesis of varicella-zoster virus infection in guinea pigs by using polymerase chain reaction. J Infect Dis 167(1):78–83
Lungu O, Annunziato PW, Gershon A, Staugaitis SM, Josefson D, LaRussa P, Silverstein SJ (1995) Reactivated and latent varicella-zoster virus in human dorsal root ganglia. Proc Natl Acad Sci USA 92(24):10980–10984
Lungu O, Panagiotidis CA, Annunziato PW, Gershon AA, Silverstein SJ (1998) Aberrant intracellular localization of Varicella-Zoster virus regulatory proteins during latency. Proc Natl Acad Sci USA 95(12):7080–7085
Mahalingam R, Wellish M, Cohrs R, Debrus S, Piette J, Rentier B, Gilden DH (1996) Expression of protein encoded by varicella-zoster virus open reading frame 63 in latently infected human ganglionic neurons. Proc Natl Acad Sci USA 93(5):2122–2124
Mahalingam R, Wellish M, Lederer D, Forghani B, Cohrs R, Gilden D (1993) Quantitation of latent varicella-zoster virus DNA in human trigeminal ganglia by polymerase chain reaction. J Virol 67(4):2381–2384
Mahalingam R, Wellish MC, Dueland AN, Cohrs RJ, Gilden DH (1992) Localization of herpes simplex virus and varicella zoster virus DNA in human ganglia. Ann Neurol 31(4):444–448
Matsunaga Y, Yamanishi K, Takahashi M (1982) Experimental infection and immune response of guinea pigs with varicella-zoster virus. Infect Immun 37(2):407–412
Merville-Louis MP, Sadzot-Delvaux C, Delree P, Piette J, Moonen G, Rentier B (1989) Varicella-zoster virus infection of adult rat sensory neurons *in vitro*. J Virol 63(7):3155–3160
Mikloska Z, Cunningham AL (2001) Alpha and gamma interferons inhibit herpes simplex virus type 1 infection and spread in epidermal cells after axonal transmission. J Virol 75 (23):11821–11826
Mikloska Z, Sanna PP, Cunningham AL (1999) Neutralizing antibodies inhibit axonal spread of herpes simplex virus type 1 to epidermal cells *in vitro*. J Virol 73(7):5934–5944
Myers MG, Duer HL, Hausler CK (1980) Experimental infection of guinea pigs with varicella-zoster virus. J Infect Dis 142(3):414–420
Myers MG, Stanberry LR, Edmond BJ (1985) Varicella-zoster virus infection of strain 2 guinea pigs. J Infect Dis 151(1):106–113
Penfold ME, Armati P, Cunningham AL (1994) Axonal transport of herpes simplex virions to epidermal cells: evidence for a specialized mode of virus transport and assembly. Proc Natl Acad Sci USA 91(14):6529–6533
Penfold ME, Armati PJ, Mikloska Z, Cunningham AL (1996) The interaction of human fetal neurons and epidermal cells *in vitro*. Cell Dev Biol Anim 32(7):420–426
Reichelt M, Zerboni L, Arvin AM (2008) Mechanisms of varicella-zoster virus neuropathogenesis in human dorsal root ganglia. J Virol 82(8):3971–3983
Ross RA, Spengler BA, Biedler JL (1983) Coordinate morphological and biochemical interconversion of human neuroblastoma cells. J Natl Cancer Inst 71(4):741–747
Sadzot-Delvaux C, Arvin AM, Rentier B (1998) Varicella-zoster virus IE63, a virion component expressed during latency and acute infection, elicits humoral and cellular immunity. J Infect Dis 178(Suppl 1):S43–S47
Sadzot-Delvaux C, Debrus S, Nikkels A, Piette J, Rentier B (1995) Varicella-zoster virus latency in the adult rat is a useful model for human latent infection. Neurology 45(12 Suppl 8):S18–S20
Sadzot-Delvaux C, Merville-Louis MP, Delree P, Marc P, Piette J, Moonen G, Rentier B (1990) An *in vivo* model of varicella-zoster virus latent infection of dorsal root ganglia. J Neurosci Res 26(1):83–89

Sato H, Callanan LD, Pesnicak L, Krogmann T, Cohen JI (2002) Varicella-zoster virus (VZV) ORF17 protein induces RNA cleavage and is critical for replication of VZV at 37 degrees C but not 33 degrees C. J Virol 76(21):11012–11023
Sato H, Pesnicak L, Cohen JI (2003) Use of a rodent model to show that varicella-zoster virus ORF61 is dispensable for establishment of latency. J Med Virol 70(Suppl 1):S79–S81
Schaap A, Fortin JF, Sommer M, Zerboni L, Stamatis S, Ku CC, Nolan GP, Arvin AM (2005) T-cell tropism and the role of ORF66 protein in pathogenesis of varicella-zoster virus infection. J Virol 79(20):12921–12933
Schmidbauer M, Budka H, Pilz P, Kurata T, Hondo R (1992) Presence, distribution and spread of productive varicella zoster virus infection in nervous tissues. Brain 115(Pt 2):383–398
Schmidt M, Kress M, Heinemann S, Fickenscher H (2003) Varicella-zoster virus isolates, but not the vaccine strain OKA, induce sensitivity to alpha-1 and beta-1 adrenergic stimulation of sensory neurones in culture. J Med Virol 70(Suppl 1):S82–S89
Smith FP (1978) Pathological studies of spinal nerve ganglia in relation to intractable intercostal pain. Surg Neurol 10(1):50–53
Somekh E, Levin MJ (1993) Infection of human fetal dorsal root neurons with wild type varicella virus and the Oka strain varicella vaccine. J Med Virol 40(3):241–243
Somekh E, Tedder DG, Vafai A, Assouline JG, Straus SE, Wilcox CL, Levin MJ (1992) Latency *in vitro* of varicella-zoster virus in cells derived from human fetal dorsal root ganglia. Pediatr Res 32(6):699–703
Stallings CL, Duigou GJ, Gershon AA, Gershon MD, Silverstein SJ (2006) The cellular localization pattern of Varicella-Zoster virus ORF29p is influenced by proteasome-mediated degradation. J Virol 80(3):1497–1512
Steain MC, Sutherland JP, Rodriguez M, Buckland M, Cunningham AL, Slobedman B, Abendroth A (2009) Comparison of naturally infected ganglia during and after herpes zoster. The 34th Annual International Herpesvirus Workshop. Ithaca, New York, USA
Theil D, Derfuss T, Paripovic I, Herberger S, Meinl E, Schueler O, Strupp M, Arbusow V, Brandt T (2003) Latent herpesvirus infection in human trigeminal ganglia causes chronic immune response. Am J Pathol 163(6):2179–2184
Turnley AM, Starr R, Bartlett PF (2002) Failure of sensory neurons to express class I MHC is due to differential SOCS1 expression. J Neuroimmunol 123(1–2):35–40
Vafai A, Murray RS, Wellish M, Devlin M, Gilden DH (1988) Expression of varicella-zoster virus and herpes simplex virus in normal human trigeminal ganglia. Proc Natl Acad Sci USA 85(7):2362–2366
van Velzen M, Laman JD, Kleinjan A, Poot A, Osterhaus AD, Verjans GM (2009) Neuron-interacting satellite glial cells in human trigeminal ganglia have an APC phenotype. J Immunol 183(4):2456–2461
Verjans GM, Hintzen RQ, van Dun JM, Poot A, Milikan JC, Laman JD, Langerak AW, Kinchington PR, Osterhaus AD (2007) Selective retention of herpes simplex virus-specific T cells in latently infected human trigeminal ganglia. Proc Natl Acad Sci USA 104(9):3496–3501
Walters MS, Kyratsous CA, Wan S, Silverstein S (2008) Nuclear import of the varicella-zoster virus latency-associated protein ORF63 in primary neurons requires expression of the lytic protein ORF61 and occurs in a proteasome-dependent manner. J Virol 82(17):8673–8686
Watson CP, Deck JH, Morshead C, Van der Kooy D, Evans RJ (1991) Post-herpetic neuralgia: further post-mortem studies of cases with and without pain. Pain 44(2):105–117
Watson CP, Morshead C, Van der Kooy D, Deck J, Evans RJ (1988) Post-herpetic neuralgia: post-mortem analysis of a case. Pain 34(2):129–138
Weller TH (1953) Serial propagation *in vitro* of agents producing inclusion bodies derived from varicella and herpes zoster. Proc Soc Exp Biol Med 83(2):340–346
Weller TH, Stoddard MB (1952) Intranuclear inclusion bodies in cultures of human tissue inoculated with varicella vesicle fluid. J Immunol 68(3):311–319
Wigdahl B, Rong BL, Kinney-Thomas E (1986) Varicella-zoster virus infection of human sensory neurons. Virology 152(2):384–399

Wroblewska Z, Valyi-Nagy T, Otte J, Dillner A, Jackson A, Sole DP, Fraser NW (1993) A mouse model for varicella-zoster virus latency. Microb Pathog 15(2):141–151

Zerboni L, Ku CC, Jones CD, Zehnder JL, Arvin AM (2005) Varicella-zoster virus infection of human dorsal root ganglia *in vivo*. Proc Natl Acad Sci USA 102(18):6490–6495

Zerboni L, Reichelt M, Arvin AM (2010) Varicella-zoster virus neurotropism in SCID mouse–human dorsal root ganglia xenografts. Curr Top Microbiol Immunol doi 10.1007/82_2009_8

Zerboni L, Reichelt M, Jones CD, Zehnder JL, Ito H, Arvin AM (2007) Aberrant infection and persistence of varicella-zoster virus in human dorsal root ganglia *in vivo* in the absence of glycoprotein I. Proc Natl Acad Sci USA 104(35):14086–14091

Molecular Characterization of Varicella Zoster Virus in Latently Infected Human Ganglia: Physical State and Abundance of VZV DNA, Quantitation of Viral Transcripts and Detection of VZV-Specific Proteins

Yevgeniy Azarkh, Don Gilden, and Randall J. Cohrs

Contents

Abstract Varicella zoster virus (VZV) establishes latency in neurons of human peripheral ganglia where the virus genome is most likely maintained as a circular episome bound to histones. There is considerable variability among individuals in the number of latent VZV DNA copies. The VZV DNA burden does not appear to exceed that of herpes simplex type 1 (HSV-1). Expression of VZV genes during latency is highly restricted and is regulated epigenetically. Of the VZV open reading frames (ORFs) that have been analyzed for transcription during latency using cDNA sequencing, only ORFs 21, 29, 62, 63, and 66 have been detected. VZV ORF 63 is the most frequently and abundantly transcribed VZV gene detected in human ganglia during latency, suggesting a critical role for this gene

Y. Azarkh, D. Gilden, and R.J. Cohrs (✉)
Departments of Neurology (Y.A., D.G., R.J.C.) and Microbiology (D.G.), University of Colorado Denver School of Medicine, 12700 E. 19th Avenue, Mail Stop B182, Aurora, CO 80045, USA
e-mail: randall.cohrs@ucdenver.edu

A.M. Arvin et al. (eds.), *Varicella-zoster Virus*,
Current Topics in Microbiology and Immunology 342, DOI 10.1007/82_2009_2

in maintaining the latent state and perhaps the early stages of virus reactivation. The inconsistent detection and low abundance of other VZV transcripts suggest that these genes play secondary roles in latency or possibly reflect a subpopulation of neurons undergoing VZV reactivation. New technologies, such as GeXPS multiplex PCR, have the sensitivity to detect multiple low abundance transcripts and thus provide a means to elucidate the entire VZV transcriptome during latency.

Abbreviations

ChIP	Chromatin immunoprecipitation
DRG	Dorsal root ganglia
HSV	Herpes simplex virus
IHC	Immunohistochemistry
LCM	Laser capture microdissection
miRNA	MicroRNA
ORF	Open reading frame
qRT-PCR	Quantitative reverse transcriptase-PCR
RT-PCR	Reverse transcriptase-PCR
TG	Trigeminal ganglia
VZV	Varicella zoster virus

1 Introduction

After primary infection, varicella zoster virus (VZV) establishes a life-long latent infection in the ganglia of the peripheral nervous system. Cranial nerve and dorsal root ganglia (DRG) are typical sites of VZV latency, although the incidence of latent infection in trigeminal ganglia (TG) is higher than in any single DRG (Gilden et al. 1983; Mahalingam et al. 1992). VZV DNA has also been detected in celiac and nodose ganglia of the autonomic nervous system (Gilden et al. 2001). Under conditions of decreased adaptive immunity (aging, HIV infection or organ transplantation), VZV can reactivate to cause herpes zoster, frequently complicated by postherpetic neuralgia, myelopathy, vasculopathy, and retinal necrosis (Arvin 1996; Nagel and Gilden 2007).

Efforts to identify new zoster therapeutics have included extensive research on the molecular events involved in VZV latency and reactivation. However, the understanding of VZV latency and reactivation lags behind that of herpes simplex type 1 (HSV-1; the prototypic neurotropic alphaherpesvirus) primarily due to the highly cell-associated nature of VZV and the lack of a suitable small-animal model. This review summarizes current knowledge of VZV DNA, RNA, and proteins in human tissue obtained at autopsy.

2 Ganglionic Cell Type Infected by VZV

Initial reports on the location of VZV within human ganglia during latency were based on the detection of viral transcripts by in situ hybridization (ISH) and led to conflicting conclusions. Using ISH, Hyman et al. (1983) found VZV RNA exclusively in neurons of TG, and Gilden et al. (1987) found VZV nucleic acids in neurons of thoracic ganglia, whereas subsequent reports by Croen et al. (1988) and Meier et al. (1993) indicated the presence of VZV transcripts exclusively in non-neuronal cells. Later, in situ PCR amplification and hybridization with digoxigenin-labeled probes revealed VZV DNA predominately in the nuclei of latently infected neurons (Dueland et al. 1995; Kennedy et al. 1998). As an alternative to ISH, a technology prone to errors arising from individual interpretation (Mahalingam et al. 1999), LaGuardia et al. (1999) used PCR to analyze VZV DNA in latently infected human ganglia after separation into neuronal and non-neuronal cell fractions by differential filtration; VZV DNA was detected in 9 of 16 samples containing an estimated 5,000 neurons, but in only 2 of 16 containing about 500,000 corresponding non-neuronal cells. In a similar experiment, Levin et al. (2003) mechanically separated ganglia into neuronal and non-neuronal cells using a micromanipulator, and again, VZV DNA was detected only in neurons; 34 of 2,226 neurons and none of 20,700 satellite cells contained VZV DNA, with an average of 4.7 copies of VZV DNA per infected neuronal cell. Wang et al. (2005), who used laser capture microdissection (LCM) and PCR to analyze a total of 1,722 neurons and 14,200 non-neuronal cells isolated from TGs of ten subjects, reported VZV DNA in 6.9% neurons and 0.06% non-neuronal cells. Overall, multiple independent studies using different methods indicate that VZV resides predominantly in neurons during latency.

3 Distribution of VZV DNA in Ganglia

In the only known study focusing on assessing the distribution of VZV DNA in human ganglia, Cohrs et al. (2005) used PCR to examine DNA extracted from 2–4 sections of fixed TG sectioned at 100-μm intervals. VZV DNA was found to be distributed evenly throughout the ganglia, but only in sections that contained neuronal nuclei.

4 Latent VZV Burden

The latent VZV DNA burden per 100,000 ganglionic cells has been variously estimated at 6–31 VZV genome copies (Mahalingam et al. 1993), 258 ± 38 copies (Pevenstein et al. 1999), and 9,046 ± 13,225 copies (Cohrs et al. 2000).

The 35-fold discrepancy between the latter two reports is curious, since both studies were based on real-time quantitative PCR, and both detected similar amounts of HSV-1 DNA in TGs: 2,902 ± 1,082 copies and 3,042 ± 3,274 copies, respectively, per 100,000 ganglionic cells. The variability in latent VZV DNA copy number most likely reflects variations in ganglia from randomly selected individuals representing an outbred human population with decades of re-exposure to the virus. Thus, while consensus on the VZV DNA burden during latency requires analysis of ganglia from additional individuals, the numbers are unlikely to greatly exceed those found for HSV-1.

5 Configuration of Latent VZV DNA

Understanding the physical state of the VZV genome in latently infected ganglionic neurons can provide information about the molecular mechanism by which VZV latency is maintained and reactivation initiated. To examine the configuration of latent VZV DNA, Clarke et al. (1995) designed PCR primers that amplified a fragment within the unique long segment of the VZV genome or, should the ends join, the junction between the genomic termini. PCR amplification using plasmid based standards and DNA extracted from ganglia and from VZV nucleocapsids revealed a ratio of products obtained by internal vs. terminal primers of ~1 for all ganglia ($n = 12$), but 15:1 for the virion DNA. Thus, genomic termini were joined in essentially all ganglia, while 5% of VZV DNA present in the virus contained inversion of the unique long genomic segment. This finding suggests that like HSV-1, VZV DNA assumes a circular configuration during latency, and that the mechanism by which both herpesviurses establish latency is analogous.

6 Epigenetic Regulation of Latent VZV Gene Transcription

Chromatin immunoprecipitation (ChIP) assays followed by PCR have been used to elucidate the epigenetic state of VZV genomes (Gary et al. 2006). Like HSV-1, VZV DNA is associated with histones at all stages of the viral life cycle, but the histone composition differs between productive infection and latency. For example, acetylated histone H3K9(Ac), indicative of a euchromatic (transcriptionally active) state, is associated with ORF62 and ORF63 promoters during both latent infection in human ganglia and lytic infection of human melanoma (MeWo) cells. ORF62 and ORF63 are regulatory proteins of the immediate-early kinetic class that are expressed during latency. In contrast, the promoters regulating transcription of VZV ORF 14 (glycoprotein C) and ORF 36 (thymidine kinase), which are not expressed in latency, are not associated with H3K9(Ac) in latent infection. Thus, chromatin remodeling appears to contribute to the restricted pattern of latent VZV gene transcription.

7 The VZV Transcriptome

Analysis of the VZV genome predicts 71 open reading frames (ORFs) encoding proteins of more than 100 amino acids (Davison and Scott 1986). Of these ORFs, 65 are present in a single copy, while the remaining 6 map within the repeated regions of the unique short segment of the virus genome and thus form three diploid ORFs. Subsequent analyzes have identified ORFs 0, 9A, and 33.5 (Kemble et al. 2000; Ross et al. 1997; Chaudhuri et al. 2008), thus revealing a total of 74 ORFs, of which 71 are unique.

Determination of VZV gene expression represents a major step in understanding virus pathogenicity, and array technology has provided a broad picture of the viral transcriptome in productively infected cells. Using macroarrays with cloned PCR fragments to target both the 5′ and 3′ ends of all ORFs, Cohrs et al. (2003b) analyzed RNA extracted from VZV (Ellen strain)-infected BSC-1 (African green monkey kidney) cells and detected transcripts mapping to all unique VZV ORFs originally identified by Davison and Scott (1986) except ORF14 (unstable when cloned). The relative expression levels of all VZV ORFs increased uniformly from days 1 to 3 postinfection with ORFs 9/9A, 33/33.5, 49, 63/70, and 64/69 being the most abundant. Kennedy et al. (2005), who used long-oligonucleotide microarrays to analyze VZV (Dumas strain) transcription in MeWo and glial cells 72 h post-infection, found that the VZV gene transcription pattern in MeWo cells was similar to that in BSC-1 cells, but markedly different from that in glial cells, where only 20 ORFs were detected. Moreover, some ORFs that were not detected by microarray were amplified by reverse transcriptase-PCR (RT-PCR), suggesting that they are expressed in low abundance in glial cells.

Whereas macroarrays can clearly detect VZV transcripts in productively infected cells in culture (Cohrs et al. 2003b), their usefulness in assessing the extent of VZV transcription in latently infected human ganglia remains problematic due to insufficient sensitivity and the low abundance of latent VZV transcripts. Preliminary experiments showed that macroarrays failed to detect VZV ORF 63 transcripts in mRNA extracted from latently infected TG in which nested-set PCR had revealed the presence of ORF 63 mRNA (Cohrs, personal communication). Subsequent use of microarrays in which six picoliters of PCR-generated target DNA were applied onto glass slides and hybridization was performed in 20 μl (compared to 20 ml for macroarrays), increased the sensitivity of detection (abundance of ~10,000 copies), but ORF 63 transcripts known to be present in human ganglionic mRNA were still undetectable (Nagel, personal communication). Efforts to identify the full extent of VZV transcription during latency continue by studying the expression of individual VZV genes and by developing a more sensitive alternative to microarrays.

8 Detection and Quantitation of Individual VZV Transcripts in Latently Infected Ganglia

VZV gene expression is highly restricted during latency. Of the 71 known VZV ORFs, only ORFs 4, 10, 18, 21, 28, 29, 40, 51, 62, 63, and 66 transcripts have been examined in latently infected human ganglia by ISH, PCR amplification of cDNA libraries or by RT-PCR (Table 1). Using ISH, Kennedy et al. (2000) detected transcripts mapping to ORFs 4, 18, 21, 29, 40, 62, and 63, but not to ORFs 28 or 61 in human TG. Transcripts mapping to ORFs 21, 29, 62, 63, and 66, but not ORFs 4, 10, 40, 51, or 61 were detected in latently infected TG by PCR amplification of cDNA libraries (Cohrs et al. 1994, 1996) and by RT-PCR (Cohrs et al. 2003b). To confirm that PCR products reflected amplification of viral cDNA, the 3′-polyadenylated termini of the transcripts were sequenced (Cohrs et al. 1996, 2003b). Based on sequence data, ORFs 21, 29, 62, 63, and 66 are transcribed in latently infected human ganglia (Table 1). Quantitative analysis demonstrated that ORF63 is the most abundant and frequently detected VZV transcript. VZV ORF 63 transcripts in individual ganglia vary over 2,000-fold from 1 to 2,785 copies per 10,000 copies of GAPdH transcript (Cohrs et al. 2000), and ORF 63 transcripts were detected in 17 of 28 ganglia with the highest copy number >29,000 copies per 1 μg of input mRNA (Cohrs and Gilden 2003). The repeated detection and high abundance of VZV ORF 63 transcripts during latency point to a critical role for this transcript in the maintenance of latency or in the early stages of virus reactivation.

9 Development of a Novel Assay to Study VZV Gene Expression During Latency

As mentioned above, current array technologies lack the sensitivity required to detect VZV transcripts in latently infected ganglia. As an alternative, Nagel et al. (2009) adapted multiplex PCR (GeXPS; Beckman Coulter) to detect the 68 VZV ORFs originally identified by Davison and Scott (1986). GeXPS technology is based on simultaneous reverse transcription of a single mRNA pool with VZV gene-specific primers followed by PCR amplification of the multiple cDNA targets using primers specific for each cDNA. The key to GeXPS analysis is the use of chimeric PCR primers that contain both gene-specific and universal DNA sequences. Amplification of the cDNA sample with optimized concentrations of these primers results in an unbiased pool of products of predetermined length, each containing a fluorescent marker on the 5′ end. Capillary electrophoresis is then used to separate the PCR products based on size and to determine the abundance of each product based on a fluorescent signal. Since the abundance of a PCR product depends on primer amplification efficiency, it is not quantitative; however, GeXPS technology allows comparison of transcript abundance in a sample-to-sample fashion, and permits screening of multiple samples in a high-throughput format.

Table 1 VZV gene expression in latently infected human ganglia

ORF	Description[a]	Findings	Transcript verified by cDNA sequencing	ORF protein detected by IHC[b]	Method	Reference
4	Transactivator, tegument protein	Transcripts found in non-neuronal cells			ISH[c]	Croen et al. (1988)
		Transcripts not found			PCR of cDNA library	Cohrs et al. (1996)
		Protein found in cytoplasm of neurons		Yes	IHC	Lungu et al. (1998)
		Transcripts found in neurons			ISH	Kennedy et al. (2000)
10	Transactivator, tegument protein	Transcripts not found			PCR of cDNA library	Cohrs et al. (1996)
18	Ribonucleotide reductase, small subunit	Transcripts found in neurons			ISH	Kennedy et al. (2000)
21	HSV-1 UL37 homolog; nucleocapsid-associated protein[d]	Transcripts found	Yes		PCR of cDNA library	Cohrs et al. (1994, 1996)
		Protein found in cytoplasm of neurons		Yes	IHC	Lungu et al. (1998), Grinfeld and Kennedy (2004)
		Transcripts found and quantified			qRT-PCR[e]	Cohrs et al. (2000)
		Transcripts found in neurons			ISH	Kennedy et al. (2000)
		Transcripts not found			qRT-PCR	Cohrs and Gilden (2007)
28	DNA polymerase	Transcripts not found			ISH	Kennedy et al. (2000)
29	Single-stranded DNA-binding protein	Transcripts found in non-neuronal cells			ISH	Croen et al. (1988), Meier et al. (1993)
		Transcripts found	Yes		PCR of cDNA library	Cohrs et al. (1996)

(*continued*)

Table 1 (continued)

ORF	Description[a]	Findings	Transcript verified by cDNA sequencing	ORF protein detected by IHC[b]	Method	Reference
		Protein found in cytoplasm of neurons		Yes	IHC	Lungu et al. (1998), Grinfeld and Kennedy (2004)
		Transcripts found and quantified			qRT-PCR	Cohrs et al. (2000), Cohrs and Gilden (2007)
		Transcripts found in neurons			ISH	Kennedy et al. (2000)
40	Major nucleocapsid protein	Transcripts not found			PCR of cDNA library	Cohrs et al. (1996)
		Transcripts found			ISH	Kennedy et al. (2000)
51	Helicase, origin-binding protein	Transcripts not found			PCR of cDNA library	Cohrs et al. (1996)
61	Transactivator, transrepressor	Transcripts not found			PCR of cDNA library	Cohrs et al. (1996)
		Transcripts not found			ISH	Kennedy et al. (2000)
62	Transactivator, tegument protein	Transcripts found in non-neuronal cells			ISH	Croen et al. (1988), Meier et al. (1993)
		Transcripts found	Yes		PCR of cDNA	Cohrs et al. (1996)
		Protein found in cytoplasm of neurons		Yes	IHC	Lungu et al. (1998), Theil et al. (2003), Grinfeld and Kennedy (2004), Hüfner et al. (2006)
		Transcripts found in neurons			ISH	Kennedy et al. (2000)
		Transcripts found and quantified			qRT-PCR	Cohrs and Gilden (2007)
63	Tegument protein	Transcripts found in neurons			ISH	Gilden et al. (1987), Kennedy et al. (2000)

		Transcripts found in non-neuronal cells			ISH	Croen et al. (1988)
		Transcripts found	Yes		PCR of cDNA	Cohrs et al. (1996)
		Protein found in cytoplasm of neurons		Yes	IHC	Mahalingam et al. (1996), Lungu et al. (1998), Grinfeld and Kennedy (2004)
		Transcripts found and quantified			qRT-PCR	Cohrs et al. (2000), Cohrs and Gilden (2007)
66	Serine–threonine protein kinase	Transcripts found	Yes		ISH, RT-PCR[f]	Cohrs et al. (2003a)
		Protein found in cytoplasm of neurons		Yes	IHC	Cohrs et al. (2003a)
		Transcripts found and quantified			qRT-PCR	Cohrs and Gilden (2007)

[a]Tyler et al. (2007)
[b]Immunohistochemistry
[c]In situ hybridization
[d]Mahalingam et al. (1998)
[e]Quantitative reverse transcriptase-PCR
[f]Reverse transcriptase-PCR

Applying GeXPS technology, Nagel et al. (2009) detected transcripts from all VZV genes in only five PCR reactions. Parallel analysis of RNA serial dilutions by GeXPS and real-time PCR showed that GeXPS multiplex analysis was sufficiently sensitive to detect as little as 20 copies of ORF 21, 29, 62, 63, and 66 transcripts. Thus, GeXPS technology can be used to determine the extent of VZV transcription in latently infected human ganglia.

10 MicroRNA Expression in Latently Infected Ganglia

MicroRNAs (miRNA) are small noncoding RNA molecules that alter transcript stability and translation when bound to their target mRNA. Deep sequencing of RNA extracted from human TG positive for VZV and HSV-1 DNA revealed several miRNAs mapping to the HSV-1 genome, but no VZV specific miRNAs (Umbach et al. 2009). Since the genetic locus containing the HSV-1 miRNA maps to the latency-associated transcript region of HSV-1, and the homologous region is deleted in VZV, the lack of VZV miRNA may reflect a basic mechanistic difference inherent in the way these neurotropic alphaherpesvirus maintain latency.

11 VZV Protein in Latently Infected Ganglia

To date, the only technique used to detect VZV proteins in latently infected ganglia has been immunohistochemistry (IHC). Proteins encoded by ORF 4, 21, 29, 62, 63, and 66 have been detected in the cytoplasm of neurons using various antibody sources (Table 1). The cytoplasmic localization of these proteins during latency, many of which are predominately nuclear during productive infection, may indicate a possible mechanism by which latency is maintained. However, the finding of proteins in ganglia without repeated detection of the respective transcript also suggests problems inherent to IHC. Just as the use of multiple independent techniques has aided the characterization of VZV transcripts in latently infected human ganglia, so too must the characterization of latently expressed VZV proteins await confirmation by independent techniques. Preliminary results suggest that Western blot analysis is not adequate to detect ORF 63 protein in ganglia containing abundant levels of ORF 63 transcripts (Cohrs, personal communication), pointing to the need for new, more sensitive technologies to detect VZV proteins during latency.

12 Future Directions

Analyses of multiple human ganglia obtained at autopsy indicate that latent VZV is predominantly, if not exclusively, located in ganglionic neurons. The VZV burden is variable, but not likely to exceed that found for HSV-1. VZV gene expression during latency is restricted, but not silenced. The studies reviewed here form a basis of our understanding of VZV latency and have helped to identify areas for future investigation:

1. VZV has been shown to establish latency in neurons (Hyman et al. 1983; Gilden et al. 1987; Cohrs et al. 2005), but is there a particular subpopulation of neurons that is selectively vulnerable to VZV infection?
2. While VZV is a remarkably stable virus, at least four distinct clades have been identified (Peters et al. 2006). Is there a correlation between specific VZV clades and the latent virus burden? Is there a demographic component to VZV latency?
3. Some latently transcribed VZV genes are regulated, in part, at the epigenetic level (Gary et al. 2006). Is this a universal feature of latent VZV gene regulation? What is the full repertoire of epigenetic markers associated with the latent virus genome? How is epigenetic regulation modified during reactivation?
4. Extensive investigation indicates that ORF 63 transcription is a hallmark of VZV latency (Cohrs and Gilden 2007); however, no complete search of the latent VZV transcriptome has been performed with the required sensitivity. Future studies await new technologies to investigate the full spectrum of VZV transcription during latency.
5. The nuclear redistribution of VZV ORF 63 protein has been detected during VZV reactivation (Lungu et al. 1998). What is the molecular basis for cytoplasmic and nuclear localization of IE63? What is the function of cytoplasmic and nuclear IE63? Does posttranslational processing of IE63 play a role in virus reactivation?
6. No VZV-specific miRNAs are detected during latency (Umbach et al. 2009); however, herpesviruses are able to modify expression of host cell miRNAs (Wang et al. 2008). Does VZV latency similarly modify the pool of miRNA in neurons?
7. VZV transcripts other than ORF 63 have been occasionally detected at low abundance during latency (Cohrs and Gilden 2007). What is the mechanism by which these transcripts are regulated? Are these transcripts required to maintain latency? Do these transcripts reflect limited virus reactivation?
8. Finally, what is the mechanism for VZV reactivation? The answer to this question will depend on the successful development of a valid experimental model of VZV latency, the most promising of which is simian varicella virus infection of monkeys (Mahalingam et al. 2007).

Answering these questions will greatly enhance our understanding of VZV molecular biology, and will require the development of exquisitely sensitive technologies to analyze the low abundance of this fascinating virus.

Acknowledgments This work was supported in part by Public Health Service grants NS032623 and AG032958 from the National Institutes of Health. The authors thank Dr. Robert Cordery-Cotter and Marina Hoffman for editorial review and Cathy Allen for preparing the manuscript.

References

Arvin A (1996) Varicella-zoster virus. Clin Microbiol Rev 9:361–381

Chaudhuri V, Sommer M, Rajamani J et al (2008) Functions of varicella-zoster virus ORF23 capsid protein in viral replication and the pathogenesis of skin infection. J Virol 82: 10231–10246

Clarke P, Beer T, Cohrs R et al (1995) Configuration of latent varicella-zoster virus DNA. J Virol 69:8151–8154

Cohrs RJ, Gilden DH (2003) Varicella zoster virus transcription in latently infected ganglia. Anticancer Res 23:2063–2070

Cohrs RJ, Gilden DH (2007) Prevalence and abundance of latently transcribed varicella-zoster virus genes in human ganglia. J Virol 81:2950–2956

Cohrs RJ, Srock K, Barbour MB et al (1994) Varicella-zoster virus (VZV) transcription during latency in human ganglia: construction of a cDNA library from latently infected human trigeminal ganglia and detection of a vzv transcript. J Virol 68:7900–7908

Cohrs RJ, Barbour M, Gilden DH (1996) Varicella-zoster virus (VZV) transcription during latency in human ganglia: detection of transcripts mapping to genes 21, 29, 62, and 63 in cDNA library enriched for VZV RNA. J Virol 70:2789–2796

Cohrs RJ, Randall J, Smith J et al (2000) Analysis of individual human trigeminal ganglia for latent herpes simplex virus type 1 and varicella-zoster virus nucleic acids using real-time PCR. J Virol 74:11464–11471

Cohrs RJ, Gilden DH, Kinchington PR et al (2003a) Varicella-zoster virus gene 66 transcription and translation in latently infected human ganglia. J Virol 77:6660–6665

Cohrs RJ, Hurley MP, Gilden DH (2003b) Array analysis of viral gene transcription during lytic infection of cells in tissue culture with varicella-zoster virus. J Virol 77:11718–11732

Cohrs RJ, LaGuardia JJ, Gilden DH (2005) Distribution of latent herpes simplex virus type-1 and varicella zoster virus DNA in human trigeminal ganglia. Virus Genes 31:223–227

Croen KD, Ostrove JM, Dragovic LJ et al (1988) Patterns of gene expression and sites of latency in human nerve ganglia are different for varicella-zoster and herpes simplex viruses. Proc Natl Acad Sci USA 85:9773–9777

Davison AJ, Scott JE (1986) The complete DNA sequence of varicella-zoster virus. J Gen Virol 67:1759–1816

Dueland AN, Ranneberg-Nilsen T, Degré M (1995) Detection of latent varicella zoster virus DNA and human gene sequences in human trigeminal ganglia by in situ amplification combined with in situ hybridization. Arch Virol 140:2055–2066

Gary L, Gilden DH, Cohrs RJ (2006) Epigenetic regulation of varicella-zoster virus open reading frames 62 and 63 in latently infected human trigeminal ganglia. J Virol 80:4921–4926

Gilden DH, Vafai A, Shtram Y et al (1983) Varicella-zoster virus DNA in human sensory ganglia. Nature 306:478–480

Gilden DH, Rozenman Y, Murray R et al (1987) Detection of varicella-zoster virus nucleic acid in neurons of normal human thoracic ganglia. Ann Neurol 22:377–380

Gilden DH, Gesser R, Smith J et al (2001) Presence of VZV and HSV-1 DNA in human nodose and celiac ganglia. Virus Genes 23:145–147

Grinfeld E, Kennedy PGE (2004) Translation of varicella-zoster virus genes during human ganglionic latency. Virus Genes 29:317–319

Hüfner K, Derfuss T, Herberger S et al (2006) Latency of α-herpes viruses is accompanied by a chronic inflammation in human trigeminal ganglia but not in dorsal root ganglia. J Neuropathol Exp Neurol 65:1022–1030

Hyman RW, Ecker JR, Tenser RB (1983) Varicella-zoster virus RNA in human trigeminal ganglia. Lancet 2:814–816

Kemble GW, Annuziato P, Lungu O et al (2000) Open reading frame S/L of varicella-zoster virus encodes a cytoplasmic protein expressed in infected cells. J Virol 74:11311–11321

Kennedy PG, Grinfeld E, Gow JW (1998) Latent varicella-zoster virus is located predominantly in neurons in human trigeminal ganglia. Proc Natl Acad Sci USA 95:4658–4662

Kennedy PGE, Grinfeld E, Bell JE (2000) Varicella-zoster virus gene expression in latently infected and explanted human ganglia. J Virol 74:11893–11898

Kennedy PGE, Grinfeld E, Craigon M et al (2005) Transcriptomal analysis of varicella-zoster virus infection using long oligonucleotide-based microarrays. J Gen Virol 86:2673–2684

LaGuardia JJ, Cohrs RC, Gilden DH (1999) Prevalence of varicella-zoster virus DNA in dissociated human trigeminal ganglion neurons and nonneuronal cells. J Virol 73:8571–8577

Levin MJ, Cai G-Y, Manchak MD et al (2003) Varicella-zoster virus DNA in cells isolated from human trigeminal ganglia. J Virol 77:6979–6987

Lungu O, Panagiotidis C, Annuziato PW et al (1998) Aberrant intracellular localization of varicella-zoster virus regulatory proteins during latency. Proc Natl Acad Sci USA 95: 7080–7085

Mahalingam R, Wellish MC, Dueland AN et al (1992) Localization of herpes simplex virus and varicella zoster virus DNA in human ganglia. Ann Neurol 31:444–448

Mahalingam R, Wellish M, Lederer D et al (1993) Quantitation of latent varicella-zoster virus DNA in human trigeminal ganglia by polymerase chain reaction. J Virol 67:2381–2384

Mahalingam R, Wellish M, Cohrs R et al (1996) Expression of protein encoded by varicella-zoster virus open reading frame 63 in latently infected human ganglionic neurons. Proc Natl Acad Sci USA 93:2122–2124

Mahalingam R, Lasher R, Wellish M et al (1998) Localization of varicella-zoster virus gene 21 protein in virus-infected cells in culture. J Virol 72:6832–6837

Mahalingam R, Kennedy PGE, Gilden DH (1999) The problems of latent varicella zoster virus in human ganglia: precise cell location and viral content. J Neurovirol 5:445–448

Mahalingam R, Traina-Dorge V, Wellish M et al (2007) Simian varicella virus reactivation in cynomologous monkeys. Virology 368:50–59

Meier JL, Holman RP, Croen KD et al (1993) Varicella-zoster virus transcription in human trigeminal ganglia. Virology 193:193–200

Nagel MA, Gilden DH (2007) The protean neurologic manifestations of varicella-zoster virus infection. Cleve Clin J Med 74:489–504

Nagel MA, Gilden D, Shade T et al (2009) Rapid and sensitive detection of 68 unique varicella zoster virus gene transcripts in five multiplex reverse transcription-polymerase chain reactions. J Virol Methods 157:62–68

Peters GA, Tyler SD, Grose C et al (2006) A full-genome phylogenetic analysis of varicella-zoster virus reveals a novel origin of replication-based genotyping scheme and evidence of recombination between major circulating clades. J Virol 80:9850–9860

Pevenstein SR, Williams RK, McChesney D et al (1999) Quantitation of latent varicella-zoster virus and herpes simplex virus genomes in human trigeminal ganglia. J Virol 73:10514–10518

Ross J, Williams M, Cohen JI (1997) Disruption of the varicella-zoster virus dUTPase and the adjacent ORF9A gene results in impaired growth and reduced syncytia formation *in vitro*. Virology 234:186–195

Theil D, Derfuss T, Paripovic I et al (2003) Latent herpesvirus infection in human trigeminal ganglia causes chronic immune response. Am J Pathol 163:2179–2184

Tyler SD, Peters GA, Grose C et al (2007) Genomic cartography of varicella-zoster virus: a complete genome-based analysis of strain variability with implications for attenuation and phenotypic differences. Virology 359:447–458

Umbach JL, Nagel MA, Cohrs RJ et al (2009) Analysis of human alphaherpesviruses microRNA expression in latently infected human trigeminal ganglia. J Virol 83:10677–10683

Wang K, Lau TY, Morales M et al (2005) Laser-capture microdissection: refining estimates of the quantity and distribution of latent herpes simplex virus 1 and varicella-zoster virus DNA in human trigeminal ganglia at the single-cell level. J Virol 79:14079–14087

Wang F-Z, Weber F, Croce C et al (2008) Human cytomegalovirus infection alters the expression of cellular microRNAspecies that affect its replication. J Virol 82:9065–9074

Neurological Disease Produced by Varicella Zoster Virus Reactivation Without Rash

Don Gilden, Randall J. Cohrs, Ravi Mahalingam, and Maria A. Nagel

Contents

Abstract Reactivation of varicella zoster virus (VZV) from latently infected human ganglia usually produces herpes zoster (shingles), characterized by dermatomal distribution pain and rash. Zoster is often followed by chronic pain (postherpetic neuralgia or PHN) as well as meningitis or meningoencephalitis, cerebellitis, isolated cranial nerve palsies that produce ophthalmoplegia or the Ramsay Hunt syndrome, multiple cranial nerve palsies (polyneuritis cranialis), vasculopathy, myelopathy, and various inflammatory disorders of the eye. Importantly, VZV reactivation can produce chronic radicular pain without rash (zoster sine herpete), as well as all the neurological disorders listed above without rash.

D. Gilden (✉), R.J. Cohrs, R. Mahalingam, and M.A. Nagel
Departments of Neurology (D.G., R.J.C., R.M., M.A.N.) and Microbiology (D.G.), University of Colorado Denver School of Medicine, 12700 E. 19th Avenue, Mail Stop B182, Aurora, CO 80045, USA
e-mail: don.gilden@ucdenver.edu

A.M. Arvin et al. (eds.), *Varicella-zoster Virus*,
Current Topics in Microbiology and Immunology 342, DOI 10.1007/82_2009_3

The protean neurological and ocular disorders produced by VZV in the absence of rash are a challenge to the practicing clinician. The presentation of these conditions varies from acute to subacute to chronic. Virological confirmation requires the demonstration of amplifiable VZV DNA in cerebrospinal fluid (CSF) or in blood mononuclear cells, or the presence of anti-VZV IgG antibody in CSF or of anti-VZV IgM antibody in CSF or serum.

Abbreviations

AIDS	Acquired immune deficiency syndrome
CSF	Cerebrospinal fluid
CT	Computerized tomography
HSV	Herpes simplex virus
Ig	Immunoglobulin
MNCs	Mononuclear cells
MRI	Magnetic resonance imaging
PORN	Progressive outer retinal necrosis
RHS	Ramsay Hunt syndrome
SSPE	Subacute sclerosing panencephalitis
VZV	Varicella zoster virus

1 Introduction

Varicella zoster virus (VZV) is an exclusively human neurotropic alphaherpesvirus. Primary infection causes varicella (chickenpox), after which virus becomes latent in cranial nerve ganglia, dorsal root ganglia, and autonomic ganglia along the entire neuraxis. Years later, as cell-mediated immunity to VZV declines with age or immunosuppression, as in patients with cancer or AIDS or organ transplant recipients, VZV reactivates to cause zoster (shingles), a syndrome characterized by pain and a vesicular rash on an erythematous base in 1–3 dermatomes. Because VZV becomes latent in any and all ganglia, zoster can develop anywhere on the body. In most patients, the disappearance of skin lesions is accompanied by decreased pain and complete resolution of pain in 4–6 weeks. Nevertheless, zoster is often followed by chronic pain (postherpetic neuralgia) as well as meningitis or meningoencephalitis, cerebellitis, isolated cranial nerve palsies that produce ophthalmoplegia or the Ramsay Hunt syndrome (RHS), multiple cranial nerve palsies (polyneuritis cranialis), vasculopathy, myelopathy and various inflammatory disorders of the eye, the most common of which is progressive outer retinal necrosis (PORN). Importantly, VZV reactivation can produce chronic radicular pain without rash (zoster sine herpete), as well as all the neurological disorders listed above without rash – the subject of this chapter.

2 Zoster Sine Herpete

The concept of zoster without rash in not new and was initially suggested as a nosological entity by Widal (1907), who described a 38-year-old man with thoracic distribution pain, hyperesthesia, a dilated left pupil, a mononuclear pleocytosis in the cerebrospinal fluid (CSF), and a negative Wasserman test for syphilis. Serologic studies for VZV were not performed. Weber (1916) later described two patients with segmental pain and weakness who ultimately developed facial and truncal zoster. In the largest series of zoster sine herpete, 120 patients were described as experiencing "zoster-type" pain without rash in a dermatome distribution distant from a dermatome with rash (Lewis 1958). CSF examination revealed the modest mononuclear pleocytosis often encountered in patients with zoster, but serologic studies were not done.

The first serologic confirmation of zoster sine herpete was reported by a physician who developed acute trigeminal distribution pain that lasted 16 days and was associated with a rise in complement-fixing antibody to VZV from 1:8 on day 15 to 1:128 on day 32 and no rise in antibody to herpes simplex virus (HSV) (Easton 1970). Schott (1998) later reported four patients, who years after trigeminal distribution zoster, developed pain without rash in the same distribution of the trigeminal nerve; unfortunately, none were studied virologically. Virologic confirmation of zoster sine herpete did not come until the analysis of two men, ages 62 and 66 years, with thoracic-distribution radicular pain that had lasted for months to years without zoster rash. An extensive search for systemic disease and malignancy was negative. VZV DNA, but not HSV DNA, was found in the CSF of the first patient 5 months after the onset of pain, and in the second patient, 8 months after pain onset (Gilden et al. 1994a). After diagnosis, both men were treated successfully with intravenous acyclovir. A third virologically confirmed case of thoracic-distribution zoster sine herpete that persisted for years included the demonstration of frequent fibrillation potentials restricted to chronically painful thoracic root segments (Amlie-Lefond et al. 1996). In that case, the patient did not improve after treatment with intravenous acyclovir and oral famciclovir.

Although the nosologic entity of zoster sine herpete as a clinical variant has now been established, its prevalence will not be known until a greater number of patients with prolonged radicular pain have been studied virologically. Analysis should include both PCR to amplify VZV DNA in CSF and in blood mononuclear cells (MNCs), as well as a search for antibody to VZV in CSF. The latter, even in the absence of amplifiable VZV DNA, has been useful to support the diagnosis of vasculopathy and myelopathy produced by VZV without rash (Gilden et al. 1998). Analysis of serum anti-VZV antibody is of no diagnostic value in patients with prolonged pain, because such antibodies persist in nearly all adults throughout life, and the presence of serum antibodies to different VZV glycoproteins and non-glycosylated proteins is variable (Vafai et al. 1988). Zoster sine herpete is essentially a disorder of the peripheral nervous system (ganglioradiculopathy) produced by VZV without rash. VZV also produces disease of the CNS without rash (described below). At the University of Colorado, School of Medicine, we have encountered more cases of VZV infection without rash in the CNS than in the peripheral nervous system.

3 Preherpetic Neuralgia

The existence of ganglionitis without rash is supported by radicular pain preceding zoster, so-called preherpetic neuralgia. We studied six patients whose pain preceded rash by 7–100 days and was located in dermatomes different from as well as in the area of eventual rash (Gilden et al. 1991). Two patients ultimately developed disseminated zoster with neurologic complications of zoster paresis and fatal zoster encephalitis; both had been taking long-term low-dose steroids. A third case of preherpetic neuralgia developed in a patient with prior metastatic carcinoma, and a fourth was in a patient with an earlier episode of brachial neuritis. Two subjects had no underlying disease. Further documentation of preherpetic neuralgia will determine whether its apparent association with steroid therapy and serious complications is statistically significant.

4 Meningitis/Meningoencephalitis

Meningitis is characterized by headache, fever, and stiff neck. Additional impairment of higher cognitive function, alterations in state of consciousness, or seizures indicate underlying parenchymal involvement (encephalitis). VZV involvement of the CNS without rash was verified by the intrathecal synthesis of antibodies to VZV in two patients with aseptic meningitis (Martinez-Martin et al. 1985) and later in four additional patients with aseptic meningitis (Echevarria et al. 1987), and in one patient with acute meningoencephalitis (Vartdal et al. 1982). Powell et al. (1995) reported a patient with meningoencephalitis without rash whose CSF contained VZV DNA, and Mancardi et al. (1987) described a patient with encephalomyelitis without rash, in whom anti-VZV antibody was detected in the CSF. Most recently, VZV meningitis with hypoglycorrhachia in the absence of rash developed in an immunocompetent woman (Habib et al. 2009).

5 Ramsay Hunt Syndrome

The strict definition of the RHS is peripheral facial nerve palsy accompanied by an erythematous vesicular rash on the ear (zoster oticus) or in the mouth. VZV causes the RHS. Some patients develop peripheral facial paralysis without ear or mouth rash associated with either a fourfold rise in antibody to VZV or the presence of VZV DNA in auricular skin, blood MNCs, middle ear fluid, or saliva (Hato et al. 2000). In a study of 32 patients with isolated peripheral facial palsy, Murakami et al. (1998) identified RHS zoster sine herpete in six (19%) of the patients based on a fourfold rise in serum antibody titer to VZV; four of these six patients were positive for VZV DNA by PCR when geniculate zone skin scrapings were studied. Further, Morgan and Nathwani (1992) found that 9.3% of Bell's palsy patients seroconverted to VZV, and Terada et al. (1998) found that blood MNCs from 4 of 17 Bell's palsy patients were positive for VZV DNA by PCR. Thus, a small proportion of "Bell's palsy patients" (idiopathic peripheral facial palsy) have RHS zoster sine herpete.

6 Polyneuritis Cranialis

There have been multiple instances of polyneuritis cranialis produced by VZV. The first report was of a 70-year-old man who seroconverted to VZV during acute disease (Mayo and Booss 1989). Another report described a 43-year-old man with acute polyneuritis cranialis who developed antibody in CSF to VZV but not to other human herpesviruses or to multiple ubiquitous paramyxoviruses or togaviruses (Osaki et al. 1995). Both men were immunocompetent. VZV-induced polyneuritis cranialis without rash has also been described in a patient with involvement of cranial nerves IX, X, and XI as well as upper cervical nerve roots without rash, and with anti-VZV antibody in the CSF (Funakawa et al. 1999).

7 Cerebellitis

While acute cerebellar ataxia (unsteadiness with or without tremor) is well-recognized as a potential complicating factor in childhood varicella (Connolly et al. 1994), there is one report of a child who became ataxic 5 days before chickenpox developed (Dangond et al. 1993). A virologically documented case of VZV cerebellitis without rash was in a 66-year-old man whose CSF at the time of presentation revealed mild lymphocytic pleocytosis and VZV DNA; 3 weeks later, anti-VZV IgG antibodies were detected in his CSF and additional analysis revealed intrathecal synthesis of anti-VZV IgG (Ratzka et al. 2006). A second case of acute cerebellitis caused by VZV in the absence of rash occurred in a middle-aged, immunocompetent woman; virological analysis of her CSF revealed VZV DNA and anti-VZV IgG antibody (Moses et al. 2006). The primary clue that led to the request for virological tests for VZV in the CSF of this patient was the patient's early age of chickenpox. Interestingly, many individuals who develop zoster before age 60 had chickenpox before age 4 years (Kakourou et al. 1998; Chiappini et al. 2002). The reason that early chickenpox predisposes to early zoster remains unknown, but is reminiscent of the observation that many patients with subacute sclerosing panencephalitis, a chronic fatal encephalitis caused by measles virus, had measles before the age of 2 years.

8 Vasculopathy

Transient ischemic attacks and focal deficit are not uncommon after VZV reactivation. A prototype fatal case of chronic progressive vasculopathy of 314 days' duration in an immunocompetent 73-year-old-man (Case Records of the Massachusetts General Hospital; Case 5-1995) was virologically verified to be caused by VZV (Gilden et al. 1996). Nau et al. (1998) reported another case of VZV vasculopathy without rash with virological confirmation. Kleinschmidt-DeMasters et al. (1998) described an HIV-positive patient with fatal encephalomyelitis and necrotizing vasculitis without rash, pathologically verified to be caused by VZV. In the past two decades, there have been numerous reports of VZV vasculopathy without rash. In fact, the largest study of virologically confirmed cases of VZV vasculopathy revealed that more than one-third of patients did not have rash. Even in a virologically confirmed case of spinal cord infarction caused by VZV, rash did not occur until 3 days after the onset of myelopathy (Orme et al. 2007).

9 Myelopathy

Acute VZV myelopathy is usually characterized by paralysis in the legs, with bladder and bowel incontinence, and occurs in the absence of rash. Heller et al. (1990) initially described a 31-year-old immunocompetent man who developed transverse myelitis with partial recovery. Disease was attributed to VZV based on the development of antibody in CSF. Later, we encountered two patients with VZV myelopathy in the absence of rash (Gilden et al. 1994b). The first patient developed zoster followed by myelopathy 5 months later, when PCR-amplifiable VZV DNA was detected in CSF. The second equally fascinating patient developed myelopathy at the time of acute zoster. The myelopathy resolved, but recurred 6 months later. Five months after the recurrence of myelopathy, the patient's CSF contained both amplifiable VZV DNA as well as antibody to VZV. Another case of recurrent VZV myelopathy revealed anti-VZV IgG in CSF with reduced serum/CSF ratios of anti-VZV IgG at the time of recurrence in the absence of rash (Gilden et al. 2009). Overall, the spectrum of VZV myelopathy is broad, ranging from acute to chronic and rarely recurrent.

10 Ocular Disorders

An emerging body of literature describes multiple ocular inflammatory disorders produced by VZV in the absence of rash. VZV is the most common cause of PORN. Multiple cases of PORN caused by VZV in the absence of rash have been described (Friedman et al. 1993; Galindez et al. 1996). A case of severe unremitting eye pain without rash was proven to be caused by VZV based on the detection of VZV DNA in nasal and conjunctival samples (Goon et al. 2000). Other cases include third cranial nerve palsies (Hon et al. 2005), retinal periphlebitis (Noda et al. 2001), uveitis (Akpel and Gottsch 2000; Hon et al. 2005), iridocyclitis (Yamamoto et al. 1995), and disciform keratitis (Silverstein et al. 1997), all without rash and all confirmed virologically to be caused by VZV.

11 Remarkable Cases of VZV Infection Without Rash

Two remarkable cases of VZV infection without rash deserve mention. The first and most extreme example of VZV infection of the nervous system that we encountered was a 77-year-old man with T cell lymphoma and no history of zoster rash who developed meningoradiculitis of the cranial nerve roots and cauda equina; he died 3 weeks after the onset of neurologic disease, confirmed pathologically and virologically to have been caused by VZV (Dueland et al. 1992). Autopsy revealed hemorrhagic inflammatory lesions with Cowdry A inclusions in the meninges and nerve roots, extending from cranial nerve roots to the cauda equina. The same lesions were present in the brain, although to a lesser extent. VZV antigen and nucleic acid, but not herpes zoster virus (HSV) or cytomegalovirus antigen or nucleic acid, were found in the infected tissue at all levels of the neuraxis. Thus, VZV should be included in the differential diagnosis of acute encephalomyeloradiculopathy, particularly since antiviral treatment is available.

The second case was an immunocompetent 45-year-old woman with a 13-month history of right facial numbness and "lightning bolt-like" pain in the maxillary distribution of the right trigeminal nerve. Initially, numbness was present over the right side of the nose adjacent to the zygoma, and a light touch over this area or the right forehead produced pain (allodynia) over the entire face. Neurological examination revealed loss of pinprick sensation in the maxillary distribution of the right trigeminal nerve. Fifth nerve motor function and the corneal reflex were intact. The CSF was acellular. Brain scanning by both computerized tomography (CT) and MRI was normal. She was treated with neurontin, 900 mg three times daily, but her pain increased. One year later, both CT and MRI scans revealed a $0.9 \times 0.9 \times 2.0$-cm homogeneously enhancing mass at the base of the right brain at the site of the trigeminal ganglion. There was never any history of zoster rash. The tumor was removed surgically. Pathological and virological analysis of the trigeminal ganglionic mass confirmed chronic active VZV-induced ganglionitis (Hevner et al. 2003).

12 Systemic Disease Produced by VZV Reactivation Without Rash

Ross et al. (1980) reported a case of fatal hepatic necrosis caused by VZV in a 64-year-old woman. She had undergone splenectomy 14 months before the fatal hepatitis. The postmortem diagnosis was based on characteristic pathological changes in the liver, the detection of herpesvirus virions in liver by electron microscopy, and serologic evidence of recent VZV infection. A report of fulminant fatal disseminated VZV infection without rash occurred in an 8-year girl undergoing chemotherapy for leukemia; virus was present in blood, lungs, liver, kidneys, and bone marrow (Grant et al. 2002). Although the nervous system was spared in both patients, these cases confirm the existence of disseminated VZV infection of multiple organs in the absence of rash.

13 Subclinical VZV Reactivation

All of the above reports support the notion that VZV can reactivate to produce neurological and systemic disease without a characteristic zosteriform rash. VZV also reactivates subclinically. Ljungman et al. (1986) demonstrated subclinical reactivation in 19 of 73 (26%) bone marrow transplant recipients, as determined by a fivefold increase in VZV-specific antibody titers or a reappearance of a specific lymphoproliferative response to VZV antigen. Only 2 of the 19 patients with VZV reactivation had signs: one experienced transient fever, and the other developed hepatitis simultaneously with increasing VZV titers. Serologic analysis has also indicated VZV as the cause of unexplained fever in three of nine metastatic breast cancer patients (Rasoul-Rockenschaub et al. 1990). In addition, VZV reactivates subclinically (without pain or rash) in astronauts (Mehta et al. 2004), including shedding of infectious virus (Cohrs et al. 2008), most likely reflecting transient stress-induced depression of cell-mediated immunity to VZV (Taylor and Janney 1992). Importantly, asymptomatic shedding of human neurotropic alphaherpsesviruses is not restricted to VZV, since HSV-1 (Kaufman et al. 2005) and HSV-2 (Koelle et al. 1992) also show subclinical reactivation. Finally, VZV DNA has been detected in saliva of patients with zoster in multiple dermatomes (Mehta et al. 2008), suggesting the simultaneous reactivation of VZV not only from geniculate ganglia but also from ganglia corresponding to the dermatome where zoster occurred. Detection of VZV DNA in saliva of such zoster patients provides a virological explanation for the classic clinical observations of Lewis (1958) who described dermatomal distribution radicular pain in areas distinct from pain with rash. Such clinical and virological phenomena are not surprising, since VZV becomes latent in cranial nerve ganglia, dorsal root ganglia, and autonomic ganglia along the entire neuraxis.

14 Diagnostic Testing

While zoster is usually diagnosed based on observation alone, all neurological complications of VZV reactivation without rash require virological confirmation. Evidence of active VZV infection is supported by any or all of the following positive tests: the presence of anti-VZV IgM antibody in serum or CSF or anti-VZV IgG antibody in CSF, or the detection of VZV DNA in blood MNCs or CSF. Because nearly all adult serum contains anti-VZV IgG antibody, serological testing for anti-VZV IgG antibody alone without other virological testing is of no value. In our experience, the most valuable tests are those that detect VZV DNA or anti-VZV IgG antibody in CSF. Much less frequently, anti-VZV IgM is found in serum or CSF, and even less often, VZV DNA is found in blood MNCs during acute neurological disease. Some, but not all, clinical hospital laboratories either perform tests for anti-VZV IgG antibody or send the specimen to an outside laboratory. If the hospital cannot conduct a test for anti-VZV IgG or IgM antibody, the referring physician should contact Dr. Don Gilden (don.gilden@ucdenver.edu) who will arrange to have the specimen tested at no cost to the patient. Because VZV DNA was found in the saliva of every zoster patient in the study by Mehta et al. (2008), it is possible that saliva may be a useful source to detect VZV in patients with neurological disease in the absence of rash. To date, definitive virological confirmation has required blood and CSF examination for VZV DNA and anti-VZV IgM and/or IgG antibody. The continuing challenge for the clinician is in establishing a diagnosis at a time when treatment will still benefit the patient.

Acknowledgements This work was supported in part by Public Health Service grants AG006127, NS032623 and AG032958 from the National Institutes of Health. Maria Nagel is supported by Public Health Service grant NS007321 from the National Institutes of Health. The authors thank Marina Hoffman for editorial review and Cathy Allen for manuscript preparation.

References

Akpel EK, Gottsch JD (2000) Herpes zoster sine herpete presenting with hypema. Ocul Immunol Inflamm 8:115–118

Amlie-Lefond C, Mackin GA, Ferguson M et al (1996) Another case of virologically confirmed zoster sine herpete, with electrophysiologic correlation. J Neurovirol 2:136–138

Case Records of the Massachusetts General Hospital (Case 5-1995) (1995) N Engl J Med 332:452–459

Chiappini E, Calabri G, Galli L et al (2002) Varicella-zoster virus acquired at 4 months of age reactivates at 24 months and causes encephalitis. J Pediatr 140:250–251

Cohrs RJ, Mehta SK, Schmid DS et al (2008) Asymptomatic reactivation and shed of infectious varicella zoster virus in astronauts. J Med Virol 80:1116–1122

Connolly AM, Dodson WE, Al P et al (1994) Course and outcome of acute cerebellar ataxia. Ann Neurol 35:673–679

Dangond F, Engle E, Yessayan L et al (1993) Pre-eruptive varicella cerebellitis confirmed by PCR. Pediatr Neurol 9:491–493

Dueland AN, Martin JR, Devlin ME et al (1992) Acute simian varicella infection: clinical, laboratory, pathologic, and virologic features. Lab Invest 66:762–773

Easton HG (1970) Zoster sine herpete causing acute trigeminal neuralgia. Lancet 2:1065–1066

Echevarria JM, Martinez-Martin P, Tellez A et al (1987) Aseptic meningitis due to varicella-zoster virus: serum antibody levels and local synthesis of specific IgG, IgM and IgA. J Infect Dis 155:959–967

Friedman SM, Mames RN, Sleasman JW et al (1993) Acute retinal necrosis after chickenpox in a patient with acquired immunodeficiency syndrome. Arch Ophthalmol 111:1607–1608

Funakawa I, Terao A, Koga M (1999) A case of zoster sine herpete with involvement of the unilateral IX, X and XI cranial and upper cervical nerves. Rinsho Skinkeigaku 39:958–960
Galindez OA, Sabates NR, Whitacre MM et al (1996) Rapidly progressive outer retinal necrosis caused by varicella zoster virus in a patient infected with human immunodeficiency virus. Clin Infect Dis 22:149–151
Gilden DH, Dueland AN, Cohrs R et al (1991) Preherpetic neuralgia. Neurology 41:1215–1218
Gilden DH, Wright RR, Schneck SA et al (1994a) Zoster sine herpete, a clinical variant. Ann Neurol 35:530–533
Gilden DH, Beinlich BR, Rubinstien EM et al (1994b) Varicella-zoster virus myelitis: an expanding spectrum. Neurology 44:1818–1823
Gilden DH, Kleinschmidt-DeMasters BK, Wellish M et al (1996) Varicella zoster virus, a cause of waxing and waning vasculitis: the New England Journal of Medicine case 5-1995 revisited. Neurology 47:1441–1446
Gilden DH, Bennett JL, Kleinschmidt-DeMasters BK et al (1998) The value of cerebrospinal fluid antiviral antibody in the diagnosis of neurologic disease produced by varicella zoster virus. J Neurol Sci 159:140–144
Gilden D, Nagel MA, Ransohoff RM et al (2009) Recurrent varicella zoster virus myelopathy. J Neurol Sci 276:196–198
Goon P, Wright M, Fink C (2000) Ophthalmic zoster sine herpete. J R Soc Med 93:191–192
Grant RM, Weitzman SS, Sherman CG et al (2002) Fulminant disseminated varicella zoster virus infection without skin involvement. J Clin Virol 24:7–12
Habib AA, Gilden D, Schmid DS et al (2009) Varicella zoster virus meningitis with hypoglycorrhachia in the absence of rash in an immunocompetent woman. J Neurovirol 15:206–208
Hato N, Kisaki H, Honda N et al (2000) Ramsay Hunt syndrome in children. Ann Neurol 48: 254–256
Heller HM, Carnevale NT, Steigbigel RT (1990) Varicella zoster virus transverse myelitis without cutaneous rash. Am J Med 88:550–551
Hevner R, Vilela M, Rostomily R et al (2003) An unusual cause of trigeminal-distribution pain and tumour. Lancet Neurol 2:567–571
Hon C, Au WY, Cheng VC (2005) Ophthalmic zoster sine herpete presenting as oculomotor palsy after marrow transplantation for acute myeloid leukemia. Haematologica 90 Suppl 12:EIM04
Kakourou T, Theodoridou M, Mostrou G et al (1998) Herpes zoster in children. J Am Acad Dermatol 39:207–210
Kaufman HE, Azcuy AM, Varnell ED et al (2005) HSV-1 DNA in tears and saliva of normal adults. Invest Ophthalmol Vis Sci 46:241–247
Kleinschmidt-DeMasters BK, Mahalingam R, Shimek C et al (1998) Profound cerebrospinal fluid pleocytosis and Froin's syndrome secondary to widespread necrotizing vasculitis in an HIV-positive patient with varicella zoster virus encephalomyelitis. J Neurol Sci 159:213–218
Koelle DM, Benedetti J, Langenberg A et al (1992) Asymptomatic reactivation of herpes simplex virus in woman after the first episode of genital herpes. Ann Intern Med 116:433–437
Lewis GW (1958) Zoster sine herpete. Br Med J 34:418–421
Ljungman P, Lönnqvist B, Gahrton G et al (1986) Clinical and subclinical reactivations of varicella-zoster virus in immunocompromised patients. J Infect Dis 153:840–847
Mancardi GL, Melioli G, Traverso F et al (1987) Zoster sine herpete causing encephalomyelitis. Ital J Neurol Sci 8:67–70
Martinez-Martin P, Garcia-Sáiz A, Rapun JL et al (1985) Intrathecal synthesis of IgG antibodies to varicella-zoster virus in two cases of acute aseptic meningitis syndrome with no cutaneous lesions. J Med Virol 16:201–209
Mayo DR, Booss J (1989) Varicella zoster associated neurologic disease without skin lesions. Arch Neurol 46:313–315
Mehta SK, Cohrs RJ, Forghani B et al (2004) Stress-induced subclinical reactivation of varicella zoster virus in astronauts. J Med Virol 72:174–179

Mehta SK, Tyring SK, Gilden DH et al (2008) Varicella-zoster virus in the saliva of patients with herpes zoster. J Infect Dis 197:654–657

Morgan M, Nathwani D (1992) Facial palsy and infection: the unfolding story. Clin Infect Dis 14:263–271

Moses H, Nagel MA, Gilden DH (2006) Acute cerebellar ataxia in a 41 year old woman. Lancet Neurol 5:984–988

Murakami S, Honda N, Mizobuchi M et al (1998) Rapid diagnosis of varicella zoster virus infection in acute facial palsy. Neurology 51:202–205

Nau R, Lantsch M, Stiefel M et al (1998) Varicella zoster virus-associated focal vasculitis without herpes zoster: recovery after treatment with acyclovir. Neurology 51:914–915

Noda Y, Nakazawa M, Takahashi D et al (2001) Retinal periphlebitis as zoster sine herpete. Arch Ophthalmol 119:1550–1552

Orme HT, Smith AG, Nagel MA et al (2007) VZV Spinal cord infarction identified by diffusion-weighted MRI (DWI). Neurology 69:398–400

Osaki Y, Matsubayashi K, Okumiya K et al (1995) Polyneuritis cranialis due to varicella-zoster virus in the absence of rash. Neurology 45:2293

Powell KF, Wilson HG, Corxson MO et al (1995) Herpes zoster meningoencephalitis without rash: varicella zoster virus DNA in CSF. J Neurol Neurosurg Psychiatry 59:198–199

Rasoul-Rockenschaub S, Zielinski CC, Müller C et al (1990) Viral reactivation as a cause of unexplained fever in patients with progressive metastatic breast cancer. Cancer Immunol Immunother 31:191–195

Ratzka P, Schlachertzki JC, Bahr M et al (2006) Varicella zoster virus cerebellitis in a 66-year-old patient without herpes zoster. Lancet 357:182

Ross JS, Fanning WL, Beautyman W et al (1980) Fatal massive hepatic necrosis from varicella-zoster hepatitis. Am J Gastroenterol 74:423–427

Schott GC (1998) Triggering of delayed-onset postherpetic neuralgia. Lancet 351:419–420

Silverstein BE, Chandler D, Neger R et al (1997) Disciform keratitis: a case of herpes zoster sine herpete. Am J Ophthalmol 123:254–255

Taylor GR, Janney RP (1992) In vivo testing confirms a blunting of the human cell-mediated immune mechanism during space flight. J Leukoc Biol 51:129–132

Terada K, Niizuma T, Kawano S et al (1998) Detection of varicella-zoster virus DNA in peripheral mononuclear cells from patients with Ramsay Hunt syndrome or zoster sine herpete. J Med Virol 56:359–363

Vafai A, Mahalingam R, Zerbe G et al (1988) Detection of antibodies to varicella-zoster virus proteins in sera from the elderly. Gerontology 34:242–249

Vartdal F, Vandvik B, Norrby E (1982) Intrathecal synthesis of virus-specific oligoclonal IgG, IgA and IgM antibodies in a case of varicella-zoster meningoencephalitis. J Neurol Sci 57:121–132

Weber FP (1916) Herpes zoster: its occasional association with a generalized eruption and its occasional connection with muscular paralysis—also an analysis of the literature of the subject. Int Clin 3:185–202

Widal (1907). Med Chiropractic Pract 78:12

Yamamoto S, Tada R, Shimomura Y et al (1995) Detecting varicella-zoster virus DNA in iridocyclitis using polymerase chain reaction: a case of zoster sine herpete. Arch Ophthalmol 113:1358–1359

Varicella-Zoster Virus Neurotropism in SCID Mouse–Human Dorsal Root Ganglia Xenografts

L. Zerboni, M. Reichelt, and A. Arvin

Contents

L. Zerboni (✉), M. Reichelt, and A. Arvin
Stanford University Medical Center, Stanford, CA, USA
e-mail: Zerboni@stanford.edu

A.M. Arvin et al. (eds.), *Varicella-zoster Virus*,
Current Topics in Microbiology and Immunology 342, DOI 10.1007/82_2009_8

Abstract Varicella-zoster virus (VZV) is a neurotropic human alphaherpesvirus and the causative agent of varicella and herpes zoster. VZV reactivation from latency in sensory nerve ganglia is a direct consequence of VZV neurotropism. Investigation of VZV neuropathogenesis by infection of human dorsal root ganglion xenografts in immunocompromised (SCID) mice has provided a novel system in which to examine VZV neurotropism. Experimental infection with recombinant VZV mutants with targeted deletions or mutations of specific genes or regulatory elements provides an opportunity to assess gene candidates that may mediate neurotropism and neurovirulence. The SCID mouse–human DRG xenograft model may aid in the development of clinical strategies in the management of herpes zoster as well as in the development of "second generation" neuroattenuated vaccines.

1 Introduction

Varicella-zoster virus (VZV) is a neurotropic human herpesvirus (subfamily *Alphaherpesviridae*) and the etiological agent of two human diseases, varicella and herpes zoster (*reviewed in Cohen et al.* (2007) *and Zerboni and Arvin* (2008)). Varicella, commonly called "chickenpox", results from primary VZV infection in susceptible individuals. In healthy children, varicella is a self-limiting infection that manifests as successive crops of pruritic lesions accompanied by fever. As a consequence of primary infection, VZV gains access to cells within the sensory cranial nerve and dorsal root ganglia (DRG) in which it appears to persist long term in a non-replicating "latent" state. Herpes zoster is the medical consequence of "reactivation" of latent VZV genomes within sensory nerve ganglion cells in which newly formed progeny virions transit in the anterograde direction to reinfect peripheral skin sites. Unlike the generalized rash of varicella, herpes zoster (commonly called "shingles") involves a localized region comprising one or more adjacent dermatomes, reflecting the sensory ganglion from which the virus has reactivated and in some cases, neighboring ganglia (Hope-Simpson 1965; Silverstein and Straus 2000). Whereas varicella resolves within 7–10 days, zoster lesions may require several weeks to heal and neuropathic pain may persist for months (Gilden et al. 2006; Silverstein and Straus 2000). In that herpes zoster is contingent upon reactivation of latent varicella within neuronal ganglia, it is a direct consequence of VZV neurotropism.

VZV is highly contagious and is maintained in the human population by shedding from vesicular lesions during varicella and herpes zoster episodes (Hope-Simpson 1965). Prior to widespread vaccination, varicella was a common childhood disease. Resolution of primary infection and herpes zoster corresponds to induction of VZV-specific T cell mediated immune responses (Arvin 2005). Immunoscenescence and immunodeficiency are risk factors for reactivation of latent VZV, and so herpes zoster is most common among the elderly and immunocompromised (Arvin 2005; Oxman et al. 2005). Live attenuated varicella vaccines have been licensed for use in preventing primary VZV infection and restoring protective cellular immune responses in the elderly (Arvin 2001; Oxman et al. 2005; Takahashi et al. 1974). Derived from serial passage of the "Oka" clinical isolate, Oka varicella vaccines retain neurotropism and can reactivate; infection with naturally circulating virus can also occur after vaccination. Therefore, vaccination does not consistently prevent VZV from establishing latency or from reactivating from sensory nerve ganglia (Arvin 2001; Takahashi et al. 1974; Zerboni et al. 2005b).

In this chapter, we review our investigation of VZV neurotropism through experimental infection of human DRG xenografts in severe compromised immunodeficient (SCID) mice (Oliver et al. 2008; Reichelt et al. 2008; Zerboni et al. 2005b, 2007). These experiments provide insight into VZV interactions with neural cells in sensory ganglia and may have potential relevance in the design of neuroattenuated varicella vaccines.

2 The SCID Mouse–Human Xenograft Model for VZV Pathogenesis

The clinical consequences of primary varicella infection and herpes zoster have been investigated for several decades; however, knowledge concerning VZV pathogenesis and virulence within the natural host is limited because infections are rarely fatal and, most importantly, no small animal exists that recapitulates disease in the human (Cohen et al. 2007). Whereas other human alphaherpesviruses readily infect rodents, rabbits, and guinea pigs, progress in the development of a small animal model for the in vivo investigation of VZV pathogenesis has been hampered by a strict restriction for human tissues (Cohen et al. 2007). Experimental infection of mice, rats, rabbits, or guinea pigs with a high potency VZV inoculant fails to produce robust infection (Cohen et al. 2007). Xenotransplanation of immunodeficient SCID mice with human tissues was first reported in 1988 (McCune et al. 1988; Mosier et al. 1988). To address the growing need for an experimental system to carry out "in vivo" infection of tissues with permissivity for VZV, we established a small animal model in which human tissue xenografts with relevance to VZV pathogenesis are infected within the SCID mouse host (Moffat et al. 1995, 1998; Zerboni et al. 2005b). SCID mice are not susceptible to VZV infection due to the host cell restriction, and so the consequences of VZV xenograft infection can be examined over a long interval, which is critical for investigation of latency and reactivation. For more than a decade, studies using human skin and T cell (thymus/liver) xenografts in SCID mice have yielded a wealth of information regarding VZV skin and T cell pathogenesis (*reviewed in Arvin* (2006)). These SCID xenograft models have also proven to be valuable experimental tools for the assessment of VZV gene function in vivo through experimental infection with recombinant viruses (*reviewed in Zerboni and Arvin* (2008)).

3 SCID Mouse–Human DRG Xenografts

Investigations of VZV neurotropism and virus–host interaction involving the various cell types that reside within sensory nerve ganglia have relied primarily upon tissues acquired from deceased individuals (Cohen et al. 2007). Tissues acquired postmortem from individuals who died during primary varicella or reactivation are exceedingly rare, and so information concerning acute VZV replication within sensory nerve ganglia has been limited. Examinations of latently infected cadaver ganglia are fraught with discrepant reports, which may be a consequence of physiological changes that occur during the interval between death and tissue acquisition (Cohen et al. 2007). Alternative in vitro model systems, such as explanted human DRG and primary neuronal cultures, do not survive long term and may also undergo physiological alterations in response to culture conditions. Optimally, a model for VZV neurotropism and neuropathogenesis would contain all cell populations that are potentially relevant to VZV neuropathobiology within their normal tissue microenvironment under physiological (in vivo) conditions. Therefore, we developed a SCID mouse–human xenograft model by establishing intact human DRG xenografts in SCID mice (Zerboni et al. 2005b). This model provides the opportunity to explore VZV infection and tropism for human neurons while preserving the interactions between neurons and non-neuronal cells within the DRG microenvironment.

3.1 Xenotransplantation of Human DRG in SCID Mice

SCID mouse–human DRG xenografts are constructed by implantation of human fetal DRG under the renal capsule of 5–8 week old male *scid/scid* homozygous CB.17 mice (Zerboni et al. 2005b). DRG are dissected from fetal spinal tissues at 18–24 gestational weeks and are $\sim$1–2 mm^3 at the time of xenotransplantation. Visual inspection at 12 weeks post xenotransplantation reveals a small (2–3 mm^3) well-vascularized graft attached to the mouse kidney (Fig. 1a). Histopathic examination reveals a neural tissue with organotypic features of human ganglia including clusters of small and large diameter neural cell bodies surrounded by satellite cells, fibroblasts, and other cell types within a dense supportive matrix comprising collagen and nerve fibers (Fig. 1b). Neural specific cell markers, such as synaptophysin and neural cell adhesion molecule (NCAM), are expressed and retain their expected localization. Human endothelial cell marker, PECAM-1, is also expressed on blood vessels supporting the graft. DRG engraftment is successful in >95% of mice and xenografts survive up to 8 months post xenotransplantation.

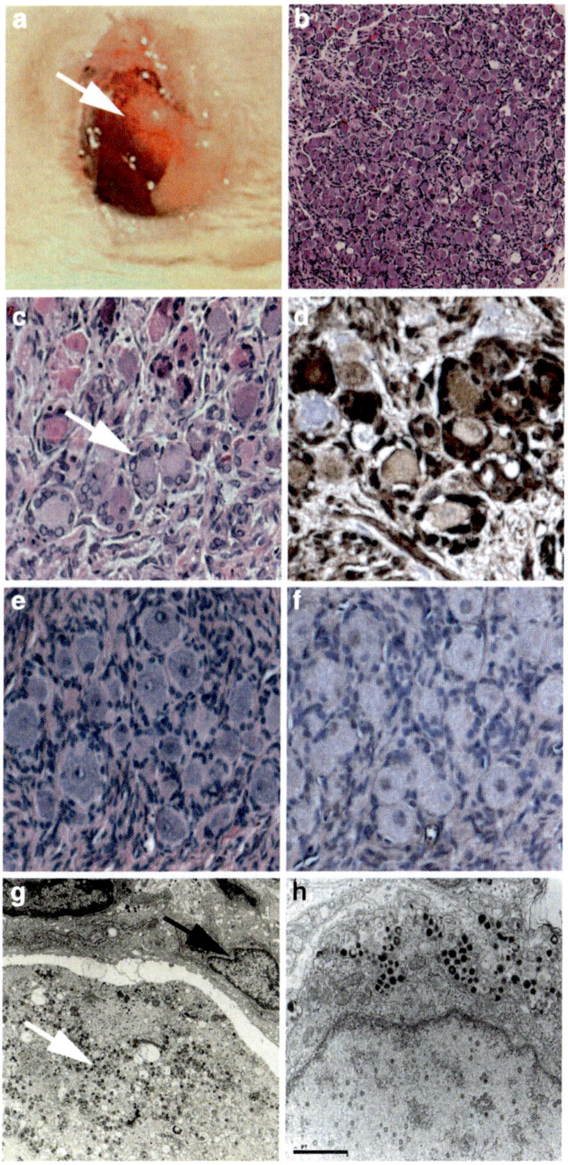

Fig. 1 VZV infection of DRG xenografts in SCID mice. (**a**) Human DRG xenograft under the renal capsule of a SCID mouse 12 weeks post xenotransplantation. (**b**) SCID DRG xenografts retain organotypic features of normal human ganglia, magnification 100×, stained with hematoxylin and eosin. (**c**) SCID DRG xenograft 14 days after infection, *white arrow* indicates cytopathology in neurons, magnification 200×, stained with hematoxylin and eosin. (**d**) SCID DRG xenograft 14 days after infection stained with rabbit polyclonal antibody to VZV IE63 (*brown*, DAB signal) and counterstained with hematoxylin, magnification 200×. (**e**) SCID DRG xenograft 56 days after VZV infection showing no evidence of productive infection, magnification 200×,

4 VZV Neurotropism in Human DRG Xenografts

4.1 Acute VZV Replication in Human DRG

As VZV is a highly cell-associated virus, inoculation of DRG xenografts is accomplished by direct injection of VZV-infected fibroblasts ($\sim$10 ul, 10^2–10^4 PFU). Unlike experimental infection of ganglia from non-permissive hosts, VZV infection of DRG is characterized by productive replication of the virus within different cell types that comprise host neuronal tissues. Cytopathic changes, such as denucleated neuronal cell bodies and cytoplasmic inclusion bodies, are apparent 10–14 days post infection (Fig. 1c) (Zerboni et al. 2005b). During this acute stage of infection, VZV regulatory proteins, such as IE63, are expressed within large numbers of neurons and satellite cells (Fig. 1d) (Zerboni et al. 2005b). Infectious virus can be recovered from productively infected DRG xenografts by mechanical disruption and coculture with a permissive cell monolayer up to 14 days post infection and correlates with detection of high viral genome copy numbers (10^8–10^9 copies/10^5 cells) in homogenized ganglia by quantitative real-time DNA PCR (Zerboni et al. 2005b).

Aggregate data from several investigations of VZV latency using cadaver ganglia implicate neurons within sensory ganglia, and, to a much lesser extent, satellite cells, as potential hosts for the latent VZV genome (Gilden et al. 1983; Kennedy et al. 1998, 1999; LaGuardia et al. 1999; Lungu et al. 1995; Mahalingam et al. 1993; Pevenstein et al. 1999; Wang et al. 2005). However, information on targets of VZV replication during acute infection and reactivation is limited. Investigation of acute VZV infection by correlative immunofluorescence-electron microscopy (IF-EM), which permits evaluation of histological and ultrastructural changes within the same cell, demonstrates a role for satellite cells as well as neurons in virion production (Reichelt et al. 2008). Ontogeny of infectious VZV requires assembly and packaging of viral capsids with a DNA core in the cell nucleus, egress of the nucleocapsid through the nuclear membrane to the cytoplasm, association with virion tegument proteins, and acquisition of the virion envelopment followed by release of infectious virus through egress pathways (Cohen et al. 2007). Intranuclear and cytoplasmic viral capsids were observed within neurons as well as satellite cells in acutely infected DRG xenografts by IF-EM, which indicates productive infection in both cell types (Fig. 1g, h; Reichelt et al. 2008). Moreover, extracellular particles were observed external to the plasma membranes of both cell

Fig. 1 (continued) hematoxylin and eosin staining. (**f**) SCID DRG xenograft 56 days after VZV infection stained with rabbit polyclonal antibody to VZV IE63 (*brown*, DAB signal) and counterstained with hematoxylin, showing absence of IE63 expression, magnification 200×. (**g**) TEM of SCID DRG xenograft 14 days after VZV infection, *white arrow* points to virions in cytoplasm of neuronal cell nucleus, *black arrow* points to satellite cell nucleus, magnification 3,800×. (**h**) TEM of SCID DRG xenograft 14 days after VZV infection, virions are present in the nucleus and cytoplasm of a satellite cell types. VZV IE proteins, the early ORF47 protein kinase, and late viral glycoproteins were expressed on satellite cells as well as neurons (Reichelt et al. 2008; Zerboni et al. 2005b). Evidence of acute replication in satellite cells is interesting given the preponderance of evidence that the neuron is the primary site of latency. It is not clear if VZV is unable to persist long term in satellite cells or if particles detected in satellite cells are non-infectious. By IF-EM, virion particles within satellite cells contained no obvious ultrastructural defects and were similar to those observed in neurons (Fig. 1h) (Reichelt et al. 2008).

4.2 *Acute VZV Infection of Human DRG Xenografts Is Associated with Polykaryon Formation*

Our examination of VZV replication in human DRG xenografts in SCID mice showed the presence of VZV DNA, viral proteins, and virion production in both neurons and satellite cells. A hallmark of VZV infection in skin lesions and in cultured cells is cell–cell fusion with resulting polykaryon formation. We examined the role of polykaryon formation during acute infection of DRG between neurons and their surrounding satellite cells, which together form a neuron-satellite cell complex (NSC) (Reichelt et al. 2008).

Sections from VZV-infected DRG (14 days after infection) were examined by staining with antibodies for NCAM and VZV IE62 protein, as a marker for productive infection. NCAM is expressed on both the neuronal membranes encapsulating the NSC and within the NSC, providing separation between the neuronal cell cytoplasm and the satellite cell cytoplasm (Fig. 2a–c). Control (uninfected) DRG were stained with antibody to NCAM and synaptophysin, which is localized to neuronal cytoplasmic vesicles. Examination by confocal IF microscopy revealed a large proportion of NSCs in VZV-infected DRG sections with evidence of polykaryon formation, as judged by the absence of NCAM-expressing membranes within NSC; whereas no fusion was detected within NSC from uninfected DRG, which maintained separation as indicated by intact NCAM-stained membranes. NCAM expression was retained on membranes that surround the NSC in both infected and uninfected sections.

To provide additional evidence for polykaryon formation, ultrathin sections of acutely infected DRG were evaluated by TEM (Fig. 2d–g). Ultrastructural analysis revealed examples of both infected NSC in which no fusion between the neuronal cell body and the adjacent satellite cells had occurred, and others in which the neural and the satellite cell nuclei shared the same cytoplasm within a polykaryon. In the absence of fusion, the plasma membranes of both the satellite cells and the neuron were clearly identifiable and a double-membrane like structure resulted when two cells, each surrounded by its own plasma membrane, were adjacent within the same NSC. However, where fusion occurred, plasma membranes separating nuclei within the same cytoplasm were not observed.

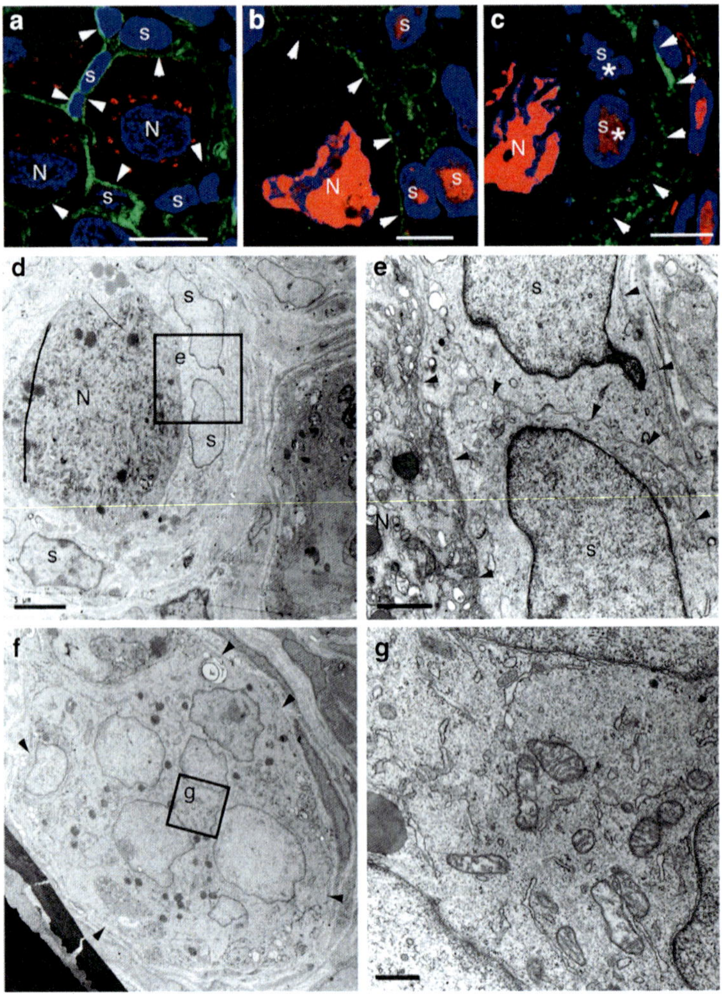

Fig. 2 Evidence of polykaryon formation in acutely infected DRG by immunofluorescence and EM analysis. Cryosections of VZV-infected DRG (**a–c**) were stained with mouse monoclonal anti-NCAM antibody (**a–c**; *green*), rabbit polyclonal anti-synaptophysin antibody (**a**; *red*), rabbit polyclonal anti-IE62 antibody (**b** and **c**; *red*) and Hoechst-stain (**a–c**; *blue*). Texas Red-labeled goat anti-rabbit or FITC-labeled goat anti-mouse antibodies were used for secondary detection (**a–c**). Nuclei of satellite cells (s) and neurons (N) are marked and nuclei of satellite cells located within a putative polykaryon are indicated with an *asterisk* (**a–c**). *White arrowheads* point to cell boundaries detected by NCAM staining (*green*). *Scale bars* are 10 μm (**a–c**). (**d–g**) Acutely infected DRG were analyzed by standard EM. *Black arrowheads* point to the cell membrane surrounding satellite cells (s) or the neuron cell body (N) or a polykaryon (**f** and **g**). *Black boxes* in (**d**) and (**f**) indicate the area that is seen at higher magnification in the images (**e**) and (**g**), respectively. Note, that despite the normal morphology of the mitochondria, the ER and of the nuclear envelope in the polykaryon (**f** and **g**), no cell membranes can be detected between the nuclei of the polykaryon. Panels for composite reprinted from (Reichelt et al. 2008) with permission from American Society for Microbiology

Consistent with observations by confocal IF microscopy in VZV-infected NSCs in which polykaryon formation was observed, NCAM expression was detected on membranes that surround the infected NSC. Notably, these NSC polykaryons were engaged in viral replication and assembly, as indicated by the presence of numerous VZV nucleocapsids, intracellular viral particles within vacuoles, and extracellular virions associated with the limiting plasma membrane of the polykaryon. Most of these polykaryons did not show any obvious damage of the nuclear envelope or intracellular organelles, despite the presence of many nucleocapsids and enveloped virions in the cytoplasm, indicating that merging of the satellite and neuronal cell cytoplasm occurred during active VZV replication in NSC and was therefore not a late effect caused by cellular necrosis in VZV-infected sensory ganglia.

A similar finding of incomplete boundaries between neurons and satellite cells during acute DRG infection was observed in an ultrastructural examination of a VZV-infected trigeminal ganglion acquired at autopsy from a patient with active infection (Esiri and Tomlinson 1972). Taken together, these observations suggest that VZV replication in a single infected neuron can result in polykaryon formation with associated satellite cells within the NSC. Satellite cell infection and polykaryon formation in NSCs provide a mechanism for amplification of VZV upon entry into neuronal cell bodies or reactivation from latency, and may contribute to VZV spread to adjacent NSCs. This may, in part, explain the neuropathic consequences of VZV ganglionitis that occur during primary infection and episodes of herpes zoster (Gilden et al. 2006).

4.3 Longterm Persistence of VZV in Human DRG Xenografts Following Acute Infection

VZV latency can be characterized as the continued presence of non-replicating VZV genomes within neuronal ganglia. Whereas the extent of VZV gene transcription and translation during latency is controversial, that the VZV genome persists in the absence of infectious virus production, is a defining criterion.

At 8 weeks post infection, several areas within VZV-infected DRG xenografts contain histopathic indicators of a prior infectious process, such as fibrosis and nodules of Nageotte, which indicate neuronal cell loss (Zerboni et al. 2005b). However, clusters of neurons and satellite cells remain that do not exhibit any cytopathic effect (Fig. 1e). At this timepoint, infectious VZV can no longer be recovered from homogenized DRG xenografts; however, VZV genomes persist (Fig. 3a). Thus, as in natural infection of human sensory ganglia, VZV infection of DRG xenografts can be characterized as a biphasic process in which productive infection transitions to a phase in which the virus persists, but in a non-infectious form (Zerboni et al. 2005b; Zerboni et al. 2007). Presumably, these latent genomes retain the capacity to replicate and cause disease upon reactivation. However, axotomized DRG xenografts do not possess any neural connection to the mouse peripheral tissues and so we refer to this stage as "persistence" rather than latency (Zerboni et al. 2005b; Zerboni and Arvin 2008).

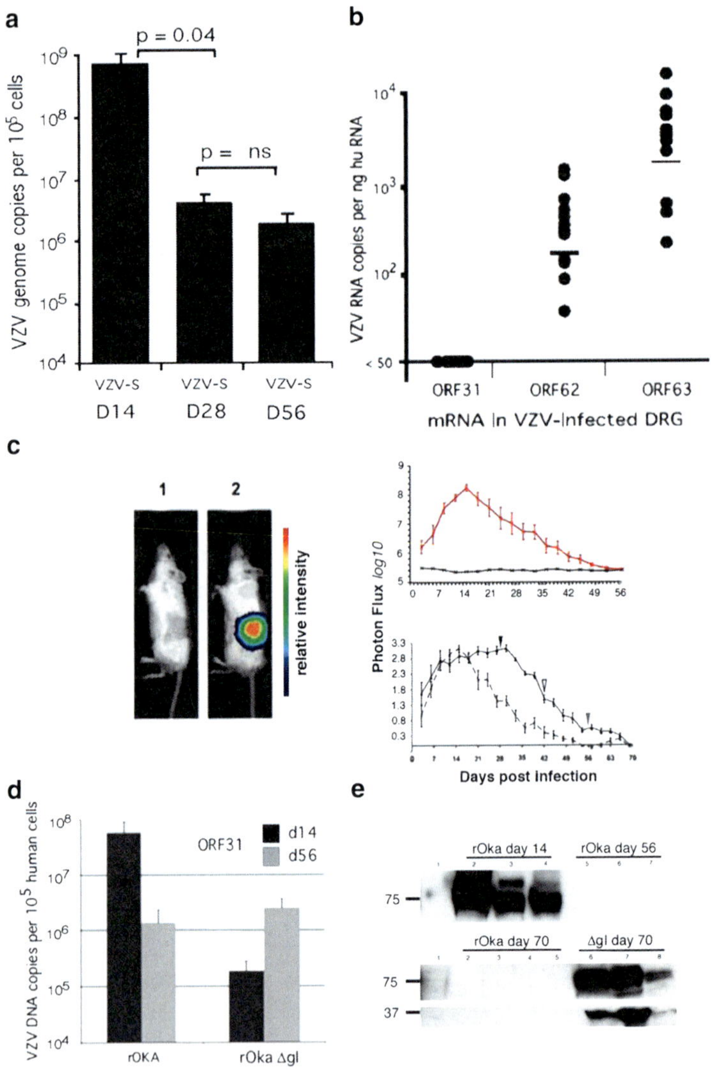

Fig. 3 Assessment of VZV-infected DRG xenografts in SCID mice. (**a**) VZV genome copy numbers in VZV-infected (VZV-S strain) DRG xenografts in SCID mice evaluated by quantitative real-time DNA PCR at 14, 28 and 56 days post-infection using a probe to ORF62. (**b**) VZV mRNA transcripts evaluated by quantitative real-time RNA PCR at 56 days post-infection using probes to ORF31 (VZV gB, a late viral glycoprotein), IE62 and IE63. (**c**) Two mice with SCID DRG xenografts, the mouse on the left (1) was infected with VZV (pOka virus), the mouse on the right (2) was infected with VZV (pOka-63/70-CBR) expressing click beetle luciferase as an IE63/70 fusion protein. (**d**) Luminescence values (Photon flux) generated by in vivo expression of IE63. (*top*) pOka, non-luminescent control virus, (*dark line*) and pOka-63/70-CBR (*grey line*). (*bottom*) pOka-63/70-CBR infection in mice provided with valacyclovir in drinking water (*solid line*) and untreated mice (*broken line*). (**e**) VZV genome copy numbers in rOka and rOka ΔgI-infected DRG xenografts at 14 and 56 days post infection. (**f**) Immunoblot of DRG extracts using human

4.4 *Viral Genomic DNA Burden in DRG Xenografts Persistently Infected with VZV*

Estimates of the VZV DNA burden in latently infected cadaver ganglia have varied according to the methodology employed. Whereas some studies have detected low VZV genome copies (6–31 per 10^5 cells), other studies have reported 258 $\pm$ 38 copies and 557–55, 543 (mean 9,046 copies) per 10^5 cells (Cohrs et al. 2000; LaGuardia et al. 1999; Levin et al. 2003; Mahalingam et al. 1993; Pevenstein et al. 1999). In a recent study, VZV DNA was detected in 1.0–6.9% of neurons with $\sim$7 genomes/cell (Wang et al. 2005).

As with latently infected cadaver ganglia, VZV DNA can be readily detected in persistently infected DRG xenografts by PCR and the viral DNA burden can be quantified by real-time PCR (Zerboni et al. 2005b; Zerboni et al. 2007). We routinely detect 10^5 to 10^7 VZV DNA copies per 10^5 human cells in DRG xenografts at 56 days post infection, a 100-fold reduction when compared with VZV genome copies detected during acute infection (Fig. 3a) (Zerboni et al. 2005b, 2007). That this estimate is higher than estimates using cadaver ganglia is not unexpected, considering that cadaver ganglia are often examined decades following primary VZV infection.

4.5 *Differential Regulation of Viral Gene Expression During VZV Persistence in Human DRG Xenografts*

Various techniques have been employed to identify VZV genes transcribed during latent infection. Transcripts corresponding to ORFs 21, 29, 62, and 63 are frequently detected; ORF63 RNAs are most consistently detected and at greater abundance (>2,000-fold) that transcripts mapping to other ORFs (Cohrs et al. 1994, 1996, 2000; Gilden et al. 1987; Hyman et al. 1983; Kennedy et al. 1999; Meier et al. 1993).

We have used quantitative real time RNA PCR to detect viral mRNAs in persistently infected DRG xenografts (Zerboni et al. 2005b, 2007). VZV RNA corresponding to ORFs 62, 63, and 31 (encoding gB) are present in acutely infected DRG xenografts, as expected for productive infection. However, transcripts corresponding to ORF63, but not ORF31, are present at 56 days post infection (Fig. 3b). Detection of IE62 mRNA at late times post infection is inconsistent. This suggests that, as observed in latently infected cadaver ganglia, persistence of

Fig. 3 (continued) polyclonal anti-VZV antibody, primarily reactive against VZV gE (*top blot*), in rOka-infected DRG at day 14 and day 56 post-infection; (*bottom blots*), immunoblot of DRG extracts using human polyclonal anti-VZV antibody (*upper*) or rabbit polyclonal antibody to IE63 (*lower blot*) in rOka and rOka ΔgI-infected DRG at 14 and 70 days post-infection. Panel (**a**) reprinted with permission from Zerboni et al. (2005b). Panels (**c**) and (**d**) reprinted with permission from Oliver et al. (2008). Panels (**e**) and (**f**) reprinted with permission from Zerboni et al. (2007)

VZV in DRG xenografts is characterized by differential regulation of VZV gene expression such that late viral glycoproteins are not expressed. A similar block on glycoprotein expression was observed during VZV infection of human neural stem cells xenotransplanted in the brain of nonobese diabetic SCID mice (Baiker et al. 2004).

4.6 *Expression of Viral Proteins During VZV Persistence in Human DRG Xenografts*

Evidence of VZV protein expression during latency in human cadaver ganglia has relied exclusively upon detection by immunohistochemical-based methods, and verification by Western blot has not been reported (Cohrs et al. 2003; Grinfeld and Kennedy 2004; Lungu et al. 1998; Mahalingam et al. 1996). In the earliest report, VZV IE63 protein was detected in rare neurons from human ganglia recovered from two of nine individuals with no evidence of VZV reactivation at the time of death (Mahalingam et al. 1996). In more recent reports, IE63 protein – as well as proteins corresponding to ORFs 21, 29, and 62 – have been reported in a high proportion of neurons (5–30%) in all ganglia examined (Grinfeld and Kennedy 2004; Lungu et al. 1998). In that these experiments depend on immunostaining of neurons within cadaver tissues, which may exhibit postmortem physiological or biochemical aberrancies, and which are known to contain an abundance of age-related pigments which can interfere with histochemical tinctorial reactions, they must be carefully controlled (Barden 1981; Double et al. 2008).

Neural pigments are absent from fetal ganglionic tissues, and persistently infected DRG xenografts can be processed and fixed immediately upon recovery of the xenograft. In our examinations of tissue sections from persistently infected DRG xenografts, we have not detected any VZV proteins by immunohistochemical-based methods or immunoblot (Figs. 1f and 3f) (Zerboni et al. 2007). No IE63 protein was detected in DRG xenografts persistently infected with VZV, examined 70 days after infection (Fig. 3f; Zerboni et al. 2007).

To further assess IE63 protein expression in DRG xenografts, we employed live animal imaging and measured IE63 expressed as a luciferase fusion protein during the course of VZV infection (Oliver et al. 2008). Upon inoculation of DRG xenografts with pOka-63/70-luciferase, the progress of VZV infection was measured over a 70-day period by imaging every 3–7 days (Fig. 3c, d). As expected, luminescence increased from day 3 to day 15 during the acute phase of infection, and then declined sharply as the virus transitioned to a longterm persistent phase. At 41 days post inoculation, luminescence had declined to a level below that observed on day 3; by day 56 the level of luminescence was equivalent to that seen in the control virus, pOka, which lacked the luciferase cassette. This experimental system provides further evidence that IE63 protein is not expressed during VZV persistence in human DRG xenografts.

4.7 Evaluation of DRG Xenografts as a System for Assessment of Antiviral Drugs

Investigation of VZV infection in human DRG xenografts by live animal imaging using luciferase tagged viral proteins may also provide an opportunity to assess the efficacy of antiviral drugs in human neuronal ganglia in vivo. We evaluated the potential for this system by delivery of the antiviral drug valacyclovir in the drinking water of SCID mice during the course of DRG xenograft VZV infection (Oliver et al. 2008). Replication of herpes viruses is susceptible to the prodrug valacyclovir, and related nucleoside analogs, through inhibition of the viral thymidine kinase (Cohen et al. 2007). Valacyclovir is a widely used therapeutic for herpes zoster infection in immunocompromised patients.

We provided valacyclovir (1 mg/ml) ad libitum in the drinking water of SCID mice immediately after VZV infection of DRG xenografts and imaged mice over a 70-day interval (Oliver et al. 2008). As previously observed, in both treated and untreated mice, the IE63-luciferase signal increased steadily from days 3 to 15, indicating progressive infection (Fig. 3d). After 15 days, the luciferase signal decreased sharply in untreated mice, as expected for the transition to viral persistence. However, in mice receiving valacyclovir, the luciferase signal plateaued until day 28, when the drug was discontinued from the drinking water. Immediately upon withdrawal of valacyclovir, the IE63-luciferase signal decreased steadily such that signal kinetics were similar to those observed in untreated mice but shifted, with a delay in transition to persistence. Viral mRNA corresponding to ORF31 (gB, a late viral glycoprotein) was detected in DRG of treated but not untreated mice at 70 days post infection, which indicates a delay in cessation of viral replication and the transition to persistence in valacyclovir treated mice. Thus, valacyclovir treatment for 28 days following infection delayed the transition to viral persistence during the treatment interval, but upon withdrawal of the drug, infection kinetics returned to normal. This observation correlates with clinical experience that valacyclovir therapy in severely immunocompromised patients with VZV reactivation often results in relapse upon cessation of drug therapy (Boeckh et al. 2006; Manuel et al. 2008).

4.8 A Role for Innate Immunity During VZV Infection of DRG Xenografts

As discussed, VZV infection of human DRG xenografts is a biphasic process in which a brief replicative phase transitions to a non-replicative phase in which the virus persists longterm. VZV persistence is characterized by cessation of infectious progeny assembly, maintenance of the VZV genome at reduced copy numbers, disappearance of late gene transcripts, and a decline but persistence of low levels of IE63 (and some IE62) gene transcripts, and absence of VZV protein synthesis (Zerboni et al. 2005b, 2007). Remarkably, this transition from productive infection to persistence is achieved in the absence of adaptive immune responses, which SCID mice lack. This suggests that VZV has evolved an intrinsic mechanism to maintain viable neuronal sites in which to establish longterm persistence. Local innate immune responses, such as the production of alpha-interferon, may contribute to this process (Jones and Arvin 2006; Ku et al. 2004). Increased production of IFN-α has been observed in epidermal cells adjacent to VZV-infected cells within skin lesions, which may provide an innate barrier to block spread of VZV within skin xenografts (Ku et al. 2004). Investigation of local innate immune responses in human DRG xenografts in response to VZV infection may provide insight into the delicate balance that must be achieved between VZV and cellular immune responses that favor longterm persistence.

4.9 VZV T Lymphotropism May Facilitate Neurotropism

We investigated the potential for hematogenous spread of VZV in vivo by transfer of infectious virus to DRG xenografts via T lymphocytes. This work was based upon observations by Ku et al. (2004), which refined the prevailing hypothesis that VZV infection resembles the dual viremic model of mousepox (infectious ectromelia) (Grose 1981; Ku et al. 2004). According to this model, primary infection is acquired by inhalation of aerosolized infectious particles, which then replicate within regional lymphoid tissues before hematogenous spread (primary viremia) to reticuloendothelial organs. Following replication in internal organs, a secondary viremic phase delivers infectious particles to deep dermal sites, where they replicate within dermal cells progressively outwards to the epidermis forming cutaneous lesions 10–14 days following initial exposure.

Experimental infection of human T-lymphocytes in thymus/liver xenografts and in vitro demonstrated that VZV exhibits marked tropism for T cells, in particular memory CD4+ tonsil T cells expressing skin homing markers (Ku et al. 2002, 2004; Moffat et al. 1995; Zerboni et al. 2000). Ku et al. (2004) postulated that localized VZV replication in mucosal epithelial cells lining tonsillar crypts may transfer VZV to tonsil T cells and, in turn, VZV-infected memory CD4+ T cells expressing skin homing markers such as cutaneous leukocyte antigen or chemokine receptor-4 may transfer virus directly to skin (without amplification in reticuloendothelial organs) (Ku et al. 2002, 2004). We hypothesized that VZV may spread directly to sensory ganglia through a hematogenous route as well, negating the requirement for skin infection which permits neuronal access via retrograde transport from dermal nerve fibrils. The observation that VZV DNA is present in ganglia recovered from immunocompromised children who died during the incubation period (before the appearance of rash) provides clinical evidence that VZV may be delivered directly to DRG during cell-associated viremia (Cohen et al. 2007).

We investigated this alternative pathway to VZV neurotropism by adoptive transfer of VZV-infected tonsil T cells into SCID DRG mice via the tail vein (Zerboni et al. 2005b). DRG xenografts were homogenized at 14 days after transfer and infectious virus was recovered from one of four xenografts by coculture of the homogenate with a permissive cell line; VZV DNA was detected in two of four DRG by quantitative PCR (8.3×10^5 and 9.5×10^6 VZV genome copies per 10^5 human cells). These experiments indicate that VZV-infected T cells can traffic to and infect human neuronal ganglion cells in vivo; this is substantial evidence that VZV T cell tropism provides a second mechanism by which the virus reaches DRG sites for longterm persistence.

4.10 VZV "Oka" Varicella Vaccines Retain Neurotropism

Clinical and epidemiological observation indicate that the live attenuated "Oka" varicella vaccine has the capacity to cause herpes zoster in vaccine recipients, but appears to do so less often than wildtype virus (Arvin 2001). Vaccine Oka (vOka) was classically attenuated by serial passage of the Oka parental isolate (pOka) in human and guinea pig embryo fibroblast cell lines (Arvin 2001). The molecular basis for vaccine Oka attenuation is unknown; vaccine formulations appear to comprise heterogeneous genomes (Quinlivan et al. 2007).

Using the SCID mouse–human DRG xenograft model, we directly assessed neurotropism and neurovirulence of vOka and pOka viruses (Zerboni et al. 2005a). Cytopathic effect of vOka and pOka viruses was indistinguishable. VZV genome copies were equivalent at acute and persistent phases of infection with the expected decrease in VZV genomes between days 14 and 28 post infection and cessation of infectious virus production. These DRG experiments suggest that attenuation of pOka has not altered tropism for neurons or neurovirulence. In previous studies in skin and T cell xenografts, we observed that vOka replicates less efficiently than pOka in skin but both viruses retain T cell infectivity (Moffat et al. 1998; Zerboni et al. 2005a). Taken together, these data suggest that attenuation of vOka limits skin infectivity which in turn may limit access of virions to nerve termini. Efforts to completely eliminate vaccine virus latency will likely require development of second generation varicella vaccines with a molecular basis for altered neurovirulence.

5 Investigation of VZV Neurovirulence Using Recombinant VZV Mutants

5.1 Generation of VZV Recombinants

VZV is a highly cell-associated virus. Lack of infectious virus release into the extracellular media of cultured cells makes plaque purification of viral mutants derived by homologous recombination impractical (Cohen et al. 2007). To overcome this hindrance, VZV cosmids that contain overlapping fragments of the VZV genome have been constructed which, when cotransfected into a permissive cell line, yield infectious virus (Mallory et al. 1997). Cosmids used to generate infectious virus can be altered by PCR-directed mutagenesis to contain mutations in genes or sequences of interest. Cosmids spanning the VZV genome have been constructed from the attenuated vaccine Oka strain as well as from the original clinical isolate (parental Oka strain) (Mallory et al. 1997; Niizuma et al. 2003). Infectious VZV derived from bacterial artificial chromosome (BAC) technology has also been utilized (Nagaike et al. 2004; Tischer et al. 2007; Zhang et al. 2007).

Experimental infection of SCID mouse–human tissue xenografts in vivo has facilitated the identification of virulence determinants for skin and T-lymphocytes. Our investigations have demonstrated that in vitro replication phenotypes of recombinant viruses are not predictive of behavior within differentiated human cells within their unique tissue microenvironments. We have often observed that VZV genes and promoter elements that are dispensable for replication in cultured cells are required for, or significantly modulate, VZV infection in vivo. Evaluation of cosmid-derived VZV mutants in SCID DRG xenografts will make it possible to determine if particular gene products or motifs within VZV proteins affect neurotropism and neurovirulence. Our lab has generated over 100 recombinant VZV mutants in various genes including viral glycoproteins, viral kinases, regulatory proteins, and promoter elements *(reviewed in Zerboni and Arvin (*2008*))*. So far, we have evaluated only a few of these viruses in SCID DRG xenografts but additional experiments are ongoing. Using these mutants, we hope to identify molecular mechanisms of VZV neuropathogenesis with the aim of defining the contributions of VZV genes and promoter elements to neuropathogenesis. These findings will have direct relevance for the design of neuroattenuated varicella vaccines.

5.2 The Role of Glycoprotein I in VZV Neurotropism

VZV gI (ORF67) and VZV gE (ORF68), encoded in the unique short (U_S) region of the VZV genome, are the major envelope glycoproteins (Cohen et al. 2007). As in other alphaherpesviruses, gE and gI form a noncovalently linked heterodimer which functions in cell–cell spread and viral envelopment. Whereas VZV gI is similar to orthologous gI proteins in human alphaherpesviruses, VZV gE is unique in that it is indispensable for virus replication (Cohen et al. 2007). VZV gE shares regions that are required for gE/gI heterodimer formation but also possesses a large, nonconserved N-terminal ectodomain. Although gI is dispensable for replication, it is critical for normal syncytia formation and spread in cultured cells (Moffat et al. 2002). Deletion of ORF67 has dramatic effects on intracellular localization of gE, causing aberrant punctate expression on cell surface membranes (Moffat et al. 2002). Virion envelopment in the *trans* Golgi network is severely affected as assessed by TEM.

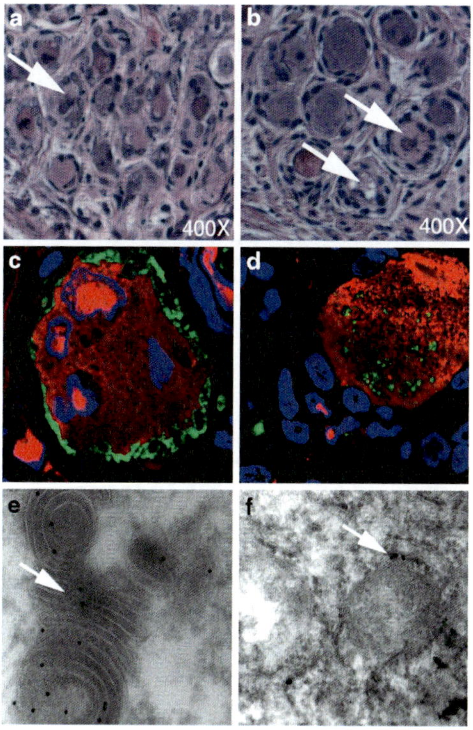

Fig. 4 Comparison of rOka-infected and rOka ΔgI-infected DRG xenografts. (**a**) Histopathic effect in rOka-infected DRG xenografts (*white arrow*) at 14 days post infection, magnification 400×, hematoxylin and eosin staining. (**b**) Rare neurons exhibiting cytopathic effect in rOka ΔgI-infected DRG xenografts (*white arrows*) at 70 days post infection, magnification 400×, hematoxylin and eosin staining. (**c**) rOka-infected DRG neuron at 14 days post-infection and rOka ΔgI-infected DRG neuron (**d**) stained by confocal IF with rabbit polyclonal antibody to IE63 (*green*) and mouse monoclonal antibody to gE (*red*) demonstrates punctate and cytoplasmic retention of gE in the absence of gI. (**e**) Immuno-EM of rOka ΔgI-infected DRG neuron with gold-labeled gE (*black dots*), *white arrow* denotes strange convolutions in Golgi structures. (**f**) Immuno-EM of rOka ΔgI-infected DRG neuron with gold-labeled gE (*black dots*), *white arrow* denotes gE retention in rough ER. Composite from panels reprinted with permission from Zerboni et al. (2007)

Whereas ORF67 is dispensable for replication in vitro, gI functions are critical for pathogenesis of VZV infection in skin and T cell xenografts in vivo (Moffat et al. 2002). As efficient spread of VZV in skin tissues requires polykarocyte formation, the requirement for gI for skin virulence indicates that other fusogenic proteins (gB, gH and gL), cannot compensate for gI functions that facilitate fusion in skin. However, VZV spread in T cells does not involve fusion, and so the requirement for gI suggests that gI functions in normal cytoplasmic envelopment of virions and egress are essential for T cell tropism.

We examined the requirement of gI in VZV neurotropism by inoculation of DRG xenografts with rOkaΔgI, which lacks ORF67 (Zerboni et al. 2007). Whereas infection with the recombinant parental strain (rOka) initiated the expected short replicative phase followed by persistence in DRGs, rOkaΔgI was markedly impaired for replication and showed no transition to persistence up to 70 days after infection (Zerboni et al. 2007). While not strictly required for replication within DRG sensory neurons, replication was significantly altered in that VZV genome copies were 100-fold lower at 14 days after inoculation than with rOka and rose appreciably over an 8-week interval instead of the typical decrease (Fig. 3e). Histopathic effect at day 70 after infection was marginal, with only a few rare cells exhibiting cytopathology (Fig. 4b). Infectious virus production (release of infectious virus from homogenized xenografts) was greater at 70 days post infection (2/4 ganglia) than at 14 days post infection (1/11 ganglia). Membrane localization of gE, which usually surrounds the infected NSC, was greatly reduced with only marginal punctate foci within the neuronal cytoplasm (Fig. 4d). Ultrastructural analysis of rOkaΔgI-infected xenografts revealed unusual Golgi stacks in the cytoplasm of those few neurons that contained virus particles (Fig. 4e). Intracellular trafficking of gE was aberrant, with evidence of retention in the rough ER by immuno-EM (Fig. 4f).

In contrast to the lethality of deletion of gI in skin and T cell xenografts, absence of gI did not alter VZV neurotropism in that rOkaΔgI was infectious for human sensory neurons. However, without gI, the virus lifecycle was significantly impaired, in particular at late stages. Although gE was present in the virion envelope, gE intracellular trafficking was aberrant, Golgi stacks were disorganized, and gE was retained in the rough ER. These experiments illustrate that genetic requirements for VZV replication in skin and T lymphocytes may be more stringent than in sensory neurons and, most importantly, that targeted mutations that limit viral replication in vitro may have unintended consequences on replication within sensory neurons in vivo.

5.3 Effect of Targeted Mutation of gI Promoter Elements on VZV Neurotropism

The activities of herpes viral gene promoters, like cellular gene promoters, can be modulated by cellular transactivating proteins through ubiquitous promoter elements. Host cell regulatory proteins recognize consensus binding sites in VZV gene promoters. Many consensus binding sites of VZV gene promoters have been mapped and examined for their role in enhancing viral transactivation via IE62 or other VZV regulatory proteins. We have examined consensus sites in the promoter of gI by generation of reporter constructs as well as targeted mutation (Ito et al. 2003). The VZV gI promoter contains consensus binding sequences for interaction with specificity factor 1 (Sp1) and upstream stimulatory factor (USF) cellular proteins. Base pair substitution of the Sp1 and USF cellular transactivating binding sites alone or in combination had a dramatic effect on VZV infection of skin xenografts and T cell xenografts (Ito et al. 2003). Disruption of both Sp1 and

USF promoter sites significantly reduced viral titers as well as plaque size in cultured cells. This dual Sp1/USF promoter mutant was unable to replicate in skin xenografts and produced only low viral titers upon infection of T cell xenografts. The overall effect of the Sp1/USF mutations were reduced gI expression, which indicate that less gI is required for VZV infection of T cells than in skin, where gI is required for fusion (Ito et al. 2003).

We examined the contributions of the Sp1 and USF binding site consensus sequences in the gI promoter on VZV neurotropism (Zerboni et al. 2007). Whereas disruption of both Sp1 and USF sequences significantly reduced skin virulence, we observed no effect on acute replication or persistence in DRG xenografts. Recombinant Oka with substitutions in the Sp1 sequence (rOka-gI-Sp1) and both the Sp1 and USF sequences (rOka-gI-Sp1/USF) replicated equally well and was equivalent to the parental rOka strain. No differences were observed in release of infectious virus or in histopathic effect. These observations demonstrate tissue-specific differences in the requirements for cellular transactivators known to enhance viral gene expression. Whereas virulence in skin and T cells depends upon functions of cellular transactivators, in this example VZ neurovirulence is independent of cellular transactivators. Understanding the requirements of tissue-specific VZV gene expression will facilitate a better understanding of the requirements for VZ neurovirulence.

6 Conclusion

The SCID mouse–human xenograft model for VZV pathogenesis has provided an opportunity to examine complex virus–host biological interactions within human tissues that are relevant to VZV pathogenesis, and within a small animal model that can be experimentally manipulated. Experimental infection of human skin, T cells, and sensory ganglia with recombinant VZV viruses that have targeted mutations of specific genes and regulatory elements enables the assessment of factors that influence virulence and tropism. In this chapter, we have focused on initial investigations of VZV infection of human DRG xenografts. We hope that our investigation of VZV neurotropism and neuropathogenesis using this model can improve our understanding of VZV neuropathogenesis with an aim to improving the clinical management of herpes zoster and the development of neuroattentuated varicella vaccines.

References

Arvin AM (2001) Varicella vaccine: genesis, efficacy, and attenuation. Virology 284:153–158
Arvin A (2005) Aging, immunity, and the varicella-zoster virus. N Engl J Med 352:2266–2267
Arvin AM (2006) Investigations of the pathogenesis of Varicella zoster virus infection in the SCIDhu mouse model. Herpes 13:75–80

Baiker A, Fabel K, Cozzio A, Zerboni L, Fabel K, Sommer M, Uchida N, He D, Weissman I, Arvin AM (2004) Varicella-zoster virus infection of human neural cells in vivo. Proc Natl Acad Sci USA 101:10792–10797

Barden H (1981) The biology and chemistry of neuromelanin. In: Sohal R (ed) Age pigments. Elsevier/North Holland Biomedical Press, Amsterdam, pp 155–180

Boeckh M, Kim HW, Flowers ME, Meyers JD, Bowden RA (2006) Long-term acyclovir for prevention of varicella zoster virus disease after allogeneic hematopoietic cell transplantation–a randomized double-blind placebo-controlled study. Blood 107:1800–1805

Cohen J, Straus S, Arvin A (2007) Varicella-zoster virus replication, pathogenesis, and management. In: Knipe D, Howley P (eds) Fields virology, vol 2. Lippincott-Williams & Wilkins, Philadelphia, pp 2774–2818

Cohrs RJ, Srock K, Barbour MB, Owens G, Mahalingam R, Devlin ME, Wellish M, Gilden DH (1994) Varicella-zoster virus (VZV) transcription during latency in human ganglia: construction of a cDNA library from latently infected human trigeminal ganglia and detection of a VZV transcript. J Virol 68:7900–7908

Cohrs RJ, Barbour M, Gilden DH (1996) Varicella-zoster virus (VZV) transcription during latency in human ganglia: detection of transcripts mapping to genes 21, 29, 62, and 63 in a cDNA library enriched for VZV RNA. J Virol 70:2789–2796

Cohrs RJ, Randall J, Smith J, Gilden DH, Dabrowski C, van Der Keyl H, Tal-Singer R (2000) Analysis of individual human trigeminal ganglia for latent herpes simplex virus type 1 and varicella-zoster virus nucleic acids using real-time PCR. J Virol 74:11464–11471

Cohrs RJ, Gilden DH, Kinchington PR, Grinfeld E, Kennedy PG (2003) Varicella-zoster virus gene 66 transcription and translation in latently infected human Ganglia. J Virol 77: 6660–6665

Double KL, Dedov VN, Fedorow H, Kettle E, Halliday GM, Garner B, Brunk UT (2008) The comparative biology of neuromelanin and lipofuscin in the human brain. Cell Mol Life Sci 65:1669–1682

Esiri MM, Tomlinson AH (1972) Herpes Zoster. Demonstration of virus in trigeminal nerve and ganglion by immunofluorescence and electron microscopy. J Neurol Sci 15:35–48

Gilden DH, Vafai A, Shtram Y, Becker Y, Devlin M, Wellish M (1983) Varicella-zoster virus DNA in human sensory ganglia. Nature 306:478–480

Gilden DH, Rozenman Y, Murray R, Devlin M, Vafai A (1987) Detection of varicella-zoster virus nucleic acid in neurons of normal human thoracic ganglia. Ann Neurol 22:377–380

Gilden D, Mahalingam R, Deitch S, Cohrs RJ (2006) Varicella-zoster virus neuropathogenesis and latency. Caister Academic, Norwich

Grinfeld E, Kennedy PG (2004) Translation of varicella-zoster virus genes during human ganglionic latency. Virus Genes 29:317–319

Grose C (1981) Variation on a theme by Fenner: the pathogenesis of chickenpox. Pediatrics 68:735–737

Hope-Simpson R (1965) The nature of herpes zoster: a long-term study and a new hypothesis. Proc R Soc Med 58:9–20

Hyman RW, Ecker JR, Tenser RB (1983) Varicella-zoster virus RNA in human trigeminal ganglia. Lancet 2:814–816

Ito H, Sommer MH, Zerboni L, He H, Boucaud D, Hay J, Ruyechan W, Arvin AM (2003) Promoter sequences of varicella-zoster virus glycoprotein I targeted by cellular transactivating factors Sp1 and USF determine virulence in skin and T cells in SCIDhu mice in vivo. J Virol 77:489–498

Jones JO, Arvin AM (2006) Inhibition of the NF-kappaB pathway by varicella-zoster virus in vitro and in human epidermal cells in vivo. J Virol 80:5113–5124

Kennedy PG, Grinfeld E, Gow JW (1998) Latent varicella-zoster virus is located predominantly in neurons in human trigeminal ganglia. Proc Natl Acad Sci USA 95:4658–4662

Kennedy PG, Grinfeld E, Gow JW (1999) Latent Varicella-zoster virus in human dorsal root ganglia. Virology 258:451–454

Ku CC, Padilla JA, Grose C, Butcher EC, Arvin AM (2002) Tropism of varicella-zoster virus for human tonsillar CD4(+) T lymphocytes that express activation, memory, and skin homing markers. J Virol 76:11425–11433

Ku CC, Zerboni L, Ito H, Graham BS, Wallace M, Arvin AM (2004) Varicella-zoster virus transfer to skin by T cells and modulation of viral replication by epidermal cell interferon-alpha. J Exp Med 200:917–925

LaGuardia JJ, Cohrs RJ, Gilden DH (1999) Prevalence of varicella-zoster virus DNA in dissociated human trigeminal ganglion neurons and nonneuronal cells. J Virol 73:8571–8577

Levin MJ, Cai GY, Manchak MD, Pizer LI (2003) Varicella-zoster virus DNA in cells isolated from human trigeminal ganglia. J Virol 77:6979–6987

Lungu O, Annunziato PW, Gershon A, Staugaitis SM, Josefson D, LaRussa P, Silverstein SJ (1995) Reactivated and latent varicella-zoster virus in human dorsal root ganglia. Proc Natl Acad Sci USA 92:10980–10984

Lungu O, Panagiotidis CA, Annunziato PW, Gershon AA, Silverstein SJ (1998) Aberrant intracellular localization of Varicella-Zoster virus regulatory proteins during latency. Proc Natl Acad Sci USA 95:7080–7085

Mahalingam R, Wellish M, Lederer D, Forghani B, Cohrs R, Gilden D (1993) Quantitation of latent varicella-zoster virus DNA in human trigeminal ganglia by polymerase chain reaction. J Virol 67:2381–2384

Mahalingam R, Wellish M, Cohrs R, Debrus S, Piette J, Rentier B, Gilden DH (1996) Expression of protein encoded by varicella-zoster virus open reading frame 63 in latently infected human ganglionic neurons. Proc Natl Acad Sci USA 93:2122–2124

Mallory S, Sommer M, Arvin AM (1997) Mutational analysis of the role of glycoprotein I in varicella-zoster virus replication and its effects on glycoprotein E conformation and trafficking. J Virol 71:8279–8288

Manuel O, Kumar D, Singer LG, Cobos I, Humar A (2008) Incidence and clinical characteristics of herpes zoster after lung transplantation. J Heart Lung Transplant 27:11–16

McCune JM, Namikawa R, Kaneshima H, Shultz LD, Lieberman M, Weissman IL (1988) The SCID-hu mouse: murine model for the analysis of human hematolymphoid differentiation and function. Science 241:1632–1639

Meier JL, Holman RP, Croen KD, Smialek JE, Straus SE (1993) Varicella-zoster virus transcription in human trigeminal ganglia. Virology 193:193–200

Moffat JF, Stein MD, Kaneshima H, Arvin AM (1995) Tropism of varicella-zoster virus for human CD4+ and CD8+ T lymphocytes and epidermal cells in SCID-hu mice. J Virol 69:5236–5242

Moffat JF, Zerboni L, Kinchington PR, Grose C, Kaneshima H, Arvin AM (1998) Attenuation of the vaccine Oka strain of varicella-zoster virus and role of glycoprotein C in alphaherpesvirus virulence demonstrated in the SCID-hu mouse. J Virol 72:965–974

Moffat J, Ito H, Sommer M, Taylor S, Arvin AM (2002) Glycoprotein I of varicella-zoster virus is required for viral replication in skin and T cells. J Virol 76:8468–8471

Mosier DE, Gulizia RJ, Baird SM, Wilson DB (1988) Transfer of a functional human immune system to mice with severe combined immunodeficiency. Nature 335:256–259

Nagaike K, Mori Y, Gomi Y, Yoshii H, Takahashi M, Wagner M, Koszinowski U, Yamanishi K (2004) Cloning of the varicella-zoster virus genome as an infectious bacterial artificial chromosome in Escherichia coli. Vaccine 22:4069–4074

Niizuma T, Zerboni L, Sommer MH, Ito H, Hinchliffe S, Arvin AM (2003) Construction of varicella-zoster virus recombinants from parent Oka cosmids and demonstration that ORF65 protein is dispensable for infection of human skin and T cells in the SCID-hu mouse model. J Virol 77:6062–6065

Oliver SL, Zerboni L, Sommer M, Rajamani J, Arvin AM (2008) Development of recombinant varicella-zoster viruses expressing luciferase fusion proteins for live in vivo imaging in human skin and dorsal root ganglia xenografts. J Virol Methods 154:182–193

Oxman MN, Levin MJ, Johnson GR, Schmader KE, Straus SE, Gelb LD, Arbeit RD, Simberkoff MS, Gershon AA, Davis LE, Weinberg A, Boardman KD, Williams HM, Zhang JH, Peduzzi

PN, Beisel CE, Morrison VA, Guatelli JC, Brooks PA, Kauffman CA, Pachucki CT, Neuzil KM, Betts RF, Wright PF, Griffin MR, Brunell P, Soto NE, Marques AR, Keay SK, Goodman RP, Cotton DJ, Gnann JW Jr, Loutit J, Holodniy M, Keitel WA, Crawford GE, Yeh SS, Lobo Z, Toney JF, Greenberg RN, Keller PM, Harbecke R, Hayward AR, Irwin MR, Kyriakides TC, Chan CY, Chan IS, Wang WW, Annunziato PW, Silber JL (2005) A vaccine to prevent herpes zoster and postherpetic neuralgia in older adults. N Engl J Med 352:2271–2284

Pevenstein SR, Williams RK, McChesney D, Mont EK, Smialek JE, Straus SE (1999) Quantitation of latent varicella-zoster virus and herpes simplex virus genomes in human trigeminal ganglia. J Virol 73:10514–10518

Quinlivan ML, Gershon AA, Al Bassam MM, Steinberg SP, LaRussa P, Nichols RA, Breuer J (2007) Natural selection for rash-forming genotypes of the varicella-zoster vaccine virus detected within immunized human hosts. Proc Natl Acad Sci USA 104:208–212

Reichelt M, Zerboni L, Arvin AM (2008) Mechanisms of varicella-zoster virus neuropathogenesis in human dorsal root ganglia. J Virol 82:3971–3983

Silverstein S, Straus S (2000) Pathogenesis of latency and reactivation. In: Arvin A, Gershon A (eds) Varicella zoster virus: virology and clinical management. Cambridge University Press, Cambridge, pp 123–141

Takahashi M, Otsuka T, Okuno Y, Asano Y, Yazaki T (1974) Live vaccine used to prevent the spread of varicella in children in hospital. Lancet 2:1288–1290

Tischer BK, Kaufer BB, Sommer M, Wussow F, Arvin AM, Osterrieder N (2007) A self-excisable infectious bacterial artificial chromosome clone of varicella-zoster virus allows analysis of the essential tegument protein encoded by ORF9. J Virol 81:13200–13208

Wang K, Lau TY, Morales M, Mont EK, Straus SE (2005) Laser-capture microdissection: refining estimates of the quantity and distribution of latent herpes simplex virus 1 and varicella-zoster virus DNA in human trigeminal ganglia at the single-cell level. J Virol 79:14079–14087

Zerboni L, Sommer M, Ware CF, Arvin AM (2000) Varicella-zoster virus infection of a human CD4-positive T-cell line. Virology 270:278–285

Zerboni L, Hinchliffe S, Sommer MH, Ito H, Besser J, Stamatis S, Cheng J, Distefano D, Kraiouchkine N, Shaw A, Arvin AM (2005a) Analysis of varicella zoster virus attenuation by evaluation of chimeric parent Oka/vaccine Oka recombinant viruses in skin xenografts in the SCIDhu mouse model. Virology 332:337–346

Zerboni L, Ku CC, Jones CD, Zehnder JL, Arvin AM (2005b) Varicella-zoster virus infection of human dorsal root ganglia in vivo. Proc Natl Acad Sci USA 102:6490–6495

Zerboni L, Reichelt M, Jones CD, Zehnder JL, Ito H, Arvin AM (2007) From the Cover: Aberrant infection and persistence of varicella-zoster virus in human dorsal root ganglia in vivo in the absence of glycoprotein I. Proc Natl Acad Sci USA 104:14086–14091

Zerboni L, Arvin A (2008) The pathogenesis of varicella-zoster virus neurotropism and infection. In: Reiss C (ed) Neurotropic viral infections. Cambridge University Press, Cambridge, pp 225–251

Zhang Z, Rowe J, Wang W, Sommer M, Arvin A, Moffat J, Zhu H (2007) Genetic analysis of varicella-zoster virus ORF0 to ORF4 by use of a novel luciferase bacterial artificial chromosome system. J Virol 81:9024–9033

Rodent Models of Varicella-Zoster Virus Neurotropism

Jeffrey I. Cohen

Contents

Abstract Inoculation of rodents with varicella-zoster virus (VZV) results in a latent infection in dorsal root ganglia with expression of at least five of the six VZV transcripts and one of the viral proteins that are reported to be expressed during latency in human ganglia. Rats develop allodynia and hyperalgesia in the limb distal to the site of injection and the resulting exaggerated withdrawal response to stimuli is reduced by treatment with gabapentin and amitryptyline, but not by antiviral therapy. Inoculation of rats with VZV mutants show that most viral genes are dispensable for latency, but that some genes (e.g., ORF4, 29, and ORF63) that are expressed during latency are important for the establishment of latency in rodents, but not for infection of rodent ganglia. The rodent model for

---

J.I. Cohen
Laboratory of Clinical Infectious Diseases, National Institutes of Health, Bldg. 10, Room 11N234, 10 Center Drive, Bethesda, MD 20892, USA
e-mail: jcohen@niaid.nih.gov

A.M. Arvin et al. (eds.), *Varicella-zoster Virus*,
Current Topics in Microbiology and Immunology 342, DOI 10.1007/82_2010_11

VZV latency allows one to study ganglia removed immediately after death, avoiding the possibility of reactivation, and helps to identify VZV genes required for latency.

Abbreviations

VZV Varicella-zoster virus
HSV Herpes simplex virus

1 Varicella-Zoster Virus Latency in Humans

Varicella-zoster virus (VZV) establishes latency in human cranial nerve and dorsal root ganglia (Mahalingam et al. 1990). Viral DNA is present in the neurons and a limited set of VZV genes are expressed: ORF4, ORF21, ORF29, ORF62, ORF63, and ORF66 (Cohrs et al. 2003; Grinfeld and Kennedy 2004; Kennedy et al. 2000; Lungu et al. 1998; Mahalingam et al. 1996). Both viral transcripts and proteins have been detected in the cells. Interestingly during latency, ORF4, 21, 29, 62, and 63 proteins have been detected in the cytoplasm of neurons (Grinfeld and Kennedy 2004; Lungu et al. 1998; Mahalingam et al. 1996), while during reactivation the proteins were detected in both the cytoplasm and the nucleus (Lungu et al. 1998). In contrast, several viral proteins, glycoprotein C (gC), gI, and ORF10 protein were not detected in latently infected human neurons (Lungu et al. 1998).

2 VZV Latency in Rats

Initial studies in adult rats showed that injection of VZV-infected cells subcutaneously along the side of the spine resulted in latent infection in the dorsal root ganglia innervating the site of injection (Sadzot-Delvaux et al. 1990). VZV nucleic acid was detected by in situ hybridization 1–9 months after inoculation in neurons in the dorsal root ganglia. VZV proteins were detected in neurons using human sera, and viral proteins were detected only in the cytoplasm of the neurons. In one set of experiments, serial repeated passages of latently infected ganglia in human fibroblasts eventually resulted in cytopathic effects consistent with VZV in fibroblasts. Similar reactivation of VZV with infectious virus ex vivo has not been reported subsequently. In a follow-up study by the same authors, VZV DNA was detected only in the dorsal root ganglia corresponding to the side of the spine injected; however, sectioning of the nerve roots before inoculation still allowed infection of the ganglia, but at a lower level of infection (Sadzot-Delvaux et al. 1995). In situ hybridization showed VZV in both neurons and nonneuronal cells.

Subsequent experiments from other researchers showed that VZV nucleic acid could be detected by in situ hybridization in dorsal root ganglia of rats inoculated in the foot pad (Annunziato et al. 1998; Kennedy et al. 2001) or subcutaneously along the skin (Annunziato et al. 1998). In one report, VZV nucleic acid was detected in both neurons and satellite cells (Annunziato et al. 1998). VZV in situ hybridization was positive in bilateral dorsal root ganglia, even though only one side of the spine or foot pad was infected, suggesting that viremia had resulted with seeding of the dorsal root ganglia. No difference was noted in the frequency of latent infection in animals vaccinated with the Oka vaccine strain and VZV strain Ellen or in animals inoculated 1 or 3 months before ganglia were obtained. In another report, VZV RNA was detected in neurons and noneuronal cells at a ratio of 3:1, and only in the nucleus of neurons (Kennedy et al. 2001). VZV RNA was detected in dorsal root ganglia at 1 week, 1 month, and 18 months after infection; similar numbers of ganglia were positive for VZV at 1 and 18 months. VZV was not detected in the liver, lung, or kidney after establishment of latency in dorsal root ganglia, but was detected in the spleen in one of three rats (Grinfeld et al. 2004).

2.1 VZV Latency in Newborn Rats and Other Rodents

Intraperitoneal inoculation of newborn rats with VZV resulted in a latent infection of the trigeminal ganglia with viral DNA and RNA detected 5–6 weeks after infection (Brunell et al. 1999). Inoculation of cotton rats with VZV-infected cells along the side of the spine resulted in latent infection of dorsal root ganglia (Sato et al. 2002b). Latency was not established in animals infected with heat-inactivated VZV-infected cells.

Corneal inoculation of mice resulted in latent infection in the trigeminal ganglia as well as viral DNA in the brainstem, kidneys, spleen, and liver by nested PCR at 1 month after infection (Wroblewska et al. 1993). In situ hybridization showed VZV RNA in trigeminal ganglia and spleen of mice 1 month after infection. Viral RNA was present at a higher frequency in nonneuronal cells than in neurons.

2.2 VZV Allodynia Model in Rats

Inoculation of VZV-infected cells into the footpad of rats resulted in allodynia and hyperalgesia in the ipsilateral limb (Fleetwood-Walker et al. 1999). Animals were tested for paw withdrawal after exposure to nylon microfilaments or to heat. The effects persisted for up to 33 days after injection. In contrast, no neurologic changes were noted in the contralateral limb. VZV IE63 protein, but not inflammatory cells, was detected in the ipsilateral dorsal root ganglia. The authors concluded that the neurological changes in the rats might serve as a model for post-herpetic neuralgia.

A follow-up study by the same research group showed that allodynia began within 3 days of inoculation of VZV-infected cells, persisted at least 30 days, and resolved within 100 days (Dalziel et al. 2004). Treatment of animals with valacyclovir for the first 10 days after infection had no effect on allodynia, implying that continued virus replication is not required for the neurologic effects. Inoculation of the footpad of rats with herpes simplex virus (HSV) resulted in hyperalgesia on days 1–4 after injection and the neurologic effects resolved by day 6. Some animals developed hindlimb paralysis after HSV infection.

Examination of dorsal root ganglia from rats infected with VZV showed increased expression of neuropeptide Y, galanin, and ATF-3, a marker of nerve injury (Garry et al. 2005). Treatment of rats with gabapentin, sodium channel blockers (mexiletine and lamotrigine), but not a nonsteroidal anti-inflammatory drug (diclofenac) reduced the paw withdrawal response to mechanical and noxious thermal stimuli. Testing of different strains of VZV showed a dose response to mechanical, but not thermal stimuli (Hasnie et al. 2007). Treatment of rats with morphine, amitryptyline, ibuprofen, a cannabinoid (Hasnie et al. 2007), a palmitoylethanolamide analogue (Wallace et al. 2007), and histamine H3 receptor antagonists (Medhurst et al. 2008) resulted in a reduced paw withdrawal in response to mechanical stimuli.

3 VZV Transcripts in Latently Infected Rats and Cotton Rats

VZV transcripts corresponding to ORF4, 29, 62, and 63 were detected by Sadzot-Delvaux et al. (1995) in adult rats latently infected with VZV (Table 1). No transcripts were detected for the thymidine kinase or glycoprotein E genes in latently infected rats. Northern blotting showed that transcripts for ORF62 and ORF63 were of the same length in latently infected animals as in infected cells in culture. VZV ORF21 and ORF63 were detected by in situ hybridization in dorsal root ganglia of rats 1 month after infection, and ORF62 and ORF63 were detected 18 months after infection (Kennedy et al. 2001). One dorsal root ganglia was positive for ORF29, and one each for ORF18 and for ORF40 (the latter two genes have not been associated with latency in humans) at 1–18 months after infection. In another study by the same authors, 39% of latently infected rat dorsal root ganglia were positive for ORF62 RNA, 35% for ORF63, 25% for ORF4, 25% for ORF 29, 24% for ORF18, 22% for ORF28, 13% for ORF21, and 0% for ORF40 (Grinfeld et al. 2004). ORF21 but not ORF40 RNA was detected by reverse transcriptase PCR in latently infected newborn rats (Brunell et al. 1999). VZV ORF63 RNA was detected in dorsal root ganglia in 25 of 36 latently infected cotton rats by reverse transcriptase-PCR 1 month after infection; in contrast, ORF40 RNA was detected in only one of 36 animals (Sato et al. 2002b).

One concern raised about detection of multiple VZV transcripts in humans is whether these represent transcripts during latency or whether the virus may reactivate between the time of death and when the ganglia are studied. To address this

Table 1 VZV genes and proteins expressed during latency in rodents

Transcript/protein	Animal	References
ORF4 RNA	Rat	Sadzot-Delvaux et al. (1995)
	Rat	Grinfeld et al. (2004)
ORF21 RNA	Neonatal rat	Brunell et al. (1999)
	Rat	Kennedy et al. (2001)
	Rat	Grinfeld et al. (2004)
ORF29 RNA	Rat	Sadzot-Delvaux et al. (1995)
	Rat	Kennedy et al. (2001)
	Rat	Grinfeld et al. (2004)
ORF62 RNA	Rat	Sadzot-Delvaux et al. (1995)
	Rat	Kennedy et al. (2001)
	Rat	Grinfeld et al. (2004)
ORF63 RNA	Rat	Sadzot-Delvaux et al. (1995)
	Rat	Kennedy et al. (2001)
	Rat	Grinfeld et al. (2004)
	Cotton rat	Sato et al. (2002b)
ORF18 RNA[a]	Rat	Grinfeld et al. (2004)
ORF28 RNA[a]	Rat	Grinfeld et al. (2004)
ORF40 RNA[a]	Rat	Grinfeld et al. (2007)
ORF63 protein	Rat	Debrus et al. (1995)
	Rat	Sadzot-Delvaux et al. (1995)
	Rat	Kennedy et al. (2001)

[a]These transcriptase have not generally been detected in latently infected human ganglia; ORF40 transcripts were absent or detected in less than 3% of rodents in other studies (Brunell et al. 1999; Sato et al. 2002a, b)

issue, Grinfeld et al. (2007) assayed VZV RNA transcripts by in situ hybridization and by nested reverse-transcriptase PCR in latently infected rat ganglia 18 months after infection at various times after death of the animal. VZV ORF63 transcripts were detected in 48, 67, and 42% of ganglia at 0, 24, and 48 h after death, respectively, while ORF40 transcripts were detected in 9–10% of animals at 0, 24, and 48 h after death. Thus, there was no increase in the percentage of ganglia expressing VZV transcripts and these results argue against VZV reactivation between the time of death and when the ganglia are examined.

4 VZV Proteins in Latently Infected Rats

VZV proteins have been detected in dorsal root ganglia from rats infected with VZV. IE63 was detected in 50–80% of neurons in dorsal root ganglia from latently infected rats by using rabbit antibody to IE63 (Debrus et al. 1995; Sadzot-Delvaux et al. 1995) (Table 1). IE63 was detected in both the nucleus and the cytoplasm of the neurons, and some noneuronal cells and axons also contained IE63. VZV IE62, ORF29 protein, and glycoprotein E could not be detected in latently infected dorsal root ganglia. IE63 was also detected in latently infected dorsal root ganglia of rats using a monoclonal antibody to IE63 at 1 and 18 months after infection

(Kennedy et al. 2001). IE63 was detected in both the nucleus and the cytoplasm of the neurons with the monoclonal antibody.

5 VZV Genes Important for Latency in Rats and Cotton Rats

The observation that infection of rats or cotton rats with VZV, either in the foot pad or along the side of the spine, results in latent infection of dorsal root ganglia has allowed many VZV mutants to be studied for their ability to establish latency (Table 2).

5.1 Role of VZV Genes Not Conserved with HSV in Establishment of Latency in Rodents

VZV apparently lacks a latency associated transcript (LAT) similar to that in HSV and many other alphaherpesviruses including pseudorabies virus, bovine herpesvirus, and equine herpesvirus. Deletion of the LAT has been associated with reduced establishment of latency in mice in some (Thompson and Sawtell 2001), but not all (Leib et al. 1989) experiments. Since VZV encodes six genes (ORF 1, 2, 13, 32, 57, and S/L) that lack homologs in HSV, we tested whether some of these genes might be important for

Table 2 VZV genes tested for latency in cotton rats or rats

VZV gene	Mutation	Phenotype in latency	References
ORF1	Stop	Dispensable	Sato et al. (2003b)
ORF2	Del	Dispensable	Sato et al. (2002b)
ORF4	Del	Impaired	Cohen et al. (2005b)
ORF10	Del	Dispensable	Sato et al. (2003b)
ORF13	Stop	Dispensable	Sato et al. (2003b)
ORF14	Del	Dispensable	Grinfeld et al. (2004)
ORF17	Del	Dispensable	Sato et al. (2002a)
ORF21	Del	Dispensable[a]	Xia et al. (2003)
ORF29	Del	Impaired	Cohen et al. (2007)
	Ectopic promoter	Impaired	Cohen et al. (2007)
ORF32	Del	Dispensable	Sato et al. (2003b)
ORF47	Stop	Dispensable	Sato et al. (2003b)
ORF57	Del	Dispensable	Sato et al. (2003b)
ORF61	Del	Dispensable	Sato et al. (2003a)
ORF63	Del	Impaired	Cohen et al. (2004)
	Carboxy terminal truncations	Dispensable or impaired	Cohen et al. (2005a)
	Phosphorylation sites	Impaired	Cohen et al. (2005a)
ORF66	Stop	Dispensable	Sato et al. (2003b)
ORF67	Stop	Dispensable	Grinfeld et al. (2004)

Stop stop codon; *del* deletion; *ectopic promoter* ORF29 expressed under a cytomegalovirus promoter
[a]Virus was not passaged in non-complementing cells before testing for latency

establishment of latency. VZV unable to express ORF1, 2, 13, 23, or 57 is not impaired for growth in cell culture (Cohen and Seidel 1993, 1995; Cox et al. 1998; Reddy et al. 1998; Sato et al. 2002b); each of the VZV mutants established latency at similar levels as parental virus (Sato et al. 2002b, 2003b). Both the number of animals with latent infection and the geometric mean of VZV genome copies for VZV-positive ganglia were similar in animals infected with the mutant and wild-type viruses.

5.2 *Role of VZV Genes Expressed During Latency in Humans in Establishment of Latency in Rodents*

VZV encodes six genes (ORF4, 21, 29, 62, 63, and 66) that are expressed during latency in humans (Cohrs et al. 2003; Grinfeld and Kennedy 2004; Kennedy et al. 2000; Lungu et al. 1998; Mahalingam et al. 1996). ORF4 encodes a transcriptional activator that is essential for replication of VZV in vitro; VZV deleted for ORF4 could be grown only in cells expressing ORF4 protein (Cohen et al. 2005b). VZV deleted for ORF4 was passaged once in noncomplementing cells so that the no ORF4 protein would be present at the time of infection. Inoculation of cotton rats with cells containing the ORF4 deleted virus resulted in latent infection of the dorsal root ganglia. The frequency of animals with latent infection was lower in those infected with the ORF4 deletion mutant than with wild-type virus. Examination of dorsal root ganglia of cotton rats 3 days after inoculation with the ORF4 deletion mutant showed that most of the animals had viral DNA in the ganglia. Therefore, VZV deleted for ORF4 was impaired for establishing latency, but not for entering ganglia.

VZV ORF21 encodes a nucleocapsid protein and ORF21 transcripts and protein are expressed in latently infected human and rodent ganglia (Grinfeld and Kennedy 2004; Kennedy et al. 2000; Lungu et al. 1998). VZV deleted for ORF21 cannot replicate in cell culture unless grown in cells expressing ORF21 (Xia et al. 2003). Inoculation of cotton rats with cells expressing ORF21 that were infected with the ORF21 deletion mutant resulted in latent infection in a similar number of animals, with a similar copy number of VZV genomes in VZV-positive ganglia, as infection of animals with wild-type virus. The ORF21 deletion mutant had been grown in ORF21 expressing cells immediately before infection of the animals and ORF21 protein may have been present when the ganglia were infected.

VZV ORF29 encodes a single-stranded DNA binding protein that is expressed during latency in human and rodent ganglia (Grinfeld and Kennedy 2004; Kennedy et al. 2000; Lungu et al. 1998). ORF29 is essential for replication of the virus in cell culture and could only be grown in cells expressing ORF29 (Cohen et al. 2007). Inoculation of cotton rats with cells infected with ORF29 deleted virus, which had been passaged once in the absence of cells expressing the protein, showed that VZV deleted for ORF29 is severely impaired for the establishment of latency in cotton rats (Cohen et al. 2007). Examination of dorsal root ganglia from animals 3 days after infection with VZV deleted for ORF29 showed that the deletion mutant was present in similar numbers of ganglia as wild-type virus; thus ORF29 is not required for entry into ganglia. Interestingly, expression of ORF29 by the human cytomegalovirus promoter, instead of its own promoter, also resulted in impaired latency in cotton rats.

ORF66 transcripts and protein have been detected in latently infected human ganglia (Cohrs et al. 2003). ORF66 encodes a viral protein kinase that phosphorylates VZV IE62, resulting in its incorporation into the virion. ORF66 is dispensable for infection in vitro (Heineman et al. 1996). ORF66 is important for growth of VZV in human T cells (Moffat et al. 1999; Soong et al. 2000), but is dispensable for the establishment of latency in cotton rats (Sato et al. 2003b).

ORF63 is located in both internal repeats of the VZV genome, inhibits expression of ORF62 (Hoover et al. 2006), and inhibits the activity of interferon alpha (Ambagala and Cohen 2007). ORF63 transcripts and protein are expressed in latently infected human ganglia (Grinfeld and Kennedy 2004; Kennedy et al. 2000; Lungu et al. 1998, Mahalingam et al. 1996). VZV deleted for both copies of ORF63 was severely impaired for replication in vitro. Inoculation of the ORF63 deletion mutant into cotton rats resulted in fewer animals with latent infection and fewer VZV genome copies in latently infected ganglia compared to wild-type virus (Cohen et al. 2004). VZV deleted for ORF63 could enter ganglia since high levels of VZV DNA were present in ganglia 3 days after infection; however, by days 6 and 10 after infection the numbers of viral genomes had rapidly declined.

A series of carboxyl terminal truncation mutations of ORF63 showed that deletion of the last 70 amino acids did not impair growth in cell culture or establishment of latency in cotton rats; in contrast, deletion of the last 108 amino acids impaired growth in vitro and latency in rodents (Cohen et al. 2005a). Substitution of five serine or threonine putative phosphorylation sites in ORF63 with alanine also impaired the growth of the virus in vitro and latency in cotton rats.

5.3 *Role of Other VZV Genes in Establishment of Latency in Rodents*

VZV ORF61 is a transcriptional activator that upregulates expression from immediate-early and putative early promoters. A deletion mutant of ORF61 is impaired for replication in vitro (Cohen and Nguyen 1998), but ORF61 is dispensable for establishment of latency (Sato et al. 2003a). VZV ORF47 encodes a viral protein kinase that phosphorylates several viral proteins and is required for replication in human T cells and skin (Moffat et al. 1999; Soong et al. 2000), but is not required for replication in vitro (Heineman and Cohen 1995) or for latency in cotton rats (Sato et al. 2003b). VZV ORF10 is a transcriptional activator that is not required for VZV replication in vitro (Cohen and Seidel 1994). While HSV-1 VP16, the homolog of ORF10, is unable to establish latency in mice (Tal-Singer et al. 1999), VZV ORF10 is dispensable for latency in cotton rats (Sato et al. 2003b). VZV ORF17 induces RNA degradation in cells (Sato et al. 2002a). While VZV deleted for ORF17 is impaired for growth at 37°C, the mutant virus was not impaired for latency in cotton rats, which have a core temperature of 39°C. Interestingly, HSV-1 deleted for vhs, the homolog of ORF17, is impaired for establishment of latency (Strelow and Leib 1995).

VZV ORF14 encodes gC, which is dispensable for replication in cultured cells, but is critical for replication in human skin (Moffat et al. 1998). VZV deleted for gC was not impaired for establishment of latency in rats (Grinfeld et al. 2004). VZV ORF67 encodes glycoprotein I, which is required for virus replication in human skin and T cells (Moffat et al. 2002); however, glycoprotein I is dispensable for latency in rats (Grinfeld et al. 2004).

5.4 Summary of Latency Studies in Rodents

Analysis of the growth of VZV mutants in vitro and their ability to establish latent infection in rodents indicates that some mutants that are severely impaired for growth in cell culture are not impaired for establishment of latency (ORF17 and ORF61 deletion mutants), while other mutants impaired for growth in culture are impaired for latency (ORF63 deletion mutant). All the mutants are able to infect ganglia and establish latency, including some that are unable to replicate in cell culture in the absence of complementing cells (ORF4 and ORF29 deletion mutants). Some (ORF4 and 29 deletion), but not all (ORF66 stop codon), mutants in genes expressed during latency are impaired for establishment of latency.

6 Advantages and Limitations of the Rodent Model of VZV Latency

These are several advantages of the rodent model for studying VZV latency (Table 3). One of the concerns of latency studies with human ganglia is that nervous system tissues are often removed hours to days after death, and reactivation of VZV may have occurred, which might be confused with latent infection. The rodent

Table 3 Advantages and limitations of rodent models for VZV latency compared with studies in humans

Advantages
1. Ganglia can removed immediately after death to avoid the possibility of reactivation post mortem
2. Virus mutants can be studied to determine the role of individual genes in latency
3. Viruses labeled with markers might be used to identify latently infected cells
Limitations
1. VZV is not a natural pathogen of rodents and viral proteins may not function the same in rodents and in humans
2. Viremia appears to be limited in rodents infected with VZV
3. VZV reactivation has not occurred in rodents in vivo

model allows ganglia to be removed immediately after death and avoid concerns about reactivation. Studies in animals allow investigation of various mutant viruses that can determine the role of individual viral genes in latency. In addition, viruses labeled with beta-galactosidase (Cohen et al. 1998), green fluorescence protein (Zerboni et al. 2000; Li et al. 2006), and luciferase (Oliver et al. 2008; Zhang et al. 2007) might be used to identify latently infected neurons.

The rodent model of VZV latency has several limitations compared with the human system (Table 3). First, VZV is not a natural pathogen in rodents and therefore the virus has not evolved with the immune system of the animal. Viral proteins that change in response to human host defenses or that are pirated from host genes may differ in their activity in rodents when compared to humans. Second, animals do not develop chickenpox after infection, and therefore viremia, with spread of virus to distant ganglia is probably more limited in rodents than in humans. However, dissemination of virus to ganglia not directly innervating the site of inoculation has been demonstrated (Annunziato et al. 1998; Brunell et al. 1999). Third, reactivation of VZV has not been detected in vivo in rodents or in explanted ganglia from these animals, with one possible exception (Sadzot-Delvaux et al. 1990). It is important to note that while VZV causes zoster in up to 50% of persons who live to age 85, reactivation has never been demonstrated in explanted human ganglia. Since reactivation in humans most frequently occurs after the sixth decade of life, and rodents are usually tested a few months to a year after infection, it is possible that insufficient time has occurred to demonstrate reactivation. However, studies with immunosuppressed animals have also failed to show reactivation (Cohen et al. unpublished data).

Acknowledgement I thank the intramural research program of the National Institute of Allergy and Infectious Diseases for support.

References

Ambagala AP, Cohen JI (2007) Varicella-Zoster virus IE63, a major viral latency protein, is required to inhibit the alpha interferon-induced antiviral response. J Virol 81:7844–7851

Annunziato P, LaRussa P, Lee P, Steinberg S, Lungu O, Gershon AA, Silverstein S (1998) Evidence of latent varicella-zoster virus in rat dorsal root ganglia. J Infect Dis 178(Suppl 1): S48–S51

Brunell PA, Ren LC, Cohen JI, Straus SE (1999) Viral gene expression in rat trigeminal ganglia following neonatal infection with varicella-zoster virus. J Med Virol 58:286–290

Cohen JI, Nguyen H (1998) Varicella-zoster virus ORF61 deletion mutants replicate in cell culture, but a mutant with stop codons in ORF61 reverts to wild-type virus. Virology 246:306–316

Cohen JI, Seidel KE (1993) Generation of varicella-zoster virus (VZV) and viral mutants from cosmid DNAs: VZV thymidylate synthetase is not essential for replication in vitro. Proc Natl Acad Sci USA 90:7376–7380

Cohen JI, Seidel KE (1994) Varicella-zoster virus (VZV) open reading frame 10 protein, the homolog of the essential herpes simplex virus protein VP16, is dispensable for VZV replication in vitro. J Virol 68:7850–7858

Cohen JI, Seidel KE (1995) Varicella-zoster virus open reading frame 1 encodes a membrane protein that is dispensable for growth of VZV in vitro. Virology 206:835–842
Cohen JI, Wang Y, Nussenblatt R, Straus SE, Hooks JJ (1998) Chronic uveitis in guinea pigs infected with varicella-zoster virus expressing Escherichia coli beta-galactosidase. J Infect Dis 177:293–300
Cohen JI, Cox E, Pesnicak L, Srinivas S, Krogmann T (2004) The varicella-zoster virus ORF63 latency-associated protein is critical for establishment of latency. J Virol 78:11833–11834
Cohen JI, Krogmann T, Bontems S, Sadzot C, Pesnicak L (2005a) Regions of the varicella-zoster virus ORF63 latency-associated protein important for efficient replication in vitro are also critical for efficient establishment of latency. J Virol 79:5069–5077
Cohen JI, Krogmann T, Ross JP, Pesnicak LP, Prikhod'ko EA (2005b) The varicella-zoster virus ORF4 latency associated protein is important for establishment of latency. J Virol 79:6969–6975
Cohen JI, Krogmann T, Pesnicak L, Ali MA (2007) Absence or overexpression of the varicella-zoster virus (VZV) ORF29 latency-associated protein impairs late gene expression and reduces latency in a rodent model. J Virol 81:1586–1591
Cohrs RJ, Gilden DH, Kinchington PR, Grinfeld E, Kennedy PG (2003) Varicella-zoster virus gene 66 transcription and translation in latently infected human Ganglia. J Virol 77:6660–6665
Cox E, Reddy S, Iofin I, Cohen J (1998) Varicella-zoster virus ORF57, unlike its pseudorabies virus UL3.5 homolog, is dispensable for replication in cell culture. Virology 250:205–209
Dalziel RG, Bingham S, Sutton D, Grant D, Champion JM, Dennis SA, Quinn JP, Bountra C, Mark MA (2004) Allodynia in rats infected with varicella zoster virus–a small animal model for post-herpetic neuralgia. Brain Res Brain Res Rev 46:234–242
Debrus S, Sadzot-Delvaux C, Nikkels AF, Piette J, Rentier B (1995) Varicella-zoster virus gene 63 encodes an immediate-early protein that is abundantly expressed during latency. J Virol 69:3240–3245
Fleetwood-Walker SM, Quinn JP, Wallace C, Blackburn-Munro G, Kelly BG, Fiskerstrand CE, Nash AA, Dalziel RG (1999) Behavioural changes in the rat following infection with varicella-zoster virus. J Gen Virol 80:2433–2436
Garry EM, Delaney A, Anderson HA, Sirinathsinghji EC, Clapp RH, Martin WJ, Kinchington PR, Krah DL, Abbadie C, Fleetwood-Walker SM (2005) Varicella zoster virus induces neuropathic changes in rat dorsal root ganglia and behavioral reflex sensitisation that is attenuated by gabapentin or sodium channel blocking drugs. Pain 118:97–111
Grinfeld E, Kennedy PG (2004) Translation of varicella-zoster virus genes during human ganglionic latency. Virus Genes 29:317–319
Grinfeld E, Sadzot-Delvaux C, Kennedy PG (2004) Varicella-zoster virus proteins encoded by open reading frames 14 and 67 are both dispensable for the establishment of latency in a rat model. Virology 323:85–90
Grinfeld E, Goodwin R, Kennedy PG (2007) Varicella-Zoster virus gene expression at variable periods following death in a rat model of ganglionic infection. Virus Genes 35:29–32
Hasnie FS, Breuer J, Parker S, Wallace V, Blackbeard J, Lever I, Kinchington PR, Dickenson AH, Pheby T, Rice AS (2007) Further characterization of a rat model of varicella zoster virus-associated pain: Relationship between mechanical hypersensitivity and anxiety-related behavior, and the influence of analgesic drugs. Neuroscience 144:1495–1508
Heineman TC, Cohen JI (1995) The varicella-zoster virus (VZV) open reading frame 47 (ORF47) protein kinase is dispensable for viral replication and is not required for phosphorylation of ORF63 protein, the VZV homolog of herpes simplex virus ICP22. J Virol 69:7367–7370
Heineman TC, Seidel K, Cohen JI (1996) The varicella-zoster virus ORF66 protein induces kinase activity and is dispensable for viral replication. J Virol 70:7312–7317
Hoover SE, Cohrs RJ, Rangel ZG, Gilden DH, Munson P, Cohen JI (2006) Downregulation of varicella-zoster virus (VZV) immediate-early ORF62 transcription by VZV ORF63 correlates with virus replication in vitro and with latency. J Virol 80:3459–3468

Kennedy PG, Grinfeld E, Bell JE (2000) Varicella-zoster virus gene expression in latently infected and explanted human ganglia. J Virol 74:11893–11898

Kennedy PG, Grinfeld E, Bontems S, Sadzot-Delvaux C (2001) Varicella-Zoster virus gene expression in latently infected rat dorsal root ganglia. Virology 289:218–223

Leib DA, Bogard CL, Kosz-Vnenchak M, Hicks KA, Coen DM, Knipe DM, Schaffer PA (1989) A deletion mutant of the latency-associated transcript of herpes simplex virus type 1 reactivates from the latent state with reduced frequency. J Virol 63:2893–2900

Li Q, Ali MA, Cohen JI (2006) Insulin degrading enzyme is a cellular receptor for varicella-zoster virus infection and for cell-to-cell spread of virus. Cell 127:305–316

Lungu O, Panagiotidis CA, Annunziato PW, Gershon AA, Silverstein SJ (1998) Aberrant intracellular localization of Varicella-Zoster virus regulatory proteins during latency. Proc Natl Acad Sci USA 95:7080–7085

Mahalingam R, Wellish M, Cohrs R, Debrus S, Piette J, Rentier B, Gilden DH (1996) Expression of protein encoded by varicella-zoster virus open reading frame 63 in latently infected human ganglionic neurons. Proc Natl Acad Sci USA 93:2122–2124

Mahalingam R, Wellish M, Wolf W, Dueland AN, Cohrs R, Vafai A, Gilden D (1990) Latent varicella-zoster viral DNA in human trigeminal and thoracic ganglia. N Engl J Med 323:627–631

Medhurst SJ, Collins SD, Billinton A, Bingham S, Dalziel RG, Brass A, Roberts JC, Medhurst AD, Chessell IP (2008) Novel histamine H3 receptor antagonists GSK189254 and GSK334429 are efficacious in surgically-induced and virally-induced rat models of neuropathic pain. Pain 138:61–69

Moffat JF, Zerboni L, Kinchington PR, Grose C, Kaneshima H, Arvin AM (1998) Attenuation of the vaccine Oka strain of varicella-zoster virus and role of glycoprotein C in alphaherpesvirus virulence demonstrated in the SCID-hu mouse. J Virol 72:965–974

Moffat JF, Zerboni L, Sommer MH, Heineman TC, Cohen JI, Kaneshima H, Arvin AM (1999) The ORF47 and ORF66 putative protein kinases of varicella-zoster virus determine tropism for human T cells and skin in the SCID-hu mouse. Proc Natl Acad Sci USA 95:11969–11974

Moffat J, Ito H, Sommer M, Taylor S, Arvin AM (2002) Glycoprotein I of varicella-zoster virus is required for viral replication in skin and T cells. J Virol 76:8468–8471

Oliver SL, Zerboni L, Sommer M, Rajamani J, Arvin AM (2008) Development of recombinant varicella-zoster viruses expressing luciferase fusion proteins for live in vivo imaging in human skin and dorsal root ganglia xenografts. J Virol Methods 154:182–193

Reddy SM, Cox E, Iofin I, Soong W, Cohen JI (1998) Varicella-zoster virus (VZV) ORF32 encodes a phosphoprotein that is posttranscriptionally modified by the VZV ORF47 protein kinase. J Virol 72:8083–8088

Sadzot-Delvaux C, Merville-Louis MP, Delrée P, Marc P, Piette J, Moonen G, Rentier B (1990) An in vivo model of varicella-zoster virus latent infection of dorsal root ganglia. J Neurosci Res 26:83–89

Sadzot-Delvaux C, Debrus S, Nikkels A, Piette J, Rentier B (1995) Varicella-zoster virus latency in the adult rat is a useful model for human latent infection. Neurology 45(12 Suppl 8): S18–S20

Sato H, Callanan LD, Pesnicak L, Krogmann T, Cohen JI (2002a) Varicella-zoster virus (VZV) ORF17 protein induces RNA cleavage and is critical for replication of VZV at 37°C, but not 33°C. J Virol 76:11012–11023

Sato H, Pesnicak L, Cohen JI (2002b) Varicella-zoster virus ORF2 encodes a membrane phosphoprotein that is dispensable for viral replication and for establishment of latency. J Virol 76:3575–3578

Sato H, Pesnicak L, Cohen JI (2003a) Use of a rodent model to show that varicella-zoster virus ORF61 is dispensable for establishment of latency. J Med Virol 70:S79–S81

Sato H, Pesnicak L, Cohen JI (2003b) Varicella-zoster Virus ORF47 protein kinase which is required for replication in human T cells, and ORF66 protein kinase which is expressed during latency, are dispensable for establishment of latency. J Virol 77:11180–11185

Soong W, Schultz JC, Patera AC, Sommer MH, Cohen JI (2000) Infection of human T lymphocytes with varicella-zoster virus: an analysis with viral mutants and clinical isolates. J Virol 74:1864–1870

Strelow LI, Leib DA (1995) Role of the virion host shutoff (vhs) of herpes simplex virus type 1 in latency and pathogenesis. J Virol 69:6779–6786

Tal-Singer R, Pichyangkura R, Chung E, Lasner TM, Randazzo BP, Trojanowski JQ, Fraser NW, Triezenberg SJ (1999) The transcriptional activation domain of VP16 is required for efficient infection and establishment of latency by HSV-1 in the murine peripheral and central nervous systems. Virology 259:20–23

Thompson RL, Sawtell NM (2001) Herpes simplex virus type 1 latency-associated transcript gene promotes neuronal survival. J Virol 75:6660–6675

Wallace VC, Segerdahl AR, Lambert DM, Vandevoorde S, Blackbeard J, Pheby T, Hasnie F, Rice AS (2007) The effect of the palmitoylethanolamide analogue, palmitoylallylamide (L-29) on pain behaviour in rodent models of neuropathy. Br J Pharmacol 151:1117–1128

Wroblewska Z, Valyi-Nagy T, Otte J, Dillner A, Jackson A, Sole DP, Fraser NW (1993) A mouse model for varicella-zoster virus latency. Microb Pathog 15:141–151

Xia D, Srinivas S, Sato H, Pesnicak L, Straus SE, Cohen JI (2003) Varicella-zoster virus ORF21, which is expressed during latency, is essential for virus replication but dispensable for establishment of latency. J Virol 77:1211–1218

Zerboni L, Sommer M, Ware CF, Arvin AM (2000) Varicella-zoster virus infection of a human CD4-positive T-cell line. Virology 270:278–285

Zhang Z, Rowe J, Wang W, Sommer M, Arvin A, Moffat J, Zhu H (2007) Genetic analysis of varicella-zoster virus ORF0 to ORF4 by use of a novel luciferase bacterial artificial chromosome system. J Virol 81:9024–9033

Simian Varicella Virus: Molecular Virology

Wayne L. Gray

Contents

Abstract Simian varicella virus (SVV) is a primate herpesvirus that is closely related to varicella-zoster virus (VZV), the causative agent of varicella (chickenpox) and herpes zoster (shingles). Epizootics of simian varicella occur sporadically in facilities housing Old World monkeys. This review summarizes the molecular properties of SVV. The SVV and VZV genomes are similar in size, structure, and gene arrangement. The 124.5 kilobase pair (kbp) SVV genome includes a 104.7 kbp long component covalently linked to a short component, which includes a 4.9 kbp unique short segment flanked by 7.5 kbp inverted repeat sequences. SVV DNA encodes 69 distinct open reading frames, three of which are duplicated within the viral inverted repeats. The viral genome is coordinately expressed, and immediate early (IE), early, and late genes have been characterized. Genetic approaches have been developed to create SVV mutants, which will be used to study the role of SVV genes in viral pathogenesis, latency, and reactivation. In addition, SVV expressing foreign genes are being investigated as potential recombinant varicella vaccines.

W.L. Gray
Department of Microbiology and Immunology, University of Arkansas for Medical Sciences, 4301 West Markham Street, Little Rock, AR 72205, USA
e-mail: graywaynel@uams.edu

A.M. Arvin et al. (eds.), *Varicella-zoster Virus*,
Current Topics in Microbiology and Immunology 342, DOI 10.1007/82_2010_27

1 Introduction

Simian varicella virus (SVV) produces a natural varicella-like disease in nonhuman primates. Simian varicella was initially reported in 1967 as an erythematous disease occurring in a colony of vervet monkeys (*Cercopithecus aethiops*) at the Liverpool School of Medicine (Clarkson et al. 1967). Since then, simian varicella outbreaks have occurred sporadically in primate facilities housing Old World monkeys, including vervet and patas (*Erythrocebus patas*) monkeys, and pig-tailed, Japanese, cynomolgous, and rhesus macaque monkeys (*Macaca sp.*) (Soike 1992; Gray 2004, 2008). In some of these outbreaks, the monkeys exhibited relatively mild clinical symptoms including fever and vesicular skin rash, similar to typical chickenpox in children. Other epizootics have been associated with a more severe disease characterized by a hemorrhagic rash, visceral dissemination, and high morbidity, and mortality rates. After the acute disease is resolved, SVV establishes a latent infection within host neural ganglia and may reactivate later in life to cause a secondary disease, analogous to varicella-zoster virus (VZV)-mediated herpes zoster (Mahalingam et al. 1991, 1992). SVV outbreaks are often initiated by viral reactivation from a latently infected monkey.

The etiologic agent is a primate herpesvirus that is genetically related to VZV. SVV is classified as *Cercopithecus herpesvirus 9*, a member of the Alphaherpesvirus family and the *Varcellovirus* genus along with VZV, equine herpesvirus types 1 and 4 (EHV-1, EHV-4), pseudorabies virus (PRV), Marek's disease virus (MDV), and bovine herpesvirus type 1 (BHV-1). The virus has previously been classified as Delta herpesvirus, Liverpool vervet virus (LVV), Medical Lake macaque virus, and patas herpesvirus based on the geographic location of the epizootic or the monkey species involved, but molecular analysis has confirmed that these are isolates of the same virus, designated as SVV (Gray and Gusick 1996).

In addition to its importance as a veterinary disease, the genetic relatedness of SVV and VZV and the similarities between simian and human varicella pathogenesis and clinical symptoms make SVV infection of nonhuman primates a useful experimental model to study the molecular basis of varicella pathogenesis and to evaluate potential antiviral agents and vaccines (Gray 2004). Recent reviews have described simian varicella epidemiology, pathogenesis, latency, and clinical aspects of disease (White et al. 2001; Dueland 1998; Gray 2003, 2004, 2008). This review will focus on the molecular properties of SVV, including SVV morphology, DNA structure, genetic content, gene expression, and mutagenesis of the SVV genome.

2 Cell-Associated Nature of SVV

A discussion of the molecular properties of SVV should include an appreciation of the cell-associated nature of the virus. SVV is propagated *in vitro* in African green monkey kidney cells (Vero, BSC-1, or CV-1). Viral-induced cytopathic effect (cpe) including rounded, swollen, and fused, multinucleated cells is evident during lytic

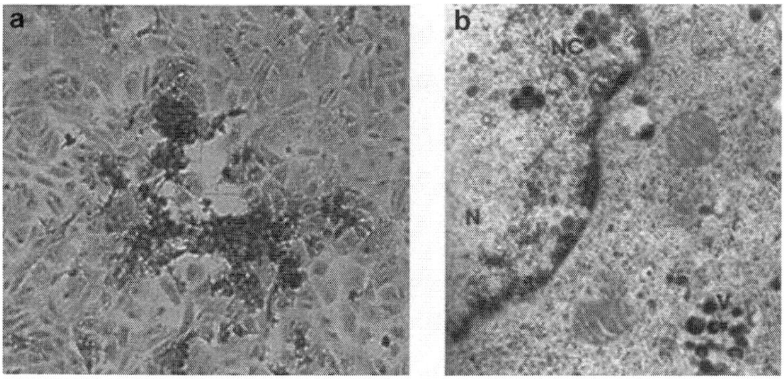

Fig. 1 SVV-infected Vero cells. (**a**) Cytopathic effect in a SVV-infected Vero cell monolayer (20×). (**b**) Electron microscopy of a SVV-infected Vero cell. SVV nucleocapsids (NC) within the cell nucleus (N) and degraded virions within cell vacuoles (V) are shown. Reprinted with permission by Springer Science and Business Media from Oakes and d'Offay (1988)

infection of cell monolayers (Fig. 1a). Cell death occurs within 24 h and infected cells detach from the monolayer. Electron microscopic analysis of infected cells reveals viral nucleocapsids within the cell nucleus (Fig. 1b). However, as virions progress into the cell cytoplasm, a majority of these accumulate within vacuoles and are degraded by lysosomal enzymes. A mixture of intact and aberrant degraded virions are released into the extracellular media, which has a relatively low titer of infectious virus (10^2–10^4 plaque forming units [pfu]/mL). Likewise, VZV infection of cultured cells is characterized by virion degradation and low infectious virus titers. This phenomenon is an unusual feature of SVV and VZV replication in cell culture, as herpes simplex virus type 1 (HSV-1) virions are not degraded in this manner in the same cell types.

Studies of SVV growth in cell culture have determined optimal conditions for generating maximal viral titers (Schmidt 1982). Yields of cell-free SVV generated from infected Vero or BSC-1 cells are generally less than 10^5 pfu/mL. Similarly, high titer VZV cell-free preparations are not feasible. By comparison, HSV-1-infected Vero cell preparations may routinely yield cell-free virus yields of $>10^8$ pfu/mL. Therefore, SVV is serially passaged in culture by transfer of infected cells rather than with cell-free virus, and virus stocks are cryopreserved at -70°C or in liquid nitrogen as infected cells. Studies on the molecular properties of SVV are limited by this inability to generate high titer virus stocks of infectious cell-free SVV and the inability to produce synchronous, high multiplicity per cell infection in cell culture.

3 SVV Morphology and Relatedness of SVV and VZV

SVV virions are isolated from infected BSC-1 or Vero cells by zonal centrifugation on glycerol gradients followed by differential centrifugation on glycerol/potassium tartrate gradients (Fletcher and Gray 1992). Purified SVV virions have a calculated buoyant density of 1.21 g/mL, identical to the density of VZV virions isolated by the same method. Electron microscopy of purified SVV virions reveals particles with typical herpesvirus morphology consisting of ≈100 nm nucleocapsid with an electron dense core containing the viral DNA genome. The nucleocapsid is surrounded by a viral membrane envelope, giving the virion an overall size of 170–200 nm, similar to the size of VZV. SVV virions consist of at least thirty polypeptide species ranging in size from 16 to >200 kilodaltons (kDa) as indicated by SDS-polyacrylamide gel electrophoresis (SDS-PAGE) analysis (Fletcher and Gray 1992). The polypeptide profiles of SVV and VZV purified virions are strikingly similar. At least four of the SVV virion polypeptides ranging in size from 113 to 46 K are antigens immunoprecipitated with immune serum from SVV-infected monkeys. SVV includes at least six glycoproteins (46–115 kDa) as indicated by immunoprecipitation analysis of infected cell lysates.

The SVV virion and infected cell proteins and glycoproteins share cross-reacting epitopes with VZV as indicated by the extensive immunoprecipitation of SVV proteins with VZV immune serum derived from herpes zoster patients (Fletcher and Gray 1992). Other studies have confirmed the antigenic relatedness of SVV and VZV. Immune serum from SVV-infected monkeys neutralizes VZV and reacts with VZV antigens as indicated by immunoflourescence, complement fixation, and immunoblot assays (Felsenfeld and Schmidt 1977, 1975; Blakely et al. 1973). In addition, while VZV inoculation of patas monkeys does not result in clinical disease, it induces cross-reacting antibodies to SVV antigens and immune protection against simian varicella disease following SVV challenge (Felsenfeld and Schmidt 1979).

A comparison of the restriction endonuclease profiles of SVV and VZV DNA confirmed that SVV and VZV are distinct herpesviruses (Gray et al. 1992). However, Southern blot DNA hybridizations performed under conditions of varied stringency demonstrated the genetic similarity of SVV and VZV and determined that the SVV and VZV genomes share 70–75% DNA homology (Gray and Oakes 1984). In addition, the SVV and VZV DNAs are colinear with respect to genome organization as indicted by hybridizations conducted with labeled SVV and VZV restriction endonuclease DNA fragments (Pumphrey and Gray 1992). These experimental findings have been confirmed by comparison of the DNA sequence of the SVV and VZV genomes (Gray et al. 2001a, b).

4 The SVV Genome

SVV DNA is purified from viral nucleocapsids isolated by centrifugation of infected Vero cell lysates through glycerol gradients (Gray et al. 1992). The molecular size of SVV DNA was initially estimated be ≈121–125 kilobases (kb) by electron microscopy, pulse field electrophoresis, and restriction endonuclease analyses (Gray et al. 1992; Clarke et al. 1992). Subsequent DNA sequence analysis has determined the SVV genome to be 124,785 bp in size (Gray et al. 2001a, b), only slightly smaller than the 124,884 bp VZV DNA. The viral DNA has a buoyant density of 1.700 g/mL, corresponding to a guansine plus cytosine (G + C%) content of 40.8%, as determined by CsCl gradient isopycnic banding (Clarke et al. 1992). DNA sequence analysis has subsequently confirmed the SVV DNA content to be 40.4% G + C, compared to 46% G + C for VZV DNA (Table 1). Transfection of Vero or CV-1 cell monolayers with purified SVV DNA yields viral plaques within 10 days indicating the infectious nature of SVV DNA (Clarke et al. 1992).

The structure of SVV DNA was resolved using a variety of approaches including electron microscopy, restriction endonuclease, lambda exonuclease, and Southern blot hybridization analyses (Gray et al. 1992; Clarke et al. 1992). The presence of inverted repeat sequences, typical for herpesvirus DNAs, was initially indicated by electron microscopy of denatured and re-annealed SVV DNA. SVV molecules consisted of a ≈7.2 kb double-stranded stem continuous with a ≈5.2 kb single-stranded loop and terminating with a long, ≈100 kb single-stranded sequence. Southern blot hybridization and restriction endonuclease analyses were used to confirm the inverted repeat sequences and to construct restriction endonuclease maps of the SVV genome. These studies indicated that the SVV genome consists of ≈100 kb long (L) component covalently linked to a ≈20 kb short (S) component. The S component includes 7.5 kb terminal and internal inverted repeat sequences (IRS and TRS), which bracket the 4.9 kb unique short (US) component (Fig. 2). The detection of 0.5 molar DNA bands in restriction endonuclease profiles, as well as PCR and hybridization analysis of SVV DNA, revealed that the L component can invert relative to the S component and that the SVV genome exists in two major isomeric forms (Gray et al. 1992; Clarke et al. 1995b). Subsequent DNA sequence analysis of the entire SVV genome more precisely defined the SVV DNA size and structure and confirmed that the SVV and VZV genomes are similar in size and DNA structure (Table 1) (Fletcher and Gray 1993; Gray et al. 1995; 2001a, b).

The only significant difference between the structure of SVV and VZV DNA occurs at the left end of the viral genomes. The left terminus of the SVV genome possesses a 666 bp terminal element that includes a 506 bp unique sequence bracketed by 80 bp inverted repeat sequences (Fig. 2; Mahalingam and Gray 2007). A portion of this inverted repeat, 65 bp, is also present at the junction of the L and S

Table 1 SVV–VZV genome comparison

	Size (bp)		G + C (%)	
	SVV	VZV[a]	SVV	VZV[a]
TRS/IRS	7,557	7319.5	65.0%	59.0%
US	4,904	5,232	39.1%	42.8%
Total S	20,018	19,871	58.6%	54.7%
TRL/IRL	65	88.5	69.3%	68.4%
UL	104,036	104,836	38.3%	44.3%
TE[b]	666	No	53.8%	–
Total L	104,767	105,013	38.3%	44.3%
Total genome	124,785	124,884	40.4%	46.0%

[a]Data derived from Davison and Scott (1986)
[b]Terminal element –SVV genome left end

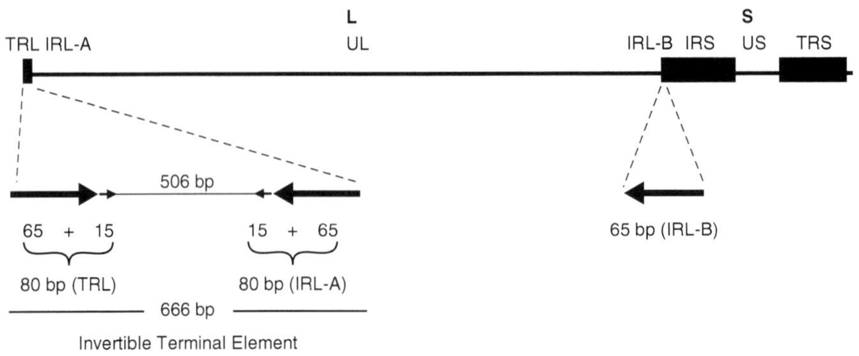

Fig. 2 Structure of SVV genome. The 124.7 kb SVV genome consists of a long (L) component covalently linked to a short (S) component. The L component includes a 104.0 kb unique long (UL) segment bracketed by repeat sequences. The S component includes a 4.9 kb unique short (US) segment flanked by 7.5 kb terminal (TRS) and internal (IRS) repeat sequences. The SVV left end has a 666 bp terminal element which includes a 506 bp unique sequence flanked by 80 bp inverted repeats (TRL and IRL-A), of which 65 bp are also present at the right end of the UL segment (IRL-B) (data revised from Mahalingam and Gray 2007)

components, such that the SVV L component is flanked by 65 bp inverted repeat sequences, similar to the 88 bp inverted repeat sequences that bracket the VZV L component. The 666 bp terminal element inverts relative to the SVV DNA UL component. Analysis of the genomes of various SVV isolates indicates that the terminal element is conserved, but may vary in size as the MLM and LVV isolates of SVV possess a 712 bp terminal element, which includes an additional 46 bp repeat within the unique sequence. The terminal element was also confirmed to exist in SVV DNA isolated from tissues of acutely and latently infected monkeys. The EHV-1, EHV-4, and PRV varicellovirus genomes also include a analogous terminal element at the left end of their genomes (Mahalingam and Gray 2007). In contrast, the VZV genome does not include a similar element at the left terminus.

Within infected cells, SVV DNA appears to exist as circular or concatemeric genomic molecules (Clarke et al. 1995b). Compared to SVV virion DNA, SVV genomes with connected termini are increased in infected BSC-1 cell DNA as indicated by PCR employing terminal primers from the left and rightward ends of the viral genome. DNA sequence analysis reveals an additional G + C base pair across the connected termini, indicating that SVV genome termini possess an unpaired 3′ nucleotide, which might enable SVV DNA circulation (Clarke et al. 1995b). Circulation may facilitate SVV DNA replication by a rolling circle mechanism that generates genomic DNA head-to-tail concatemers. While the structure of SVV DNA during latency is unknown, the termini of VZV genomes are connected in latently infected ganglia, suggesting that latent VZV DNA exists in an extra-chromosomal, circular configuration (Clarke et al. 1995a).

Tandem direct repeat elements, which are typical for herpesvirus genomes, are present in SVV DNA as they are in VZV DNA. In the SVV genome, these repeat

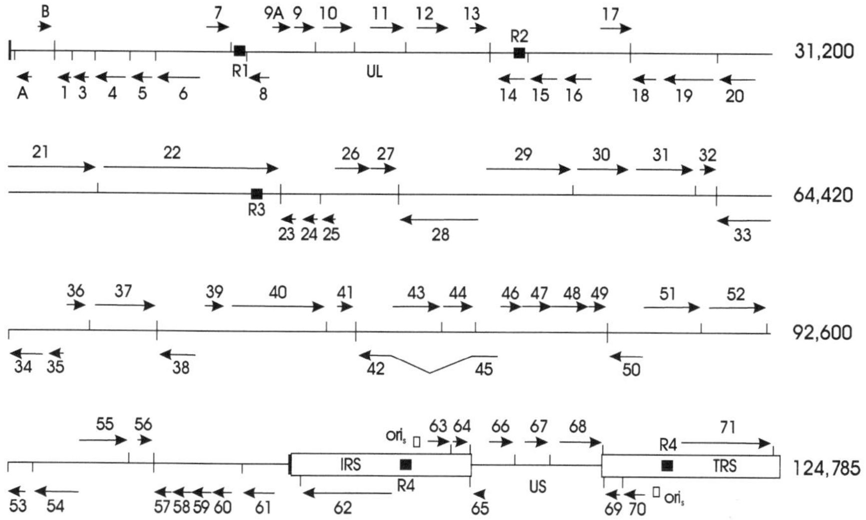

Fig. 3 The SVV gene map. SVV ORFs are designated as *horizontal arrows* indicating gene location on each DNA strand. Potential polyadenylation sites are shown as *vertical lines*. The putative origins of replication (oriS) are indicated as *open boxes*. The R1, R2, R3, and R4 repeat elements are shown as *black boxes*

elements are designated as R1, R2, R3, and R4 (Fig. 3; Gray et al. 2001a, b). Located between ORFs 7 and 8, SVV R1 is an A + T rich 37 bp element that is repeated three times. SVV R2, a G + C-rich 83 bp element that is repeated twice, is present within the viral glycoprotein C (gC) gene (ORF 14). The most complex and largest, the SVV tandem repeats is SVV R3, which includes eight 12 bp repeats and seventeen 9 bp repeats found within SVV ORF 22. SVV R4 is a G + C-rich 16 bp sequence that is repeated seven times and is present in a noncoding region between ORFs 62 and 63. The VZV genome has reiterations corresponding in location to the SVV R2, R3, and R4 repeat elements, although the DNA sequences of the SVV and VZV tandem repeats are different (Davison and Scott 1986). VZV DNA does not have a reiteration corresponding in location to the SVV R1, but instead the VZV R1 is located within ORF 11. VZV DNA also includes R5, an 88 bp element located between ORF 60 and 61, that is not present in SVV DNA. The function of these conserved repeat elements in SVV and other herpesvirus genomes is not understood.

The SVV and VZV genomes are colinear with respect to gene organization. The SVV genome encodes 69 distinct ORFs, three of which are duplicated within the S component inverted repeat sequences (Table 2; Fig. 3; Gray et al. 2001a, b). Corresponding SVV and VZV genes share extensive homology, with predicted proteins having from 27% to 75% amino acid identity. The only significant difference in SVV and VZV gene content occurs in the left terminus of the viral DNAs (Mahalingam et al. 2000). SVV DNA lacks a homolog to the VZV ORF 2, a gene that is nonessential for *in vitro* replication and dispensible for the virus to establish

Table 2 SVV open reading frames

ORF	Start	Stop	Size (aa)	Size (kDa)[a]	VZV % Homol.[b]	VZV Size (aa)[c]	HSV-1 Homolog	Putative function[d]/notes
A	1,807	926	293	33.5	–	–	UL54	Truncated homolog of ORF 4
B	2,327	2,671	114	12.7	29.9	157–224	None	Homolog of VZV ORF S/L
1	3,036	2,730	101	11.7	27.3	108	None	Membrane protein
2	–	–	–	–	–	238	None	SVV DNA does not include a homolog of VZV ORF 2
3	3,852	3,301	183	19.7	63.7	179	UL55	Virion assembly
4	5,533	4,121	470	54.3	43.2	452	UL54	Transcriptional activator, immediate early protein 2
5	6,656	5,643	337	42.0	59.3	340	UL53	Glycoprotein K
6	9,906	6,661	1,081	123.3	37.7	1,083	UL52	Component of DNA helicase-primase complex
7	9,917	10,612	231	25.3	72.9	259	UL51	Virion phosphoprotein
8	12,068	10,881	395	44.9	38.2	396	UL50	DeoxyUTPase
9A	12,037	12,300	87	9.7	68.8	87	UL49A	Glycoprotein N
9	12,405	13,310	301	33.2	59.4	302	UL49	Tegument protein
10	13,537	14,757	406	46.6	62.5	410	UL48	Transcriptional activator, tegument protein
11	15,076	17,004	642	71.7	50.8	819	UL47	Tegument protein
12	17,150	19,117	655	73.4	60.1	661	UL46	Tegument protein
13	19,243	20,130	295	33.9	71.1	301	None	Thymidylate synthetase
14	21,841	20,219	540	60.6	43.8	560	UL44	Glycoprotein C
15	23,259	21,994	421	47.3	36.2	406	UL43	Membrane protein
16	24,493	23,336	385	42.8	46.8	408	UL42	Associated with DNA polymerase
17	24,719	26,134	471	53.6	55.0	455	UL41	Host shutoff virion protein
18	27,117	26,182	311	36.2	72.6	306	UL40	Ribonucleotide reductase, small subunit
19	29,471	27,120	783	88.2	68.7	775	UL39	Ribonucleotide reductase, large subunit
20	31,068	29,665	467	52.9	60.5	483	UL38	Capsid protein
21	31,243	34,368	1,041	116.8	52.0	1,038	UL37	Tegument protein
22	34,559	42,520	2,653	294.8	48.9	2,763	UL36	Tegument protein
23	43,870	43,184	228	23.8	47.3	235	UL35	Capsid protein
24	44,596	43,952	214	23.9	61.5	269	UL34	Membrane phosphoprotein
25	45,337	44,876	153	17.4	45.2	156	UL33	Viral DNA cleavage/ packaging
26	45,249	46,967	572	65.1	61.3	585	UL32	

(*continued*)

Table 2 (continued)

ORF	Start	Stop	Size (aa)	Size (kDa)[a]	VZV % Homol.[b]	VZV Size (aa)[c]	HSV-1 Homolog	Putative function[d]/notes
								DNA cleavage/ packaging
27	46,894	47,826	310	35.4	74.4	333	UL31	Nuclear phosphoprotein
28	51,268	47,750	1,172	113.2	65.7	1,194	UL30	DNA polymerase
29	51,454	55,038	1,194	132.0	71.9	1,204	UL29	Single-stranded DNA binding protein
30	55,125	57,407	760	86.5	61.3	770	UL28	Viral DNA cleavage/ packaging
31	57,266	60,016	916	104.0	75.4	868	UL27	Glycoprotein B
32	60,152	60,559	135	15.0	49.6	143	None	Phosphoprotein
33	62,412	60,646	588	65.1	64.4	605	UL26	Protease, capsid assembly protein
34	64,182	62,443	579	65.7	61.1	579	UL25	Viral DNA cleavage/ packaging
35	64,992	64,246	248	28.7	49.7	258	UL24	Membrane protein
36	65,018	66,031	337	37.9	52.3	341	UL23	Thymidine kinase
37	66,204	68,762	852	96.8	55.5	841	UL22	Glycoprotein H
38	70,414	68,813	533	59.8	60.3	541	UL21	Virion protein
39	70,693	71,364	223	25.4	53.1	240	UL20	Envelope protein, viral egress
40	71,553	75,731	1,392	155.9	73.3	1,396	UL19	Major capsid protein
41	75,826	76,773	315	34.3	70.5	316	UL18	Capsid protein
42/45	78,005	76,821	744	84.1	67.7	747	UL15	Spliced product
	82,424	81,372						Viral terminase
43	78,037	80,073	678	75.9	47.3	676	UL17	Viral DNA cleavage/ packaging
44	80,217	81,299	360	39.8	69.3	363	UL16	Virion protein
46	82,536	83,135	199	22.5	58.2	199	UL14	Tegument protein
47	82,988	84,511	507	57.7	64.8	510	UL13	Protein Kinase
48	84,475	86,004	509	57.8	56.2	551	UL12	Deoxyribonuclease
49	86,004	86,252	82	9.2	50.0	81	UL11	Myristylated virion protein
50	87,656	86,337	439	49.5	57.5	435	UL10	Glycoprotein M
51	87,665	90,115	816	92.8	53.7	835	UL9	Origin binding protein
52	90,232	92,529	765	86.0	50.4	771	UL8	Component of DNA helicase-primase complex
53	93,512	92,598	304	34.5	56.6	331	UL7	Gamma-1 protein
54	95,610	93,403	735	83.5	59.5	769	UL6	Viral DNA cleavage/ packaging
55	95,645	98,254	869	97.8	75.2	881	UL5	Component of DNA helicase-primase complex
56	98,326	98,889	187	21.3	37.7	244	UL4	Gamma-2 protein
57	99,137	98,919	72	8.2	39.5	71	None	Nonessential VZV protein
58	99,745	99,134	203	22.9	42.0	221	UL3	Phosphoprotein
59	100,661	99,759	300	34.6	55.4	305	UL2	

(*continued*)

Table 2 (continued)

ORF	Start	Stop	Size (aa)	Size (kDa)[a]	VZV % Homol.[b]	VZV Size (aa)[c]	HSV-1 Homolog	Putative function[d]/notes
								Uracil DNA glycosylase
60	101,092	100,565	175	20.2	43.5	159	UL1	Glycoprotein L
61	103,781	102,270	503	54.1	42.8	467	RL2	Transcriptional activator, repressor, immediate early protein 1
62	108,458	104,619	1,279	136.8	58	1,310	RS1	Transcriptional activator, immediate early protein 3
63	109,844	110,629	261	29.3	52	278	US1	Transcriptional activator, immediate early protein 4
64	110,851	111,414	187	21.1	56	180	US10	Tegument phosphoprotein
65	111,948	111,715	77	9.0	49	102	US9	Tegument phosphoprotein
66	112,388	113,425	345	38.9	66	393	US3	Protein kinase
67	113,656	114,717	353	40.5	37	354	US7	Glycoprotein I
68	114,937	116,751	604	67.6	47	623	US8	Glycoprotein E
69	117,408	116,845	187	21.1	56	180	US10	Duplicate of ORF 64
70	118,415	117,629	261	29.3	52	278	US1	Duplicate of ORF 63
71	119,801	123,640	1,279	136.8	58	1,310	ICP4	Duplicate of ORF 62

[a]Predicted size based on amino acid (AA) sequence
[b]Based on % amino acid identity with the homologous VZV protein
[c]Derived from Davison and Scott (1986)
[d]Based on known VZV function or function of HSV-1 homolog

latent infection of neural ganglia (Sato et al. 2002). The SVV DNA left terminus includes an 882 bp ORF A that is not present in the VZV genome. The SVV ORF A encodes a truncated homolog of the SVV ORF 4 and the VZV ORF 4, a viral transactivator of viral gene expression (Perera et al. 1994), although the SVV ORF A has not been demonstrated to transactivate SVV promoters (Gray, unpublished data). The SVV ORF B corresponds to the VZV ORF S/L located at the VZV DNA left terminus, which encodes a cytoplasmic viral protein of unknown function (Kemble et al. 2000).

Sensitive techniques have been developed for detection of SVV DNA. A PCR-based assay readily detects SVV DNA in skin rash specimens, peripheral blood lymphocytes, and other tissues of acutely infected monkeys (Gray et al. 1998). In addition, SVV DNA can be detected in neural ganglia of latently infected animals. The SVV PCR assay is specific, detecting DNA derived from various SVV isolates, but does not cross-react with DNA derived from VZV, HSV-1, or other primate herpesviruses. Real-time quantitative PCR assay may be used to determine SVV copy number in infected tissues (White et al. 2002a). These sensitive assays for SVV DNA detection are useful for rapid diagnosis of simian varicella in primate facilities and for investigating SVV pathogenesis.

5 SVV Gene Expression

Investigation of SVV gene expression is hampered by the cell-associated nature of SVV and the resulting inability to generate high titer infectious virus stocks for synchronous infection in cell culture. However, SVV gene expression, like that of other herpesviruses, is considered to be coordinately regulated into immediate early (IE), early, and late phases with the IE genes expressing regulatory proteins, the early genes encoding enzymes involved in viral DNA synthesis, and late genes expressing structural proteins and glycoproteins.

The SVV IE ORF 62, homolog of the HSV-1 ICP 4, encodes a regulatory protein that serves as a major transactivator of viral genes. The IE62 protein stimulated the SVV ORF 28 and ORF 29 early gene promoters by over 100-fold using a luciferase reporter gene assay (Ou and Gray 2006). A cellular transcription factor, the upstream stimulatory factor (USF), appears to play a critical role, as a USF DNA binding sequence (5′-CACGTG) within the ORF 28/29 bidirectional promoter is necessary for efficient IE62-mediated transactivation. The SVV IE62 also transactivates its own promoter, other SVV immediate early genes (ORFs 4 and 61), other early genes (ORF 21), and late genes (ORF 68) (Mahalingam et al. 2006, Gray, unpublished data).

The SVV ORFs 4, 61, and 63 are putative IE genes, based upon their VZV and HSV-1 homologs. The SVV ORF 61 protein, homolog to the HSV-1 ICP0, transactivates its own promoter, the ORF 62 IE promoter, early promoters (ORFs 28 and 29), and ORF 68 (gE) late gene promoter in transfected Vero cells (Gray et al. 2007). However, IE61-mediated transactivation is modest compared to IE62-mediated induction, and ORF 61 is nonessential for replication in cell culture (Gray et al. 2007). The IE61 protein includes a RING finger motif at the amino-terminus and a nuclear localization signal sequence (NLS) at the carboxy-terminus, both of which are required for transactivation. The SVV ORF 63 IE gene product did not transactivate the SVV ORF 21 early promoter upon co-transfection of Vero and Mewo cells (Mahalingam et al. 2006). The IE63 down-regulated IE62-mediated transactivation of the ORF 21 promoter in these cells, but augmented IE62-transactivation of this promoter in neuronal cell culture, indicating that the IE63 has a differential effect on the ORF 21 promoter depending on the cell type (Mahalingam et al. 2006). In limited studies to date, the SVV ORF 4 and its truncated homolog, ORF A, have not been demonstrated to transactivate SVV genes (Gray, unpublished data).

Expression of several SVV early and late genes has been characterized. The SVV thymidine kinase (ORF 36), uracil DNA glycosylase (ORF 59), and deoxyuridine nucleotidohydrolase (dUTPase, ORF 8) genes are demonstrated to express functional enzymes involved in viral DNA synthesis and repair (Pumphrey and Gray 1996; Ashburn and Gray 1999; Ward et al. 2009). SVV expresses several glycoproteins, which are incorporated into the viral envelope and the surface and inner membranes of infected cells. The expression and molecular properties of the SVV gB, gC, gE, gH, and gL have been characterized, although the specific roles of these glycoproteins in viral attachment and penetration has not been studied (Pumphrey and Gray 1994, 1995; Gray and Byrne 2003; Gray et al. 2001a, b; Ashburn and Gray 2001). Unlike HSV-1 and other alphaherpesviruses, SVV and VZV do not express a gD homolog.

All regions of the SVV genome are transcriptionally active during lytic infection of Vero cell monolayers. At least 53 distinct RNA species ranging in size from 9.2 to 0.8 kb were detected by Northern blot hybridization analysis using SVV restriction endonuclease probes representing the entire SVV genome (Gray et al. 1993). A SVV transcription map was constructed, which revealed similarities between SVV and VZV gene expression. More recently, microarray assay was used to analyze viral gene transcription from 70 predicted SVV ORFs during lytic infection (Deitch et al. 2006). Using cloned DNA fragments generated from the 5′ and 3′ ends of each SVV ORF and polyadenylated RNA isolated from infected Vero cells at maximal cytopathic effect (3 days postinfection), viral transcripts mapping to each ORF were identified, and their relative abundance was determined. The most abundant transcript mapped to SVV ORF 9. Interestingly, a similar array analysis of VZV transcription also identified the VZV ORF 9 as the most abundant transcript in infected BSC-1 cells (Cohrs et al. 2003). The VZV ORF 9 encodes a tegument protein that is essential for replication in cell culture and interacts with IE62 in the cytoplasm of infected cells (Cilloniz et al. 2007). Putative splice sites are predicted for SVV ORF 42/45 and ORF B (S/L) transcripts, but splicing of transcripts is unusual for processing of SVV and VZV RNA (Gray et al. 2001a, b). Analysis of SVV promoter sequences have revealed conservation of the *cis* acting elements, which control SVV and VZV gene expression (Fletcher and Gray 1994; Pumphrey and Gray 1994, 1996).

SVV gene expression has been analyzed in tissues of African green monkeys with acute simian varicella disease (day 10–12 postinfection). SVV IE, early, and late transcripts and viral antigens were detected in the skin, lung, liver, spleen, and neural ganglia, indicating expression of viral genes from across the viral genome and active viral replication in tissues of acutely infected monkeys (Gray et al. 2002). A persistent infection was described in African green monkeys experimentally infected by intratracheal inoculation with IE, early, and late transcripts detected for as long as 12 months postinfection (White et al. 2002b).

In contrast, viral gene expression was restricted in tissues of African green monkeys confirmed to be latently infected following intratracheal and subcutaneous SVV inoculation followed by a second SVV immunization (Ou et al. 2007a, b). A latency associated transcript (LAT)-oriented antisense to the SVV ORF 61 mRNA was consistently detected in neural ganglia, but not in other tissues, derived from latently infected monkeys (Ou et al. 2007a, b). Similarly, gene expression of HSV-1 and other neurotropic varicelloviruses, including EHV-1 and BHV-1, is limited to expression of a LAT antisense to an ICP0 homolog during latent infection. In contrast, several VZV transcripts (ORF 4, 21, 29, 62, 63, 66) may be detected in latently infected human ganglia (Cohen et al. 2006).

6 Genetic Manipulation of the SVV Genome-SVV Mutants and Recombinant Viruses

Development of genetic approaches to insert site-specific mutations into the SVV genome is important to elucidate the roles of viral genes in replication and pathogenesis. The inability to generate high titer infectious virus due the cell-associated nature of SVV creates challenges for genetic manipulation of the SVV genome. The initial approach for generating SVV mutants was based on the insertion of a reporter gene encoding the green fluorescent protein (GFP) into the SVV genome by homologous recombination. A SVV gC deletion mutant was constructed by insertional mutagenesis using the GFP gene (Gray and Byrne 2003). A cassette consisting of the GFP gene (under control of the human cytomegalovirus [HCMV] IE promoter) flanked by SVV gC sequences was transfected into SVV-infected BSC-1 cells, and the recombinant SVV gC^{-}/GFP mutant was selected by plaque purification of fluorescent cells. The SVV gC^{-}/GFP mutant replicated efficiently in cell culture, indicating that the SVV gC is nonessential for *in vitro* replication. Similarly, a SVV mutant was constructed by homologous recombination insertion of the GFP reporter gene into the SVV genome between SVV ORFs 66 and 67 (Mahalingam et al. 1998). This recombinant SVV replicated efficiently in Vero cells and in lung tissues of an infected African green monkey. Although successful, genetic manipulation of the SVV genome by homologous recombination and insertion of a reporter gene is laborious, requiring extensive plaque purification. In addition, the presence of the reporter gene may complicate evaluation of the SVV mutants in studies of viral pathogenesis, and so additional steps are required to remove the foreign gene from the viral genome.

The development of a cosmid-based recombination system was an advance in the ability to genetically manipulate the SVV genome (Gray and Mahalingam 2005). Following a genetic approach used to alter VZV DNA (Cohen and Seidel 1993), the entire SVV genome in four overlapping DNA fragments, each 32–38 kilobase pairs (kbp) in size, was cloned into cosmid vectors. Co-transfection of all four of the SVV cosmids (Cos A, Cos B, Cos C, and Cos D) into Vero or CV-1 cells results in homologous recombination of the overlapping DNA ends and generation of infectious virus plaques within 9–12 days post-transfection (Fig. 4). A SVV gC^{-} mutant was generated by insertion of a 30 bp oligonucleotide containing translational stop codons into a unique KpnI site within ORF 14 of Cos A, followed by

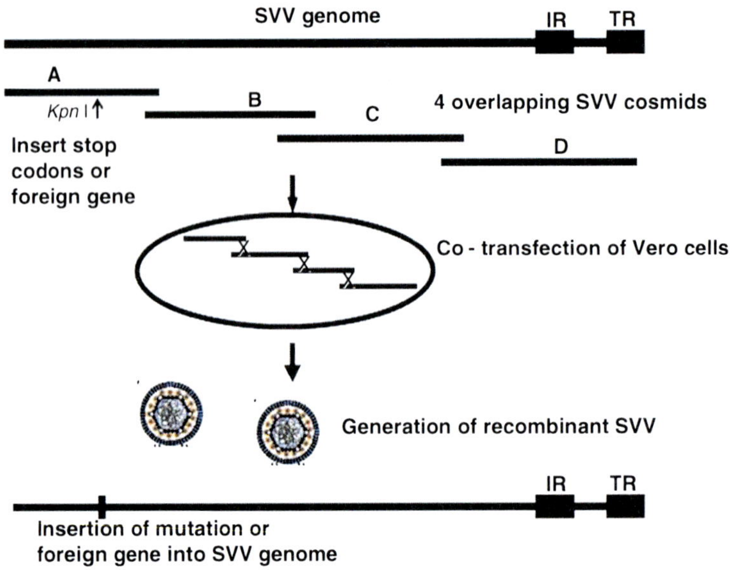

Fig. 4 Creation of SVV mutants and recombinant viruses using the SVV cosmid-based recombination system. The relative genomic location of each 32–38 kb SVV fragment cloned into cosmid vectors is indicated. As an example, the *arrow* indicates the location for insertion of translational stop codons or a foreign gene into the unique KpnI site of ORF 14 (gC gene) within Cos A. Following co-transfection of Vero cells, genetic recombination occurs resulting in the generation of infectious recombinant virus harboring the specified mutation or foreign gene

co-transfection of a Vero cell monolayer with Cos A/gC$^-$, Cos B, Cos C, and Cos D. Analysis of the infectious SVV gC$^-$ virus that was generated verified the mutation within ORF 14 and confirmed that the SVV gC is nonessential for replication in cell culture. A more versatile approach for inserting gene-specific mutations within SVV cosmids employs rec A-assisted restriction endonuclease (RARE) cleavage (Ferrin and Camerini-Otero 1991). RARE-mediated deletion of 1.3 kb from the SVV gene 61 within Cos D followed by co-transfection of Vero cells with the other three cosmids generated a SVV ORF 61 deletion mutant. This SVV ORF61*del* multiplied only slightly less efficiently than wild-type SVV in CV-1 cells, indicating that SVV 61 transactivation is not critical for *in vitro* replication (Gray et al. 2007). SVV mutants defective in dUTPase (ORF 8) and UDG (ORF 59) were also generated using RARE mutagenesis and the SVV cosmid recombination system (Ward et al. 2009). While the SVV dUTPase and UDG are nonessential for replication in cell culture, these enzymes may be important for viral DNA synthesis and repair in neuronal cells and play a role in SVV latency or reactivation.

The SVV cosmid recombination system has also been used to insert foreign genes into the SVV genome. A recombinant SVV expressing simian immunodeficiency virus (SIV) antigens was constructed by insertion of the SIV env and gag genes along with the HCMV IE promoter into the SVV genome at a unique KpnI site within ORF 14 (gC) of Cos A followed by co-transfection of CV-1 cells with the other three cosmids. Immunization of African green monkeys with the rSVV/SIVenv, gag recombinant virus-induced antibody and cellular immune responses to SIV gag and env antigens as well as to SVV antigens (Ou et al. 2007a, b). An additional study generated a recombinant SVV expressing the respiratory syncytial virus (RSV) G and M antigens and the rSVV/RSVG,M recombinant virus-induced antibody responses to the RSV antigens in immunized monkeys (Ward et al. 2008). These studies provide support for the potential use of recombinant VZV vaccines expressing antigens of other pathogens.

Most recently, the entire SVV genome has been cloned into a bacterial artificial chromosome (BAC) permitting stable maintenance of SVV DNA in *Escherichia coli* (Gray, unpublished data). The SVV BAC was constructed by insertion of the BAC vector into Cos A within the intergenic sequence between SVV ORFs 12 and 13, followed by co-transfection of CV-1 cells with Cos A-BAC along with Cos B, Cos C, and Cos D. Generation of the SVV BAC represents an advance over the SVV cosmid system as it avoids the necessity of transfecting four large independent cosmids into the same cell for the required recombination. The BAC sequence is flanked by *lox*P sites, so that the BAC vector can be excised from the SVV genome by Cre recombinase. The SVV BAC system permits genetic manipulation (site-specific point mutations, insertions, and deletions) of SVV DNA using efficient techniques including Red-mediated recombination (Tischer et al. 2006). This approach was recently used to generate a SVV mutant with a deletion in ORF 10, which encodes a homolog of the HSV-1 VP16 transactivator protein (Gray, unpublished data). These studies employing SVV mutants will provide a foundation to elucidate the molecular basis of SVV pathogenesis, latency, and reactivation.

Acknowledgements Research conducted by the author was supported by Public Health Service Grant RO1 AI052373 of the National Institutes of Health.

References

Ashburn CV, Gray WL (1999) Identification and characterization of the simian varicella virus uracil DNA glycosylase. Arch Virol 144:2161–2172

Ashburn CV, Gray WL (2001) Expression of the simian varicella virus glycoprotein L and H. Arch Virol 147:335–348

Blakely GA, Lourie B, Morton WG, Evans HH, Kaufmann AF (1973) A varicella-like disease in macaque monkeys. J Infect Dis 127:617–625

Cilloniz C, Jackson W, Grose C, Czechowski D, Hay J, Ruyechan WT (2007) The varicella-zoster virus (VZV) ORF 9 protein interacts with the IE62 major VZV transactivator. J Virol 81:761–774

Clarke P, Rabkin SD, Inman MV, Mahalingam R, Cohrs R, Wellish M, Gilden DH (1992) Molecular analysis of simian varicella virus DNA. Virology 190:597–605

Clarke P, Beer T, Cohrs R, Gilden DH (1995a) Configuration of latent varicella-zoster virus DNA. J Virol 69:8151–8154

Clarke P, Beer T, Gilden DH (1995b) Configuration and terminal sequences of the simian varicella virus genome. Virology 207:154–159

Clarkson MJ, Thorpe E, McCarthy K (1967) A virus disease of captive vervet monkeys (*Cercopithicus aethiops*) caused by a new herpes virus. Arch Gesamte Virusforsch 22: 219–234

Cohen JI, Seidel KE (1993) Generation of varicella-zoster virus (VZV) and viral mutants from cosmid DNAs: VZV thymidylate synthetase is not essential for replication *in vitro*. Proc Natl Acad Sci USA 90:7376–7380

Cohen JI, Straus SE, Arvin AM (2006) Varicella-zoster virus replication, pathogenesis, and management. In: Knipe DM et al (eds) Field's Virology. Lippincott Williams & Wilkins, Philadelphia, pp 2774–2840

Cohrs RJ, Hurley MP, Gilden DH (2003) Array analysis of viral gene transcription during lytic infection of cells in tissue culture with varicella-zoster virus. J Virol 77:11718–11732

Davison AJ, Scott JE (1986) The complete DNA sequence of varicella-zoster virus. J Gen Virol 67:1759–1816

Deitch SB, Gilden DH, Wellish M, Smith J, Cohrs R, Mahalingam R (2006) Array analysis of simian varicella virus gene transcription in productively infected cell in tissue culture. J Virol 79:5315–5325

Dueland AN (1998) Simian varicella virus infection in primates as a model for human varicella zoster virus infection. Herpes 5:76–78

Felsenfeld AD, Schmidt NJ (1975) Immunological relationship between delta herpesvirus of patas monkeys and varicella-zoster virus of humans. Infect Immun 12:261–266

Felsenfeld AD, Schmidt NJ (1977) Antigenic relationships among several simian varicella-like viruses and varicella-zoster virus. Infect Immun 15:807–812

Felsenfeld AD, Schmidt NJ (1979) Varicella-zoster virus immunizes patas monkeys against simian varicella-like disease. J Gen Virol 42:171–178

Ferrin LJ, Camerini-Otero RD (1991) Selective cleavage of human DNA: RecA-assisted restriction endonuclease (RARE) cleavage. Science 254:1494–1497

Fletcher TM III, Gray WL (1992) Simian varicella virus: characterization of virion and infected cell polypeptides and the antigenic cross-reactivity with varicella-zoster virus. J Gen Virol 73:1209–1215

Fletcher TM, Gray WL (1993) DNA sequence and genetic organization of the unique short (Us) region of the simian varicella virus genome. Virology 193:762–773

Fletcher TM III, Gray WL (1994) Transcriptional analysis of two simian varicella virus glycoprotein genes which are homologous to varicella-zoster virus gpl (gE) and gpIV (gl). Virology 205:352–359

Gray WL (2003) Pathogenesis of simian varicella virus. J Med Virol 70:S4–S8

Gray WL (2004) Simian varicella: a model for human varicella-zoster virus infections. Rev Med Virol 14:363–381

Gray WL (2008) Simian varicella in Old World monkeys. Comp Med 58:22–30

Gray WL, Byrne BH (2003) Characterization of the simian varicella virus glycoprotein C, which is nonessential for *in vitro* replication. Arch Virol 148:537–545

Gray WL, Gusick NJ (1996) Viral isolates derived from simian varicella epizootics are genetically related but distinct from other primate herpesviruses. Virology 224:161–166

Gray WL, Mahalingam R (2005) A cosmid-based system for inserting mutations and foreign genes into the simian varicella virus genome. J Virol Meth 130:89–94

Gray WL, Oakes JE (1984) Simian varicella virus DNA shares homology with human varicella-zoster virus DNA. Virology 136:241–246

Gray WL, Pumphrey CY, Ruyechan WT, Fletcher TM (1992) The simian varicella virus and varicella zoster virus genomes are similar in size and structure. Virology 186:562–572

Gray WL, Gusick N, Fletcher TM, Pumphrey CY (1993) Characterization and mapping of simian varicella virus transcripts. J Gen Virol 74:1639–1643

Gray WL, Gusick NJ, Ek-kommonen C, Kempson SE, Fletcher TM (1995) The inverted repeat regions of the simian varicella virus and varicella-zoster virus genomes have a similar genetic organization. Virus Res 39:181–193

Gray WL, Williams RJ, Soike KF (1998) Rapid diagnosis of simian varicella using the polymerase chain reaction. Lab Anim Sci 48:45–49

Gray WL, Mullis LB, Soike KF (2001a) Expression of the simian varicella virus glycoprotein E. Virus Res 79:27–37

Gray WL, Starnes B, White MW, Mahalingam R (2001b) The DNA sequence of the simian varicella virus genome. Virology 284:123–130

Gray WL, Mullis LB, Soike KF (2002) Viral gene expression during acute simian varicella virus infection. J Gen Virol 83:841–846

Gray WL, Davis K, Ou Y, Ashburn C, Ward TM (2007) Simian varicella virus gene 61 encodes a viral transactivator but is non-essential for *in vitro* replication. Arch Virol 152:553–563

Kemble GW, Annunziato PW, Lungu O, Winter RE, Cha T, Silverstein SJ, Spaete RR (2000) Open reading frame S/L of varicella-zoster virus encodes a cytoplasmic protein expressed in infected cells. J Virol 74:11311–11321

Mahalingam R, Gray WL (2007) The simian varicella virus genome contains an invertible 665 base pair terminal element that is absent in the varicella-zoster virus genome. Virology 366:387–393

Mahalingam R, Smith D, Wellish M, Wolf W, Dueland AN, Cohrs R, Soike K, Gilden D (1991) Simian varicella virus DNA in dorsal root ganglia. Proc Natl Acad Sci USA 88:2750–2752

Mahalingam R, Clarke P, Wellish M, Dueland AN, Soike KF, Gilden DH, Cohrs R (1992) Prevalence and distribution of latent simian varicella virus DNA in monkey ganglia. Virology 188:193–197

Mahalingam R, Wellish M, White T, Soike K, Kleinschmidt-DeMasters BK, Gilden DH (1998) Infectious simian varicella virus expressing the green fluorescent protein. J Neurovirol 4:438–444

Mahalingam R, White T, Wellish M, Gilden D, Gray WL (2000) Sequence analysis of the leftward end of simian varicella virus (EcoRI- I fragment) reveals the presence of an 8-bp repeat flanking the unique long segment and an 881-bp open reading frame that is absent in the varicella-zoster virus genome. Virology 274:420–428

Mahalingam R, Gilden DH, Wellish M, Pugazhenthi S (2006) Transactivation of the simian varicella virus (SVV) open reading fram (ORF) 21 promoter by SVV ORF 62 is upregulated in neuronal cells but downregulated in non-neuronal cells by SVV ORF 63 protein. Virology 345:244–250

Oakes JE, d'Offay JM (1988) Virus diseases in laboratory and captive animals. Kluwer Academic/Plenum Publishers, Boston, pp 163–174

Ou Y, Gray WL (2006) The simian varicella virus gene 28 and 29 promoters share a common USF binding site and are induced by IE62 transactivation. J Gen Virol 87:1501–1508

Ou Y, Davis KA, Traina-Dorge V, Gray WL (2007a) Simian varicella virus expresses a latency associated transcript that is antisense to ORF 61 (ICP0) mRNA in neural ganglia of latency infected monkeys. J Virol 81:8149–8156

Ou Y, Traina-Dorge V, Davis KA, Gray WL (2007b) Recombinant simian varicella vaccines induce immune responses to simian immunodeficiency virus (SIV) antigens in imunized vervet monkeys. Virology 364:291–300

Perera LP, Kaushal S, Kinchington PR, Mosca JD, Hayward GS, Straus SE (1994) Varicella-zoster virus open reading frame 4 encodes a transcriptional activator that is functionally distinct from that of herpes simplex virus homolog ICP27. J Virol 68:2468–2477

Pumphrey CY, Gray WL (1992) The genomes of simian varicella virus and varicella zoster virus are colinear. Virus Res 26:255–266

Pumphrey CY, Gray WL (1994) DNA sequence and transcriptional analysis of the simian varicella virus glycoprotein B gene. J Gen Virol 75:3219–3227

Pumphrey CY, Gray WL (1995) DNA sequence of the simian varicella virus (SVV) glycoprotein H (gH) gene and analysis of the SVV and varicella-zoster virus (VZV) gH transcripts. Virus Res 38:55–70

Pumphrey CY, Gray WL (1996) Identification and analysis of the simian varicella virus thymidine kinase gene. Arch Virol 141:43–55

Sato H, Pesnicak L, Cohen JI (2002) Varicella-zoster virus open reading frame 2 encodes a membrane phosphoprotein that is dispensible for viral replication and for establishment of latency. J Virol 76:3575–3578

Schmidt NJ (1982) Improved yields and assay of simian varicella virus, and a comparison of certain biological properties of simian and human varicella viruses. J Virol Meth 5:229–241

Soike KF (1992) Simian varicella virus infection in African and Asian monkeys. The potential for development of antivirals for animal diseases. Ann NY Acad Sci 653:323–333

Tischer BK, von Einem J, Kaufer B, Osterrieder N (2006) Two-step Red-mediated recombination for versatile high-efficiency markerless DNA manipulation in *Escherichia coli*. Biotechniques 40:191–196

Ward TM, Traina-Dorge V, Davis KA, Gray WL (2008) Recombinant simian varicella viruses expressing respiratory syncytial virus antigens are immunogenic. J Gen Virol 89:741–740

Ward TM, Williams MV, Traina-Dorge V, Gray WL (2009) The simian varicella virus uracil glycosylase and dUTPase genes are expressed *in vivo*, but are non-essential for replication in cell culture. Virus Res 142:78–84

White TM, Gilden DH, Mahalingam R (2001) An animal model of varicella virus infection. Brain Pathol 11:475–479

White TM, Mahalingam R, Traina-Dorge V, Gilden DH (2002a) Persistence of simian varicella virus DNA in CD4+ and CD8+ blood mononuclear cells for years after intratracheal inoculation of African green monkeys. Virology 303:192–198

White TM, Mahalingam R, Traina-Dorge V, Gilden DH (2002b) Simian varicella virus DNA is present and transcribed months after experimental infection of adult African green monkeys. J Neurovirol 8:191–205

Simian Varicella Virus Pathogenesis

Ravi Mahalingam, Ilhem Messaoudi, and Don Gilden

Contents

Abstract Because varicella zoster virus (VZV) is an exclusively human pathogen, the development of an animal model is necessary to study pathogenesis, latency, and reactivation. The pathological, virological, and immunological features of

R. Mahalingam (✉)
Departments of Neurology, University of Colorado Denver School of Medicine, 12700 E. 19th Avenue, Mailstop B182, Aurora, CO 80045, USA
e-mail: ravi.mahalingam@ucdenver.edu

I. Messaoudi
Vaccine and Gene Therapy and Division of Pathobiology and Immunology, Oregon National Primate Research Center, Oregon Health and Science, 505 NW 185th Avenue, Beaverton, OR 97006, USA

D. Gilden
Departments of Neurology, University of Colorado Denver School of Medicine, 12700 E. 19th Avenue, Mailstop B182, Aurora, CO 80045, USA
Departments of Microbiology, University of Colorado Denver School of Medicine, 12700 E. 19th Avenue, Mailstop B182, Aurora, CO 80045, USA

A.M. Arvin et al. (eds.), *Varicella-zoster Virus*,
Current Topics in Microbiology and Immunology 342, DOI 10.1007/82_2009_6

simian varicella virus (SVV) infection in nonhuman primates are similar to those of VZV infection in humans. Both natural infection of cynomolgus and African green monkeys as well as intrabronchial inoculation of rhesus macaques with SVV provide the most useful models to study viral and immunological aspects of latency and the host immune response. Experimental immunosuppression of monkeys latently infected with SVV results in zoster, thus providing a new model system to study how the loss of adaptive immunity modulates virus reactivation.

Abbreviations

CMI	Cell-mediated immunity
MNCs	Mononuclear cells
ORF	Open reading frame
SVV	Simian varicella virus
VZV	Varicella zoster virus

1 Introduction

During primary infection, varicella zoster virus (VZV) causes chickenpox in children, becomes latent in cranial nerve ganglia, dorsal root ganglia, and autonomic ganglia along the entire neuraxis and reactivates decades later to produce zoster. With advancing age, a natural decline in cell-mediated immunity (CMI) to VZV causes reactivation of latent VZV resulting in zoster and postherpetic neuralgia, as well as stroke from uni- or multifocal vasculopathy, myelitis, zoster paresis, and even pain without rash (zoster sine herpete). All of these neurologic complications of zoster are increased in the rapidly expanding aging and immunocompromised population, especially in AIDS patients. Thus, a better understanding of VZV reactivation is essential.

VZV causes disease only in humans. Development of an experimental animal model that recapitulates the pathogenesis of VZV seen in humans has been a goal of several laboratories. Important criteria for any animal model of VZV latency include: (1) presence of virus nucleic acids in ganglia, but not in non-ganglionic tissues; (2) presence of virus exclusively in neurons; (3) limited transcription of virus genes; and (4) ability to reactivate the virus.

2 Animal Models of VZV Infection

There have been several attempts to generate an animal model of VZV infection (Table 1). Subcutaneous inoculation of VZV into the breast of a chimpanzee produced a mild rash near the site of infection with mild fever, but latency was

Table 1 Results of attempts to produce VZV-induced disease in animals

Species	Route of inoculation	Sero-conversion	Rash (dpi)	Viremia	Time of sacrifice	Detection of SVV nucleic acids in tissue: ganglionic: hyb[1]	ganglionic: PCR	ganglionic: neuronal	ganglionic: Non-neuronal	lung/liver	References
chimpanzee	subcutaneous	+	10	+	nd	nd	nd	nd	nd	nd	Cohen et al. 1996
guinea pig	intranasal/ corneal	+	4	nd	nd	nd	nd	nd	nd	nd	Matsunaga et al. 1982
guinea pig	intramuscular	+	9–25	+	23 d	+[2]	nd	nd	nd	+	Myers et al. 1985
guinea pig	intratracheal	+	–	nd	60 d	nd	nd	nd	nd	nd	Walz-Cicconi et al. 1986
guinea pig	occular	nd[3]	–	nd	21 d	nd	nd	nd	nd	nd	Pavan-Langstan and Dunkel 1989
guinea pig	intramuscular	+	4–7	+	nd	na[4]	na	na	na	na	Myers et al. 1991
guinea pig	subcutaneous	nd	+[5]	+	80 d	+[2]	+	nd	nd	nd	Lowry et al. 1993
guinea pig	corneal	nd	nr[6]	nd	35 d	+[7]		+	–	nd	Tenset and Hyman 1987
rat	subcutaneous	+	none	nd	9 m	+[7]	nd	+	nd	nd	Sadzot-Delvaux et al. 1990
rat	subcutaneous	nd	nr	nd	30 d	nd	nd	+	+	nd	Debrus et al. 1995
rat	subcutaneous	nd	nr	nd	18 m	+[7]	+	+	+	nd	Kennedy et al. 2001
rat	subcutaneous	+	none	–	1–3 m	+[7]		+	+	nd	Annunziato et al. 1998
rat	intramuscular	nd	nr	nd	1 m		+	nd	nd	nd	Sato et al. 2003
mouse	subcutaneous	+	nr	nd	4–9 d	nd	nd	nd	nd	nd	Wroblewska et al. 1982
mouse	corneal	+	none	nd	33 d	+[7]	+	+	+	nd	Wrobliwska et al. 1993
SCID-hu-mouse	skin implants	na	na	nd	nd	nd	nd	nd	nd	nd	Moffat et al. 1995
SCID-hu-mouse	ganglia implants	na	na	na	14 d	nd	+	+	–	nd	Zerboni et al. 2005

[1]hybridization; [2]dot biot; [3]not done; [4]not applicable; [5]dpi not reported; [6]not reported; [7]insitu

not studied in this model (Cohen et al. 1996). Seroconversion in the absence of clinical signs was demonstrated by experimental inoculation of VZV into rabbits, mice, and rats (Myers et al. 1980, 1985; Matsunaga et al. 1982; Wroblewska et al. 1982; Walz-Cicconi et al. 1986). Viral DNA was found in both ganglionic neurons and nonneuronal cells, as well as in non-ganglionic tissues 1 month after corneal inoculation of VZV in mice (Wroblewska et al. 1993). A papular exanthem without vesicles was seen in guinea pigs inoculated intramuscularly with VZV (Myers et al. 1991). VZV DNA was detected by PCR in ganglia of guinea pigs 80 days after subcutaneous inoculation (Lowry et al. 1993). Although VZV RNA was detected in ganglia of guinea pigs by in situ hybridization at 5 weeks after ocular inoculation (Tenser and Hyman 1987), the absence of viral nucleic acids in non-ganglionic tissues has not been confirmed. Although no clinical signs developed, virus nucleic acids and proteins were detected in dissociated rat ganglionic neurons up to 9 months after experimental infection; however, although ganglia were cultured for 3–12 days, *in vitro* reactivation could not be excluded (Sadzot-Delvaux et al. 1990). Latent VZV infection was reported in rats inoculated via footpad and sacrificed 1 month later (Debrus et al. 1995; Kennedy et al. 2001), but the validity of the rat model is debatable since VZV DNA was not detected in ganglia 1–3 months after footpad-inoculation (Annunziato et al. 1998). Importantly, reactivation of latent VZV has not been demonstrated in any rodent species (Chen et al. 2003; Sadzot-Delvaux et al. 1990).

A SCID-humanized (SCID-hu) mouse model was developed in which human thymus and liver implants were introduced under the kidney capsule followed by subcutaneous introduction of skin implants from the same donor (Ku et al. 2005). Direct inoculation of VZV into the skin implants results in virus infection, as evidenced by the detection of virus proteins for 3 weeks after infection in $CD4^+$ and $CD8^+$ T cells of these mice (Moffat et al. 1995). These skin implants also become infected upon intravenous inoculation with VZV-infected human T cells (Ku et al. 2004). Because of the partially immunodeficient status of the mice, it is difficult to use this model to study the immune response of the host to VZV infection. Similar studies using human ganglionic implants have been used to demonstrate virus infection in SCID-hu mice (Zerboni et al. 2005).

3 Simian Varicella Virus

Simian varicella virus (SVV, *Cercopithecine herpesvirus 7*) causes chickenpox in nonhuman primates. SVV has been isolated during natural outbreaks in African green and Patas monkeys (Soike et al. 1984a). Clinical, pathological, immunological, and virological features of SVV infection in monkeys resemble those of human VZV infection (Wenner et al. 1977; Felsenfeld and Schmidt 1977, 1979; Padovan and Cantrell 1986; Myers and Connelly 1992; Dueland et al. 1992).

3.1 Clinical and Pathological Features

During primary SVV infection, infectious virus can be recovered at the peak of viremia from blood mononuclear cells (MNCs; Clarkson et al. 1967; Wolf et al. 1974; Soike et al. 1984a). SVV-induced rash is often hemorrhagic and disseminated (Soike 1992). Like disseminated VZV in immunosuppressed patients, lung and liver are the most severely affected organs (Roberts et al. 1984). Histological examination of skin and viscera reveals foci of hemorrhagic necrosis, inflammation, and eosinophilic intranuclear inclusions (Clarkson et al. 1967; Wolf et al. 1974). Analysis of DNA extracted from multiple tissues 5 to 60 days post-infection (dpi) from intravenously inoculated monkeys demonstrated the time course and route of virus spread into sensory ganglia (Mahalingam et al. 2001). SVV DNA was detected in ganglia 6 dpi, before the appearance of rash. Intravenous inoculation produced more SVV DNA-positive ganglia (63%) than after intratracheal inoculation (13%), supporting the notion that like other organs, ganglia become infected by hematogenous spread of virus (Mahalingam et al. 2001).

3.2 Immunological Features

Like VZV infection in humans, recovery from SVV infection in monkeys correlates with both humoral and cellular immune responses. The antibody response to SVV has been investigated in several nonhuman primate species, including Patas, African green monkeys, and rhesus macaques. In Patas monkeys, SVV-specific IgM antibodies were detected using a high sensitivity double sandwich ELISA 5–8 days after subcutaneous inoculation, with a peak titer 10–15 dpi, a decline in titer 17–22 dpi, and eventual disappearance within 2 months (Iltis et al. 1984). IgG antibodies appeared 10–12 dpi, reaching a plateau 17–19 dpi and remained stable for at least 2 months (Iltis et al. 1984; Achilli et al. 1984). Viremia preceded detectable IgM antibody by 2 days, and the appearance of IgG coincided with the end of viremia (Iltis et al. 1984). Neutralizing antibodies develop in St. Kitts vervet monkeys 10–14 dpi and increase in titer until 21 dpi thereafter remaining at stable levels for at least 2 months (Gray et al. 1998). The antibody response to SVV is directed against several polypeptides, some of which are glycosylated (Gray et al. 1995). The exact identity of these polypeptides has not been elucidated, and the kinetics of the neutralizing antibody response remains to be determined. The immune response to SVV plays a critical role in protection against disease, as evidenced by resistance to reinfection and viremia after SVV challenge in animals that recover from SVV infection (Gray et al. 1995; Messaoudi et al. 2009).

3.3 Virological Features

SVV and VZV have similar size, structure, and genomic organization, with an estimated 70–75% DNA homology (Gray et al. 1992, 2001; Pumphrey and

Gray 1992). The two viruses encode polypeptides that are antigenically related. Immunization of monkeys with VZV protects monkeys from SVV infection (Felsenfeld and Schmidt 1979). In serum neutralization and complement fixation tests, SVV-specific antibodies cross-react with human VZV (Felsenfeld and Schmidt 1979; Soike et al. 1987; Fletcher and Gray 1992).

3.4 Features of Latency and Reactivation

LIke VZV infection in humans, SVV becomes latent in ganglionic neurons at multiple levels of the neuraxis (Mahalingam et al. 2002; Kennedy et al. 2004) after primary infection. The pattern of SVV transcription in latently infected monkey ganglia is similar to that of VZV transcription in latently infected human ganglia, with minor differences such as the presence of sense and antisense SVV ORF 61 transcripts in monkey ganglia but their absence in human ganglia during latency (Messaoudi et al. 2009).

Latent SVV reactivates in naturally infected monkeys exposed to social and environmental stress (Soike et al. 1984b). Epizootic SVV infections have been associated with transportation of monkeys or introduction of new monkeys into an existing colony (Clarkson et al. 1967; McCarthy et al. 1968; Soike 1992). Between 1966 and 1989, several SVV outbreaks were observed in primate centers in the USA and UK, most of which were attributed to reactivation of latent SVV (Gray 2004). Unlike VZV reactivation in humans, zoster in primates often appears as a whole-body rash lasting less than 1 week, albeit obscured by fur. Although neither VZV nor SVV can be isolated from blood in otherwise healthy immunocompetent seropositive humans or primates, SVV has been isolated from skin vesicles after reactivation (Soike et al. 1984a) and also during primary infection and reactivation in the same monkey (Gray and Gusick 1996).

4 Experimental SVV Infection

Three models of experimental SVV infection have been generated. The first is intratracheal inoculation of SVV into African green and cynomolgus monkeys, which results in persisting viremia for months to years (White et al. 2002 a, b). The second involves simulated natural infection, which produces latency and which has been used to demonstrate experimental reactivation (Mahalingam et al. 2002). The last model is intrabronchial inoculation of SVV into rhesus macaques, which provides a novel model to analyze viral and immunological mechanisms of varicella latency (Messaoudi et al. 2009). The latter two models are the best-suited for future studies involving SVV latency and reactivation. Each one of the three models is discussed in detail below.

4.1 Model 1: Persistent Viremia

Experimental intratracheal inoculation of 10^4 pfu of SVV in African green and Cynomolgus monkeys produces vesicular skin rash 7–10 dpi. Viremia is detectable 3 dpi, peaks 5 dpi, and disappears by 11 dpi, indicating hematogenous spread of virus. At the peak of rash, SVV produces hepatitis and pneumonia. Resolution of rash correlates with the detection of virus-specific antibody by 12 dpi (Wenner et al. 1977; Iltis et al. 1982; Soike et al. 1984a; Dueland et al. 1992; Gray et al. 1998; Gray 2003). During acute infection, virus antigens and nucleic acids have been found in multiple organs, including lung, liver, spleen, adrenal gland, kidney, lymph node, bone marrow, and in ganglia at all levels of the neuraxis (Wenner et al. 1977; Roberts et al. 1984; Padovan and Cantrell 1986; Dueland et al. 1992; Gray et al. 2002). In African green monkeys, serum-neutralizing antibodies are commonly detected by 14 dpi and reach peak levels 21 dpi (Soike et al. 1984b; Gray et al. 1995). Levels of neutralizing antibodies remain stable for at least 4 months after SVV infection.

SVV DNA continues to persist for months to years in several tissues, including ganglia, liver, and blood MNCs in monkeys inoculated intratracheally with SVV (White et al. 2002a). Multiple regions of SVV DNA are found in blood MNCs from SVV-infected monkeys 7 dpi and 10 months post-infection (mpi) (White et al. 2002b). This is best explained by infection of MNCs that traffic through infected tissue in which SVV DNA persists. Virus could not be recovered 14 mpi, from monkey kidney cells that had been co-cultivated with blood MNCs, even after multiple sub-cultivations (White et al. 2002b). At 14 mpi, SVV DNA was found in CD4+ and CD8+ cells, but not in CD14+ or CD20+ cells (White et al. 2002b). Earlier studies in VZV demonstrated that during chickenpox, virus DNA and antigens are present in human T and B cells, monocytes, and macrophages, but infectious VZV was recovered from macrophages and T lymphocytes (Arbeit et al. 1982; Koropchak et al. 1989; Soong et al. 2000). Multiple attempts to infect MNCs *in vitro* with VZV suggest that these cells are only semi-permissive to VZV (Arbeit et al. 1982; Gilden et al. 1987; Koropchak et al. 1989; Soong et al. 2000; Zerboni et al. 2000). In this model, Grinfeld and Kennedy (2007) showed that SVV DNA was present in both neuronal as well as nonneuronal satellite cells 9–10 mpi and only in neurons 2 years post-infection. Transcripts specific for immediate-early, early, and late SVV genes are present in lung, liver, and ganglia (White et al. 2002a). Ou et al. (2007) reported the detection of transcripts specific for SVV ORFs 21, 29, and 63, as well as both sense and antisense RNA with respect to SVV ORF 61, in ganglia of vervet monkeys that were inoculated intratracheally multiple times with SVV.

4.2 Model 2: Latency

Intrabronchial inoculation of rhesus macaques with SVV produced rash, viremia, and both humoral and cell-mediated immune responses to SVV in all monkeys

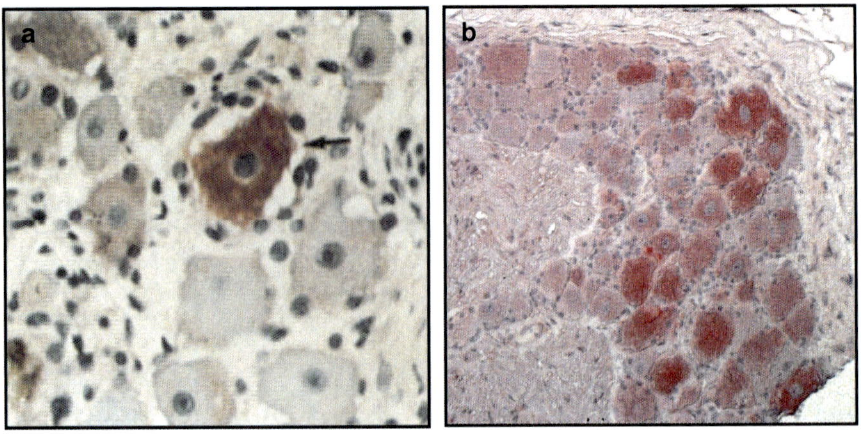

Fig. 1 Detection of VZV and SVV ORF 63 proteins in the cytoplasm of neurons in ganglia of a human latently infected with VZV and a rhesus macaque latently infected with SVV. Paraformaldehyde-fixed, paraffin-embedded sections of thoracic ganglia from a VZV seropositive 46-year-old man (**a**) (Mahalingam et al. 1996) and from a rhesus macaque latently infected with SVV (**b**) (Messaoudi et al. 2009) were analyzed by immunohistochemistry using rabbit anti-VZV ORF 63. Both VZV and SVV ORF 63 proteins are located exclusively in the cytoplasm of neurons in the respective ganglia. The *arrows* indicate the location of ORF 63 protein in the neuronal cytoplasm. Figure 1a reprinted with permission of National Academy of Sciences, USA (Mahalingam et al. (1996); Copyright 1996 National Academy of Sciences, USA); and Fig. 2b PLoS pathogens (Messaoudi et al. 2009)

(Messaoudi et al. 2009). SVV-specific IgG were first detected 10–12 days after intrabronchial inoculation, and the titers peaked 7 days later. Analysis of the kinetics and magnitude of the T cell response was based on measuring changes in T cell proliferation, IFNγ/TNFα secretion, and granzyme B (a marker for T cells with cytotoxic potential) production in peripheral blood as well as bronchial alveolar lavage (BAL) and revealed a proliferative burst of T cells in both peripheral blood and BAL first detected 7 dpi that peaked 14 dpi. Similarly, CD4+ T cells that produce IFNγ/TNFα in response to SVV viral lysate stimulation ex vivo were first detected 7 dpi as determined by intracellular cytokine staining. The frequency of SVV-specific CD4+ T cells peaked 14 dpi, after which it decreased and remained stable up to 73 dpi. Prevalence of granzyme B-expressing T cells among effector memory CD4+ and CD8+ T cells increased 7 dpi and remained high until 28 dpi before returning to pre-infection frequencies, indicating that SVV infection elicits a T cell response with cytotoxic potential.

Months after resolution of varicella, SVV DNA was detected only in ganglia but not in lung or liver. Furthermore, like VZV, SVV displayed limited transcriptional activity. We detected transcripts corresponding to SVV ORFs 21, 62, 63, and 66 but not 40 (a late gene) in latently infected ganglia. However, unlike in VZV latency, we detected sense as well as the much more abundant antisense transcripts specific to SVV ORF 61 (Messaoudi et al. 2009) consistent with the observations of Ou et al. (2007) in vervet monkeys inoculated intratracheally multiple times with SVV. Detection of sense or antisense SVV ORF 61 transcript monkey ganglia warrants more detailed analysis of latently infected human ganglia for VZV ORF 61 transcripts. Finally, like VZV latency in human ganglia, SVV ORF 63 protein was detected exclusively in the cytoplasm of monkey ganglia latently infected with SVV (Fig. 1). Overall, intratracheal inoculation of SVV into rhesus macaques results in the establishment of latency (Messaoudi et al. 2009).

4.3 Model 3: Latency and Reactivation

We developed a model of latent SVV infection in both African green and cynomolgus monkeys by exposing seronegative monkeys to others that had been inoculated intratracheally with SVV (Mahalingam et al. 2002). Naturally exposed monkeys develop varicella 10–12 days after exposure. SVV infection was confirmed by DNA PCR analysis of skin scrapings during varicella. SVV DNA was occasionally present in blood MNCs, and SVV DNA was detected 6–8 weeks after resolution of rash in multiple ganglia along the entire neuraxis, but not in lung or liver, indicating the establishment of latency. Latent SVV was localized exclusively in ganglionic neurons (Kennedy et al. 2004). While natural infection resulted in latency in sensory ganglia, seroconversion was random, and viremia was not found in most animals.

Although spontaneous reactivation of SVV has been observed in naturally infected monkeys in primate centers around the world (Treuting et al. 1998; Gray 2004), experimental reactivation has not been attempted until recently. We observed subclinical reactivation of latent SVV in an irradiated rhesus macaque that resulted in disseminated varicella in another monkey housed in the same colony (Kolappaswamy et al. 2007). In addition, papulovesicular dermatitis caused by reactivation of latent SVV in immunosuppressed rhesus macaques has been reported (Schoeb et al. 2008).

Based on the spontaneous development of zoster in AIDS patients as well as organ transplant recipients and cancer patients treated with X-irradiation, immunosuppressive drugs, and steroids (Dolin et al. 1978), we used the natural infection paradigm to establish latent SVV infection in four cynomolgus monkeys, and 3 months after varicella, the monkeys were irradiated and treated with tacrolimus and prednisone. Of the four latently infected monkeys that were immunosuppressed and subjected to the stress of transportation and isolation, one developed zoster and three developed subclinical reactivation in the absence of rash. A non-immunosuppressed latently infected monkey subjected to the same stress showed features of subclinical reactivation. SVV reactivation was confirmed not only by the occurrence of zoster in one monkey, but also by the presence of SVV RNA specific to late capsid proteins (ORFs 40 and 9) in ganglia, the presence of SVV DNA in non-ganglionic tissue, and the detection of SVV antigens in skin, ganglia (including axons), and lung (Mahalingam et al. 2007). Recently, we found that

SVV reactivates in monkeys treated with tacrolimus with or without exposure to irradiation (unpublished observations).

5 Conclusions

Intrabronchial inoculation of rhesus macaques and simulated natural infection of African green and cynomolgus monkeys are best-suited to study SVV latency and reactivation. Although zoster in humans is directly related to a decline in CMI to VZV, virus-specific T cells are not seen in ganglia harboring latent VZV or SVV (Verjans et al. 2007; Messaoudi et al. unpublished observations). Experimental reactivation of SVV in primates can be used to examine the role of T cells in the maintenance of latent infection in ganglia. While VZV downregulates MHC I surface expression and its retention in golgi bodies (Cohen 1998; Abendroth et al. 2001; Eisfeld et al. 2007), varicella latency may also be regulated by an innate immune response involving cytokines or chemokines.

SVV infection in the monkeys will, for the first time, allow analysis of the cascade of cellular, immune, and viral factors involved in reactivation. This is important since reactivation in elderly and immunocompromised individuals can produce serious, often chronic, and sometimes fatal neurological disease. Furthermore, an SVV reactivation model will be useful in testing preventive vaccines, antiviral drugs, SVV strain differences, age factors, as well as other variables.

Acknowledgments This work was supported in part by Public Health Service grants AG006127, NS032623 and AG032958 from the National Institutes of Health. The authors thank Marina Hoffman for editorial review and Cathy Allen for preparing the manuscript.

References

Abendroth A, Lin I, Slobedman B et al (2001) Varicella-zoster virus retains major histocompatibility complex class I proteins in the Golgi compartment of infected cells. J Virol 75:4878–4888

Achilli G, Sarasini A, Gerna G et al (1984) Antibody response of patas monkeys to experimental infection with Delta herpesvirus. Eur J Clin Microbiol 3:158–159

Annunziato P, LaRussa P, Lee P et al (1998) Evidence of latent varicella-zoster virus in rat dorsal root ganglia. J Infect Dis 178(Suppl 1):S48–S51

Arbeit RD, Zaia JA, Valerio MA et al (1982) Infection of human peripheral blood mononuclear cells by varicella-zoster virus. Intervirology 18:56–65

Chen JJ, Gershon AA, Li ZS et al (2003) Latent and lytic infection of isolated guinea pig enteric ganglia by varicella zoster virus. J Med Virol 70(Suppl 1):S71–S78

Clarkson MJ, Thorpe E, McCarthy K (1967) A virus disease of captive vervet monkeys (Cercopithecus aethiops) caused by a new herpesvirus. Arch Gesamte Virusforsch 22:219–234

Cohen JI (1998) Infection of cells with varicella-zoster virus down-regulates surface expression of class I major histocompatibility complex antigens. J Infect Dis 177:1390–1393

Cohen JI, Moskal T, Shapiro M et al (1996) Varicella in chimpanzees. J Med Virol 50:289–292

Debrus S, Sadzot-Delvaux C, Nikkels AF et al (1995) Varicella-zoster virus gene 63 encodes an immediate-early protein that is abundantly expressed during latency. J Virol 69:3240–3245

Dolin R, Reichman RC, Mazur MH et al (1978) NIH conference. Herpes zoster-varicella infections in immunosuppressed patients. Ann Intern Med 89:375–388

Dueland AN, Martin JR, Devlin ME et al (1992) Acute simian varicella infection: clinical, laboratory, pathologic, and virologic features. Lab Invest 66:762–773

Eisfeld AJ, Yee MB, Erazo A et al (2007) Downregulation of class I major histocompatibility complex surface expression by varicella-zoster virus involves open reading frame 66 protein kinase-dependent and -independent mechanisms. J Virol 81:9034–9049

Felsenfeld AD, Schmidt NJ (1977) Antigenic relationships among several simian varicella-like viruses and varicella-zoster virus. Infect Immun 15:807–812

Felsenfeld AD, Schmidt NJ (1979) Varicella-zoster virus immunizes patas monkeys against simian varicella-like disease. J Gen Virol 42:171–178

Fletcher TM, Gray WL (1992) Simian varicella virus: characterization of virion and infected cell polypeptides and the antigenic cross-reactivity with varicella-zoster virus. J Gen Virol 73:1209–1215

Gilden DH, Hayward AR, Krupp J et al (1987) Varicella-zoster virus infection of human mononuclear cells. Virus Res 7:117–129

Gray WL (2003) Pathogenesis of simian varicella virus. J Med Virol 70(Suppl 1):S4–S8

Gray WL (2004) Simian varicella: a model for human varicella-zoster virus infections. Rev Med Virol 14:363–381

Gray WL, Gusick NJ (1996) Viral isolates derived from simian varicella epizootics are genetically related but are distinct from other primate herpesviruses. Virology 224:161–166

Gray WL, Pumphrey CY, Ruyechan WT et al (1992) The simian varicella virus and varicella zoster virus genomes are similar in size and structure. Virology 186:562–572

Gray WL, Gusick NJ, Fletcher TM et al (1995) Simian varicella virus antibody response in experimental infection of African green monkeys. J Med Primatol 24:246–251

Gray WL, Williams RJ, Chang R et al (1998) Experimental simian varicella virus infection of St. Kitts vervet monkeys. J Med Primatol 27:177–183

Gray WL, Starnes B, White MW et al (2001) The DNA sequence of the simian varicella virus genome. Virology 284:123–130

Gray WL, Mullis L, Soike KF (2002) Viral gene expression during acute simian varicella virus infection. J Gen Virol 83:841–846

Grinfeld E, Kennedy PG (2007) The pattern of viral persistence in monkeys intra-tracheally infected with simian varicella virus. Virus Genes 35:289–292

Iltis JP, Arrons MC, Castellano GA et al (1982) Simian varicella virus (delta herpesvirus) infection of patas monkeys leading to pneumonia and encephalitis. Proc Soc Exp Biol Med 169:266–279

Iltis JP, Achilli G, Madden DL et al (1984) Serologic study by enzyme-linked immunosorbent assay of the IgM antibody response in the patas monkey following experimental simian varicella virus infection. Diagn Immunol 2:137–142

Kennedy PG, Grinfeld E, Bontems S et al (2001) Varicella-zoster virus gene expression in latently infected rat dorsal root ganglia. Virology 289:218–223

Kennedy PGE, Grinfeld E, Traina-Dorge V et al (2004) Neuronal localization of simian varicella virus DNA in ganglia of naturally infected African green monkeys. Virus Genes 28:273–276

Kolappaswamy K, Mahalingam R, Traina-Dorge V et al (2007) Disseminated simian varicella virus infection in an irradiated rhesus macaque (Macaca mulatta). J Virol 81:411–415

Koropchak CM, Solem SM, Diaz PS et al (1989) Investigation of varicella-zoster virus infection of lymphocytes by in situ hybridization. J Virol 63:2392–2395

Ku CC, Zerboni L, Ito H et al (2004) Varicella-zoster virus transfer to skin by T cells and modulation of viral replication by epidermal cell interferon-alpha. J Exp Med 200:917–925

Ku CC, Besser J, Abendroth A et al (2005) Varicella-zoster virus pathogenesis and immunobiology: new concepts emerging from investigations with the SCIDhu mouse model. J Virol 79:2651–2658

Lowry PW, Sabella C, Koropchak CM et al (1993) Investigation of the pathogenesis of varicella-zoster virus infection in guinea pigs by using polymerase chain reaction. J Infect Dis 167:78–83

Mahalingam R, Wellish M, Cohrs R et al (1996) Expression of protein encoded by varicella-zoster virus open reading frame 63 in latently infected human ganglionic neurons. Proc Natl Acad Sci USA 93:2122–2124

Mahalingam R, Wellish M, Soike K (2001) Simian varicella virus infects ganglia before rash in experimentally infected monkeys. Virology 279:339–342

Mahalingam R, Traina-Dorge V, Wellish M et al (2002) Naturally acquired simian varicella virus infection in African green monkeys. J Virol 76:8548–8550

Matsunaga Y, Yamanishi K, Takahashi M (1982) Experimental infection and immune response of guinea pigs with varicella-zoster virus. Infect Immun 37:407–412

McCarthy K, Thorpe E, Laursen AC et al (1968) Exanthematous disease in patas monkeys caused by a herpes virus. Lancet 2:856–857

Messaoudi I, Barron A, Wellish M et al (2009) Simian varicella virus infection of rhesus macaques recapitulates essential features of varicella zoster virus infection in humans. PLoS Pathog 5(11):1–14

Moffat JF, Stein MD, Kaneshima H et al (1995) Tropism of varicella-zoster virus for human CD4+ and CD8+ T lymphocytes and epidermal cells in SCID-hu mice. J Virol 69:5236–5242

Myers MG, Connelly BL (1992) Animal models of varicella. J Infect Dis 166(Suppl 1):S48–S50

Myers MG, Duer HL, Hausler CK (1980) Experimental infection of guinea pigs with varicella-zoster virus. J Infect Dis 142:414–420

Myers MG, Stanberry LR, Edmond BJ (1985) Varicella-zoster virus infection of strain 2 guinea pigs. J Infect Dis 151:106–113

Myers MG, Connelly BL, Stanberry LR (1991) Varicella in hairless guinea pigs. J Infect Dis 163:746–751

Ou Y, Davis KA, Traina-Dorge V et al (2007) Simian varicella virus expresses a latency-associated transcript that is antisense to open reading frame 61 (ICP0) mRNA in neural ganglia of latently infected monkeys. J Virol 81:8149–8156

Pavan-Langston D, Dunkel EC (1989) Ocular varicella-zoster virus infection in the guinea pig: a new *in vivo* model. Arch Ophthalmol 107:1058–1072

Padovan D, Cantrell CA (1986) Varicella-like herpesvirus infections of nonhuman primates. Lab Anim Sci 36:7–13

Pumphrey CY, Gray WL (1992) The genomes of simian varicella virus and varicella zoster virus are colinear. Virus Res 26:255–266

Roberts ED, Baskin GB, Soike K et al (1984) Pathologic changes of experimental simian varicella (Delta herpesvirus) infection in African green monkeys (Cercopithecus aethiops). Am J Vet Res 45:523–530

Sadzot-Delvaux C, Merville-Louis MP, Delree P et al (1990) An *in vivo* model of varicella-zoster virus latent infection of dorsal root ganglia. J Neurosci Res 26:83–89

Sato H, Pesnicak L, Cohen JI (2003) Use of a rodent model to show that varicella-zoster virus ORF61 is dispensable for establishment of latency. J Med Virol 70(Suppl 1):S79–S81

Schoeb TR, Eberle R, Black DH et al (2008) Diagnostic exercise: papulovesicular dermatitis in rhesus macaques (Macaca mulatta). Vet Pathol 45:592–594

Soike KF (1992) Simian varicella virus infection in African and Asian monkeys. The potential for development of antivirals for animal diseases. Ann NY Acad Sci 653:323–333

Soike KF, Baskin G, Cantrell C et al (1984a) Investigation of antiviral activity of 1-beta-D-arabinofuranosylthymine (ara-T) and 1-beta-D-arabinofuranosyl-E-5-(2-bromovinyl)uracil (BV-ara-U) in monkeys infected with simian varicella virus. Antiviral Res 4:245–257

Soike KF, Rangan SR, Gerone PJ (1984b) Viral disease models in primates. Adv Vet Sci Comp Med 28:151–199

Soike KF, Keller PM, Ellis RW (1987) Immunization of monkeys with varicella-zoster virus glycoprotein antigens and their response to challenge with simian varicella virus. J Med Virol 22:307–313

Soong W, Schultz JC, Patera AC et al (2000) Infection of human T lymphocytes with varicella-zoster virus: an analysis with viral mutants and clinical isolates. J Virol 74:1864–1870

Tenser RB, Hyman RW (1987) Latent herpesvirus infections of neurons in guinea pigs and humans. Yale J Biol Med 60:159–167

Treuting PM, Johnson-Delaney C, Birkebak TA (1998) Diagnostic exercise: vesicular epidermal rash, mucosal ulcerations, and hepatic necrosis in a Cynomolgus monkey (Macaca fascicularis). Lab Anim Sci 48:384–386

Verjans GM, Hintzen RQ, van Dun JM et al (2007) Selective retention of herpes simplex virus-specific T cells in latently infected human trigeminal ganglia. Proc Natl Acad Sci USA 104:3496–3501

Walz-Cicconi MA, Rose RM, Dammin GJ et al (1986) Inoculation of guinea pigs with varicella-zoster virus via the respiratory route. Arch Virol 88:265–277

Wenner HA, Abel D, Barrick S et al (1977) Clinical and pathogenetic studies of Medical Lake macaque virus infections in Cynomolgus monkeys (simian varicella). J Infect Dis 135:611–622

White TM, Mahalingam R, Traina-Dorge V et al (2002a) Simian varicella virus DNA is present and transcribed months after experimental infection of adult African green monkeys. J Neurovirol 8:191–203

White TM, Mahalingam R, Traina-Dorge V et al (2002b) Persistence of simian varicella virus DNA in CD4(+) and CD8(+) blood mononuclear cells for years after intratracheal inoculation of African green monkeys. Virology 303:192–198

Wolf RH, Smetana HF, Allen WP et al (1974) Pathology and clinical history of Delta herpesvirus infection in patas monkeys. Lab Anim Sci 24:218–221

Wroblewska Z, Devlin M, Reilly K et al (1982) The production of varicella zoster virus antiserum in laboratory animals. Arch Virol 74:233–238

Wroblewska Z, Valyi-Nagy T, Otte J et al (1993) A mouse model for varicella-zoster virus latency. Microb Pathog 15:141–151

Zerboni L, Sommer M, Ware CF et al (2000) Varicella-zoster virus infection of a human CD4-positive T-cell line. Virology 270:278–285

Zerboni K, Ku CC, Jones CD et al (2005) Varicella-zoster virus infection of human dorsal root ganglia in vivo. Proc Natl Acad Sci USA 102:6490–6495

Varicella-Zoster Virus Vaccine: Molecular Genetics

D. Scott Schmid

Contents

Abstract The genetic differences that potentially account for the attenuation of the Oka vaccine VZV preparation are more clearly defined than for perhaps any other vaccine in current use. This is due in large part to the small number of differences between the vaccine and the parental strain from which it was derived, and to the high level of genomic conservation that characterizes VZV. This information has been used with great success to develop methods that discriminate vaccine from wild-type strains, to begin determining which specific vaccine markers contribute to the attenuated phenotype, to improve evaluations of vaccine efficacy and safety,

---

The findings and conclusions in this report are those of the author and do not necessarily represent the views of the funding agency.

D.S. Schmid
Herpesvirus Team and National VZV Laboratory, MMRHLB, Centers for Disease Control and Prevention, 1600 Clifton Rd/Bldg 18/Rm 6-134/ MS G-18, Atlanta, GA 30333, USA
e-mail: SSchmid@cdc.gov

A.M. Arvin et al. (eds.), *Varicella-zoster Virus*,
Current Topics in Microbiology and Immunology 342, DOI 10.1007/82_2010_14

and to observe the behavior of the live, attenuated preparation as it becomes more prevalent through widespread immunization.

1 Introduction

1.1 Development of Varicella Vaccine

Varicella is a highly infectious disease caused by varicella-zoster virus (VZV), with characteristically high secondary attack rates among susceptible household contacts. Primary infection with VZV typically produces in lifelong immunity and, in otherwise healthy persons, subsequent re-exposure usually does not lead to symptomatic illness. As such, varicella was regarded as a practical target for prevention through immunization.

The Oka varicella vaccine was produced through the serial tissue culture passage of VZV obtained from a child with uncomplicated varicella in guinea pig cells and human diploid cells (Takahashi et al. 1974). The vaccine was also adapted to growth at 34°C. The cold-adapted, live-attenuated vaccine preparation, given intramuscularly, did not cause clinical disease in immunized children and conferred protection against varicella (Takahashi et al. 1974). A commercial preparation of this vaccine (Varivax, Merck & Co., Inc., Whitehouse Station, NJ) was licensed in the US in 1995 and recommended for routine immunization of children aged 12–18 months, susceptible adolescents, and susceptible adults at high risk of exposure (CDC 1996, 1999a).

Three commercial varicella vaccine preparations are now in use to varying degrees in countries representing all the populated continents. Although all three are derived from the same seed virus reported in 1974 (Takahashi et al. 1974), minor differences in strain content have been observed between the products (Quinlivan et al. 2004; Sauerbrei et al. 2004; Loparev et al. 2007b). Although acceptance of the vaccine appears to be increasing in developed countries, global vaccine coverage remains very limited, and only a few nations currently include varicella vaccination on their routine childhood immunization schedules.

In 2007, a vaccine preparation with roughly 14 times as much of the same attenuated virus used to prevent varicella was licensed and approved in the US for the prevention of herpes zoster.

1.2 Prevention of Varicella in the United States

Prior to routine varicella immunization, there were four million cases of varicella every year, with 100–150 deaths and 11,000–15,000 hospitalizations annually (Wharton 1996). Approximately 90% of varicella cases occurred in children under 15 years of age, with the majority developing in 5-years-old or younger children (Wharton 1996; Finger et al. 1994; Yawn et al. 2003). Severe varicella (hospitalizations and deaths) were more common among adults with varicella than among children.

The implementation of routine single-dose varicella immunization of children has led to striking reductions in varicella morbidity and mortality in the US compared with the pre-1995 levels. Single-dose vaccine coverage in the US is currently estimated at 90%, resulting in an overall reduction in varicella incidence of $>$ 80% by 2005, with corresponding reductions in hospitalizations and deaths (Guris et al. 2008). In one study, varicella incidence in adults had reduced by 74% in active surveillance sites, suggesting herd immunity (Seward et al. 2002).

In spite of dramatic reductions in the incidence of varicella through routine immunization, day care and elementary school outbreaks remain common and, in that setting, about one in five varicella vaccine recipients have been susceptible to breakthrough varicella (Lopez et al. 2006; CDC 1999a, b). While breakthrough cases are typically mild, usually afebrile and with $<$ 50 lesions, the virus is transmissible from such cases (Lopez et al. 2006). The extent to which susceptibility of immunized children reflects the immunogenicity of single-dose vaccine or primary vaccine failure remains controversial, but the observation led to the recommendation of a second dose of varicella vaccine for children in 2007 (Marin et al. 2007).

Another controversial issue arising from the use of varicella vaccine has come from modeling studies of herpes zoster (Brisson et al. 2000). Some have predicted that the reduction of varicella incidence, since fewer adult re-exposures will occur in that environment, may cause immunity to varicella to wane at an earlier age, leading to a younger mean onset age for herpes zoster. However, there are a number of considerations that make the interpretation of the model predictions difficult. First, in the US, a marked increasing trend in incidence was observed prior to the implementation of routine varicella immunization. In addition, it is not clear whether periodic asymptomatic reactivation of VZV may compensate for the lack of exogenous re-exposure, particularly among younger adults whose immune systems may be sufficiently robust to prevent symptomatic reactivation. Finally, the licensure and recommendation of herpes zoster vaccine in the US could render the issue moot, particularly if the recommendation is extended to younger adults aged $\geq$ 50. The zoster vaccine was shown to reduce the risk of zoster in persons aged $\geq$ 60 by 50%, with a two-thirds reduction in risk for severe zoster-associated pain and post-herpetic neuralgia (Oxman et al. 2005).

2 Genetic Markers in the Vaccine

2.1 Vaccine-Associated Single Nucleotide Polymorphisms

The genomic integrity of VZV is among the best-conserved human viral pathogens. VZV has a double-stranded DNA genome of approximately 125,000 bp and

	ORF6 5745	ORF9A 10900	ORF10 12779	ORF21 31732	ORF31 58595	ORF31 59287	ORF39 71252	ORF50 87306	ORF51 89734	ORF52 90535	ORF54 94167	ORF55 97748	ORF59 101089	ORF62 105310	ORF62 105544	ORF62 105705	ORF62 107136	ORF62 107252	ORF62 108838
Oka Biken	G	Y	Y	Y	R	R	Y	Y	G	R	C	R	R	R	G	C	C	C	R
Oka Merck	R	T	C	Y	R	A	Y	Y	A	R	Y	R	R	R	R	C	Y	C	A
Oka GSK lot A2*	A	T	Y	C	G	-	T	C	A	A	T	A	A	A	G	T	T	Y	A
Oka GSK lot A3	G	T	Y	C	G	-	C	T	R	R	C	A	G	G	G	C	C	C	A

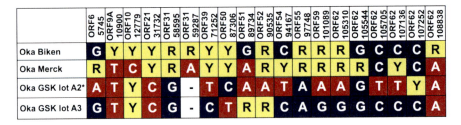

Fig. 1 Differences in the expression of vaccine-associated SNP among the three commercial preparations of Oka vaccine. *Oka GSK lot A3 and Oka GSK lot A2 were produced in 1988 and 1999, respectively. *White squares with hyphens* indicate that marker was not evaluated. *Blue boxes* indicate the vaccine-associated base, *red boxes* represent the wild-type base, *yellow boxes* represent mixed bases at that locus, *Y* pyrimidine, *R* purine

encodes about 70 genes. The gene products of ORF62, ORF63 and ORF64 located in the internal short repeat region (ISR) are duplicated in the terminal short repeat region (TSR) as ORF71, ORF69, and ORF68, respectively (Fig. 1). Currently 18 complete genome sequences are now available for wild-type VZV strains, in addition to sequences for all three commercial preparations of the Oka vaccine (vOka) and the parental Oka strain (pOka) (Loparev et al. 2009). Excluding the short direct repeat regions, R1 through R5, and the origin of replication sites (OriS), the two most diverged clades of VZV (Clade 1 and Clade 2, see VZV Molecular Epidemiology, Breuer J, this issue) display only 188 total site differences at the nucleotide level, or 99.83% identity (Peters et al. 2006). Nine wild-type Clade 1 isolates have between 99.93 and 99.97% identity at the nucleotide level when compared with the sequence of the Clade 1 strain Dumas (Peters et al. 2006). As such, the DNA genome of VZV is very highly conserved.

It was not surprising that very few differences in genome sequence were observed in the attenuated varicella vaccine compared with its parental strain. The complete genomic sequences for vOka and pOka were published in 2002 (Gomi et al. 2002). Excluding the TRS region (a duplication of the IRS), the tandem repeat regions R1 to R5, and the ORIs regions, a total of 42 single nucleotide polymorphism (SNP) were identified between the pOka and vOka sequences. Of these, 5 occurred in noncoding regions, 17 were silent mutations, and 20 caused amino acid substitutions. The vaccine-associated SNP are broadly scattered across the entire genome, occurring in 23 open reading frames; curiously, 15 of 42 SNP (36%) are localized to a single open reading frame (ORF62) and to the sequences immediately flanking that gene. This finding was unexpected, since ORF62 encodes the major transactivating protein for VZV, which is indispensible for growth and responsible for upregulating the expression of genes in every temporal class (immediate-early, early, and late) (Cohrs et al. 2006). As such, selective pressure would have been expected to work against extensive changes to ORF62. ORF62 protein produced by the strains in the vaccine remains functional, but appears to be less efficient than that produced by pOka (Gomi et al. 2000, 2001). Minor sequence variation is present between vOka and pOka in the R1 and

R4 tandem repeat elements and in OriS, and heterogeneity was seen for both pOka and vOka R3 repeat element (Gomi et al. 2002). These differences are not considered likely to contribute to the attenuated phenotype of the vaccine. The vaccine-associated SNP occurring at position 560, originally reported as noncoding, has since been shown to occur in ORF0, also known as ORF S/L; gathering evidence from several sources suggests that ORF0 encodes a protein involved in VZV virulence (Zhang et al. 2007; Kemble et al. 2000; Koshizuka et al. 2009). Recent studies indicate that the vOka SNP at position 560 eliminates a stop codon for the short glycoprotein encoded by ORF0, resulting in a longer protein with evidence of variable glycosylation (Koshizuka et al. 2009). The modified gene product produced by vOka strains appears to negatively affect viral growth.

2.2 *Insights from Vaccine Adverse Event Isolates*

Thirty-one (74%) of the vaccine-associated SNP occur in the Biken vaccine preparation as mixtures, and at least three additional SNP are present as mixtures in the Varivax vaccine preparation (Gomi et al. 2002; Loparev et al. 2007b). Thus, the vaccine comprises a mixture of probably many strains. Two groups have evaluated the vaccine-associated SNP profiles of viral DNA isolated from cases of PCR-confirmed varicella and herpes zoster (Loparev et al. 2007b; Quinlivan et al. 2004). Although a relatively small number of VZV isolates were examined in the two studies (approximately 120), the degree of concordance between the studies was remarkable (Table 1). Amplimers of segments from ORFs 1, 6, 9A, 10, 11, 14, 21, 31, 39, 50, 51, 52, 54, 55, 59, 62, and 64, together comprising almost all of the SNP observed between vaccine Oka and the pOka, were completely sequenced. Most of these markers were examined by both laboratories. Only three vOka-associated markers were universally present as an unmixed base for all of the isolated viruses at positions 106262, 107252, and 108111 (locus identifiers based on the reference sequence for Dumas strain). At positions that were analyzed by both laboratories, the frequencies at which the SNP was present in adverse event isolates as the wild-type marker were strikingly similar (Table 1). Among 27 vaccine-associated SNP evaluated by both laboratories, 13 loci were present as the wild-type base in $\geq$ 70% of the isolates examined. Nine loci were present as the wild-type marker in 100% of isolates examined by one of the laboratories. The vaccine-associated markers located at position 31732 (ORF 21) and 107797 (ORF62) were present as the wild-type base in 100% of the isolates examined by the two laboratories. The information derived from these studies should provide critical insight into which of the vaccine-associated SNP are likeliest to account for the attenuated phenotype of the vaccine. For example, among the 39 isolates evaluated in the US (Loparev et al. 2007b) and the 80 isolates examined on specimens obtained both in the UK and the US (Quinlivan et al. 2004), all carried the wild-type associated base at position 107797. The frequency of this vaccine marker in strains present in the vaccine preparation is substantially lower. The SNP at

Table 1 Frequency of vaccine-associated SNP present as the wild-type marker in clinical isolates from vaccine-attributable varicella and herpes zoster

ORF	Position[a]	% Isolates with wt marker (Loparev et al.)[b]	% Isolates with wt marker (Quinlivan et al.)
21	31732	100	100
62	107797	100	100
62/63	109200	100	95
10	12779	100	90
52	90535	88 (5)	100
62	105331	n.d.	100
62	106710	n.d.	100
64	111650	60 (10)	100
62	108838	50 (4)	100
62	107599	95	92
14	19431	n.d.	97
55	97796	95	80
62	108030	n.d.	92
9A	10900	80 (10)	90
51	89734	95	75
50	87306	92	80
62	105169	n.d.	90
62	105724	n.d.	82
31	59287	80	n.d.
62	105356	n.d.	80
62	109137	80	70
50	87815	n.d.	80
55	97748	65 (10)	80
62	105356	n.d.	80
62	109137	80	70
59	101089	40 (38)	75
31	58595	75	25
62	105310	40 (10)	70
6	5745	60 (12)	60
1	763	n.d.	60
39	71252	35 (25)	50
1	703	n.d.	50
62	107136	50 (15)	30
1	560	n.d.	30
54	94167	15 (12)	15
62	105544	25 (10)	7
62	105705	0	3
62	106262	0	0
62	107252	0	0
62	108111	0	0

[a]Positions are based on the published sequence for Dumas strain
[b]Percentages in parentheses represent the isolates that had mixed markers at that position

107797 leads to an amino acid substitution (proline in the vaccine strains; leucine in wild type strains); the wild-type leucine at this position restores an A2-restricted epitope that may play a role in the predisposition to rash illness. The locus lies in a region of IE62 that is involved in IE63 and cellular transcription factor TBP, all of which participate in the expression of glycoprotein I. Glycoprotein I, IE62, and

IE63 have all been shown to be essential for normal VZV replication in skin, suggesting that the wild-type base at 107797 may be important to viral pathogenicity. Current molecular methods should facilitate the generation of VZV strains bearing single vaccine-associated bases in a parental Oka background, thus permitting the independent assessment of individual markers for their contribution to VZV attenuation. Efforts to produce these viruses are currently underway in several laboratories.

One important difference observed between the two studies involved the clonality (or absence thereof) of skin isolates obtained from varicella vaccine adverse events. The study conducted on skin lesion isolates from both the UK and the US found that, in every instance, viruses from cases of both varicella and herpes zoster were clonal based on multiple SNP analysis (Quinlivan et al. 2004). In contrast, all 21 isolates from varicella cases and 11 of 18 isolates from herpes zoster cases were mixtures of vaccine viruses in the study of skin lesions in the US (Loparev et al. 2007b). There are several potential explanations for this disparity: (1) the isolates examined in the Quinlivan study were all viable isolates, and clonality could have arisen through selection in tissue culture. A more recent study of viable isolates obtained from a patient with DiGeorge syndrome revealed that virus recovered from skin lesions was clonal, while virus isolated from endotrachial aspirates were mixtures (Quinlivan et al. 2006); (2) The isolates studied in the Loparev study were all nonviable, collected from multiple sites. As such, it is possible that some specimens were submitted as pooled samples from multiple lesions. This cannot account for all the mixed viruses observed, however, since some were known to have been obtained from single lesions or from single scabs collected from individual crusted lesions; (3) Finally, differences in sensitivity for detecting mixed bases have been characterized among alternative sequencing techniques, and could account for some of the discrepancy in these two studies (Neve et al. 2002; Silvertsson et al. 2002; Ogino et al. 2005). In general, the issue of whether or not virus obtained from skin lesions is typically clonal vs. mixed remains unresolved and requires further study.

2.3 Composition of Commercial Varicella Vaccine Preparations

All of the marketed preparations of Oka varicella vaccine have been shown to comprise mixtures of multiple vaccine strains, and probably no individual strains in the mixture carries a complete complement of the identified vaccine-specific markers (Gomi et al. 2000; Gomi et al. 2001, 2002; Cohrs et al. 2006; Quinlivan et al. 2004; Sauerbrei et al. 2004). A few vaccine strains have been successfully isolated in culture and reduced levels of IE62 transactivation have been observed in several studies (Gomi et al. 2001, 2002; Cohrs et al. 2006), although one of these studies detected only a slight reduction in IE62 transactivation capacity of vaccine-derived IE62 compared with wild-type parental Oka IE62 (Cohrs et al. 2006). Although all commercial varicella vaccines are prepared from the same seed lot of vaccine Oka, multiple differences have been observed among these preparations in terms of vaccine-associated SNP content (Quinlivan et al. 2004; Sauerbrei et al. 2004; Loparev et al. 2007b) (Fig. 1). Lot to lot variations in the SNP profile of the same vaccine preparation have also been observed (Sauerbrei et al. 2004). This variation in SNP content likely reflects differences in the manufacturing processes used by the three companies; in addition, the limit of detection for virus mixtures using conventional sequencing is approximately 10%, so that the observed differences probably underscore relative frequencies of vaccine strains among various vaccine preparations rather than the complete absence of selected vaccine markers. This notion is supported by the observation that all the commercially available vaccine preparations provide comparable levels of protection against varicella and result in comparable rates of vaccine adverse events. However, it is noteworthy that such differences emerge among the commercial vaccine preparations and suggests that routine monitoring of varicella vaccine lots for vaccine-associated SNP content is warranted.

3 Discrimination of Oka Vaccine from Wild-Type Strains

3.1 Overview

Following the implementation of routine childhood varicella vaccination in the US and Japan, methods that reliably discriminate the vOka from wild-type VZV became critical for monitoring vaccine impact and for identifying varicella vaccine adverse events. Early approaches included the identification of point mutations in ORF38 and ORF54 and determining variation in copy number in the repeat elements in VZV, particularly the R2 repeat located in ORF14, which encodes glycoprotein C (Hondo et al. 1989; LaRussa et al. 1992; Mori et al. 1998; Hawrami and Breuer 1997; Hawrami et al. 1997). These methods employed wild-type markers that failed to distinguish wild-type strains similar to Oka; this was not problematic at the time, particularly for the US since Clade 2 strains were not in circulation. The identification of vOka-specific SNP awaited the determination of the DNA sequence of ORF62 for the vOka and the pOka in 2000 (Gomi et al. 2000). Soon thereafter, methods were published that evaluated a SNP at position 106262 in ORF62 that cleanly distinguished all vaccine preparations from wild-type strains (Argaw et al. 2000; Loparev et al. 2000a, b); subsequent surveys of vaccine adverse event isolates have revealed the selection of this ORF62 marker to be extremely fortuitous, since it is one of only three vaccine-associated SNP that has invariably been identified in vaccine strains (positions 106262, 107252, and 108111) (Quinlivan et al. 2004; Loparev et al. 2007b; Breuer and Schmid 2008). Although several refinements to these methods have been reported in the interim, most investigators have relied on the several markers to discriminate vaccine from wild-type strains of VZV.

3.2 PCR Methods for Discriminating Vaccine from Wild Type

The earliest reported methods used to distinguish vOka from wild-type VZV used restriction endonuclease fingerprinting of whole viral DNA (Hayakawa et al. 1984, 1986; Shiraki et al. 1991). By digesting viral DNA with either HpaI or EcoRI, the vaccine strain produced the unique fragments K or P, respectively. The use of this technique required the recovery of viable virus to obtain suitable preparations of DNA, and this together with the laborious nature of the protocol made it impractical for routine use in the identification of vaccine virus. Early methods incorporating PCR amplification of targeted regions in the VZV genome included evaluation of length polymorphisms in the VZV repeat elements R1–R5, particularly R2 and R5 (Mori et al. 1998; Hawrami et al. 1997), and amplification of regions of ORF38 and ORF54 that included a PstI and BglI restriction site, respectively, which were useful for distinguishing vOka strain from most wild-type strains (LaRussa et al. 1992; Hawrami et al. 1997) The BglI/R2 method (Mori et al. 1998) was reliable in Japan because the combined evaluation of the ORF54 marker plus R2 analysis effectively discriminated almost all wild-type strains, since they rarely carried both the BglI mutation in ORF54 as well as the characteristic R2 configuration of the vOka. However, a more recent study revealed that R2 polymorphism is evident even among strains within the vaccine, indicating that this method may also be unreliable (Sauerbrei et al. 2007). In the United States, the majority of circulating VZV strains lacked the BglI restriction site in ORF54 present in the vaccine, and virtually all circulating strains in the US carried the PstI restriction site in ORF38 that is missing in the vOka (La Russa et al. 1998). As such, even though the targeted markers in this assay were essentially wild-type VZV SNP that served to discriminate various parental Oka subgenotype of viruses circulating in Japan, as long as such strains were not circulating in the US, this method could reliably discriminate vOka from wild-type viruses. In 2000, the 3′ terminal 34 Kb of the VZV genome was sequenced for both the vOka and the pOka (Argaw et al. 2000). These data included comparative sequence data for the region comprising ORFs 48 through 68 including, importantly, ORF62. The largest number of differences was detected in that open reading frame, and the group settled on a vaccine-specific marker at position 106262, principally because a SmaI restriction site was introduced into the vaccine through the substitution of C for T at that position. A PCR method incorporated SmaI digestion and restriction fragment length polymorphism analysis on the basis of this marker. Later the same year, this method was enhanced first by the improvement of primers (Loparev et al. 2000a) and then further refined into a real-time, closed system, fluorophore-based PCR method that substantially shortened the time required to perform the assay and reduced the risk of contamination by eliminating the need to manipulate amplimer (Loparev et al. 2000b). Shortly thereafter, the same method was adapted to discriminate the vaccine-specific marker at position 107252 and the markers in ORF38 and ORF54 (Lopez et al. 2008). A number of alternative real-time PCR protocols have been described using various combinations of the same three SNP in ORF38, 54, and 62 (Tipples et al. 2003;

Parker et al. 2006; Toi and Dwyer 2008; Harbecke et al. 2009); each of these methods varies principally by the equipment platform and reagents used, and by the number of target SNP evaluated. The latter aspect is the most important: at a minimum the fixed vaccine-specific marker at ORF62 should be evaluated, ideally two fixed ORF62 markers (any two of positions 106262, 107252, or 108111) together with the two vaccine-associated wild-type markers in ORF38 and ORF54 should all be evaluated for maximal reliability in the confirmation of vaccine adverse events. That testing algorithm is particularly important for the evaluation of nonviable specimens taken from skin lesions or saliva, where the amount and integrity of viral DNA can be highly variable. Focusing on one or two SNP can result in false negative results on specimens with partially degraded or small amounts of viral DNA.

3.3 Impact of Changing VZV Epidemiology

Several studies in the early 2000s indicated that VZV genotypes have distinctive global distributions (Faga et al. 2001; Wagenaar et al. 2003; Muir et al. 2002; Quinlivan et al. 2002; Loparev et al. 2004). While each of these groups devised a different system for genotyping VZV, it was clear almost immediately that many of the same genotypes were being described by all three laboratories, and that similar patterns of distribution were being observed. This distribution of strains appears to have stabilized globally over the course of many centuries, in the 12,000 year period following the expansion of human populations across the habitable globe, during which relatively little long distance travel or intermixing of populations occurred. Modern industrialized technology together with political unrest in much of the developing world has made global travel routine and has led to marked increases in immigration into nations that have relatively liberal policies for accepting foreign nationals. One direct consequence of this appears to be an ongoing redistribution of VZV genotypes in nations with large number of immigrants. For example, large number of Indian and Bangladashi immigrants have moved to east London in recent decades; the percentage of $BglI^{+}$ strains, relatively uncommon in most of England, increased from 10% in the 1980s to more than 30% in the 1990s (Hawrami et al. 1997). A study of clinical isolates collected in eastern Australia and New Zealand revealed that five of seven identified VZV genotypes are circulating in Australia, which has received a substantial number of immigrants (Loparev et al. 2007a). Clade 2 strains, including a wild-type Oka-like Clade 2 strain, accounted for 13% of the strains circulating in Australia. In contrast, in New Zealand, which has received relatively few immigrants, all the strains identified were the Clade 1, Clade 3, and Clade 5 strains that have typically predominated in countries with a history of European settlement. A recent study of cases of VZV-attributable encephalitis/ meningitis revealed that a similar increase in circulating Clade 2 strains is occurring on the west coast of North America, where Asian immigration is common (Pahud et al. manuscript in preparation); 5 of 32 clinical VZV isolates (15%) were Clade 2, only one of which was vOka. Of the remainder, most would have been mistaken for vOka if only the ORF38 and ORF54 markers had been evaluated. Another study of clinical isolates from France and Spain revealed increasing numbers of Clade VI genotype isolates, which predominate in tropical Africa, presumably accounted for by an increase in immigrants to those countries from Algeria and other African nations (Loparev et al. 2006). By comparison, countries in Europe in which immigration remains uncommon still have Clade 1 and Clade 3 VZV strains as the predominant genotypes (Loparev et al. 2009; Koskiniemi et al. 2007). Finally, a survey of VZV isolates from cases of herpes zoster in the UK revealed that, while most white, native-born citizens are infected with typical European genotype strains of VZV, immigrants from Asia and Africa are more likely to be infected with genotypes that predominate in tropical climates (Sengupta et al. 2007). Taken together, these observations indicate that the global distribution of VZV strains is currently in a state of flux. Of particular interest is the increase in J genotype strains in countries that were previously almost exclusively circulating Clade 1, Clade 3, and Clade 5 strains. Approximately 30% of the circulating strains in Japan have the same ORF38/ORF54 profile as the vOka, and would be mistakenly designated as vaccine virus if only those markers are evaluated.

3.4 Specimen Collection

Successful identification of VZV and vaccine:wild type discrimination is highly dependent on the type of specimen available for testing. The best specimens are vesicular swabs obtained by unroofing a vesicle and vigorously swabbing the base of the lesion and scabs obtained from crusted vesicles. Viral DNA levels are typically quite high in both types of lesions (1×10^4–1×10^9 genome equivalents per ml; sample extracted into 0.5 ml total volume) (Lopez et al. 2008). Varicella and herpes zoster due to the vOka are typically mild diseases, and lesions may not progress to the vesicular stage. Scrapings of maculopapular lesions may contain detectable levels of VZV DNA but need to be collected rather aggressively by roughly abrading the lesion with, for example, the edge of a glass microscope slide. At best, scrapings of maculopapular lesions often test VZV negative; it is unclear to what extent this may reflect misdiagnosis of varicella or herpes zoster. VZV DNA can often be detected in saliva, cheek swabs, and throat swabs if they are collected during the acute phase of disease, but detection is far less likely by the time lesions have crusted over (Leung et al. manuscript in preparation). VZV DNA can be detected in peripheral blood or urine only about 60% and 40% of the time, respectively, during acute disease, and substantially less often after the last lesions have crusted (Leung et al. manuscript in preparation).

An interesting recent observation has been that environmental specimens may be collected in the living space where cases have occurred and can provide useful data for epidemiologic investigations. An outbreak of varicella developed in long-term-care facility in West Virginia in early 2004, mostly among elderly patients who were noncommunicative (Lopez et al. 2008). By the time investigators arrived on the scene to evaluate the outbreak, the putative index case of zoster had completely resolved. Wet swabs of a variety of surfaces (e.g., bed frame, chair, light fixture) were obtained, extracted, and tested as for conventional specimens. Samples were obtained from case rooms and control rooms. VZV DNA was detectable in environmental specimens from all case rooms; moreover, sufficient DNA was present to permit extensive genotypic analysis, which made it possible to link the strain recovered from the putative index case samples with both clinical and environmental samples taken from other case rooms. The same strain was responsible for all of the cases. This approach has proven useful in subsequent outbreak investigations as well (Schmid, unpublished observations). This includes the confirmation of one case of vOka varicella in a recently vaccinated adult where only environmental specimens were available.

3.5 *The Role of Recombination in Varicella Vaccine Surveillance*

Evidence for interstrain genetic recombination has been observed for most of the human herpesviruses, including cytomegalovirus (Haberland et al. 1999), Epstein–Barr virus (Midgley et al. 2000; Walling et al. 1994), human herpesvirus 8 (Poole et al. 1999), herpes simplex virus (Norberg et al. 2007; Umene et al. 2008), and VZV (Barrett-Muir et al. 2003; Norberg et al. 2006; Peters et al. 2006; Sauerbrei et al. 2007; Sauerbrei and Wutzler 2007; Schmidt-Chanasit et al. 2008; Wagenaar et al. 2003). In less than a decade, VZV has gone from being one of the least well studied herpesvirus genomes to one of the best studied, at least in terms of DNA sequence variation. Seven VZV genotypes have been identified thus far, five of which were confirmed as stably circulating phylogenetic groups both through targeted sequence analysis and through determination of complete genomic sequence (Clade 1, Clade 2, Clade 3, Clade 5, and Clade 4) (Barrett-Muir et al. 2003; Loparev et al. 2004, 2006, 2007a; Peters et al. 2006). The provisional genotypes Clade VII and Clade VI are currently based on limited sequence analysis and will need to be confirmed with at least two complete genome sequences each (Sergeev et al. 2006; Loparev et al. 2006). Preliminary phylogenetic analysis of complete genomic sequences available in 2006 suggested that the Clade 5 and Clade 4 strains were generated through recombination events between Clade 1 and Clade 2 strains (Norberg et al. 2006). A more recent analysis taking advantage of 18 available VZV complete genomic sequences concludes that Clade 1 arose through recombination between Clade 3 and Clade 4 viruses; limited complete genome sequences for Clade 2 and Clade 5 strains precluded speculation about recombination history, but suggested that both genotypes represent deeper lineage viruses (McGeoch 2009). Several groups have speculated on the potential for recombination between vaccine and wild-type VZV among vaccinated persons who experience breakthrough infection (Barrett-Muir et al. 2003; Peters et al. 2006; Sauerbrei and Wutzler 2007). Recently, 15 of 134 isolates (11%) from previously laboratory-confirmed vaccine varicella and herpes zoster were found to be Clade 1 or Clade 3 viruses (rather than the Clade 2 profile of the vaccine). All but two of these viruses came from cases of presumed vaccine zoster (Schmid et al. 2008). Two isolated studies evaluated in detail, including the determination of 84 Kb of genomic sequence, were demonstrated to be clonal (Schmid et al. 2008), and direct evidence for recombination was obtained by evaluating recombination arrays and performing pairwise homoplasy index analysis. Clinically apparent differences in pathogenicity have not been demonstrated among circulating wild-type VZV viruses, suggesting that perhaps recombination with an attenuated vaccine virus is unlikely to produce in viruses with enhanced pathogenicity, although that possibility cannot be ruled out. Even so, there is currently no evidence that vaccine:wild type recombination will necessitate a change in vaccine recommendations or policy. Perhaps more likely is the restoration of wild-type pathogenicity through recombination, which could in turn result in the circulation of strains bearing both vaccine and non-J wild type markers, which will greatly complicate the conduct of varicella vaccine surveillance. Simply relying on the analysis of four vaccine-associated SNP in ORF62, ORF38, and ORF54, 11% of isolates submitted to the National CDC Laboratory were mistakenly identified as vOka and were revealed to be Clade 1 and Clade 3 viruses by ORF21, ORF22, ORF50 sequencing. We also established that Clade 1 and Clade 3 viruses are absent in the vaccine preparation through the evaluation of more than 600 TA clones for two lots of the Varivax vaccine. As such, the simple analysis of one or even several vaccine SNP is no longer sufficient for confirming a vaccine adverse event. Extensive SNP analysis together with VZV genotyping by one of the established methods is required to reliably attribute disease to vOka. Finally, the occurrence of vaccine:wild type recombination will likely increase with the licensure and recommendation of zoster vaccine (Harpaz et al. 2008), which contains 14 times as much vaccine virus as the childhood varicella vaccine, and is being used to immunize persons known to have preexisting latent VZV infections.

4 Summary

The decision to routinely immunize children with Oka varicella vaccine in the United States, Japan, and other countries has driven the development of methods for distinguishing the vaccine preparation from circulating wild-type viruses. Such methods are essential to the process of monitoring vaccine impact and for confirming the occurrence of vaccine adverse events. Evidence that emerged in the last decade suggests that the global distribution of circulating wild-type strains is changing, particularly in countries in which immigration is common. Of particular concern is the apparent increase of Clade 2 strains that are similar to vOka, which has made the assessment of fixed vaccine-specific markers located in ORF62 a necessary component of identifying the vOka strain. In addition, the detection of clinical isolates resulting from recombination between the vaccine strain and wild-type superinfecting strains now makes it advisable to routinely genotype all potential vaccine adverse event isolates.

References

Argaw T, Cohen JI, Klutch M, Lekstrum K, Yoshikawa T, Asano Y, Krause PR (2000) Nucleotide sequences that distinguish Oka vaccine from parental Oka and other varicella-zoster virus isolates. J Infect Dis 181:1153–1157

Barrett-Muir W, Scott FT, Aaby P, John J, Matondo P, Chaudry QL, Siquera M, Poulsen A, Yamanishi K, Breuer J (2003) Genetic variation of varicella-zoster virus: evidence for geographical separation of strains. J Med Virol 70(Suppl 1):S42–S47

Breuer J, Schmid DS (2008) Vaccine Oka variants and sequence variability in vaccine-related skin lesions. J Infect Dis 197(Suppl 2):S54–S57

Brisson M, Edmunds WJ, Gay NJ, Law B, De Serres G (2000) Modeling the impact of immunization on the epidemiology of varicella-zoster virus. Epidemiol Infect 125:651–669

CDC (1996) Prevention of varicella: Recommendations of the Advisory Committee on Immunization Practices (ACIP). MMWR 45 (RR-11)

CDC (1999a) Prevention of varicella: updated recommendations of the Advisory Committee on Immunization Practices (ACIP). M.M.W.R. Recomm. Rep. 48 (No. RR-6)

CDC (1999b) Outbreak of varicella among vaccinated children – Michigan, 2003. MMWR 53:389–392

Cohrs RJ, Gilden DH, Gomi Y, Yamanishi K, Cohen JI (2006) Comparison of virus transcription during lytic infection of the Oka parental and vaccine strains of varicella-zoster virus. J Virol 80:2076–2082

Faga KI, Maury W, Bruckner DA, Grose C (2001) Identification and mapping of single nucleotide polymorphisms in the varicella-zoster virus genome. Virology 280:355–363

Finger R, Hughes JP, Meade BJ, Pelletier AR, Palmer CT (1994) Age specific incidence of chickenpox. Public Health Rep 109:750–755

Gomi Y, Imagawa T, Takahashi M, Yamanishi K (2000) Oka varicella vaccine is distinguishable from its parental virus in DNA sequence of open reading frame 62 and its transactivation activity. J Med Virol 61:497–503

Gomi Y, Imagawa T, Takahashi M, Yamanishi K (2001) Comparison of DNA sequence and transactivation of open reading frame 62 of Oka varicella vaccine and its parental viruses. Arch Virol (Suppl 17):49–56

Gomi Y, Sunamachi H, Mori Y, Nagaike K, Takahashi M, Yamanishi K (2002) Comparison of the complete DNA sequences of the Oka varicella vaccine and its parental virus. J Virol 76:11447–11459

Guris D, Jumaan AO, Mascola L, Watson BM, Zhang JX, Chaves SS, Gargiullo P, Perella D, Civen R, Seward JF (2008) Changing varicella epidemiology in active surveillance sites – United States, 1995-2000. J Infect Dis 197(Suppl 2):S71–S75

Haberland M, Meyer-König U, Hufert FT (1999) Variation within the glycoprotein B gene of human cytomegalovirus is due to homologous recombination. J Gen Virol 80:1495–1500

Harbecke R, Oxman MN, Arnold BA, Ip C, Johnson GR, Levin MJ, Gelb LD, Schmader KE, Strauss SE, Wang H, Wright PF, Pachucki CT, Gershon AA, Arbeit RD, Davis LE, Simberkoff MS, Weinberg A, Williams HM, Cheney C, Petrukhin L, Abraham KG, Shaw A, Manoff S, Antonello JM, Green T, Wang Y, Tan C, Keller PM, Shingles Prevention Study Group (2009) A real-time PCR assay to identify and discriminate among wild-type and vaccine strains of varicella-zoster virus and herpes simplex virus in clinical specimens, and comparison with the clinical diagnoses. J Med Virol 81:1310–1322

Harpaz R, Ortega-Sanchez IR, Seward JF, Advisory Committee on Immunization Practices (ACIP), Centers for Disease Control and Prevention (CDC) (2008) Prevention of herpes zoster: recommendations of the advisory committee on immunization practices. MMWR Recomm Rep 57(RR-5):1–30

Hawrami K, Breuer J (1997) Analysis of United Kingdom wild type strains of varicella-zoster virus: differentiation from the Oka vaccine strain. J Med Virol 53:60–62

Hawrami K, Hart LJ, Pereira F, Argent S, Bannister B, Bovill B, Carrington D, Ogilvie M, Rawstorne S, Tryhorn Y, Breuer J (1997) Molecular epidemiology of varicella-zoster virus in East London, England, between 1971 and 1995. J Clin Microbiol 35:2807–2809

Hayakawa Y, Torigoe S, Shiraki K, Yamanishi K, Takahashi M (1984) Biologic and biophysical markers of a live varicella vaccine strain (Oka): identification of clinical isolates from vaccine recipients. J Infect Dis 149:956–963

Hayakawa Y, Yamamoto T, Yamanishi K, Takahashi M (1986) Analysis of varicella-zoster virus DNAs of clinical isolates by endonuclease *Hpa*I. J Gen Virol 67:1817–1829

Hondo R, Yogo Y, Yoshida M, Fujima A, Itoh S (1989) Distribution of varicella-zoster virus strains carrying a Pst-site-less mutation in Japan and DNA change responsible for the mutation. Jpn J Exp Med 59:233–237

Kemble GW, Annunziato P, Lungu O, Winter RE, Cha T-A, Silverstein SJ, Spaete RR (2000) Open reading frame S/L of varicella-zoster virus encodes a cytoplasmic protein expressed in infected cells. J Virol 74:11311–11321

Koskiniemi M, Lappalainen M, Schmid DS, Rubtcova E, Loparev VN (2007) Genotypic analysis of varicella-zoster virus and its seroprevalence in Finland. Clin Vaccine Immunol 14:1057–1061

Koshizuka T, Yamanishi K, Mori Y (2009) Comparison and characterization of VZV encoded ORF0 of the parental and vaccine strains. 34th International Herpesvirus Workshop, Abstract 1.70

LaRussa P, Lungu O, Hardy I, Gershon A, Steinberg SP, Silverstein S (1992) Restriction fragment length polymorphism of polymerase chain reaction products from vaccine and wild type varicella-zoster virus isolates. J Virol 66:1016–1020

La Russa P, Steinberg S, Arvin A, Dwyer D, Burgess M, Menegus M, Rekrut K, Yamanishi K, Gershon A (1998) Polymerase chain reaction and restriction fragment length polymorphism analysis of varicella-zoster virus isolates from the United States and other parts of the world. J Infect Dis 178(Suppl 1):S64–S66

Loparev VN, Argaw T, Krause PR, Takayama M, Schmid DS (2000a) Improved identification and differentiation of varicella-zoster virus (VZV) wild-type strains and an attenuated varicella vaccine strain using a VZV open reading frame 62-based PCR. J Clin Microbiol 38:3156–3160

Loparev VN, McCaustland K, Holloway BP, Krause PR, Takayama M, Schmid DS (2000b) Rapid genotyping of varicella-zoster virus vaccine and wild-type strains with fluorophore-labeled hybridization probes. J Clin Microbiol 38:4315–4319

Loparev VN, Gonzalez A, Deleon-Carnes M, Tipple G, Fickenscher H, Torfason EG, Schmid DS (2004) Global identification of three major genotypes of varicella-zoster virus: longitudinal clustering and strategies for genotyping. J Virol 78:8349–8358

Loparev V, Martro E, Rubtcova E, Rodrigo C, Piette JC, Caumes E, Vernant JP, Schmid DS, Fillet AM (2006) Toward universal varicella-zoster virus genotyping: diversity of VZV strains from France and Spain. J Clin Microbiol 45:559–563

Loparev VN, Rubtcova EN, Bostik V, Govil D, Birch CJ, Druce JD, Schmid DS, Croxson MC (2007a) Identification of five major and two minor genotypes of varicella-zoster virus strains: a practical two-amplicon approach used to genotype clinical isolates in Australia and New Zealand. J Virol 81:12758–12765

Loparev VN, Rubtcova E, Seward JF, Levin MJ, Schmid DS (2007b) DNA sequence variability in isolates recovered from patients with postvaccination rash or herpes zoster caused by Oka varicella vaccine. J Infect Dis 195:502–510

Loparev VN, Rubtcova EN, Bostik V, Tzaneva V, Sauerbrei A, Robo A, Sattler-Dornbacher E, Hanovcova I, Stepanova V, Splino M, Eremin V, Koskiniemi M, Vankova OE, Schmid DS (2009) Distribution of varicella-zoster virus (VZV) wild type genotypes in northern and Southern Europe: evidence for high conservation of circulating genotypes. Virology 383:216–225

Lopez AS, Guris D, Zimmerman L, Gladden L, Moore T, Haselow DT, Loparev VN, Schmid DS, Jumaan AO, Snow SL (2006) One dose of varicella vaccine does not prevent school outbreaks: is it time for a second dose? Pediatrics 117:e1070–e1077

Lopez AS, Burnett-Hartman A, Nambiar R, Ritz L, Owens P, Loparev VN, Guris D, Schmid DS (2008) Transmission of a newly characterized strain of varicella-zoster virus from a patient with herpes zoster in a long-term-care facility, West Virginia, 2004. J Infect Dis 197:646–653

Marin M, Guris D, Chaves SS, Schmid S, Seward JF (2007) Prevention of varicella: recommendations of the advisory committee on immunization practices (ACIP). MMWR Recom Rep 56 (RR-4):1–37

McGeoch D (2009) Lineages of varicella-zoster virus. J Gen Virol 90:963–969

Midgley RS, Blake NW, Yao QY, Croom-Carter D, Cheung ST, Leung SF, Chan ATC, Johnson PJ, Huang D, Rickenson AB, Lee SP (2000) Novel intertypic recombinants of Epstein Barr virus in the Chinese population. J Virol 74:1544–1548

Mori C, Takahara R, Toriyama T, Nagai T, Takahashi M, Yamanishi K (1998) Identification of the Oka strain of the live attenuated varicella vaccine from other clinical isolates by molecular epidemiologic analysis. J Infect Dis 178:35–38

Muir WB, Nichols R, Breuer J (2002) Phylogenetic analysis of varicella-zoster virus: evidence of intercontinental spread of genotypes and recombination. J Virol 76:1971–1979

Neve B, Froguel P, Corset L, Valliant E, Vatin V, Boutin P (2002) Rapid SNP allele frequency determination in genomic DNA pools by pyrosequencing. Biotechniques 32:1138–1142

Norberg P, Kasubi MJ, Haarr L, Bergström T, Liljeqvist JA (2007) Divergence and recombination of clinical herpes simplex virus type 2 isolates. J Virol 81:13158–13167

Norberg P, Liljeqvist J-Å, Bergström T, Sammons S, Schmid DS, Loparev VN (2006) Complete-genome phylogenetic approach to varicella-zoster virus evolution: genetic divergence and evidence for recombination. J Virol 80:9569–9576

Ogino S, Kawasaki T, Brahmandam M, Yan L, Cantor M, Namgyal C, Mino-Kenudson M, Lauwers GY, Loda M, Fuchs CS (2005) Sensitive sequencing method for KRAS mutation detection by pyrosequencing. J Mol Diagn 7:413–421

Oxman MN, Levin MJ, Johnson GR, Schmader KE, Straus SE, Gelb LD, Arbeit RD, Simberkoff MS, Gershon AA, Davis LE, Weinberg A, Boardman KD, Williams HM, Zhang JH, Peduzzi PN, Beisel CE, Morrison VA, Guatelli JC, Brooks PA, Kauffman CA, Pachucki CT, Neuzil KM, Betts RF, Wright PF, Griffin MR, Brunell P, Soto NE, Marques AR, Keay SK, Goodman RP, Cotton DJ, Jr Gnann JW, Loutit J, Holodniy M, Keitel WA, Crawford GE, Yeh SS, Lobo Z, Toney JF, Greenberg RN, Keller PM, Harbecke R, Hayward AR, Irwin MR, Kyriakides TC, Chan CY, Chan IS, Wang WW, Annunziato PW, Silber JL, Shingles Prevention Study Group (2005) A vaccine to prevent herpes zoster and postherpetic neuralgia in older adults. N Engl J Med 352:2271–2284

Parker SP, Quinlivan M, Taha Y, Breuer J (2006) Genotyping of varicella-zoster virus and the discrimination of Oka vaccine strains by TaqMan real-time PCR. J Clin Microbiol 44:3911–3914

Peters GA, Tyler SD, Grose C, Severini A, Gray MJ, Upton C, Tipples GA (2006) A full genome phylogenetic analysis of varicella-zoster virus reveals a novel origin of replication-based genotyping scheme and evidence of recombination between major circulating clades. J Virol 80:9850–9860

Poole LJ, Zong J-C, Ciufo DM, Alcendor DJ, Cannon JS, Ambinder R, Orenstein JM, Reitz MS, Hayward GS (1999) Comparison of genetic variability at multiple loci across the genomes of the major subtypes of Kaposi's sarcoma-associated herpesvirus reveals evidence for recombination and for two distinct types of open reading frame K15 alleles at the right-hand end. J Virol 73:6646–6660

Quinlivan M, Hawrami K, Barrett-Muir W, Aaby P, Arvin A, Chow VT, John TJ, Matondo P, Peiris M, Poulson A, Siqueira M, Takahashi M, Talukder Y, Yamanishi K, Leedham-Cross M, Scott FT, Thomas SL, Breuer J (2002) The molecular epidemiology of varicella-zoster virus: evidence for geographic segregation. J Infect Dis 186:888–894

Quinlivan ML, Gershon AA, Steinberg SP, Breuer J (2004) Rashes occurring after immunization with a mixture of viruses in the Oka vaccine are derived from single clones of virus. J Infect Dis 190:793–796

Quinlivan MA, Gershon AA, Nichols RA, LaRussa P, Steinberg SP, Breuer J (2006) Vaccine Oka varicella-zoster virus genotypes are monomorphic in single vesicles and polymorphic in respiratory tract secretions. J Infect Dis 193:927–930

Sauerbrei A, Wutzler P (2007) Different genotype pattern of varicella-zoster virus obtained from patients with varicella and zoster in Germany. J Med Virol 79:1025–1031

Sauerbrei A, Rubtcova E, Wutzler P, Schmid DS, Loparev VN (2004) Genetic profile of an Oka vaccine variant isolated from an infant with zoster. J Clin Microbiol 42:5604–5608

Sauerbrei A, Zell R, Wutzler P (2007) Analysis of the repeat units in the R2 among different Oka varicella-zoster virus vaccine strains and wild type strains in Germany. Intervirology 50:40–44

Schmid DS, V Bostik, J Folster, N Jensen, P Norberg, B Bankamp, K Hummel (2008) 33rd international herpesvirus workshop. Evidence for recombination between Oka vaccine and wild type strains among clinical varicella-zoster virus (VZV) isolates. Abstr 803

Schmidt-Chanasit J, Olschläger S, Günther S, Jaeger G, Bleymehl K, Schäd SG, Heckel G, Ulrich RG, Doerr HW (2008) Molecular analysis of varicella-zoster virus strains circulating in Tanzania demonstrating the presence of genotype Clade 5. J Clin Microbiol 46:3540–3543

Sengupta N, Taha Y, Scott FT, Leedham-Green ME, Quinlivan M, Breuer J (2007) Varicella zoster genotypes in East London: a prospective study in patients with herpes zoster. J Infect Dis 196:1014–1020

Sergeev N, Rubtcova E, Chizikov V, Schmid DS, Loparev VN (2006) New mosaic subgenotype of varicella-zoster virus in the USA: VZV detection and genotyping by oligonucleotide-microarray. J Virol Methods 136:8–16

Seward JF, Watson BM, Peterson CL, Mascola L, Pelosi JW, Zhang JX, Maupin TJ, Goldman GS, Tabony LJ, Brodovicz KG, Jumaan AO, Wharton M (2002) Varicella disease after introduction of varicella vaccine in the United States, 1995–2000. JAMA 287:606–611

Shiraki K, Horiuchi K, Asano Y, Yamanishi K, Takahashi M (1991) Differentia-tion of Oka varicella vaccine strain from wild varicella zoster virus strains isolated from vaccinees and household contact. J Med Virol 33:128–132

Silvertsson Å, Plantz A, Hansson J, Lundeberg J (2002) Pyrosequencing as an alternative to single strand conformation polymorphism analysis for detection of N-ras mutations in human melanoma metastases. Clin Chem 48:2164–2170

Takahashi M, Otsuka T, Okuno Y, Asano Y, Yazaki T (1974) Live vaccine used to prevent the spread of varicella in children in hospital. Lancet 2:1288–1290

Tipples GA, Safronetz D, Gray M (2003) A real-time PCR assay for the detection of varicella-zoster virus DNA and differentiation of vaccine, wild-type and control strains. J Virol Methods 113:113–116

Toi CS, Dwyer DE (2008) Differentiation between vaccine and wild-type varicella-zoster virus genotypes by high-resolution melt analysis of single nucleotide polymorphisms. J Clin Virol 43:18–24

Umene K, Oohashi S, Yoshida M, Fukumaki Y (2008) Diversity of the a sequence of herpes simplex virus type 1 developed during evolution. J Gen Virol 89:841–852

Wagenaar T, Chow VTK, Buranathai C, Thawasupha P, Grose C (2003) The out of Africa model of varicella-zoster virus evolution: single nucleotide polymorphisms and private alleles distinguish Asian clades from European/North American clades. Vaccine 21:1072–1081

Walling DM, Perkins AG, Webster-Cyriaque J, Resnick L, Raab-Traub N (1994) The Epstein-Barr virus EBNA-2 gene in oral hairy leukoplakia: strain variation, genetic recombination, and transcriptional expression. J Virol 68:7918–7926

Wharton M (1996) The epidemiology of varicella-zoster virus infections. Infect Dis Clin North Am 10:571–581

Yawn BP, Yawn RA, Lydick E (2003) Community impact of childhood varicella infections. J Pediatr 130:759–765

Zhang Z, Rowe J, Wang W, Sommer M, Arvin A, Moffat J, Zhu H (2007) Genetic analysis of varicella-zoster virus ORF0 to ORF4 by use of a novel luciferase bacterial artificial chromosome system. J Virol 81:9024–9033

VZV T Cell-Mediated Immunity

Adriana Weinberg and Myron J. Levin

Contents

Abstract Primary varicella-zoster virus (VZV) infection (varicella) induces VZV-specific antibody and VZV-specific T cell-mediated immunity. T cell-mediated immunity, which is detected within 1–2 weeks after appearance of rash, and consists of both CD4 and CD8 effector and memory T cells, is essential for recovery

A. Weinberg (✉)
University of Colorado Denver, Research Complex 2, Mail Stop 8604, 12700 E. 19th Ave., Aurora, CO 80045, USA
e-mail: Adriana.Weinberg@ucdenver.edu

M.J. Levin
University of Colorado Denver, Bldg. 401, Mail Stop C227, 1784 Racine St., Room R09-108, Aurora, CO 80045, USA
e-mail: Myron.Levin@ucdenver.edu

A.M. Arvin et al. (eds.), *Varicella-zoster Virus*,
Current Topics in Microbiology and Immunology 342, DOI 10.1007/82_2010_31

from varicella. Administration of a varicella vaccine also generates VZV-specific humoral and cellular immune responses. The memory cell responses that develop during varicella or after vaccination contribute to protection following re-exposure to VZV. These responses are subsequently boosted either by endogenous re-exposure (silent reactivation of latent virus) or exogenous re-exposure (environmental). VZV-specific T cell-mediated immunity is also necessary to maintain latent VZV in a subclinical state in sensory ganglia. When these responses decline, as occurs with aging or iatrogenic immune suppression, reactivation of VZV leads to herpes zoster. Similarly, the magnitude of these responses early after the onset of herpes zoster correlates with the extent of zoster-associated pain. These essential immune responses are boosted by the VZV vaccine developed to prevent herpes zoster.

1 Introduction

Varicella (chickenpox) is the manifestation of primary infection with the varicella-zoster virus (VZV). VZV-specific antibodies and T cell-mediated immune responses protect against secondary infections. The VZV-protective immune responses appear to be life-long in immune competent persons, as suggested by life-long protection against subsequent varicella after re-exposure.

Varicella also results in life-long latency of VZV in neurons in cranial nerve and dorsal root ganglia (Levy et al. 2003). The immune responses induced by varicella may be boosted throughout life as a consequence of environmental exposures to VZV (from incident varicella or herpes zoster [HZ]) or as a consequence of endogenous, subclinical reactivation of latent VZV. Subclinical reactivation of latent VZV may occur throughout life, but clinical consequences of this reactivation are prevented by VZV-specific T cell-mediated immunity. VZV-specific T cell-mediated immunity is decreased by immune suppressive diseases or therapies and also wanes with increasing age, especially after age 40 years (see below). Although the VZV-specific T cell-mediated immunity that follows varicella remains adequate to prevent varicella after re-exposure, it may decline to a level that is insufficient to maintain latency of VZV, with the result that the reactivated latent VZV propagates and causes a ganglionitis with associated dermatomal neuropathic pain and subsequent painful dermatomal vesicular rash.

Vaccination of naïve individuals with varicella vaccine also induces VZV-specific antibody and T cell-mediated immune responses. However, these responses sometimes fail to develop (primary vaccine failure) or are less robust than those following varicella (Michalik et al. 2008), and thus approximately 15% of vaccinees develop a mild form of varicella ("breakthrough" disease) after exposure to varicella (White et al. 1992; Vessey et al. 2001). Typical varicella is generally prevented in vaccinees.

2 VZV-Specific T Cell Immunity Acquired from Wild Type VZV Infection

2.1 Characteristics of VZV-Specific T Cell-Mediated Immunity

Recovery from varicella is associated with the development of VZV-specific T cell-mediated immunity (Kumagai et al. 1980; Arvin et al. 1986a). Both effector and memory VZV-specific T cells have been identified (Asanuma et al. 2000; Diaz et al. 1989; Rotbart et al. 1993; Patterson-Bartlett et al. 2007; Sadaoka et al. 2008). These cells recognize epitopes in immediate early IE62, IE63, ORF4, and ORF10 gene products and in late gB, gC, gE, gH, and gI surface glycoproteins (Arvin et al. 1986b, 1991, 2002; Bergen et al. 1991; Diaz et al. 1988; Giller et al. 1989). There is some, albeit limited, information regarding the amino acid sequences of the epitopes recognized by VZV-specific T cells (Frey et al. 2003; Hayward 1990; Jones et al. 2006; Malavige et al. 2007). *In vitro* stimulation of T cells from VZV-experienced subjects with VZV antigens elicits production of multiple cytokines, including IL2, IFNγ, TNFα, IL10, IL12, IL4, and IL5, all of which can be demonstrated by intracellular staining or analysis of culture supernatants (Arvin et al. 1986a; Patterson-Bartlett et al. 2007; Hayward et al. 1989, 1998; Zhang et al. 1994; Jenkins et al. 1998). These cytokine profiles are consistent with the participation of T cells in Th1 and Th2 responses. Both CD4 and CD8 T cells have cytolytic activity against VZV-infected targets, representing the ultimate defensive function performed by T cells (Diaz et al. 1989; Frey et al. 2003; Hayward et al. 1989, 1990). Most VZV-specific CD4 and CD8 T cells also express the CLA homing receptor (Patterson-Bartlett et al. 2007) that facilitates their infiltration of the skin, where the virus replicates during symptomatic infection. This permits recognition and rapid destruction of infected cells at these sites. VZV-specific regulatory T cells or Th17 cells have not been described or thoroughly investigated. Studies of these relatively recently described T cell populations in VZV infection may enhance our understanding of the VZV-specific T cell-mediated immune response.

The frequency of VZV-specific T cell effectors in the peripheral blood of healthy VZV-immune adults is approximately 0.1–0.2%, which represents one order of magnitude less than the frequency of effectors specific for other herpes viruses, such as Epstein–Barr virus or cytomegalovirus (Asanuma et al. 2000; Bhaduri-McIntosh et al. 2008). This finding is somewhat unexpected, since VZV, like these other herpes viruses, causes a systemic primary infection, establishes latency, periodically reactivates, and VZV-specific T cell immunity is boosted by exogenous re-exposures. The last observation is particularly relevant, since the original studies describing the frequency of VZV-specific T cells were done while varicella was circulating in communities, that is, before the inclusion of the varicella vaccine in the childhood immunization schedule.

VZV and herpes simplex virus (HSV) are α-herpeviruses that establish latency in sensory neurons. The detailed mechanisms by which T cell-mediated immunity maintains latency of these two viruses is not known but appear to be quite different. There is a high abundance of HSV-specific CD8 T cells in ganglia latently infected with HSV, but VZV-specific CD8 T cells are not frequent in ganglia containing latent VZV (Verjans et al. 2007). Moreover, HSV-specific T cells near the peripheral termini of nerves containing HSV contribute to maintaining the clinical latency of HSV (Zhu et al. 2007). The significance of the mechanisms underlying the differences in the T cell-mediated immune response to VZV compared with other herpes viruses deserve to further study, as they may uncover important aspects of immune activation and regulation.

The first cell-mediated immune response to viral infections is orchestrated by the innate immune system. Very little is known about NK-mediated responses to VZV, except that subjects with NK or NKT cell defects are at risk of developing severe varicella (Levy et al. 2003; Vossen et al. 2005). Among the functions of NK cells in other viral infections is the transfer of information to dendritic cells that instruct T cells and initiate the adaptive immune response. The contribution of dendritic cells to the VZV-specific innate or adaptive immune response has not been completely defined, but individuals with plasmacytoid dendritic cell defects develop severe manifestations of VZV and other herpes viral infections (Kittan et al. 2007). VZV infection of dendritic cells *in vitro* impairs their MHC class I and II surface expression, which may represent a mechanism of immune evasion (Morrow et al. 2003). More studies are needed to define the interplay among NK, dendritic cells, and the adaptive immune responses to VZV, since this might be manipulated by therapeutic interventions.

2.2 Kinetics of VZV-Specific T Cell-Mediated Immunity After Varicella

After primary VZV infection, VZV-specific memory CD4 T cells are not detected in the blood of the host within 24 h of the onset of varicella rash, but are demonstrable 3–7 days later (Kumagai et al. 1980). These CD4 T cell memory responses peak between 1 and 2 weeks after rash onset, and then gradually decrease during up to 3 years of follow up (Rotbart et al. 1993). A comparison of VZV-specific CD4 T cell memory responses of children who received acyclovir during the acute phase of the disease with that of untreated children showed similar VZV responder cell frequencies in the two groups, suggesting that antiviral treatment does not interfere with the development of VZV T cell-mediated immunity (Rotbart et al. 1993). A study of the kinetics of VZV-specific memory CD4 T cells in children who developed asymptomatic primary VZV infection after receiving acyclovir for post-exposure prophylaxis (7–9 days after exposure) showed peak responses 3–5 weeks after exposure (Kumagai et al. 1999), delineating the time required to generate the VZV-specific memory T cells after infection. These results are in accordance with the data obtained in studies of the T cell-mediated immune response to the live attenuated varicella vaccine (see below).

The frequency of IFNγ-expressing VZV-specific T cell effectors in children with varicella peaked during the first week of rash, followed by a precipitous decrease over the subsequent 2–3 weeks and a gradual decrease thereafter (the maximum follow up in this study was 52 weeks) (Vossen et al. 2004). There is no information on the clonality or epitope-specificity of the T cells that expand and contract during the maturation of the immune response to VZV. This information could be utilized to optimize immunization strategies against VZV throughout life.

2.3 *Kinetics of VZV-Specific T Cell-Mediated Immunity After Herpes Zoster*

HZ, the symptomatic manifestation of VZV reactivation, is associated with a boost of VZV-specific T cell-mediated immunity (Weinberg et al. 2009). VZV-specific effector T cell responses measured by IFNγ ELISPOT rise quickly and peak between 1 and 3 weeks after the onset of HZ rash. This is followed by a rapid decay between 3 and 6 weeks after onset. A steady state level is maintained at least for the subsequent 2 years (Fig. 1). VZV-specific memory T cells rise more slowly, attaining their peak between 3 and 6 weeks after the onset of rash, after which they decay during the first year after onset, subsequently establishing a plateau for the following 2 years or longer.

Asymptomatic reactivations of VZV also boost the VZV-specific T cell immunity. This phenomenon was clearly demonstrated in hematopoietic stem cell recipients with asymptomatic VZV viremia detected by PCR (Wilson et al. 1992). Exogenous re-exposure to VZV has a similar boosting effect on VZV-specific CD4 effector and memory T cells, as well as CD8 T and NK cells (Vossen et al. 2004; Arvin et al. 1983). Both endogenous and exogenous re-exposures to VZV may contribute to the maintenance of VZV-specific T cell immunity over time.

2.4 *Effect of Age on VZV-Specific T Cell-Mediated Immunity*

After primary infection, the frequency of VZV-specific memory CD4 T cells is significantly influenced by age (Burke et al. 1982; Levin et al. 2006). We observed maximum frequencies of VZV-specific memory CD4 T cells during early adulthood (34 years of age) with lower frequencies at both ends of the age spectrum (Fig. 2). Hence, young adults had higher frequencies of VZV-specific memory T cells than children in spite of the fact that they were more remote than children from their primary VZV infection. This observation, taken together with kinetics studies showing that VZV-specific memory CD4 T cell frequency continuously decreased during 3 years after varicella, support the notion that exogenous and, perhaps, endogenous re-exposures boost VZV-specific immune responses of young adults. However, after mid-adulthood, the intensity and the quality of the antigenic

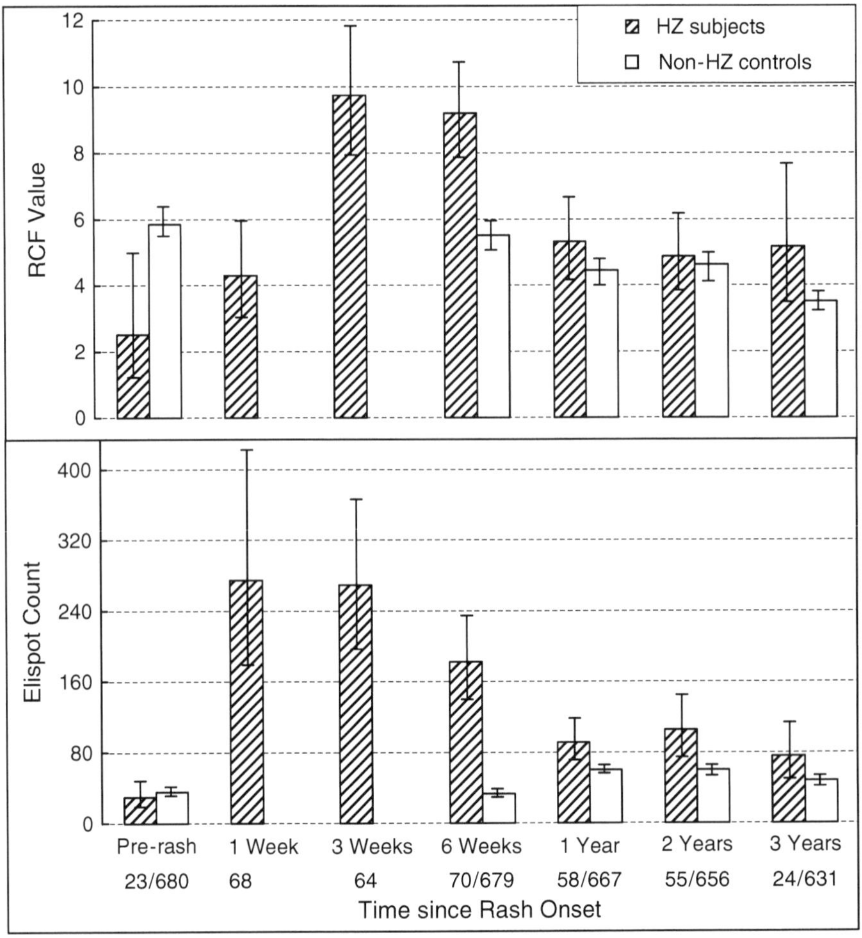

Fig. 1 Comparison of VZV-specific immune responses of individuals with and without HZ. Data were derived from 73 subjects who developed HZ and 680 who did not during a herpes zoster vaccine study. All subjects who contributed data for this graph received placebo. Bars indicate geometric means and 95% CI of RCF values measured as responder cells/10^5 PBMC and ELISPOT counts, measured as SFC/10^6 PBMC. Pre-rash levels for non-HZ subjects were those measured at enrollment. Levels at other time points were measured following HZ rash onset in the subjects with HZ or after enrollment in the non-HZ subjects. *N* indicates the number of subjects contributing samples at each time point. Subjects who developed HZ had significantly lower RCF ($p = 0.001$), but not ELISPOT values at the last visit before HZ compared with subjects who did not develop HZ. HZ significantly increased all immune responses. © 2009 Infectious Diseases Society of America

stimulation provided by the re-exposures and asymptomatic reactivations are not sufficient to maintain the VZV-specific T cell-mediated immunity of the host. The frequency of T cells that upregulate the CD69 activation marker upon *in vitro* stimulation with VZV is actually higher in elderly than in young adults

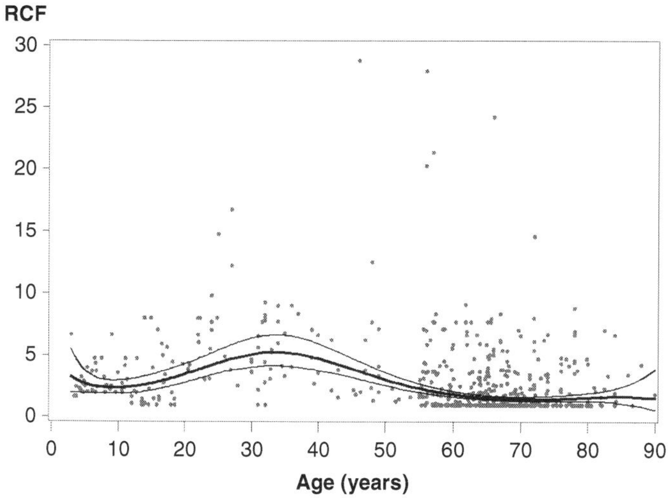

Fig. 2 Reference curve of VZV-RCF on age. Data were derived from 540 individuals with an antibody-confirmed history wild-type varicella infection and without history of herpes zoster. Lines indicate the regression curve of VZV-RCF on age and 95% confidence intervals. The highest RCF values were in mid-adulthood: mean of 5.2 responders/10^5 PBMC (95% CI of 4.1–6.6 responders/10^5 PBMC) at 34 years of age. © 2010 Infectious Diseases Society of America

(Patterson-Bartlett et al. 2007), suggesting that elderly individuals have a higher proportion of VZV-primed cells. This may reflect the life experience of repeated endogenous and environmental boosting events. However, the proportion of VZV-specific early effectors and cytokine-producing CD4 and CD8 T cells is significantly lower in elderly compared with young adults (Patterson-Bartlett et al. 2007), suggesting that a functional defect of the T cells of the elderly adults may attenuate the protection conferred by T cells against VZV.

2.5 *Clinical Evidence of Protection Against Varicella Conferred by VZV-Specific T Cell Mediated Immunity*

VZV-specific T cell-mediated immunity is essential for protection against and recovery from varicella and HZ. There is an association between the tempo and the intensity of the VZV-specific memory and effector T cell-mediated immune response and severity of varicella (Malavige et al. 2008). Furthermore, individuals with T cell immune deficits have more severe disease than normal hosts (Arvin et al. 1986a; Gershon and Steinberg 1979; Gershon et al. 1997). A single episode of varicella generates sufficient VZV-specific humoral and cellular immunity to protect normal hosts against second episodes. Passive transfer of VZV-specific antibodies can prevent development of disease after exposure to varicella, demonstrating the role of humoral immunity in preventing primary VZV infection. Passive transfer of VZV-specific T cells is not readily feasible or practical and, therefore, has not been studied as post-exposure prophylaxis. The evidence that VZV-specific T cell-mediated immunity alone can protect against varicella comes from the natural history of VZV infection in individuals with agammaglobulinemia. These individuals were protected by their primary VZV infection against second episodes of varicella despite their inability to develop VZV-specific antibodies, indicating that VZV-specific T cell-mediated immune responses are sufficient to prevent varicella (Good and Zak 1956).

2.6 Clinical Evidence of Protection Against HZ Conferred by VZV-Specific T Cell Mediated Immunity

VZV-specific cell-mediated immunity is necessary and sufficient to prevent HZ. The essential role of VZV-specific T cell-mediated immunity is strongly suggested by clinical observations. These include (1) the marked increase in incidence of HZ in association with all diseases and therapies associated with immune suppression (Buchbinder et al. 1992); (2) the correlation of the incidence of HZ in patients with lymphoma or after stem cell transplantation with their residual VZV-specific T cell-mediated immunity but not with residual VZV antibody (Arvin et al. 1978; Onozawa et al. 2006); (3) the correlation of protection with VZV-specific T cell-mediated immunity, and not VZV antibody, in human stem cell recipients who received an inactivated VZV vaccine (Hata et al. 2002). Most importantly, HZ incidence increases progressively with age in proportion to the age-related decline in VZV-specific T cell-mediated immunity. The severity of HZ also increases with age and disease-related immune suppression (Yawn et al. 2007; Kost and Straus 1996). The unique role of VZV-specific T cell-mediated immunity, rather than humoral immunity, in maintaining latency of VZV is also demonstrated by experience with human stem cell transplantation, where total ablation of immunity is partially restored, with intravenous immunoglobulin containing high levels of VZV antibody; nevertheless, a high incidence of HZ occurs in this setting.

In the elderly, the age-related loss of VZV-specific functional T cell-mediated immunity coincides with an increase in incidence of HZ (Hope-Simpson 1965). VZV-specific T cell-mediated immunity is also essential for optimal recovery from HZ. We recently demonstrated an inverse association of the magnitude of the VZV-specific T cell-mediated immune response in the first week after the onset of HZ rash, with the severity of HZ illness and the risk of developing post-herpetic neuralgia (Weinberg et al. 2009). These data, taken together, indicate that VZV T cell-mediated immunity plays a crucial role in the protection against both the occurrence and the morbidity of HZ.

3 VZV-Specific T Cell Immunity Acquired from Vaccination

3.1 Varicella Vaccine

The licensed varicella vaccines (Okavax–Biken; Varivax – Merck; Varilrix – SmithKlineBeecham) contain 1,350 pfu of the Oka strain of live attenuated VZV or its minor modifications (Takahashi 1984; Takahashi et al. 2008; Merck Varivax Package Insert 2009).The vaccine virus is administered subcutaneously, which differs from the airborne route by which varicella infection occurs. The vaccine virus propagates in varicella-naïve persons at the injection site, as evidenced by occasional local skin lesions containing vaccine virus. Vaccination must also occasionally cause a viremia, as indicated by the appearance of VZV in skin lesions distant from the injection site in a minor proportion of vaccinees (Ngai et al. 1996). Vaccine strain VZV also gains access to peripheral ganglia, either by ascending the peripheral nerve that innervates the injection site or as a consequence of viremia, as evidenced by the subsequent late appearance of HZ caused by Oka strain VZV (Galea et al. 2008; Levin et al. 2008a).

Varicella vaccine mimics natural infection, in that both VZV-specific antibody and T cell-mediated immunity are induced. Eighty percent of subjects studied at 2 weeks after vaccination by lymphoproliferation assays (LPA) using whole VZV antigen had a positive test; 95% were positive at 6 weeks, and 65–81% were positive in an assay that incorporated VZV glycoproteins B, E, and H (Diaz et al. 1988; Watson et al. 1990). The timing of these responses was similar to natural infection. VZV-specific T cell-mediated immunity measured by LPA was detectable at least 6 years after vaccination (Watson et al. 1995a). Vaccination induced cytotoxic responses against selected VZV proteins – glycoproteins B, C, E, H, and I, and against two of the immediate early proteins (Diaz et al. 1989; Sharp et al. 1992). Cytotoxic responses, which were mediated by both CD4 and CD8 cells, were detected for at least 4 years after vaccination at levels comparable to those present in adults with prior varicella. Skin test reactivity in response to intradermal inoculation of VZV antigens was present in vaccine recipients 20 years after vaccination (Asano et al. 1994). However, there are two potential problems in interpreting the accumulated data on vaccine-induced VZV-specific T cell-mediated immunity: (1) vaccine used in some studies had a content of live virus greater than that of the licensed vaccines, and virus titer is known to influence at least the peak antibody response (Varis and Vesikari 1996); (2) the results were obtained prior to widespread use of a licensed vaccine when there was significant varicella occurring in the community, which may have contributed to the maintenance of the VZV-specific immune responses.

Higher peak VZV-specific T cell-mediated responses after vaccination are achieved by children compared with adults (Nader et al. 1995), which may contribute to the greater vaccine efficacy in children. Two doses provided a higher peak value than a single dose (Watson et al. 1995a; Nader et al. 1995), consistent with the initial recommendations that susceptible adults receive two doses of vaccine.

While there are no immunologic data on long-term persistence of VZV T cell immunity after vaccination in the current era (without circulating VZV), there is a suggestion that efficacy of a single dose of varicella vaccine may decline after a decade (Chaves et al. 2007). It is likely, but not proven, that the persistence of higher VZV-specific T cell-mediated immunity will be longer when two doses of vaccine are administered, consistent with the current recommendation that all susceptible children also receive two doses of varicella vaccine (ACIP 2007). Two doses of varicella vaccine in an early clinical trial were more efficacious than a single dose (Kuter et al. 2004).

Varicella vaccine induces VZV-specific T cell-mediated immunity in carefully selected populations of children being treated for leukemia, other malignancies, liver transplantation, post-stem cell transplantation, and HIV (Gershon et al. 1986; Levin et al. 2008b). In HIV-infected children, T cell-mediated immune response after vaccination are similar to those that follow varicella (Levin et al. 2006). The efficacy of vaccination has been established only for leukemic children, but there are preliminary data that varicella vaccination also protects HIV-infected children (Son et al. 2008).

3.2 Herpes Zoster Vaccine

Given the essential role of VZV-specific T cell-mediated immunity in maintaining latency of VZV, an obvious approach to prevention of HZ would be to boost this immunity above the presumed threshold required to maintain latency. This rationale is supported by the observation that older individuals who developed HZ uncommonly experience a second episode of HZ, suggesting that the VZV antigenic load of HZ was sufficient to boost their immunity to maintain latency subsequently. The improved efficacy of a second dose of varicella vaccine in individuals who are varicella-naïve, demonstrates the capacity of this Oka strain VZV to boost VZV-specific T cell-mediated immunity (Watson et al. 1995a, b). Pilot experiments in people at least 60 years of age demonstrated that Oka strain VZV, administered in doses as high as ninefold higher than the licensed varicella vaccine, could boost VZV-specific CD4+ memory T cells approximately twofold in peripheral blood of elderly people who previously had natural immunity (varicella) (Trannoy et al. 2000; Levin et al. 1992). VZV-CMI was measured by responder cell frequency assays (RCF), which enumerated CD4+ memory cells (Hayward et al. 1994a). These experiments demonstrated the safety of such an approach in elderly people, and also indicated that the half-life of the boost in VZV-specific T cell-mediated immunity was at least 5 years (Levin et al. 1994, 1998). Vaccination of varicella-immune individuals also increased CD8+ T lymphocytes as measured by intracellular cytokine flow cytometry (Frey et al. 2003).

One potential problem that had to be overcome in the development of a HZ vaccine is the immunosenescence that progresses with aging (Cambier 2005), as demonstrated by the decline in VZV-specific T cell-mediated immunity in spite of environmental re-exposure to VZV. This was at least partially compensated for by increasing the viral content of the zoster vaccine to 19,400 pfu/dose at expiry, which is 14-fold higher than the varicella vaccine. A phenotypic analysis of mononuclear cells after exposure to VZV antigen *in vitro*, which compared young adults vs. individuals older than 60 years, indicated that elderly subjects have (1) fivefold less CD4 cells that produce IFNγ or IL 4&5, but no differences in the cytokine expression of the CD8 pool; (2) significantly less CD4 early effectors (CD45R0+62L-); and (3) significantly less CD8 effector memory (CD45RA+62L-) cells and early effectors (Patterson-Bartlett et al. 2007). The HZ vaccine reversed many of these observations in the older vaccinees: (1) CD4 and CD8 Th1 cells were increased; (2) CD4 and CD8 effector memory T cells and CD8 early effector T cells were increased; and (3) CD4 Th2 cells were increased, but not quite significantly. Most importantly, these vaccine-induced changes largely abolished the VZV-specific T cell differences that were present between the older and younger subjects prior to vaccination.

The pivotal trial of the licensed HZ vaccine was undertaken in subjects who were at least 60 years of age (45% were at least 70 years old) (Oxman et al. 2005). The baseline immunologic measurements from this trial confirmed that VZV-specific T cell-mediated immunity, as measured by RCF and an ELISPOT assay, continuously declined with advancing age (Levin et al. 2008b). The ELISPOT responses peaked at 2 weeks after vaccination (Lange et al. 2006), as they did after cases of HZ (Weinberg et al. 2009). Responses, which were measured in the trial at 6 weeks post-vaccination, were approximately twofold higher in vaccine-recipients than in placebo-recipients (Fig. 3). Values subsequently fell during the first year after vaccination, and then remained approximately 50% higher than pre-vaccine levels for the next 3 years. The persisting boost in VZV-specific T cell-mediated immunity for this duration mirrors the persisting efficacy in elderly vaccines (Levin et al. 2008b), which has been maintained for at least 5 years in an ongoing study (Oxman MN, unpublished). The magnitude of the boost was greatest in the youngest subjects (Fig. 4). This is consistent with the efficacy of the vaccine in preventing HZ, which was greater in those 60–69 years of age than in those $\geq$70 years of age. However, the efficacy against the overall morbidity in subjects and against post-herpetic neuralgia was not greatly affected by age. This suggests that the boost in immunity induced by the vaccine was not sufficient in many older subjects to prevent HZ, but the memory responses elicited even in these older subjects were sufficient to limit viral replication and subsequent damage. This paradigm is similar to that of two other vaccines for the elderly – pneumococcal and influenza vaccines. The boost in VZV-specific T cell-mediated immunity is similar in timing and magnitude to that which occurs with HZ (Weinberg et al. 2009). The administration of a second dose of HZ vaccine to older individuals at an interval of 6 weeks did not significantly increase the peak response (Lange et al. 2006).

VZV-specific T cell-mediated immunity can also be stimulated in elderly subjects by the administration of an inactivated Oka strain VZV vaccine (Hayward et al. 1994b). The RCF responses at 3 months and 1 year after vaccination were similar to those induced by a matched live virus vaccine. An inactivated vaccine was shown to be partially efficacious in human stem cell transplant recipients, with

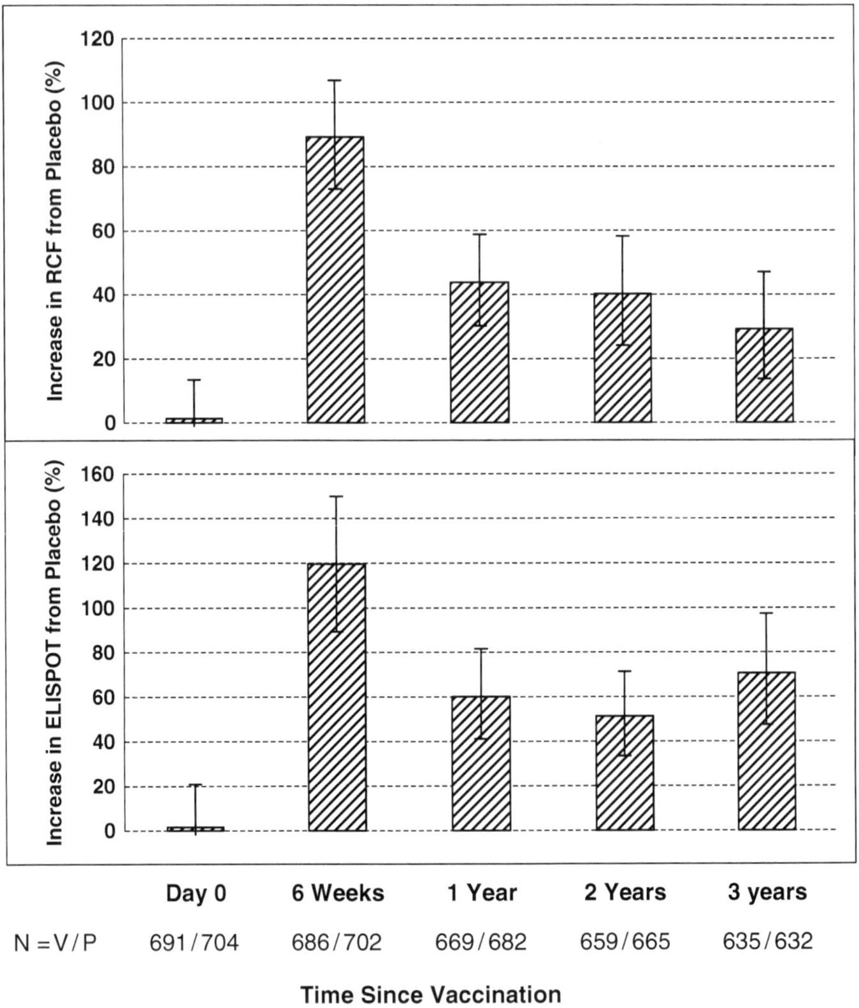

Fig. 3 Estimated percent increase of VZV-specific T cell-mediated immune responses in vaccine recipients relative to placebo recipients by time after vaccination. (1) Increase in immune response from placebo (%) at each time is the estimated geometric mean increase (%) from the placebo for each assay. The estimated increase is adjusted by age, gender, and study site. (2) ELISPOT counts = spot-forming cells per million PBMC; RCF value = responding cells per 100,000 PBMC. (3) Error bars indicate 95% confidence intervals of the geometric mean. (4) *N* is the number of subjects with blood drawn within the time window. (5) Data from subjects who developed HZ were censored from subsequent time points. (6) All responses of the vaccine recipients differed significantly from those of the placebo recipients ($p \leq 0.001$–0.0014). © 2008 Infectious Diseases Society of America

the efficacy correlating with the VZV-specific T cell response as measured by LPA (Hata et al. 2002). This vaccine is being studied in additional subjects with immune suppressive diseases.

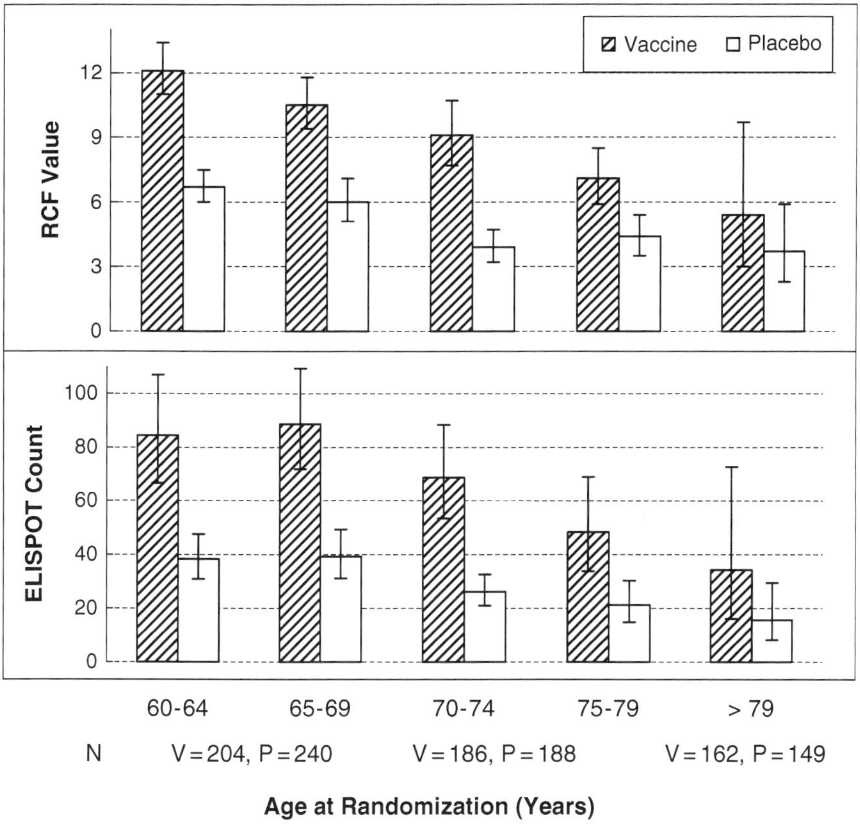

Fig. 4 VZV-specific T cell-mediated immune responses at 6-weeks after vaccination by age group. (1) ELISPOT counts = spot-forming cells per million PBMC; RCF value = responding cells per 100,000 PBMC. Error bars indicate 95% confidence intervals of the geometric mean. (2) *N* is the number of subjects with blood drawn within the time window. (3) There is a significant age effect (age slope) for the all net responses at week 6 ($p < 0.001$ for RCF and ELISPOT). © 2008 Infectious Diseases Society of America

3.3 Forward-Looking Statement

Some important questions have yet to be adequately answered:

1. Will the VZV-specific immune responses elicited by the current two dose strategy of varicella vaccination in childhood confer lifelong protection against varicella (especially typical varicella)?
2. Will this strategy lead to an increase in vaccine strain HZ as the vaccinated population ages?
3. Will this strategy, by virtue of changing the epidemiology of varicella (and associated environmental boosting), lead to earlier appearance of HZ in individuals who had varicella?

4. How long will the efficacy of the current HZ vaccine persist following vaccination?
5. How can we develop a cell-mediated immune response correlate of protection against HZ?
6. Can a HZ vaccine strategy be implemented that will overcome immune senescence – specifically to improve the efficacy of the vaccine in individuals of advanced age?
7. Can we succeed in defining immunosuppressive states where VZV vaccines can be safely used?
8. Will a non-live vaccine (alone or followed by a live vaccine) be valuable in protecting some immunosuppressed patients? Investigating this question in immunosuppressive states with high baseline levels of HZ might be one approach for developing a correlate of protection.

References

Advisory Committee on Immunization Practices, C.f.D.C.a.P. (2007) Recommendations of the ACIP on varicella immunization. http://www.cdc.gov/mmwr/preview/mmwrhtml/rr5604a1.htm

Arvin AM et al (1978) Selective impairment of lymphocyte reactivity to varicella-zoster virus antigen among untreated patients with lymphoma. J Infect Dis 137(5):531–540

Arvin AM, Koropchak CM, Wittek AE (1983) Immunologic evidence of reinfection with varicella-zoster virus. J Infect Dis 148(2):200–205

Arvin AM et al (1986a) Early immune response in healthy and immunocompromised subjects with primary varicella-zoster virus infection. J Infect Dis 154(3):422–429

Arvin AM et al (1986b) Immunity to varicella-zoster viral glycoproteins, gp I (gp 90/58) and gp III (gp 118), and to a nonglycosylated protein, p 170. J Immunol 137(4):1346–1351

Arvin AM et al (1991) Equivalent recognition of a varicella-zoster virus immediate early protein (IE62) and glycoprotein I by cytotoxic T lymphocytes of either CD4+ or CD8+ phenotype. J Immunol 146(1):257–264

Arvin AM et al (2002) Memory cytotoxic T cell responses to viral tegument and regulatory proteins encoded by open reading frames 4, 10, 29, and 62 of varicella-zoster virus. Viral Immunol 15(3):507–516

Asano Y et al (1994) Experience and reason: twenty-year follow-up of protective immunity of the Oka strain live varicella vaccine. Pediatrics 94(4 Pt 1):524–526

Asanuma H et al (2000) Frequencies of memory T cells specific for varicella-zoster virus, herpes simplex virus, and cytomegalovirus by intracellular detection of cytokine expression. J Infect Dis 181(3):859–866

Bergen RE et al (1991) Human T cells recognize multiple epitopes of an immediate early/tegument protein (IE62) and glycoprotein I of varicella zoster virus. Viral Immunol 4(3):151–166

Bhaduri-McIntosh S et al (2008) Repertoire and frequency of immune cells reactive to Epstein-Barr virus-derived autologous lymphoblastoid cell lines. Blood 111(3):1334–43

Buchbinder SP et al (1992) Herpes zoster and human immunodeficiency virus infection. J Infect Dis 166(5):1153–1156

Burke BL et al (1982) Immune responses to varicella-zoster in the aged. Arch Intern Med 142(2):291–293

Cambier J (2005) Immunosenescence: a problem of lymphopoiesis, homeostasis, microenvironment, and signaling. Immunol Rev 205:5–6

Chaves SS et al (2007) Loss of vaccine-induced immunity to varicella over time. N Engl J Med 356(11):1121–1129

Diaz PS et al (1988) Immunity to whole varicella-zoster virus antigen and glycoproteins I and p170: relation to the immunizing regimen of live attenuated varicella vaccine. J Infect Dis 158(6):1245–1252

Diaz PS et al (1989) T lymphocyte cytotoxicity with natural varicella-zoster virus infection and after immunization with live attenuated varicella vaccine. J Immunol 142(2):636–641

Frey CR et al (2003) Identification of CD8+ T cell epitopes in the immediate early 62 protein (IE62) of varicella-zoster virus, and evaluation of frequency of CD8+ T cell response to IE62, by use of IE62 peptides after varicella vaccination. J Infect Dis 188(1):40–52

Galea SA et al (2008) The safety profile of varicella vaccine: a 10-year review. J Infect Dis 197 (Suppl 2):S165–S169

Gershon AA, Steinberg SP (1979) Cellular and humoral immune responses to varicella-zoster virus in immunocompromised patients during and after varicella-zoster infections. Infect Immun 25(1):170–174

Gershon AA, Steinberg SP, Gelb L (1986) Live attenuated varicella vaccine use in immunocompromised children and adults. Pediatrics 78(4 Pt 2):757–762

Gershon AA et al (1997) Varicella-zoster virus infection in children with underlying human immunodeficiency virus infection. J Infect Dis 176(6):1496–1500

Giller RH, Winistorfer S, Grose C (1989) Cellular and humoral immunity to varicella zoster virus glycoproteins in immune and susceptible human subjects. J Infect Dis 160(6):919–928

Good RA, Zak SJ (1956) Disturbances in gamma globulin synthesis as experiments of nature. Pediatrics 18(1):109–149

Hata A et al (2002) Use of an inactivated varicella vaccine in recipients of hematopoietic-cell transplants. N Engl J Med 347(1):26–34

Hayward AR (1990) T-cell responses to predicted amphipathic peptides of varicella-zoster virus glycoproteins II and IV. J Virol 64(2):651–655

Hayward A, Giller R, Levin M (1989) Phenotype, cytotoxic, and helper functions of T cells from varicella zoster virus stimulated cultures of human lymphocytes. Viral Immunol 2(3):175–184

Hayward AR, Zerbe GO, Levin MJ (1994a) Clinical application of responder cell frequency estimates with four years of follow up. J Immunol Methods 170(1):27–36

Hayward AR, Buda K, Levin MJ (1994b) Immune response to secondary immunization with live or inactivated VZV vaccine in elderly adults. Viral Immunol 7(1):31–36

Hayward AR et al (1998) Cytokine production in varicella-zoster virus-stimulated cultures of human blood lymphocytes. J Infect Dis 178(Suppl 1):S95–S98

Hope-Simpson RE (1965) The nature of herpes zoster: a long-term study and a new hypothesis. Proc R Soc Med 58:9–20

Jenkins DE et al (1998) Interleukin (IL)-10, IL-12, and interferon-gamma production in primary and memory immune responses to varicella-zoster virus. J Infect Dis 178(4):940–948

Jones L et al (2006) Persistent high frequencies of varicella-zoster virus ORF4 protein-specific CD4+ T cells after primary infection. J Virol 80(19):9772–9778

Kittan NA et al (2007) Impaired plasmacytoid dendritic cell innate immune responses in patients with herpes virus-associated acute retinal necrosis. J Immunol 179(6):4219–4230

Kost RG, Straus SE (1996) Postherpetic neuralgia–pathogenesis, treatment, and prevention. N Engl J Med 335(1):32–42

Kumagai T et al (1980) Development and characteristics of the cellular immune response to infection with varicella-zoster virus. J Infect Dis 141(1):7–13

Kumagai T et al (1999) Varicella-zoster virus-specific cellular immunity in subjects given acyclovir after household chickenpox exposure. J Infect Dis 180(3):834–837

Kuter B et al (2004) Ten year follow-up of healthy children who received one or two injections of varicella vaccine. Pediatr Infect Dis J 23(2):132–137

Lange J et al (2006) Immunogenicity, kinetics of VZV-specific CD4+ T-cell γ-IFN production and safety of a live attenuated Oka/Merck zoster vaccine in healthy adults $\geq$ 60 years of age.

Abstract #857, in 46th Interscience Conference on Antimicrobial Agents and Chemotherapy (ICAAC). Toronto, Canada
Levin MJ et al (1992) Immune response of elderly individuals to a live attenuated varicella vaccine. J Infect Dis 166(2):253–259
Levin MJ et al (1994) Immune responses of elderly persons 4 years after receiving a live attenuated varicella vaccine. J Infect Dis 170(3):522–526
Levin MJ et al (1998) Use of a live attenuated varicella vaccine to boost varicella-specific immune responses in seropositive people 55 years of age and older: duration of booster effect. J Infect Dis 178(Suppl 1):S109–S112
Levin MJ et al (2006) Administration of live varicella vaccine to HIV-infected children with current or past significant depression of CD4(+) T cells. J Infect Dis 194(2):247–255
Levin MJ et al (2008a) Herpes zoster with skin lesions and meningitis caused by 2 different genotypes of the Oka varicella-zoster virus vaccine. J Infect Dis 198(10):1444–1447
Levin MJ et al (2008b) Varicella-zoster virus-specific immune responses in elderly recipients of a herpes zoster vaccine. J Infect Dis 197(6):825–835
Levy O et al (2003) Disseminated varicella infection due to the vaccine strain of varicella-zoster virus, in a patient with a novel deficiency in natural killer T cells. J Infect Dis 188(7):948–953
Malavige GN et al (2007) Rapid effector function of varicella-zoster virus glycoprotein I-specific CD4+ T cells many decades after primary infection. J Infect Dis 195(5):660–664
Malavige GN et al (2008) Viral load, clinical disease severity and cellular immune responses in primary varicella zoster virus infection in Sri Lanka. PLoS One 3(11):e3789
Merck Varivax Package Insert (2009) www.merck.com/product/USA/pi_circular/v/varivax/varivax_pi.pdf
Michalik DE et al (2008) Primary vaccine failure after 1 dose of varicella vaccine in healthy children. J Infect Dis 197(7):944–949
Morrow G et al (2003) Varicella-zoster virus productively infects mature dendritic cells and alters their immune function. J Virol 77(8):4950–4959
Nader S et al (1995) Age-related differences in cell-mediated immunity to varicella-zoster virus among children and adults immunized with live attenuated varicella vaccine. J Infect Dis 171(1):13–17
Ngai AL et al (1996) Safety and immunogenicity of one vs. two injections of Oka/Merck varicella vaccine in healthy children. Pediatr Infect Dis J 15(1):49–54
Onozawa M et al (2006) Relationship between preexisting anti-varicella-zoster virus (VZV) antibody and clinical VZV reactivation in hematopoietic stem cell transplantation recipients. J Clin Microbiol 44(12):4441–4443
Oxman MN et al (2005) A vaccine to prevent herpes zoster and postherpetic neuralgia in older adults. N Engl J Med 352(22):2271–2284
Patterson-Bartlett J et al (2007) Phenotypic and functional characterization of ex vivo T cell responses to the live attenuated herpes zoster vaccine. Vaccine 25(41):7087–7093
Rotbart HA, Levin MJ, Hayward AR (1993) Immune responses to varicella zoster virus infections in healthy children. J Infect Dis 167(1):195–199
Sadaoka K et al (2008) Measurement of varicella-zoster virus (VZV)-specific cell-mediated immunity: comparison between VZV skin test and interferon-gamma enzyme-linked immunospot assay. J Infect Dis 198(9):1327–1333
Sharp M et al (1992) Kinetics and viral protein specificity of the cytotoxic T lymphocyte response in healthy adults immunized with live attenuated varicella vaccine. J Infect Dis 165(5):852–858
Son M et al (2008) Vaccination of children with perinatal HIV infection protects against varicella and zoster. In: Pediatric Academic Societies and Asian Society for Pediatric Research joint meeting. Honolulu, Hawaii
Takahashi M (1984) Development and characterization of a live varicella vaccine (Oka strain). Biken J 27(2–3):31–36
Takahashi M et al (2008) Development of varicella vaccine. J Infect Dis 197(Suppl 2):S41–S44

Trannoy E et al (2000) Vaccination of immunocompetent elderly subjects with a live attenuated Oka strain of varicella zoster virus: a randomized, controlled, dose-response trial. Vaccine 18(16):1700–1706

Varis T, Vesikari T (1996) Efficacy of high-titer live attenuated varicella vaccine in healthy young children. J Infect Dis 174(Suppl 3):S330–S334

Verjans GM et al (2007) Selective retention of herpes simplex virus-specific T cells in latently infected human trigeminal ganglia. Proc Natl Acad Sci USA 104(9):3496–3501

Vessey SJ et al (2001) Childhood vaccination against varicella: persistence of antibody, duration of protection, and vaccine efficacy. J Pediatr 139(2):297–304

Vossen MT et al (2004) Development of virus-specific CD4+ T cells on reexposure to Varicella-Zoster virus. J Infect Dis 190(1):72–82

Vossen MT et al (2005) Absence of circulating natural killer and primed CD8+ cells in life-threatening varicella. J Infect Dis 191(2):198–206

Watson B et al (1990) Cell-mediated immune responses after immunization of healthy seronegative children with varicella vaccine: kinetics and specificity. J Infect Dis 162(4):794–799

Watson B et al (1995a) Humoral and cell-mediated immune responses in healthy children after one or two doses of varicella vaccine. Clin Infect Dis 20(2):316–319

Watson B et al (1995b) Safety and cellular and humoral immune responses of a booster dose of varicella vaccine 6 years after primary immunization. J Infect Dis 172(1):217–219

Weinberg A et al (2009) Varicella-Zoster Virus-Specific Immune Responses to Herpes Zoster in Elderly Participants in a Trial of a Clinically Effective Zoster Vaccine. J Infect Dis 200 (7):1068–1077

White CJ et al (1992) Modified cases of chickenpox after varicella vaccination: correlation of protection with antibody response. Pediatr Infect Dis J 11(1):19–23

Wilson A et al (1992) Subclinical varicella-zoster virus viremia, herpes zoster, and T lymphocyte immunity to varicella-zoster viral antigens after bone marrow transplantation. J Infect Dis 165(1):119–126

Yawn BP et al (2007) A population-based study of the incidence and complication rates of herpes zoster before zoster vaccine introduction. Mayo Clin Proc 82(11):1341–1349

Zhang Y et al (1994) Cytokine production in varicella zoster virus-stimulated limiting dilution lymphocyte cultures. Clin Exp Immunol 98(1):128–133

Zhu J et al (2007) Virus-specific CD8+ T cells accumulate near sensory nerve endings in genital skin during subclinical HSV-2 reactivation. J Exp Med 204(3):595–603

Perspectives on Vaccines Against Varicella-Zoster Virus Infections

Anne A. Gershon and Michael D. Gershon

Contents

Abstract Primary infection with varicella-zoster virus (VZV) results in varicella which, in populations where immunization is not used, occurs mostly in children. Varicella is a generalized rash illness with systemic features such as fever and malaise. During varicella, VZV becomes latent in sensory ganglia of the individual, and in 70% it remains asymptomatic for their lifetime. The remaining 30% develop reactivation from latency, resulting in herpes zoster (HZ). HZ usually occurs in persons over the age of 50, and is manifested by a painful unilateral rash that usually lasts about 2 weeks and then may be followed by a chronic pain syndrome called post-herpetic neuralgia (PHN). VZV infections are notoriously more severe in immunocompromised hosts than in healthy individuals. Despite gaps in our understanding of the details of immunity to VZV, successful vaccines have been developed against both varicella and zoster.

A.A. Gershon (✉) and M.D. Gershon
Departments of Pediatrics and Pathology and Cell Biology, Columbia University, College of Physicians and Surgeons, 630 West 168th Street, New York, NY 10032, USA
e-mail: aag1@columbia.edu; mdg4@columbia.edu

A.M. Arvin et al. (eds.), *Varicella-zoster Virus*,
Current Topics in Microbiology and Immunology 342, DOI 10.1007/82_2010_12

1 Varicella-Zoster Virus

VZV produces at least 12 glycoproteins (gps) during lytic infection when all of its 71 genes are expressed. These gps are present on the surface of the virion and also on the surfaces of infected cells. The gps are immunogenic and at least some are thought to play important roles in viral spread by facilitating cell adherence and possibly transfer of the virus from an infected cell to an uninfected one. The gps are designated by letters; the major one of VZV is gE. Other highly significant gps are gB, gH, and gI. In addition, a number of other internal viral antigens are expressed in lytic infection, some of which are also expressed in latent infection. It is not known which, if any or even all, of these proteins are responsible for stimulating immunity to VZV. Because of its abundance, gE is thought to be the major protective antigen (Arvin and Cohen 2007). Little is known about the roles of internal antigens in generating immunity, but ORF 4p and ORF 63p have recently been postulated to be important, based on immunologic responses demonstrated in experiments using tetramer technology to demonstrate cellular immunity (CMI) to the virus (Jones et al. 2006, 2007).

During lytic infection, VZV expresses all of its 71 genes, but during latent infection only six are known to be consistently expressed. Viral latency of VZV is quite distinctive from that of its close relative herpes simplex virus (HSV). In VZV latency, genes 4, 21, 29, 62, 63, and 66 are expressed (Gershon et al. 2008a). The proteins of these genes are segregated only in the cytoplasm of the cell and may be prevented from entering the cell nucleus by ORF61p (Hay and Ruyechan 1994). When this ORF is introduced into a cell harboring latent VZV infection, the virus reactivates and infection becomes lytic (Gershon et al. 2008a; Chen et al. 2003).

VZV remains an important pathogen; there is no useful animal model of VZV disease, and so it has been difficult to study protection that develops after vaccination in any degree of detail. Both varicella vaccine and zoster vaccine have, however, been demonstrated to have significant clinical effectiveness in large populations, and they have been used on a worldwide basis in over 50 million people (Gershon et al. 2008b; Levin 2008).

Exactly how adaptive immunity to VZV is mediated after varicella, during latency, and before and after zoster is not entirely understood. Clearly participation of humoral and CMI is important, and there may be some redundancy to assure a high degree of protection of the host. Although it is well known that patients with agammaglobulinemia are not at particular risk to develop varicella more than once, specific antibodies are thought to play a major role in protection against developing varicella. There is a very high correlation, for example, between demonstration of VZV antibodies in serum using a sensitive and specific assay and failure to develop clinical varicella after an intimate exposure to the virus (Michalik et al. 2008). It is possible that when cell-free VZV spread by the airborne route attempts to gain a foothold in an already experienced host, it is promptly eliminated from spread within the host by specific antibodies in the blood. While it is possible that VZV antibodies are surrogate markers of immunity, elderly individuals who have no demonstrable CMI to VZV but usually have high antibody titers are not at risk to develop varicella (although they are at risk to develop zoster). After a varicella-susceptible individual is exposed to VZV, infection can be prevented or modified by prompt passive immunization with preformed VZV antibodies such as varicella-zoster immune globulin (VZIG) (Gershon et al. 2008b; Ampofo et al. 2002). Thus the role of antibodies in prevention of varicella is not exactly straightforward, and so it seems most likely that there is redundancy between humoral and CMI to best protect the host.

It is generally agreed, however, that CMI responses are critical to recovery from VZV infections. The most severe (and even fatal) cases of varicella are seen in patients with absent or poor CMI to VZV, and administration of antibodies to combat active infections are ineffective (Gershon et al. 2008b; Ampofo et al. 2002). At this point, VZV is spreading within the host from one cell to another with little or no spread as cell-free virions (Gershon et al. 2008a). Zoster develops in patients with low CMI to VZV, despite high antibody titers, and protection is correlated with high levels of CMI following vaccination (Oxman et al. 2005). Finally, while asymptomatic infection with VZV is thought to be unusual (about 5%), it is not uncommon for persons with prior varicella who are exposed to VZV to develop a boost in immunity without symptoms (Gershon et al. 2008b; Ampofo et al. 2002). Data are also beginning to accumulate, which suggest that subclinical reactivation of latent VZV can also occur (Cinque et al. 1997; Mehta et al. 2004; Wilson et al. 1992).

2 Varicella Vaccine

The live attenuated Oka vaccine strain of VZV was first developed about 40 years ago by Takahashi et al. (1974) Originally, the vaccine was projected to be useful to prevent varicella ("chickenpox"), which is the primary infection with VZV. It was soon shown that immunocompromised vaccinated patients were not only protected against varicella (Gershon et al. 1984) but they were also less likely to develop zoster after natural chickenpox (Hardy et al. 1991). Roughly 15 years later it was realized that the Oka strain in a larger dose might be employed as a therapeutic vaccine to prevent HZ (Oxman et al. 2005).

At present, two live attenuated VZV Oka vaccines are licensed in the United States, both manufactured by Merck and Co., one against varicella (1,350 plaque forming units [pfu] per dose) and one against zoster (20,000 pfu per dose). The zoster vaccine is about 15 times as strong as varicella vaccine. The wide difference in dosage is of great interest because the low dose varicella vaccine is effective in preventing chickenpox in VZV naïve children. The much larger dosage of virus in zoster vaccine for adults (over age 60 years) who have already had varicella and are at high risk to develop zoster is required to achieve a significant degree of vaccine efficacy (Levin 2001). It is postulated that, in elderly individuals, immune senescence is not only responsible for the increased risk of zoster but also for the relatively refractory response to the vaccine, therefore requiring a much higher dose than that administered to children to prevent the primary infection with VZV (varicella).

Since its licensure in the US in 1995, the incidence of varicella has fallen by $> 80\%$ in both vaccine recipients and also in the unvaccinated, indicating herd immunity (Seward et al. 2002). Varicella-associated hospitalizations, moreover, have decreased by 88% (Zhou et al. 2005), and age-adjusted mortality from varicella has decreased by 66% (Nguyen et al. 2005) (See Fig. 1).

When initial clinical trials with varicella vaccine were first carried out, there was considerable controversy surrounding the vaccine. It was hypothesized that immunity from it might not be durable and that there might be severe adverse events resulting from its use, ranging from neurological problems to cancer. The first major trials were in immunocompromised children, because in the late 1970s as many as 80% of children with leukemia were being cured of this disease only to be at high risk to die of varicella during their anti-cancer chemotherapy (Feldman et al. 1975; Feldman and Lott 1987). Therefore, children with leukemia in remission were thought to have an appropriate risk-benefit ratio for clinical trials with this new vaccine, and they became the major participants in trials of the varicella vaccine in Japan and in the United States (Gershon et al. 2008b; Takahashi et al. 1975). The availability of antiviral therapy with acyclovir (ACV) beginning in the

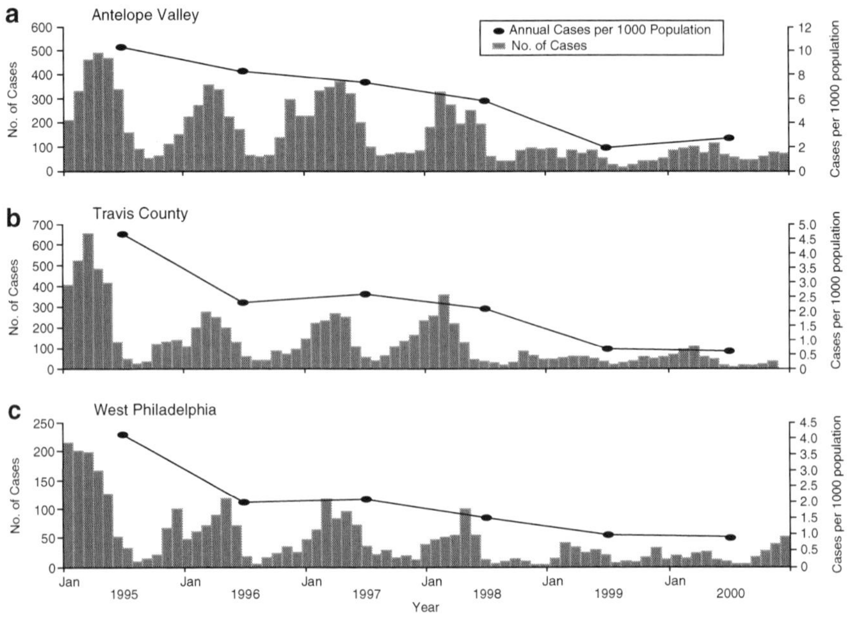

Fig. 1 Incidence of varicella in three sentinel counties in the United States from 1995, when varicella vaccine was licensed, until 2000, indicating decrease in the incidence of varicella (Gershon et al. 2008b)

mid-1970s made it possible to conduct these trials. In a large collaborative study conducted in the United States between 1979 and 1990, over 550 children with leukemia that was in remission for at least 1 year were immunized with the Oka vaccine. Most of these children had received maintenance anti-leukemic therapy for at least 1 year; this was withheld for 1 week before and 1–2 weeks after immunization. About 25% of the vaccinees developed a rash due to Oka VZV with more than 50 skin lesions, and they were given ACV to prevent further multiplication of the vaccine virus. The efficacy of the vaccine in preventing chickenpox after the usual two dose regimen for these children was demonstrated to be 85% (Gershon et al. 1984, 1996). Months to years later, some 15% of vaccinees developed what was termed "breakthrough" chickenpox after an exposure to wild-type (WT) VZV; however, the cases were uniformly mild even in leukemic children and did not require antiviral therapy. Many of these vaccinees remain alive and well today almost 30 years later, and are not known to have developed any of the feared theoretical complications that were originally projected regarding the live varicella vaccine.

When varicella vaccine was licensed in 1995, only one dose was recommended for children aged 1–12 years, and it was expected that over 90% would be fully immune to varicella. Two doses were to be given to adolescents and adults, because one dose was not clinically as effective in adults as in children (Gershon et al. 2008b). As will be discussed further, in 2006, hoping to improve vaccine efficacy, a second dose of varicella vaccine was recommended by the Centers for Disease Control and Prevention (CDC) for all vaccinated children and adults (Centers-for-Disease-Control 2007a).

3 Vaccine Safety

Varicella vaccine has proven to be remarkably safe. In the United States, about four million children have been vaccinated annually over roughly the past 10 years. As will be detailed, very few serious complications of vaccination have been reported. As with all vaccines, some reported problems may not have been due to varicella vaccine but were coincidental temporally related occurrences rather than cause and effect. Those that have occurred and were clearly due to varicella vaccine are probably less frequent than complications that are seen in individuals who have experienced natural varicella. Routine universal immunization of infants is now administered in Canada, Uruguay, Sicily, Germany, South Korea, Qatar, Taiwan, Germany, Israel, and Australia (Gershon et al. 2008b). The vaccine has also been found to be safe and effective in children with underlying HIV infection, as long as they are relatively immunologically normal when immunized (Gershon et al. 2009a; Levin et al. 2001, 2006; Son 2008). As in other studies in immunocompromised children, not only is varicella significantly prevented in HIV-infected children, but also is zoster (Hardy et al. 1991; Son 2008). Not surprisingly, data now also indicate that zoster is decreased in healthy recipients of varicella vaccine as well as in immunocompromised vaccinees. The risk of developing zoster in healthy vaccinated children less than 10 years old was found to be decreased by a factor between 4 and 12 times, compared with similar children who had experienced natural infection (Civen et al. 2009). In another study, the incidence rate of zoster in 172, 163 healthy children after varicella vaccine was demonstrated to be extremely low (Tseng et al. 2009).

The main adverse event reported in the US after immunization of healthy children is a mild rash due to the Oka vaccine in the first 6 weeks after immunization. Other problems are of somewhat greater concern but are far less frequent in occurrence. In reports to the vaccine adverse event reporting system (VAERS) in the US, from 1995 to 2005, the rate of serious adverse events (SAEs) after varicella vaccine was 2.6/100,000 distributed doses (Chaves et al. 2008). Additionally, generalized severe Oka infections in children who were found to be immunocompromised either before (in which case the underlying immunodeficiency was not yet recognized) or just following vaccination have been reported in less than 10 vaccinated children, since 1995 (Chaves et al. 2008; Galea et al. 2008; Jean-Philippe et al. 2007; Levy et al. 2003; Levin et al. 2003). These children had underlying conditions such as HIV infection with almost no detectable CD4 T lymphocytes, congenital immunodeficiency diseases such as adenosine deaminase deficiency and natural killer (NK) cell deficiency, high dose steroid therapy, and anti-cancer chemotherapy. Most of these children were treated with antiviral therapy and recovered.

As has been mentioned, the incidence of zoster after varicella vaccine is very low in healthy children (Civen et al. 2009; Tseng et al. 2009). Two rare reported SAEs following vaccination were cases of zoster with meningitis caused by Oka VZV (Chaves et al. 2008). Two additional cases of meningitis due to Oka zoster have also been reported (Iyer et al. 2009; Levin et al. 2008). One of these patients was a 1-year-old boy vaccinated just prior to the diagnosis of neuroblastoma, who developed chronic zoster that became drug resistant after prolonged unsuccessful courses of antiviral therapy. His zoster remitted after he was no longer receiving anti-cancer therapy (Levin et al. 2003). The other occurred in an immunocompetent child; he recovered with antiviral therapy (Iyer et al. 2009). The occurrence of meningitis associated with zoster due to WT VZV is, in contrast, not at all uncommon (Amlie-Lefond and Jubelt 2009; Gilden et al. 2000). Zoster is thus unusual after vaccination against chickenpox, and based on what we know today, meningitis is likely to be less common from Oka zoster than from WT VZV zoster, although this is an area that deserves further scrutiny. No long-term sequelae from these rare cases have been reported.

Why zoster after varicella vaccination is less common than after natural infection is not fully understood; the Oka strain is capable of causing latent infection in an in vitro neuronal model (Chen et al. 2003). It may be that the burden of viral latency after vaccination is lower than it is with WT VZV or that the Oka strain is attenuated and therefore less likely to reactivate. It is possible to rapidly and accurately distinguish between WT VZV and the Oka strain by polymerase chain reaction (PCR) (LaRussa et al. 1992). Data from our group have indicated that Oka

VZV can be demonstrated in the dorsal root ganglia (DRG), cranial nerve ganglia (CNG), and enteric ganglia (EG) of vaccinated children who have died suddenly and have therefore come to autopsy (Gershon et al. 2009b). We hypothesize that vaccinated children have a low incidence of zoster due to lower viral loads from latent Oka VZV than latent WT virus after natural varicella, secondary to the lower rate of skin involvement after vaccination compared to that following WT varicella.

Transmission of Oka VZV to others has been extremely rare and is estimated to occur in 1 recipient per 10 million doses of vaccine (Gershon et al. 2008b; Galea et al. 2008). Transmission to others has occurred only when an individual develops skin lesions due to the Oka strain (Gershon et al. 2008b; Tsolia et al. 1990; Brunell 2000).

A combination of varicella vaccine with other viral vaccines specifically as measles–mumps–rubella–varicella vaccine (MMRV) would be extremely useful clinically. In 2005, this combination vaccine was licensed in the United States, but it was soon noted that it produced a doubling of the rate of febrile seizures as a complication of vaccination (Centers-for-Disease-Control 2008; Jacobsen et al. 2009). Between days 7 and 14 after MMR vaccine, the rate of febrile seizures was 4/10,000, while after MMRV it was 9/10,000 ($p = <0.001$) At present, after much deliberation, the CDC recommended that a choice of either MMR plus monovalent varicella vaccine in a different body site or MMRV should be offered to parents for use in their 1-year-old infants who are receiving their first immunization against these viruses. Parents are to be carefully counseled about the rare possibility of their infant developing a febrile seizure 1–2 weeks after the vaccination with MMRV for the first immunizing dose. There is also a precaution based on the increased risk of febrile seizures in children with an immediate family history of epilepsy, or febrile seizures, or past personal history of seizures. In that case, for the first dose, MMR and varicella vaccines should be administered, not MMRV. For the second dose, MMRV is preferred because the complication of febrile seizures is rarely seen after a second dose of MMRV, one reason being that children receiving the second dose of MMRV are usually older and beyond the age when febrile seizures are most common (Centers-for-Disease-Control 2008; Jacobsen et al. 2009).

4 Effectiveness of Varicella Vaccine

Case–control studies of the effectiveness of the licensed varicella vaccine in healthy children indicate that there is about 85% protection against developing any form of chickenpox following one dose of vaccine (Vazquez et al. 2001, 2004). It was originally anticipated that protection would be higher, approaching 95%, based on serologic data from studies in which a glycoprotein ELISA assay was used as a surrogate to evaluate immunity to varicella (Provost et al. 1989, 1991). Soon after vaccine licensure in the US, however, a great number of outbreaks in day care facilities and schools where most children had been vaccinated began to be described (Gershon et al. 2008b). Infection rates from these outbreaks were used to try to calculate vaccine effectiveness, but there was a great deal of variability from study to study (44–100% effectiveness). Interestingly, however, the average effectiveness was about 85%.

Whether immunity wanes with time after vaccination (known as secondary vaccine failure) is controversial. One observational study purported to show waning of protection against varicella with time, but this was purely an epidemiologic study with no laboratory component regarding specific viral diagnosis of breakthrough varicella (Chaves et al. 2007). It is not always easy to recognize breakthrough varicella and therefore laboratory confirmation of illness is important. Moreover, the existence of primary vaccine failure, which could appear to be waning immunity, was not considered in the study (Gershon et al. 2007a). In the already mentioned case–control studies, the incidence of varicella in vaccinees did not increase in time, over a period of 2–8 years after vaccination; there was no indication of secondary vaccine failure (Vazquez et al. 2004) (See Fig. 2). In this study, laboratory confirmation of breakthrough varicella was obtained, mostly using PCR (Vazquez et al. 2001, 2004). Additional studies regarding whether immunity to varicella wanes significantly with time would be extremely useful for understanding how varicella vaccine behaves and planning whether and when booster doses of vaccine should be recommended. Based on what we now know, waning immunity may occur, but it does not seem to be a serious problem after varicella vaccination. Primary vaccine failure is more likely to be problematic (see below) but is being addressed by giving two doses of vaccine to all children.

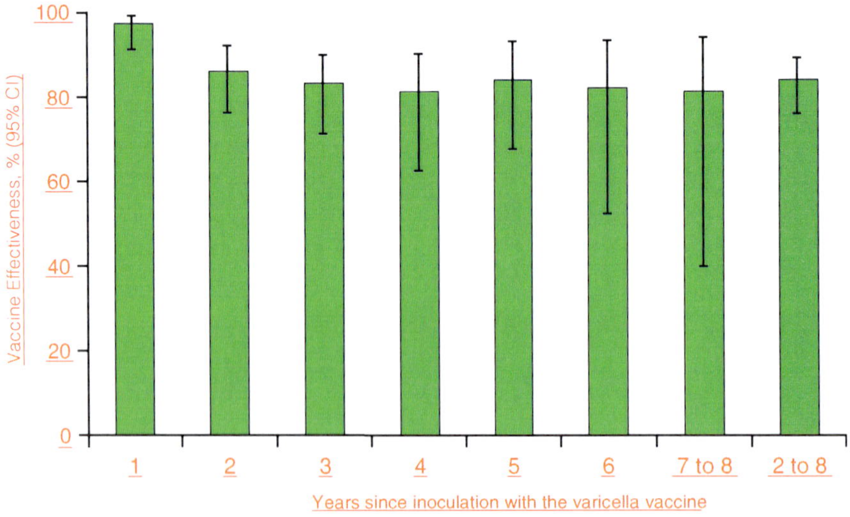

Fig. 2 Case–control study indicating the effectiveness of varicella vaccine over time (Vazquez et al. 2004)

Based on the clinical and laboratory observations mentioned in 2006, the CDC recommended that all children receive two doses of varicella vaccine rather than one dose as originally proposed (Centers-for-Disease-Control 2007b). Dramatic boosts in humoral and CMI to VZV have been observed following a second dose of vaccine, which suggests that two doses will provide increased protection against chickenpox (Centers-for-Disease-Control 2007b). Although it remains to be proven whether two doses will result in a further decline in incidence of varicella, it is projected to do so (Kuter et al. 2004). It was recently reported, however, that in one study, two doses did not decrease the attack rate of varicella in an outbreak situation in a school where most children had received two doses (Gould et al. 2009). Breakthrough varicella, which was usually very mild, was not proven by laboratory testing in these children; many may not have had varicella but rather were suffering from insect bites. Bites were apparently recognized to be common during the period when the outbreak was occurring. Data from our laboratory, a case–control study in which varicella was proven by PCR testing, (Vazquez et al.; presented at the Annual Conference of the Infectious Diseases of America in October 2009) indicates that two doses provides a high degree of protection against chickenpox. In 62 cases of varicella, 62 (100%) had one dose of vaccine and 0 had 2 doses. In 133 controls without varicella, 115 (85.8%) had one dose and 18 (13.4%) had two doses. Vaccine effectiveness of two doses compared to unvaccinated children was 98% ($p < 0.001$). These studies are ongoing.

Additional research on the subject of protection against varicella after two doses of vaccine compared to one dose is extremely important. Currently, the CDC recommends that second doses of MMRV be administered between the ages of 4 and 6 years old. Because of vaccine failures from monovalent vaccine, however, it might be preferable to administer the second dose of MMRV at age 2 or 3 years of age. On the other hand, since the dose of VZV in MMRV is about 17,000 pfu, the rate of vaccine failure (if there is one) to the VZV component is unknown and may be lower than that identified with monovalent varicella vaccine. As an aside, it has been suggested that robust immune responses to varicella vaccine are genetically determined (Klein et al. 2007). Additional information on these issues would be extremely practical and useful to have in planning when to vaccinate children.

Varicella is one of the most contagious diseases known. After a household exposure to the virus, the attack rate of chickenpox in susceptibles has been reported to be as high as 90% (Ross et al. 1962). The best laboratory indication for immunity to varicella is the fluorescent antibody to membrane antigen (FAMA) assay (Fig. 3). Extensive clinical validation indicates that persons with positive FAMA titers have a protection rate against varicella of over 98% after a close exposure to VZV. Negative FAMA titers are highly correlated with developing varicella after exposure to the virus, particularly in the unvaccinated (Michalik et al. 2008). Our laboratory is one of very few laboratories worldwide that perform the FAMA test; it is considered the "gold standard" for measuring immunity to varicella (Gershon et al. 2007b; Williams et al. 1974). This assay is the best available immune correlate for determining whether an individual is or is not immune to varicella.

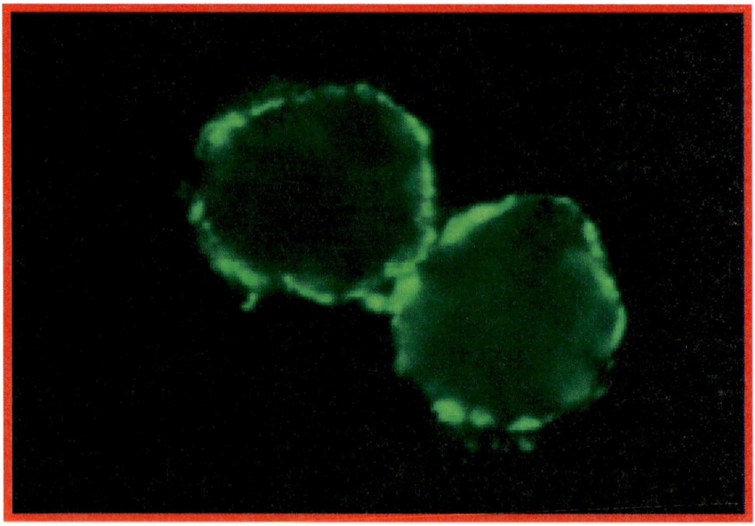

Fig. 3 A FAMA assay, indicating the presence of antibodies to VZV in serum. Unfixed tissue culture cells are incubated first with diluted serum and subsequently with anti-human globulin labeled with fluorsecein isothyocyanate and viewed by fluorescence microscopy (Gershon et al. 2007b)

5 Vaccination Against Zoster

Zoster, which is due to reactivation of VZV that becomes latent in the nervous system during varicella, develops when immunity to VZV declines, usually as a result of aging or being immunocompromised (Hope-Simpson 1965). About 30 years ago, it was realized that individuals over 50 years of age are at increased risk to develop zoster and that, while they have detectable antibodies to VZV, they often have no demonstrable CMI to the virus (Berger et al. 1981; Burke et al. 1982). Interestingly, not all persons with low CMI to VZV develop zoster; most do not. Most individuals over the age of 50 no longer have demonstrable CMI to VZV (Berger et al. 1981; Burke et al. 1982). In the United States about one-third of individuals will go on to develop zoster during their lives; this amounts to about one million annual cases of zoster (Oxman et al. 2005). Despite the high dosage of VZV in the zoster vaccine, it is only about 50–60% effective in preventing zoster, and protection also declines with age. When lower vaccinating doses were tried, stimulation of VZV CMI was deemed too low to lead to durable protection against zoster (Levin 2001). Thus the dose of 20,000 pfu of live VZV was utilized, and it proved to be effective (roughly 50–60%) in a double blind placebo controlled study involving over 38,000 healthy adults over age 60 (Oxman and Levin 2008). Vaccinees were more likely to increase their CMI responses to VZV than placebo recipients (Oxman and Levin 2008). Despite the apparent correlation of zoster with VZV CMI, there are no specific immune markers that indicate protection against zoster as there are for varicella. Identification of a specific immune pattern indicating protection against zoster would be extremely useful for deciding whom to vaccinate at present and for testing of newly developed molecular vaccines of the future.

The zoster vaccine used in the United States has proven to be extremely safe, despite the large dose of virus. No common SAEs emerged from the double-blind controlled trial, nor have any been identified since the vaccine was licensed in 2005. Major questions regarding vaccination against zoster concern the best age at which to administer it and whether booster doses might improve the effectiveness as seen with varicella vaccine. Currently, one dose is recommended for individuals over 60 years of age who are relatively healthy. A small study in 1,122 subjects aged 50–59 at vaccination indicated that it was immunogenic and safe (Sutradhar et al. 2009). It might therefore make sense, for example, to vaccinate younger individuals who are projected to have solid organ transplants in the near future. In addition, inactivated forms of zoster vaccines have proven to safely decrease zoster in immunocompromised patients, and this vaccine might prove be very useful if it were licensed and available (Hata et al. 2002) Subunit vaccines are also being studied for efficacy in preventing zoster (ClinicalTrial.gov 2009a, b).

The disease burden of zoster in the elderly and the immunocompromised is considerable, and therefore improved vaccines and refinement of the use of the existing vaccine are very important for individuals and for public health. While zoster is rarely fatal, the morbidity associated with this disease is considerable. By decreasing the incidence of zoster, thereby decreasing spread of VZV, we gain greater protection from this viral infection. Despite the modeling studies that suggest that zoster might become more common for some 30 years after widespread use of varicella vaccine in a population(Brisson et al. 2003), this possibility, should it occur, can be managed by judicious use of zoster vaccine. The aim of vaccination against varicella and zoster is to decrease transmission of the virus and lower the overall viral burden in the population.

Clearly, vaccines against VZV have been extremely successful, despite the realization that there remains much to be learned about them immunologically. Although VZV vaccines are proven to be highly effective, there are a number of questions associated with their use. These need to be addressed, but this research can be ongoing as we proceed to immunize more and more of our population against varicella and HZ by using these safe and effective immunizing agents.

Acknowledgments Supported by NIH grants AI27187, and AI24021.

References

Amlie-Lefond C, Jubelt B (2009) Neurologic manifestations of varicella zoster virus infections. Curr Neurol Neurosci Rep 9:430–434

Ampofo K, Saiman L, LaRussa P, Steinberg S, Annunziato P, Gershon A (2002) Persistence of immunity to live attenuated varicella vaccine in healthy adults. Clin Infect Dis 34:774–779

Arvin AM, Cohen J (2007) Varicella-zoster virus. In: Knipe DM, Howley PM (eds) Virology, 5th edn. Raven, NY, Philadelphia, pp 2773–2818

Berger R, Florent G, Just M (1981) Decrease of the lympho-proliferative response to varicella-zoster virus antigen in the aged. Infect Immun 32:24–27

Brisson M, Edmunds WJ, Gay NJ (2003) Varicella vaccination: impact of vaccine efficacy on the epidemiology of VZV. J Med Virol 70(Suppl 1):S31–37

Brunell PA (2000) Chickenpox attributable to a vaccine virus contracted from a vaccinee with zoster. Pediatrics 106:e28–e29

Burke BL, Steele RW, Beard OW, Wood JS, Cain TD, Marmer DJ (1982) Immune responses to varicella-zoster in the aged. Arch Intern Med 142:291–293

Centers-for-Disease-Control (2007) Prevention of varicella: recommendations of the advisory committee on immunization practices (ACIP). MMWR 56:1–40

Centers-for-Disease-Control (2007) Prevention of varicella. MMWR 56:1–55

Centers-for-Disease-Control (2008) Update: recommendations from the advisory committee on immunization practices (ACIP) regarding administration of combination MMRV vaccine. MMWR 57:258–260

Chaves SS, Gargiullo P, Zhang JX et al (2007) Loss of vaccine-induced immunity to varicella over time. N Engl J Med 356:1121–1129

Chaves SS, Haber P, Walton K et al (2008) Safety of varicella vaccine after licensure in the United States: experience from reports to the vaccine adverse event reporting system, 1995–2005. J Infect Dis 197(Suppl 2):S170–S177

Chen J, Gershon A, Silverstein SJ, Li ZS, Lungu O, Gershon MD (2003) Latent and lytic infection of isolated guinea pig enteric and dorsal root ganglia by varicella zoster virus. J Med Virol 70:S71–S78

Cinque P, Bossolasco S, Vago L et al (1997) Varicella-zoster virus (VZV) DNA in cerebrospinal fluid of patients infected with human immunodeficiency virus: VZV disease of the central nervous system or subclinical reactivation of VZV infection? Clin Infect Dis 25:634–639

Civen R, Chaves SS, Jumaan A et al (2009) The incidence and clinical characteristics of herpes zoster among children and adolescents after implementation of varicella vaccination. Pediatr Infect Dis J 28:954–959

ClinicalTrial.gov (2009) Immunogenicity and safety study of GSK biologicals' herpes zoster vaccine with various formulations in adults $\geq$ 50 years

ClinicalTrial.gov (2009) Safety and immunogenicity of the zoster vaccine GSK1437173A in elderly subjects

Feldman S, Lott L (1987) Varicella in children with cancer: impact of antiviral therapy and prophylaxis. Pediatrics 80:465–472

Feldman S, Hughes W, Daniel C (1975) Varicella in children with cancer: 77 cases. Pediatrics 80:388–397

Galea SA, Sweet A, Beninger P et al (2008) The safety profile of varicella vaccine: a 10-year review. J Infect Dis 197(Suppl 2):S165–S169

Gershon AA, Steinberg S, Gelb L, NIAID-Collaborative-Varicella-Vaccine-Study-Group (1984) Live attenuated varicella vaccine: efficacy for children with leukemia in remission. JAMA 252:355–362

Gershon A, LaRussa P, Steinberg S (1996) Varicella vaccine: use in immunocompromised patients. In: White J, Ellis R (eds) Infectious disease clinics of North America, vol 10. Saunders, Philadelphia, pp 583–594

Gershon AA, Arvin AM, Shapiro E (2007) Varicella vaccine. N Engl J Med 356:2648–2649

Gershon A, Chen J, LaRussa P, Steinberg S (2007b) Varicella-zoster virus. In: Murray PR, Baron E, Jorgensen J, Landry M, Pfaller M (eds) Manual of Clinical Microbiology, 9th edn. ASM, Washington, DC, pp 1537–1548

Gershon AA, Chen J, Gershon MD (2008a) A model of lytic, latent, and reactivating varicella-zoster virus infections in isolated enteric neurons. J Infect Dis 197(Suppl 2):S61–S65

Gershon A, Takahashi M, Seward J (2008b) Live attenuated varicella vaccine. In: Plotkin S, Orenstein W, Offit P (eds) Vaccines, 5th edn. WB Saunders, Philadelphia, pp 915–958
Gershon AA, Levin MJ, Weinberg A et al (2009a) A Phase I-Ii study of live attenuated varicella-zoster virus vaccine to boost immunity in human immunodeficiency virus-infected children with previous varicella. Pediatr Infect Dis J 28:653–655
Gershon AA, Chen J, Davis L, Krinsky C, Cowles R, Gershon MD (2009) Distribution of latent varicella zoster virus in sensory ganglia and gut after vaccination and wild-type infection: evidence for viremic spread. In: 34th International Herpesvirus Workshop, Ithaca, NY
Gilden DH, Kleinschmidt-DeMasters BK, LaGuardia JJ, Mahaliingham R, Cohrs RJ (2000) Neurologic complications of the reactivation of varicella-zoster virus. N Engl J Med 342:635–645
Gould PL, Leung J, Scott C et al (2009) An outbreak of varicella in elementary school children with two-dose varicella vaccine recipients–Arkansas, 2006. Pediatr Infect Dis J 28:678–681
Hardy IB, Gershon A, Steinberg S, LaRussa P et al (1991) The incidence of zoster after immunization with live attenuated varicella vaccine. A study in children with leukemia. N Engl J Med 325:1545–1550
Hata A, Asanuma H, Rinki M et al (2002) Use of an inactivated varicella vaccine in recipients of hematopoietic- cell transplants. N Engl J Med 347:26–34
Hay J, Ruyechan WT (1994) Varicella-zoster virus: a different kind of herpesvirus latency? Semin Virol 5:241–248
Hope-Simpson RE (1965) The nature of herpes zoster: a long term study and a new hypothesis. Proc R Soc Med 58:9–20
Iyer S, Mittal MK, Hodinka RL (2009) Herpes zoster and meningitis resulting from reactivation of varicella vaccine virus in an immunocompetent child. Ann Emerg Med 53:792–795
Jacobsen SJ, Ackerson BK, Sy LS et al (2009) Observational safety study of febrile convulsion following first dose MMRV vaccination in a managed care setting. Vaccine 27(34):4656–4661
Jean-Philippe P, Freedman A, Chang MW et al (2007) Severe varicella caused by varicella-vaccine strain in a child with significant T-cell dysfunction. Pediatrics 120:e1345–e1349
Jones L, Black AP, Malavige GN, Ogg GS (2006) Persistent high frequencies of varicella-zoster virus ORF4 protein-specific CD4+ T cells after primary infection. J Virol 80:9772–9778
Jones L, Black AP, Malavige GN, Ogg GS (2007) Phenotypic analysis of human CD4+ T cells specific for immediate-early 63 protein of varicella-zoster virus. Eur J Immunol 37:3393–3403
Klein NP, Fireman B, Enright A, Ray P, Black S, Dekker CL (2007) A Role for Genetics in the Immune Response to the Varicella Vaccine. Pediatr Infect Dis J 26:300–305
Kuter B, Matthews H, Shinefield H et al (2004) Ten year follow-up of healthy children who received one or two injections of varicella vaccine. Pediatr Infect Dis J 23:132–137
LaRussa P, Lungu O, Hardy I, Gershon A, Steinberg S, Silverstein S (1992) Restriction fragment length polymorphism of polymerase chain reaction products from vaccine and wild-type varicella-zoster virus isolates. J Virol 66:1016–1020
Levin MJ (2001) Use of varicella vaccines to prevent herpes zoster in older individuals. Arch Virol (Suppl 17):151–160
Levin MJ (2008) Zoster vaccine. In: Plotkin S, Orenstein W, Offit PA (eds) Vaccines, 5th edn. WB Saunderws, Philadelphia, pp 1057–1068
Levin MJ, Gershon AA, Weinberg A et al (2001) Immunization of HIV-infected children with varicella vaccine. J Pediatr 139:305–310
Levin MJ, Dahl KM, Weinberg A, Giller R, Patel A (2003) Development of resistance to acyclovir during chronic Oka strain varicella-zoster virus infection in an immunocompromised child. J Infect Dis 188:954–959
Levin MJ, Gershon AA, Weinberg A, Song LY, Fentin T, Nowak B (2006) Administration of live varicella vaccine to HIV-infected children with current or past significant depression of CD4(+) T cells. J Infect Dis 194:247–255

Levin MJ, DeBiasi RL, Bostik V, Schmid DS (2008) Herpes zoster with skin lesions and meningitis caused by 2 different genotypes of the Oka varicella-zoster virus vaccine. J Infect Dis 198:1444–1447
Levy O, Orange JS, Hibberd P et al (2003) Disseminated varicella infection due to vaccine (Oka) strain varicella-zoster virus in a patient with a novel deficiency in natural killer cells. J Infect Dis 188:948–953
Mehta SK, Cohrs RJ, Forghani B, Zerbe G, Gilden DH, Pierson DL (2004) Stress-induced subclinical reactivation of varicella zoster virus in astronauts. J Med Virol 72:174–179
Michalik DE, Steinberg SP, LaRussa PS et al (2008) Primary vaccine failure after 1 dose of varicella vaccine in healthy children. J Infect Dis 197:944–949
Nguyen HQ, Jumaan AO, Seward JF (2005) Decline in mortality due to varicella after implementation of varicella vaccination in the United States. N Engl J Med 352:450–458
Oxman MN, Levin MJ (2008) Vaccination against Herpes Zoster and Postherpetic Neuralgia. J Infect Dis 197(Suppl 2):S228–S236
Oxman MN, Levin MJ, Johnson GR et al (2005) A vaccine to prevent herpes zoster and postherpetic neuralgia in older adults. N Engl J Med 352:2271–2284
Provost P, Krah D, Miller W et al (1989) Comparative sensitivities of assays for antibodies induced by live varicella vaccine. Interscience conference on antimicrobial agents and chemotherapy, Houston, TX
Provost PJ, Krah DL, Kuter BJ et al (1991) Antibody assays suitable for assessing immune responses to live varicella vaccine. Vaccine 9:111–116
Ross AH, Lencher E, Reitman G (1962) Modification of chickenpox in family contacts by administration of gamma globulin. N Engl J Med 267:369–376
Seward JF, Watson BM, Peterson CL et al (2002) Varicella disease after introduction of varicella vaccine in the United States, 1995–2000. JAMA 287:606–611
Son M (2008) varicella vaccine protects children with perinatal HIV infection against zoster. In: 56th Annual Meeting of IDSA. Washington, DC
Sutradhar SC, Wang WW, Schlienger K et al (2009) Comparison of the levels of immunogenicity and safety of Zostavax in adults 50 to 59 years old and in adults 60 years old or older. Clin Vaccine Immunol 16:646–652
Takahashi M, Otsuka T, Okuno Y, Asano Y, Yazaki T, Isomura S (1974) Live vaccine used to prevent the spread of varicella in children in hospital. Lancet 2:1288–1290
Takahashi M, Okuno Y, Otsuka T et al (1975) Development of a live attenuated varicella vaccine. Biken J 18:25–33
Tseng HF, Smith N, Marcy SM, Sy LS, Jacobsen SJ (2009) Incidence of herpes zoster among children vaccinated with varicella vaccine in a prepaid health care plan in the United States, 2002–2008. Pediatr Infect Dis J 28(12):1069–1072
Tsolia M, Gershon A, Steinberg S, Gelb L (1990) Live attenuated varicella vaccine: evidence that the virus is attenuated and the importance of skin lesions in transmission of varicella-zoster virus. J Pediatr 116:184–189
Vazquez M, LaRussa P, Gershon A, Steinberg S, Freudigman K, Shapiro E (2001) The effectiveness of the varicella vaccine in clinical practice. N Engl J Med 344:955–960
Vazquez M, LaRussa PS, Gershon AA et al (2004) Effectiveness over time of varicella vaccine. JAMA 291:851–855
Williams V, Gershon A, Brunell P (1974) Serologic response to varicella-zoster membrane antigens measured by indirect immunofluorescence. J Infect Dis 130:669–672
Wilson A, Sharp M, Koropchak CM, Ting SF, Arvin AM (1992) Subclinical varicella-zoster virus viremia, herpes zoster, and T lymphocyte immunity to varicella-zoster viral antigens after bone marrow transplantation. J Infect Dis 165:119–126
Zhou F, Harpaz R, Jumaan AO, Winston CA, Shefer A (2005) Impact of varicella vaccination on health care utilization. JAMA 294:797–802

Index